BIOLOGY

The Unity and Diversity of Life

General Advisors/Contributors

JOHN ALCOCK
Arizona State University

AARON BAUER
Villanova University

ROBERT COLWELL
University of Connecticut

GEORGE COX
San Diego State University

DANIEL FAIRBANKS
Brigham Young University

EUGENE KOZLOFF
University of Washington

WILLIAM PARSON
University of Washington

CLEON ROSS
Colorado State University

SAMUEL SWEET
University of California, Santa Barbara

Developmental Editor

BEVERLY McMILLAN
Gloucester Point, Virginia

BIOLOGY
The Unity and Diversity of Life

SIXTH EDITION

CECIE STARR
Belmont, California

RALPH TAGGART
Michigan State University

Wadsworth Publishing Company
Belmont, California
A Division of Wadsworth, Inc.

BIOLOGY PUBLISHER: *Jack C. Carey*

EDITORIAL ASSISTANT: *Kathryn Shea*

ART DIRECTOR AND DESIGNER: *Stephen Rapley*

PRODUCTION EDITOR: *Mary Forkner Douglas*

COPY EDITOR: *Carolyn McGovern*

PRODUCTION COORDINATOR: *Jerry Holloway*

MANUFACTURING: *Randy Hurst*

MARKETING: *Todd Armstrong, Karen Culver*

EDITORIAL PRODUCTION: *Scott Alkire, John Douglas, Kathy Hart, Gloria Joyce, Ed Serdziak, Karen Stough*

PERMISSIONS: *Marion Hansen*

PHOTO RESEARCH: *Marion Hansen, Stuart Kenter*

ARTISTS: *Susan Breitbard, Lewis Calver, Joan Carol, Raychel Ciemma, Robert Demerest, Ron Erwin, Enid Hatton, Darwin Hennings, Vally Hennings, Joel Ito, Robin Jensen, Keith Kasnot, Julie Leech, Laszlo Mezoly, Leonard Morgan, Palay/Beaubois (Phoebe Gloeckner, Lynne Larson, Betsy Palay), Victor Royer, Jeanne Schreiber, Kevin Somerville, John Waller, Judy Waller, Jennifer Wardrip*

DESIGN CONSULTANT: *Gary Head*

COVER DESIGN: *Stephen Rapley*

COVER PHOTOGRAPH: © *Thomas D. Mangelsen*

COMPOSITOR: *G&S Typesetters, Inc.: Bill M. Grosskopf, Merry Finley, Pat Molenaar, Beverly Zigal, Maurine Zook*

COLOR SEPARATOR: *H&S Graphics, Inc.: Tom Andersen, Nancy Dean, Roger Tillander, Marty O'Dean, Dennis Schnell*

PRINTING: *R. R. Donnelley & Sons Company/Willard*

3 4 5 6 7 8 9 10   96 95 94 93 92

Library of Congress Cataloging in Publication Data
Starr, Cecie.
    Biology: the unity and diversity of life / Cecie Starr, Ralph Taggart—6th ed.
        p.   cm.
    Includes bibliographical references and index.

    1. Biology.   I. Taggart, Ralph.   II. Title.
QH308.2.S72 1992
574—dc20   91-44997
                                                            CIP

PREFACE

Ask people at random to comment on the photograph on the cover of this book and you might hear something like, *What sweet little birds!* (as we did).

Those "sweet little birds" are highly specialized predators, the African bee-eaters of the family Meropidae. They swoop preferentially after bees and wasps in midair, using their long bill to catch safely, hold onto, and crush the stinging types. Then they *WHAP WHAP WHAP* the crushed body against the side of a branch or some other hard surface until the stinger protrudes and its venom drips out. After tossing the body several times in the air, the bee-eater gulps it down, head first, stinger last.

All textbooks for survey courses in biology include such descriptions of events in the natural world. All put on a parade of representatives from the five great kingdoms of organisms. They all describe the structure and function of those organisms at different levels of biological organization. The descriptions are useful, in that they help students build a working vocabulary about the parts and processes of life.

And yet, if textbooks are to convey accurately the nature of biological science, they must be more than a collection of observations. They must help students become familiar with the approach that biologists take to answering questions about what it is they see. With this approach, for example, the photograph of bee-eaters may open a door of inquiry, with all sorts of questions tumbling out.

. . . I've read about the aggressive African bees and what happened after they were introduced to South America. If the bees are so aggressive, why don't they defend themselves? Are they color-blind? How else could such brilliant-blue birds sneak up on them in midair? Maybe bees don't see blue. (Come to think of it, they must not be color-blind—I see them all the time on yellow flowers.) How could I test this idea? If it's true, then bees should pass up blue flowers in favor of yellow ones. How might I set up an experiment to test this prediction? For one thing, I'd better use odorless plastic flowers . . . .

Biologists ask questions, make educated guesses (hypotheses) about possible answers, then devise ways to rigorously test predictions that will hold true if the hypotheses are correct. In broad outline, their approach is that simple. Yet it has proved to be one of the most useful of all tools for explaining the world around us. Students can use the biological approach to satisfy curiosity about bees and bee-eaters. They can use this approach to pick their way logically through today's environmental, medical, and social landmines. Finally, they can use it to understand the past and predict possible futures for ourselves and all other organisms.

OBJECTIVES FOR THIS EDITION

Biology: The Unity and Diversity of Life has been evolving for eighteen years. More than 2 million students have used this book, and each revision becomes more refined in response to their experience with it. More than 1,500 dedicated teachers and researchers have shared insights with us during the years of refinements. They are our guardians of reading level and depth of coverage, of currency and accuracy.

As with previous editions, we mapped out major objectives to guide us in our approach to the sixth edition: First, write in a clear, engaging style, without being patronizing. Second, give enough examples of problem solving and experiments to provide familiarity with a scientific approach to interpreting the world. Third, identify the key concepts and select topics that reflect current research in all major fields, then present this material in light of two major themes in biology—evolution and energy flow. Fourth, use interesting, informative applications to stimulate student interest. Fifth, create easy-to-follow line art and select informative photographs.

REVISION HIGHLIGHTS

Writing Style

Years ago, we thought students could be enticed into the biological sciences with lively writing, memorable analogies, and engaging bits of natural history. However, the prevailing view was that such an approach somehow would be inappropriate for a textbook dealing with biological science. And so, for the past fifteen years, we focused primarily on making the writing clear and the science accurate.

Today, students often pick up biology textbooks with apprehension. If the words do not engage them, they sometimes end up hating the book *and* the subject. Instructors still ask for a scientifically accurate book—but now they also ask for one that puts the life back in.

We could not be more pleased. Because we devoted so many years to writing about biology with confidence, precision, and objectivity, we knew where the writing in this new edition *could* be loosened up. Interrupting a description of, say, the mechanisms of mitosis with a dithering of words will do the struggling student no good. Plunking humorous anecdotes into a chapter on the correlation between geologic and organismic evolution trivializes a magnificent story. Taking up valuable reading time with bits of natural history is pointless—unless those bits lead students to the big concepts.

By contrast, it certainly is appropriate to liven up a paragraph on, say, the structure of the nuclear envelope (page 61), the role of mitosis in growth and development (page 146), and the functions of skin (page 621). When you look through this new edition, you will see stunning line art and photographs. Don't let them distract you from the line-by-line judgment calls made with respect to the writing in every chapter. Improving and livening up the writing was our major objective.

Vignettes

What authors say in a preface sometimes bears no apparent resemblance to what they did in the book. As corroboration of what we did with the writing, look at the chapter introductions, each a short story that leads into the chapter's key concepts. Some provide glimpses into the natural history of an organism, others show how biological science applies to human affairs, and many do both.

If the Chapter 9 vignette rings true, it is because Beverly McMillan sat quietly beside the Alagnak at dawn, watching life end for a female salmon. If the body language of a bulldog making his contribution to cardiology provides a light touch (Chapter 38), this is Margaret Warner's offering. Remembering a journal article from her graduate school days, she rummaged through her university's archives and found those photographs for us. If you wonder whether a confrontation with a tornado will work in class, this is Fred Delcomyn's story (the Chapter 32 vignette) and it works for him. If the Chapter 24 vignette on daisies as supermarkets seems accurate yet refreshing, Edward Ross has been thinking about this for a long time.

Applications

This new edition features a greatly increased number of applications, all indexed on the back endpapers for easy reference. Some examples enrich the text. Others are boxed illustrations or *Commentaries* that provide in-depth information for the interested student but do not interrupt the text flow.

Many examples convey the importance of biological science in general, as when basic concepts in population ecology are applied to the prospects and problems of 5.4 billion of us now living on this planet (pages 797 and 875). Others bring home the impact of biological research on individual lives, as when students are asked to think about the effects of crack cocaine (page 562), anabolic steroids (page 637) or the implications of human gene therapy (page 257), which to a limited extent is already under way.

Starting in February, 1992, Wadsworth will be publishing an *Annual Newsletter* on important new applications that may be used to supplement those already incorporated into the sixth edition.

Doing Science

Earlier editions included many examples of biologists at work as a way to help students develop their own understanding of critical thinking. The entire chapter on DNA structure and function has been especially successful in this respect. So have the descriptions of experimental evidence for the concepts being discussed, as in the chapter on plant growth and development. (See also the index entry, Experiments.) This edition builds on our base of science in action.

At John Alcock's suggestion, we added *Doing Science* essays. In one, students will follow Molly Lutcavage's line of questioning in her studies of the leatherback turtle, a species on the brink of extinction (page 712). In one of the essays that John drafted himself, they will see how DNA fingerprinting was used to help explain self-sacrificing behavior—not among insects, but among a fascinating group of mammals (page 918).

Illustrations

One of us (Cecie Starr) has for eighteen years been obsessed with writing and creating illustrations simultaneously. It takes her almost as much time to research, develop, and integrate art with the text as to write and rewrite manuscripts. The obsession extends to positioning art and text references on the same two-page spread, no page-flipping required. Chapter 8, an obvious example, shows how layouts make it easier to study glycolysis, the Krebs cycle, electron transport phosphorylation, and anaerobic pathways.

Icons (pictorial representations) next to the main art show students where pathways or structures occur in a cell, multicelled body, or some other system. Zoom-sequence illustrations, from the macroscopic to microscopic, serve a similar purpose. Simple color-coded diagrams help students interpret micrographs.

Careful use of color helps students track information on hard-to-visualize topics. Throughout the book, for instance, proteins are color-coded green, carbohydrates pink, lipids yellow and gold, DNA blue, and RNA orange. Full-color anatomical paintings help give students a sense of the splendid internal complexity of organisms.

Often we incorporated written summaries *within* diagrams to make concepts easier to grasp. Where possible, we broke down information into a series of steps that are far less threatening than one large, complex diagram. Students find this approach useful, particularly with respect to art on mitosis, meiosis, and protein synthesis. It works just as effectively for such topics as antibody-mediated immunity and neural functioning.

Consider also the pedagogical impact of illustration size, as Starr did for every page. One photograph (page 885) conveys the magnitude of tropical rain forest destruction; a small photograph of a patch of burning trees could never do this. Probably few students gasp in wonder over a diagram of biomes—but ask them to use that diagram to interpret the spectacular photograph preceding Chapter 1. Pieced together from thousands of satellite images, it reveals the sweep of the Sahara, the collective green of boreal forests, and other features of the earth's surface.

STUDY AIDS

New to this edition is a *list of key concepts* following the vignette for each chapter. We increased the number of summary statements of concepts within the text itself to help keep readers on track. Several end-of-chapter study aids reinforce the key concepts. Each chapter has a *summary* in list form, *review questions*, a *self-quiz*, *selected key terms*, and *recommended readings*. Page numbers tie each review question and key term to the relevant text page.

Numerous *genetics problems* help students grasp the principles of inheritance. The *glossary* includes pronunciation guides and origins of words, when such information will make formidable words less so. The *index* is comprehensive; students find a door to the text more quickly through finer divisions of topics. The first appendix has *metric-English conversion charts*. The second is a *classification scheme* that students can use for reference purposes. The third has *detailed answers* to the genetics problems; and the fourth, *answers* to self-quizzes. The final appendix shows structural formulas for *major metabolic pathways* for interested students and instructors who prefer the added detail.

The appendixes and glossary are printed on paper of different tints to preclude frustrating searches for where one ends and the next starts.

SUPPLEMENTS

Twenty supplements are available. *Full-color transparencies* and *35mm slides* of almost all illustrations from the book are labeled with large, boldface type. A *Test Items* booklet has 5,000 questions by outstanding test writers. Questions are available in electronic form on IBM, Apple IIe, and Macintosh.

An *Instructor's Resource Manual* has, for each chapter, an outline, objectives, list of boldface or italic terms, and a detailed lecture outline. It also includes suggestions for lecture presentations, classroom and laboratory demonstrations, suggested discussion questions, research paper topics, and annotations for filmstrips and videos. *Lecture outlines* in the Instructor's Resource Manual are available on a data disk for those who wish to modify the material. A *Videodisc Correlation Directory and Barcode Guide* correlates the text with popular videodiscs. Software—*HyperCard Stacks for Videodiscs*—correlates the text with the same videodiscs.

A new, active-oriented *Study Guide and Workbook* asks students to respond to almost all questions by writing in the guide. Questions are arranged by chapter section. Each chapter also has a set of critical thinking questions. The *chapter objectives* of the Study Guide are available on disk as part of the testing file for those who wish to modify or select portions of the material. An *electronic study guide* consists of multiple-choice questions different from those in the test-item booklet. Students get feedback on why their answer is correct or incorrect. A 100-page *Answer Booklet* has answers to the book's end-of-chapter review questions.

A special version of *STELLA II*, a software tool for developing critical thinking skills, is available to users of the book, together with a workbook.

Approximately 400 *flashcards* with 1,000 glossary items are available. There are four anthologies. *Contemporary Readings in Biology* has articles on applications of interest to students. *Science and the Human Spirit: Contexts for Writing and Learning* helps students learn to write effectively about biology. *Ethical Issues in the New Reproductive Technologies* discusses some major issues of our time. *The Game of Science* gives students a realistic view of what science is and what scientists do.

A new *Laboratory Manual* has 38 experiments and exercises. It now contains hundreds of labeled photographs, and all illustrations are in full color. Many experiments are divided into distinct parts that can be assigned individually, depending on time available. All have objectives, discussion (introduction, background, and relevance), a list of materials for each part of an experiment, procedural steps, pre-lab questions, and post-lab questions. An *Instructor's Manual* accompanies the Laboratory Manual. It covers quantities, procedures for preparing reagents, time requirements for each portion of the exercise, hints to make the lab a success, and vendors of materials with item numbers.

IMPROVEMENTS IN CONTENT AND ORGANIZATION

Although the following paragraphs are by no means inclusive, they convey the magnitude of the revision.

INTRODUCTION We streamlined the first two chapters of the preceding edition. Now one chapter provides an overview of key biological concepts and a revised treatment of scientific methods, livened up by John Alcock. Simple examples introduce the pertinent points of evolution by natural selection, but the history of evolutionary thought now is the stage-setting chapter for the evolution unit (III).

UNIT I. CELLULAR BASIS OF LIFE More concise writing and greatly improved art make the chapters on cell structure and biochemistry more accessible. Chapter 2 has a new *Commentary* on radioisotopes. Chapter 3 includes an improved discussion of carbohydrates, a new *Commentary* on cholesterol and atherosclerosis, and a better diagram of hemoglobin structure. Chapter 4 has a simpler description of the cytomembrane system. Notice the cell icons in the illustrations of organelles. We updated the classification of membrane proteins (Chapter 5). Better diagrams of freeze-fracturing accompany the *Doing Science* essay (page 78). Chapter 6 presents a simpler overview of basic metabolism.

The plant in the zoom-sequence of chloroplast structure (Chapter 7) is now the same species used in Chapter 28, which continues the story by showing translocation. David Fisher helped develop these illustrations. Chapter 8 is reorganized—first the aerobic pathway, then anaerobic pathways. To keep text concepts uncluttered, details of ATP formation in Chapters 7 and 8 are presented in boxed illustrations.

UNIT II. PRINCIPLES OF INHERITANCE Chapters 9 through 11 already are effective in the classroom. We sharpened the writing, included new examples (such as Labrador coat color), but left the organization much the same. Robert Robbins suggested the vignette for Chapter 10 and the *Commentary* on HeLa cells (page 147). Chromosomal inheritance and human genetics are combined in one chapter (12). Morgan's fruit fly experiments are in an optional, boxed illustration. Students should enjoy the new *Commentary* on sex determination (page 186).

The Chapter 13 vignette provides background for the Watson-Crick story and reminds students that

science proceeds as a community effort (more or less). The organization of DNA in chromosomes is now described and illustrated in this chapter (page 212). The Chapter 14 vignette actually makes the idea of reading about protein synthesis nonthreatening. A simple overview and improved art make the chapter easier to follow. Early studies of gene function are described in an optional boxed illustration. Gene mutation is now introduced in this chapter, with a *Commentary* on its role in evolution.

Chapter 15 (gene regulation) is updated, with better delineation between prokaryotic and eukaryotic mechanisms. The vignette on control of cell division sets the stage for the *Commentary* on cancer. Daniel Fairbanks and Lisa Starr made solid contributions to Chapter 16, which provides a reorganized, updated, and more accurate picture of recombinant DNA technology and genetic engineering. The *Doing Science* essay gives interested students simple descriptions of gel electrophoresis of DNA and DNA sequencing methods.

UNIT III. PRINCIPLES OF EVOLUTION The evolution and diversity units now immediately follow the genetics unit. We overhauled the content and added spectacular illustrations (see page 312). Chapter 17 provides the historical background. We polished the chapter on microevolution (18) and added an in-depth look at a current study of speciation (page 288).

Chapter 19 has crisper descriptions of the evidence for the origin of the earth and life. Events and mechanisms underlying large-scale evolutionary patterns and rates of change are described succinctly. Macroevolutionary patterns dominating each major geologic era are sketched out. Figure 19.16 graphically emphasizes a central concept—that changes in the environment have been a profound force in the evolution of life.

Aaron Bauer wrote a new chapter on systematics (20) and, amazingly, made cladistics understandable (page 324). We moved the case study on human evolution (Chapter 21) here to make the unit self-contained. It includes a *Doing Science* essay on mitochondrial DNA and recent human ancestry. A section on classification outlines the five-kingdom scheme.

In Chapter 20 and elsewhere, we remind students that boundaries between taxa are not real; we impose them on a continuum of evolutionary lines. Taxonomists take the impositions seriously, and possibly our decision to classify the red, brown, and green algae as plants will make some of them cranky. However, we did not make the decision lightly. It reflects the overwhelming preference of hundreds of teachers who responded to a questionnaire on this issue.

UNIT IV. EVOLUTION AND DIVERSITY We reworked the diversity unit extensively. Introductory texts (our earlier editions included) tend to slight the microbial world. Notice the expanded, richly illustrated coverage of viruses, bacteria, and protistans in Chapter 22. The chapter also has *Commentaries* on infectious diseases (page 352), eukaryotic origins (page 361), and the beginnings of multicellularity (page 369). Fungi now have their own chapter (23) that conveys the diversity in this often-ignored kingdom.

We clarified the Chapter 24 survey of evolutionary trends among plants. More applications are woven into descriptions of the major divisions, as on page 388. Once again, Eugene Kozloff guided us through the maze of invertebrates and helped refine the chapter (25). Vertebrates are described in a separate chapter (26). In both chapters, icons serve as effective roadmaps.

UNIT V. PLANT STRUCTURE AND FUNCTION We made the writing easier to follow, added applications, and improved the art and page layouts. Chapter 28 has better coverage of root nodules (page 489) and superior diagrams for absorption (491), transpiration (492), and translocation (496).

The vignette in Chapter 29 gives new meaning to the word chocolate, the *Commentary* (page 506) provides vivid examples of pollination, and a *Doing Science* essay asking why some plants produce so many flowers (page 512) reminds students of what it means to think critically.

UNIT VI. ANIMAL STRUCTURE AND FUNCTION Extensive rewriting and many more applications make this inherently complex unit approachable. The new art speaks for itself.

We updated tissue classification and micrographs and explained homeostasis with tangible examples (Chapter 31). Neurobiologists helped update Chapter 32, and our teacher reviewers helped make it accessible. Chapter 33 better describes the evolution of nervous systems and the neural wiring of vertebrates.

The endocrine chapter (34) has less abstract examples and art. Chapter 35 provides a more accurate picture of sensory function. It has new material on echolocation, pain, and vision, including a *Commentary* on eye disorders (612).

We packed Chapter 36 with applications that should hold student interest. The sections on muscle function and energy metabolism make better sense. An integrative diagram at the start of Chapters 37, 38, and 39 helps students visualize how systems are integrated. We expanded the material on human nutrition and included a *Commentary* on eating disorders (650). Chapters 38 through 42 underwent major reorganization and updating. Whether assigned or not, Chapter 43 on human reproduction and development is one that students read closely, and we took special care to provide them with accurate and current information.

UNIT VII. ECOLOGY AND BEHAVIOR We worked closely with Robert Colwell and George Cox to reorganize and update the ecology chapters. Growth equations in Chapter 44 are described more clearly and the section on human population growth is expanded. Chapter 45 has refined definitions for habitat, niche, and species richness. Charles Krebs suggested an update for the Canadian lynx-hare story (pages 811–812). Jane Lubchenko's study of predation and competition is included (page 817). There is a new *Commentary* on species introduction (page 818).

The Chapter 46 vignette on a major environmental issue may leave students with a sense that things *can* change when we put our minds to it. A new carbon cycle diagram (page 837) leads into the *Commentary* on global warming (838). Chapter 47 provides a tighter overview of factors shaping climate, hence ecosystems. Photographs are large enough to show biome features. The text on lake ecosystems and intertidal zonation is more straightforward. Ernest Benfield provided material on stream ecosystems (864). Tropical reefs are now illustrated in this chapter (868). The Chapter 48 vignette describes tropical rain forests, then the text conveys the magnitude and pace of their destruction (874 and 884). Our friend Tyler Miller, Jr., helped us update this important chapter.

John Alcock's interest in teaching students how to think critically is evident in his two chapters (49 and 50), starting with the vignette on nest-building behavior. He updated and reorganized both.

JONES, PATRICIA, *Stanford University*
JUILLERAT, FLORENCE, *Indiana University–Purdue University*
KAYE, GORDON, *Albany Medical School*
KEIM, MARY, *Seminole College*
KELLY, DOUGLAS, *University of Southern California*
KELSEN, STEVEN, *Temple University Hospital*
KEYES, JACK, *Linfield College, Portland*
KIGER, JOHN, *University of California, Davis*
KIMBALL, JOHN, *Tufts University*
KIRK, HELEN, *University of Western Ontario*
KNUTTGEN, HAROLD, *Boston University*
KREBS, CHARLES, *University of British Columbia*
KURIS, ARMAND, *University of California, Santa Barbara*
KUTCHAI, HOWARD, *University of Virginia Medical School*
LATIES, GEORGE, *University of California, Los Angeles*
LATTA, VIRGINIA, *Jefferson State Junior College*
LEFEVRE, GEORGE, *California State University, Northridge*
LEVY, MATTHEW, *School of Medicine, City University of New York*
LEWIS, LARRY, *Bradford University*
LINDE, RANDY, *Palo Alto Medical Foundation*
LINDSEY, JERRI, *Tarrant County Junior College*
LITTLE, ROBERT, *Medical College of Georgia*
LOCKE, MICHAEL, *University of Western Ontario*
MACKLIN, MONICA, *Northeastern State University*
MADIGAN, MICHAEL, *Southern Illinois University, Carbondale*
MAJUMDAR, S. K., *Lafayette College*
MALLOCH, DAVID, *University of Toronto*
MANN, ALAN, *University of Pennsylvania*
MARGULIES, MAURICE, *Rockville, Maryland*
MARGULIS, LYNN, *University of Massachusetts, Amherst*
MARTIN, JAMES, *Reynolds Community College*
MATHEWS, ROBERT, *University of Georgia*
MATSON, RONALD, *Kennesaw State College*
MATTHAI, WILLIAM, *Tarrant County Junior College*
MAXSON, LINDA, *Pennsylvania State University*
MAXWELL, JOYCE, *California State University, Northridge*
MCCLINTIC, J. ROBERT, *California State University, Fresno*
MCEDWARD, LARRY, *University of Florida*
MCKEAN, HEATHER, *Eastern Washington State University*
MCKEE, DOROTHY, *Auburn University, Montgomery*
MCNABB, ANNE, *Virginia Polytechnic Institute and State University*
MCREYNOLDS, JOHN, *University of Michigan Medical School*
MERTENS, THOMAS, *Ball State University*
MEYER, NANCY, *University of Michigan, Dearborn*
MIMMS, CHARLES, *University of Georgia*
MITZNER, WAYNE, *Johns Hopkins University*
MOCK, DOUG, *University of Washington*
MOHRMAN, DAVID, *University of Minnesota*
MOISES, HYLAND, *University of Michigan Medical School*
MOORE-LANDECKER, ELIZABETH, *Glassboro State University*
MORBECK, MARY ELLEN, *University of Arizona*
MORRISON, WILLIAM, *Shippensburg University*
MORTON, DAVID, *Frostburg State University*
MOUNT, DAVID, *University of Arizona*
MURPHY, RICHARD, *University of Virginia Medical School*
MURRISH, DAVID, *State University of New York, Binghamton*
MYERS, NORMAN, *Oxford University, England*
MYRES, BRIAN, *Cypress College*
NAGLE, JAMES, *Drew University*
NEMEROFSKY, ARNOLD, *State University of New York, New Paltz*
NICHOLS-KIRK, HELEN, *University of Western Ontario*
NORRIS, DAVID, *University of Colorado*
O'BRIEN, ELINOR, *Boston College*
OJANLATVA, ANSA, *Sacramento, California*
OLSON, MERLE, *University of Texas Health Science Center*
ORR, CLIFTON, *University of Arkansas*
ORR, ROBERT, *California Academy of Sciences*
PAI, ANNA, *Montclair State College*
PALMBLAD, IVAN, *Utah State University*
PAPPENFUSS, HERBERT, *Boise State University*
PARSONS, THOMAS, *University of Toronto*
PAULY, JOHN, *University of Arkansas for Medical Sciences*
PECHENIK, JAN, *Tufts University*

PERRY, JAMES, *Frostburg State University*
PETERSON, GARY, *South Dakota State University*
PIERCE, CARL, *Washington University*
PIKE, CARL, *Franklin and Marshall College*
PLEASANTS, BARBARA, *Iowa State University*
POWELL, FRANK, *University of California, San Diego*
RALPH, CHARLES, *Colorado State University*
RAWN, CARROLL, *Seton Hall University*
REEVE, MARIAN, *Emeritus, Merritt Community College*
RICKETT, JOHN, *University of Arkansas, Little Rock*
RIEDER, CONLY, *Wadsworth Center for Laboratories and Research*
ROMANO, FRANK, *Jacksonville State University*
ROSEN, FRED, *Harvard University School of Medicine*
ROSS, EDWARD, *California Academy of Sciences*
ROSS, GORDON, *University of California Medical Center, Los Angeles*
ROSS, IAN, *University of California, Santa Barbara*
ROST, THOMAS, *University of California, Davis*
RUIBAL, RUDOLFO, *University of California, Riverside*
SACHS, GEORGE, *University of California, Los Angeles*
SACKETT, JAMES, *University of California, Los Angeles*
SALISBURY, FRANK, *Utah State University*
SCHAPIRO, HARRIET, *San Diego State University*
SCHECKLER, STEPHEN, *Virginia Polytechnic Institute and State University*
SCHIMEL, DAVID, *NASA Ames Research Center*
SCHLESINGER, WILLIAM, *Duke University*
SCHMID, RUDI, *University of California, Berkeley*
SCHMOYER, IRVIN, *Muhlenberg College*
SCHNERMANN, JURGEN, *University of Michigan School of Medicine*
SEARLES, RICHARD, *Duke University*
SEHGAL, PREM, *East Carolina University*
SHARP, ROGER, *University of Nebraska, Omaha*
SHEPHERD, JOHN, *Mayo Medical University*
SHERMAN, PAUL, *Cornell University*
SHOEMAKER, DAVID, *Emory University School of Medicine*
SHONTZ, NANCY, *Grand Valley State University*
SHOPPER, MARILYN, *Johnson County Community College*
SILK, WENDY, *University of California, Davis*
SLATKIN, MONTGOMERY, *University of California, Berkeley*
SMILES, MICHAEL, *State University of New York, Farmingdale*
SMITH, RALPH, *University of California, Berkeley*
SOLOMON, TRAVIS, *Kansas City V.A. Medical Center*
STARR, LISA, *Scripps Clinic and Research Foundation*
STEIN-TAYLOR, JANET, *University of Illinois, Chicago*
STEINERT, KATHLEEN, *Bellevue Community College*
STITT, JOHN, *John B. Pierce Foundation Laboratory*
SULLIVAN, LAWRENCE, *University of Kansas*
SUMMERS, GERALD, *University of Missouri*
SUNDBERG, MARSHALL, *Louisiana State University*
SWABY, JAMES, *United States Air Force Academy*
TERHUNE, JERRY, *Jefferson Community College, University of Kentucky*
THAMES, MARC, *Medical College of Virginia*
TIFFANY, LOIS, *Iowa State University*
TIZARD, IAN, *Texas A & M University*
TORSTVEIT, ELINOR, *Concordia College*
TRAMMELL, JAMES, JR., *Arapahoe Community College*
TUTTLE, JEREMY, *University of Virginia, Charlottesville*
VALENTINE, JAMES, *University of California, Santa Barbara*
VALTIN, HEINZ, *Dartmouth Medical School*
WAALAND, ROBERT, *University of Washington*
WADE, MICHAEL, *University of Chicago*
WALSH, BRUCE, *University of Arizona*
WARING, RICHARD, *Oregon State University*
WARMBRODT, ROBERT, *University of Maryland*
WEIGL, ANN, *Winston-Salem State University*
WEISBRODT, NORMAN, *University of Texas Medical School, Houston*
WELKIE, GEORGE, *Utah State University*
WHEELIS, MARK, *University of California, Davis*
WHIPP, BRIAN, *University of California, Los Angeles*
WHITTOW, G. CAUSEY, *University of Hawaii School of Medicine*
WINICUR, SANDRA, *Indiana University, South Bend*
WISE, ROBERT, *Francis Scott Key Medical Center*
WOMBLE, MARK, *University of Michigan Medical School*
ZIHLMAN, ADRIENNE, *University of California, Santa Cruz*

A COMMUNITY EFFORT

One, two, or a smattering of authors can write accurately and often very well about their field of interest, but it takes more than this to deal with the full breadth of biological sciences. For us, it takes an educational network that extends through the United States and on into Canada, England, Germany, France, Sweden, Australia, and elsewhere. We continually track down respected researchers, teachers, and photographers. For a few rarified topics, we invite resource manuscripts from specialists who have never turned their back on the call to teach a new generation of students. We integrate such material into our own manuscript, rewriting or graphically shaping it according to our strong convictions about what an introductory book must be.

John Alcock, Aaron Bauer, Rob Colwell, George Cox, Daniel Fairbanks, Eugene Kozloff, Bill Parson, Cleon Ross, and Sam Sweet have been exceptional in their commitment to our efforts. They have responded with grace to phone calls and faxes, to queries concerning the odd fact.

Many reviewers were content specialists, others were diary reviewers who evaluated the fifth edition's effectiveness in their own classrooms. Collectively, their comments helped us shape the revision. Our advisors assisted us in evaluating reviewer comments and in suggesting improvements in the new manuscripts and art. Katherine Denniston, Pamela Hanratty, William Hess, John Jackson, Bernard Frye, Greig Rose, and Nancy Meyer advised us on every chapter for level and clarity. Jackson and Denniston helped us develop new self-quiz questions.

If the writing now seems fresher, this is largely because of our creative interaction with Beverly McMillan, the developmental editor for this edition.

She also researched material and drafted several vignettes, *Doing Science* essays, and *Commentaries*. We treasure her as a friend and gifted writer.

Dick Greenberg, Jack Carey, Kathie Head, Stephen Rapley, and Randy Hurst at Wadsworth never let the users of this book down. This time they assembled the best production team and manufacturers in the business, starting with Mary Douglas, who has our unequivocal respect and friendship. Mary has the talent, toughness, sensitivity, compulsiveness, and oblique sense of humor required to shepherd text and art manuscripts of this complexity.

Because of Jerry Holloway and Kathryn Shea, we kept smiling instead of bashing our head against the wall. Marion Hansen would have liked to bash our head against the wall, but instead she persevered through one more edition and, with Stuart Kenter, collected exquisite photographs. Carolyn McGovern, Gloria Joyce, and Ed Serdziak took care of complicated editorial functions. Todd Armstrong, Karen Culver, and Debbie Dennis have been most supportive. Ryan Carey was the captive student reader. Barbara Odone, John Douglas, and Verbal Clark kept the paper flowing.

Besides designing a memorable cover for the book, Stephen Rapley worked out its interior design in consultations with Gary Head. Raychel Ciemma, Bob Demerest, Darwin Hennings, Vally Hennings, Len Morgan, and Betsy Palay did much of the outstanding new art. Susan Breitbard, Natalie Hill, Carole Lawson, Joan Olson, Jill Turney, and Kathryn Werhane worked directly with us as resident artists. Tom Anderson once again was responsive to picky requests on color separations. G&S is in a league by itself.

Jack Carey, would you have believed eighteen years ago that you would be publisher of such a widely used textbook in biology? You charted our course, and made it happen.

CONTENTS IN BRIEF

DETAILED
TABLE OF CONTENTS

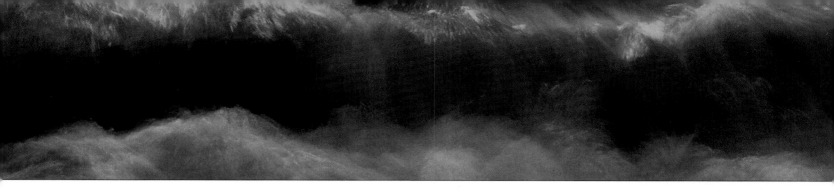

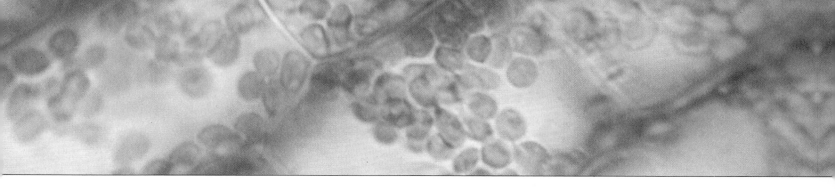

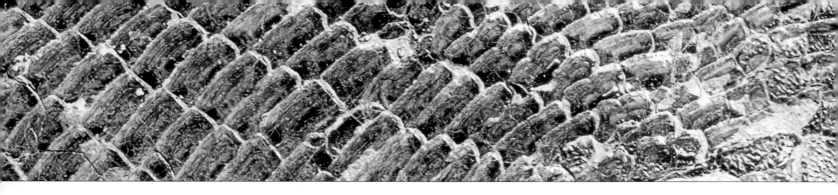

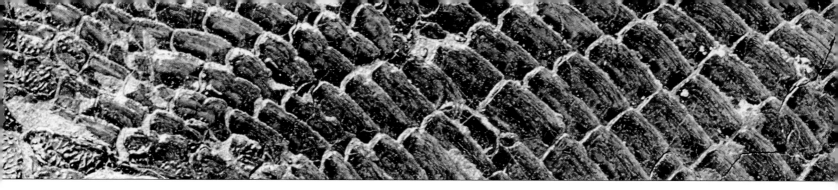

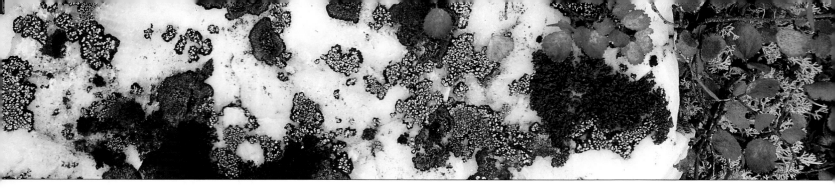

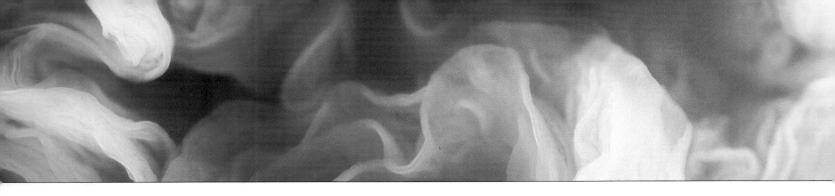

Current configurations of the earth's oceans and land masses—
the geologic stage upon which life's drama continues to unfold.
Thousands of separate images were pieced together to create
this remarkable photograph of our planet.

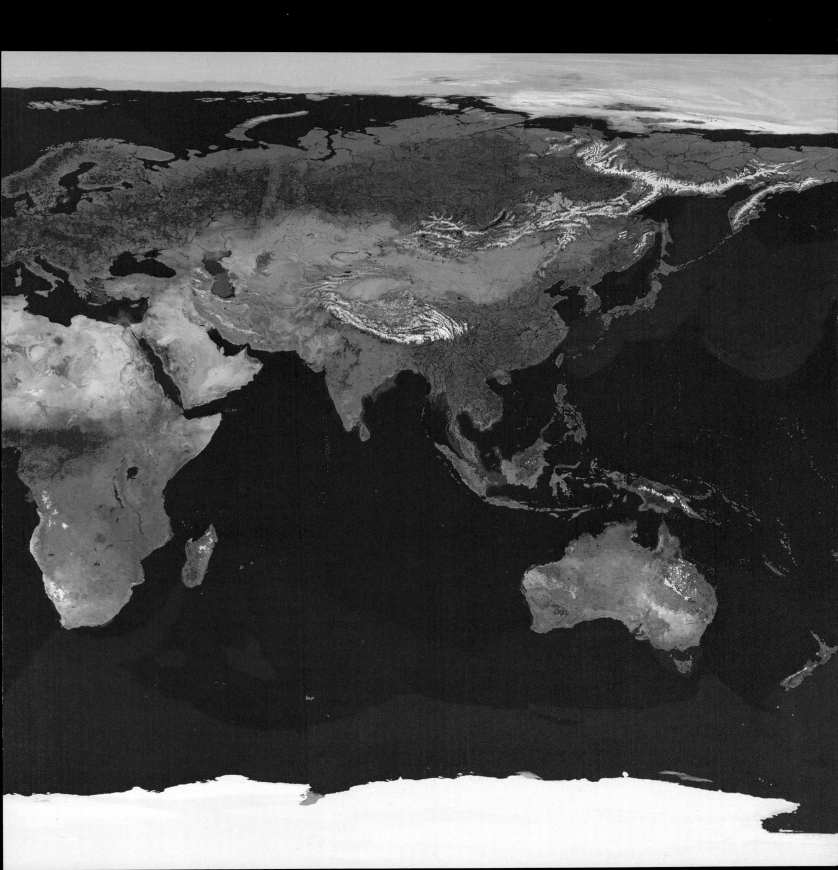

1 METHODS AND CONCEPTS IN BIOLOGY

Biology Revisited

Buried somewhere in that mass of nerve tissue just above and behind your eyes are memories of first encounters with the living world. Still in residence are sensations of discovering your own two hands and feet, your family, the change of seasons, the smell of rain-drenched earth and grass. In that brain are traces of early introductions to a great disorganized parade of insects, flowers, friends, and furred things, mostly living, sometimes dead. There, too, are memories of questions—*"What is life?"* and, inevitably, *"What is death?"* There are memories of answers, some satisfying, others less so.

Observing, asking questions, accumulating answers—in this manner you have acquired a store of knowledge about the world of life. As you have grown older, experience and education have been refining your questions, and no doubt the answers are more difficult to come by. What *is* life? What defines the living state? The answers you get may vary, depending, for example, on whether they come from a physician, a court of law, or the parents of a severely injured girl who is being maintained by mechanical life support systems because her brain no longer functions at all.

Yet despite the changing character of your questions, the world of living things persists much as it did before. New leaves still unfurl during the spring rains. Animals are born, they grow, reproduce, and die even as new individuals of their kind replace them. The world of life has not changed in these respects. It is just that your perceptions about them have deepened.

It is scarcely appropriate, then, for a book to claim that it is your introduction to biology—the study of life—when you have been studying life ever since awareness of the world began penetrating your brain. The subject is the same familiar world you have already thought about for many years. That is why this book claims only to be biology *revisited*, in ways that may help carry your thoughts about life to deeper, more organized levels of understanding.

Let us return to the question, What is life? Offhandedly, you might reply that you know it when you see it. Yet there is no simple answer, for the question opens up a story that has been unfolding in countless directions for several billion years! To biologists, "life" is an outcome of ancient events that led to the assembly of nonliving materials into the first organized, living cells. "Life" is a way of capturing and using energy and materials. "Life" is a way of sensing and responding to specific changes in the environment. "Life" is a capacity to reproduce; it is a capacity to follow programs of growth and development. And "life" evolves, meaning that details in the body plan and functions of each kind of organism can change through successive generations.

Clearly, a short description only hints at the meaning of life. Deeper insight requires wide-ranging study of life's characteristics.

Throughout this book you will encounter many different examples of living things—how they are constructed, how they function, where they live, what they do. The examples provide evidence in support of certain concepts which, when taken together, will give you a sense of what "life" is. The next few pages will provide you with a brief overview of those basic concepts. As you continue your reading in subsequent chapters, you may find it useful now and then to return to this overview as a way of reinforcing your grasp of details.

Figure 1.1 Think back on all you have known and seen, and this is the foundation for your deeper probes into the world of life.

1. All organisms are alike in these respects: Their structure, organization, and interactions arise from the properties of matter and energy. They obtain and use energy and materials from their environment, and they make controlled responses to changing conditions. They grow and reproduce, and instructions for traits that they pass on from one generation to the next reside in their DNA.

2. Organisms show great diversity in their structure, function, and behavior, largely as a result of evolution by means of natural selection.

3. The theories of science are based on systematic observations, hypotheses, predictions, and relentless testing. The external world, not internal conviction, is the testing ground for scientific theories.

SHARED CHARACTERISTICS OF LIFE

DNA and Biological Organization

Picture a frog on a rock, busily croaking. Without even thinking about it, you know that the frog is alive and the rock is not. At a much deeper level, however, the difference between them blurs. Frogs, rocks, and all other living or nonliving things are composed of the same particles (protons, electrons, and neutrons). The particles are organized into atoms, in every case according to the same physical laws. At the heart of those laws is something called **energy**—a capacity to make things happen, to do work. Energetic interactions bind atom to atom in predictable patterns, giving rise to the structured bits of matter we call molecules. Energetic interactions among molecules hold a rock together—and they hold a frog together.

It takes a special type of molecule called deoxyribonucleic acid, or **DNA**, to set living things apart from the nonliving world. No chunk of granite or quartz has it. DNA molecules contain the instructions for assembling each new organism from carbon, hydrogen, and a few

3

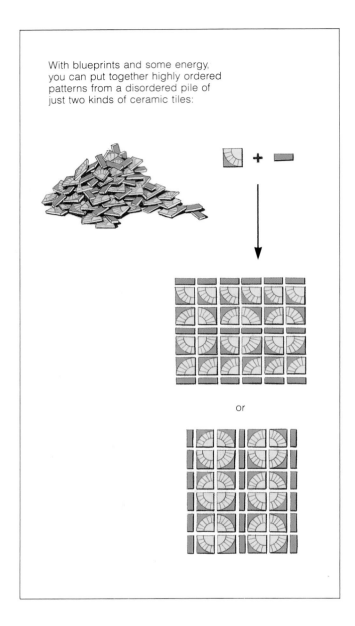

With blueprints and some energy, you can put together highly ordered patterns from a disordered pile of just two kinds of ceramic tiles:

or

Figure 1.2 Emergence of organized patterns from disorganized beginnings. Two ceramic tile patterns are shown here. You probably can visualize other possible patterns using the same two kinds of tiles. Similarly, the organization characteristic of life emerges from pools of simple building blocks, given energy sources and specific DNA "blueprints."

other kinds of "lifeless" molecules. By analogy, think of what you can do with just two kinds of ceramic tiles in a crafts kit. With a little effort, you can glue the tiles together according to the kit's directions, so that you can produce many organized patterns of tiles (Figure 1.2). Similarly, the organization of life emerges from lifeless matter with DNA "directions," some raw materials, and energy.

Look carefully at Figure 1.3, which outlines the levels of organization in nature. The quality of "life" actually emerges at the level of cells. A *cell* is the basic living unit. This means it has the capacity to maintain itself as an independent unit and to reproduce, given appropriate sources of energy and raw materials. Amoebas and many other single-celled organisms lead such independent lives.

A *multicelled organism* is more complex, with specialized cells typically arranged into tissues, organs, and often organ systems. Its cells depend on the integrated activities of one another, but each generally retains the capacity for independent existence. How do we know this? Individual cells that have been removed from humans and other multicelled organisms can be kept alive under controlled laboratory conditions.

The next, more inclusive level of organization is the *population*: a group of single-celled or multicelled organisms of the same kind occupying a given area. A congregation of penguins at a rookery in Antarctica is an example. Moving on, the populations of whales, seals, fishes, and all other organisms living in the same area as the penguins make up a *community*.

The next level, the *ecosystem*, includes the community *and* its physical and chemical environment. The most inclusive level of organization is the *biosphere*. The biosphere includes all regions of the earth's waters, crust, and atmosphere in which organisms live.

The structure and organization of nonliving *and* living things arise from the fundamental properties of matter and energy.

The structure and organization *unique* to living things starts with instructions contained in DNA molecules.

Metabolism

You never, ever will find a rock engaged in metabolic activities. Only living cells can do this. **Metabolism** refers to the cell's capacity to (1) *extract and transform energy* from its surroundings and (2) *use energy* and so maintain itself, grow, and reproduce. In essence, metabolism means "energy transfers" within the cell.

A growing rice plant nicely illustrates this aspect of life. Like other plants, it has cells that engage in *photosynthesis*. The cells convert sunlight energy to chemical energy, which is then parceled out to the tasks of building sugars, starch, and other good things from simple raw materials in the environment. (Chemical energy is remarkable stuff. Cells use it to build large molecules out of smaller bits. They also use it to split molecules apart and liberate various bits.) In photosynthesis, energy from sunlight drives the attachment of a bit of phosphate to a certain molecule, which thereby becomes known as ATP. ATP is a generous molecule. It readily transfers chemical energy to other molecules that function as metabolic workers (enzymes), building blocks, or energy reserves.

In rice plants, energy reserves are especially concentrated in starchy seeds—rice grains—from which more rice plants may grow. The energy reserves in countless trillions of rice grains also provide energy for billions of rice-eating humans around the world. How? In rice plants, humans, and most other organisms, stored chemical energy can be tapped for use by way of another metabolic process, called *aerobic respiration*. Later chapters will describe the splendid metabolic jugglings of photosynthesis and aerobic respiration. For now, the point to keep in mind is this:

Living things show metabolic activity: Their cells acquire and use energy to stockpile, tear down, build, and eliminate materials in ways that promote survival and reproduction.

Interdependency Among Organisms

With few exceptions, a flow of energy from the sun maintains the great pattern of organization in nature. Plants and some other photosynthetic organisms are the entry point for this flow—the food "producers" for the world of life. Animals are "consumers." Directly or indirectly, they feed on the energy stored in plant parts. For example, zebras tap directly into the stored energy when they nibble on grass, and lions tap into it indirectly when they nibble on zebras. Certain bacteria and fungi are "decomposers." When they feed on the tissues or remains of other organisms, they break down complex molecules to simple raw materials—which can be recycled back to the producers.

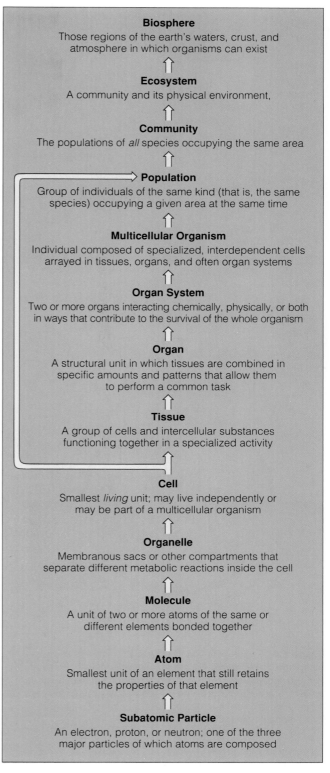

Figure 1.3 Simplified picture of the levels of organization in nature, starting with the subatomic particles that serve as the fundamental building blocks of all organisms.

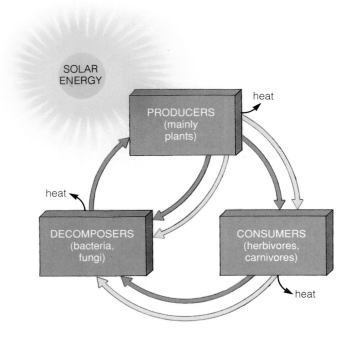

Figure 1.4 Energy flow and the cycling of materials in the biosphere. Here, grasses of the African savanna are producers that provide energy directly for zebras (herbivores) and indirectly for lions and vultures (carnivores). The wastes and remains of all these organisms are energy sources for decomposers, which cycle nutrients back to the producers.

Figure 1.4 is a generalized picture of energy flow and the cycling of materials through the world of life. Figure 1.5 is a specific example of the resulting interdependencies among organisms.

Energy flows to, within, and from single cells and multicelled organisms. It flows within and between populations, communities, and ecosystems. As you will see, interactions among organisms are part of the cycling of carbon and other substances on a global scale. They also have profound influence on the earth's energy "budget." Understand the extent of those interactions and you will gain insight into the greenhouse effect, acid rain, and many other modern-day problems.

All organisms are part of webs of organization in nature, in that they depend directly or indirectly on one another for energy and raw materials.

Homeostasis

It is often said that only living organisms "respond" to the environment. Yet a rock also responds to the environment, as when it yields to gravity and tumbles downhill or when it changes shape slowly under the battering of wind, rain, or tides. The real difference is this: *Organisms have the cellular means to sense environmental changes and make controlled responses to them.* They do so with the help of diverse **receptors**, which are molecules and structures that can detect specific information about the environment. When cells receive information from receptors, their activities become adjusted in ways that bring about an appropriate response.

a

b

c

d

Your body, for example, can withstand only so much heat or cold. It must rid itself of harmful substances. Certain foods must be available to it, in certain amounts. Yet temperatures shift, harmful substances may be encountered, and food is sometimes plentiful and sometimes scarce.

Even so, your body usually can adjust to the variations and so maintain internal operating conditions for its cells. **Homeostasis** refers to a state in which conditions in this "internal environment" are being maintained within a tolerable range. Homeostasis, too, is a common attribute of living things.

Think about what happens after you eat and simple sugar molecules make their way into your bloodstream. Certain cells detect the rising level of sugar in your blood and cause molecules of insulin, a hormone, to be released. Most of your body's cells have receptors for insulin, which prods the cells into taking up sugar molecules. With this uptake, the blood sugar level returns to normal. Now suppose you can't eat when you should and your blood sugar level falls. Then, a different hormone prods cells in your liver and elsewhere to dig into

Figure 1.5 How organisms interact through their requirements for energy and raw materials. This example makes the point, even though the cast of characters seems of a most improbable sort.

First we have the adult male elephant of the African savanna (**a**). It stands almost two stories high at the shoulder and weighs more than eight tons. This grazing animal eats large quantities of plants, the remains of which leave its body as droppings of considerable size (**b**). Appearances to the contrary, locked in the droppings are substantial stores of unused nutrients. With resource availability being what it is, even waste products from one kind of organism are food for another.

And so we next have little dung beetles rushing to the scene almost simultaneously with the uplifting of the elephant's tail. With great precision they carve out fragments of the dung into round balls (**c**). The dung balls are rolled off and buried underground in burrows, where they serve as compact food supplies. In these balls the beetles lay eggs, a reproductive behavior that assures the forthcoming offspring of a food supply (**d**). Also assured is an uncluttered environment. If the dung were to remain aboveground, it would dry out and pile up beneath the hot African sun. Instead, the surface of the land is tidied up, the beetle has its resource, and the remains of the dung are left to decay in burrows—there to enrich the soil that nourishes the plants that sustain (among others) the elephants.

Figure 1.6 "The insect"—a continuum of developmental stages, with new adaptive properties emerging at each stage. Shown here: the development of a silkworm moth, from egg (**a**) to larval stage (**b**), to pupal form (**c**), to emergence of the splendid adult form (**d, e**).

a   b

their storehouses of energy-rich molecules. Those molecules are broken down into simple sugars, which are released into the bloodstream—and again the blood sugar level returns to normal.

All organisms respond to changing conditions through use of homeostatic controls, which help maintain their internal operating conditions.

Reproduction

We humans tend to think we enter the world rather abruptly and are destined to leave it the same way. Yet we and all other organisms are more than this. *We are part of an immense, ongoing journey that began billions of years ago.* Think about the first cell of a new human individual, which is produced when a sperm joins with an egg. The cell would not even exist if the sperm and egg had not been formed earlier, according to DNA instructions that were passed down through countless generations. With time-tested DNA instructions, a new human body develops in ways that will prepare it, ultimately, for *reproduction*. With reproduction, the journey of life continues.

If someone asked you to think of a moth, would you simply picture a winged insect? What of the tiny fertilized egg deposited on a branch by a female moth (Figure 1.6)? The egg contains all the instructions necessary to become an adult. By those instructions, the egg first develops into a caterpillar, a larval form adapted for rapid feeding and growth. The caterpillar eats and increases in size until an internal "alarm clock" goes off. Then its body enters a so-called pupal stage of development, which requires wholesale remodeling. Some cells

die, while other cells multiply and become organized in different patterns. Now the adult moth emerges. It is equipped with organs in which eggs or sperm develop. Its wings are brightly colored and flutter at a frequency that can attract a potential mate. In short, the adult stage is adapted for reproduction.

None of these stages is "the insect." "The insect" is a series of organized stages from one fertilized egg to the next, each vital for the ultimate production of new moths. The instructions for each stage were written into moth DNA long before each moment of reproduction—and so the ancient moth story continues.

Each organism arises through *reproduction* (the production of offspring by one or more parents).

Each organism is part of a reproductive continuum that extends back through countless generations.

Mutation and Adapting to Change

The word *inheritance* refers to the transmission, from parents to offspring, of structural and functional patterns characteristic of each kind of organism. In living cells, hereditary instructions are encoded in molecules of DNA. Those instructions have two striking qualities. They assure that offspring will resemble their parents— and they also permit *variations* in the details of their traits. By "trait" we mean some aspect of an organism's body, functioning, or behavior. For example, having five fingers on each hand is a human trait. Yet some humans are born with six fingers on each hand instead of five! Variations in traits arise through **mutations**, which are changes in the structure or number of DNA molecules.

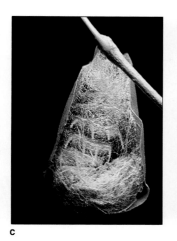

c

d

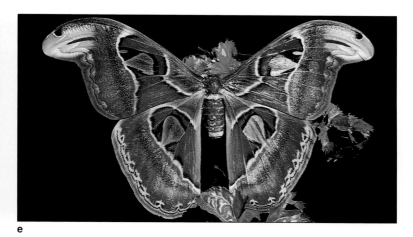

e

Many mutations are harmful, for the separate bits of information in DNA are part of a coordinated whole. A single mutation in a tiny segment of human DNA may lead to a genetic disorder such as hemophilia, in which blood cannot clot properly in response to a cut or bruise. Yet on rare occasions, a mutation may prove to be harmless, even beneficial, under prevailing conditions. One type of mutation in light-colored moths leads to dark-colored offspring. What happens when a dark moth rests on a soot-covered tree? Bird predators simply do not see it. If most trees are soot-covered (as in industrial regions), light moths are more likely to be seen and eaten—so the dark form has a better chance of living long enough to reproduce. Under such conditions, the mutated form of the trait is more adaptive.

An **adaptive trait** simply is one that helps an organism survive and reproduce under a given set of environmental conditions.

In all organisms, DNA is the molecule of inheritance: Its instructions for reproducing traits are passed on from parents to offspring.

Mutations introduce variations in heritable traits.

Although most mutations are harmful, some give rise to variations in form, function, or behavior that turn out to be adaptive under prevailing conditions.

LIFE'S DIVERSITY

Five Kingdoms, Millions of Species

Until now, we have focused on the unity of life—on characteristics shared by all organisms. Superimposed on this shared heritage is immense diversity. Many mil-lions of different kinds of organisms, or **species**, inhabit the earth. And many millions more existed in the past and became extinct. Early attempts to make sense of life's diversity led to a classification scheme in which each species was assigned a two-part name. The first part designates the **genus** (plural, genera). It encompasses all the species having perceived similarities to one another. The second part designates a particular species within that genus.

For instance, *Quercus alba* is the scientific name of the white oak. *Quercus rubra* is the name of the red oak. (Once the genus name has been spelled out, subsequent uses of it in the same document can be abbreviated—for example, to *Q. rubra*.)

Life's diversity is classified further by using more inclusive groupings. For example, similar genera are placed in the same *family*, similar families into the same *order*, then similar orders into the same *class*. Similar classes are placed into a *division* or *phylum* (plural, phyla). In turn, phyla are assigned to a *kingdom*. Today, most biologists recognize the following five kingdoms:

Monera	*Bacteria. Single cells of relatively little internal complexity. Producers or decomposers.*
Protista	*Protistans. Single cells of considerable internal complexity. Producers or consumers.*
Fungi	*Fungi. Mostly multicelled. Decomposers.*
Plantae	*Plants. Mostly multicelled. Mostly producers.*
Animalia	*Animals. Multicelled. Consumers.*

Figure 1.7 Representatives of the five kingdoms of life.

Kingdom Monera. (**a**) A bacterium, seen with the aid of a microscope. The single-celled bacteria making up this kingdom live nearly everywhere, including in or on other organisms. The ones in your gut and on your skin outnumber the cells making up your body.

Kingdom Protista. (**b**) A parasitic trichonomad, from a termite gut. This kingdom of single-celled organisms has poorly defined boundaries; many lineages seem to have evolutionary connections with plants, fungi, and animals.

Kingdom Fungi. (**c**) A stinkhorn fungus. Many fungi are major decomposers. Even a single elm tree can shed 400 pounds of leaves in one season. Without decomposers, communities would gradually be buried in their own garbage.

Kingdom Plantae. (**d**) A grove of California coast redwoods. Like nearly all members of the plant kingdom, redwoods produce their own food through photosynthesis. (**e**) Flower of a plant called a composite. Its intricate pattern guides bees to the flower's nectar. The bees get food and the plant gets pollinated. Many organisms are locked in such mutually beneficial interactions.

Kingdom Animalia. (**f**) Male bighorn sheep competing for females. Like all animals, they cannot produce their own food and so must eat other organisms. Like most animals, they move about far more than the adult organisms of other kingdoms (most of which do not move about at all).

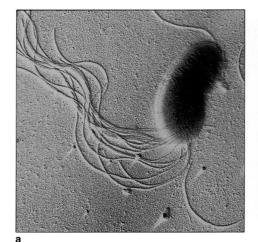
a

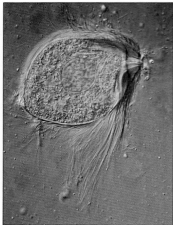
b

d

Table 1.1   Characteristics of Organisms in All Five Kingdoms

1. Complex structural organization based on instructions contained in DNA molecules.

2. Directly or indirectly, dependence on other organisms for energy and material resources.

3. Metabolic activity by the single cell or multiple cells composing the body.

4. Use of homeostatic controls that maintain favorable operating conditions in the body despite changing conditions in the environment.

5. Reproductive capacity, by which the instructions for heritable traits are passed from parents to offspring.

6. Diversity in body form, in the functions of various body parts, and in behavior. Such traits are adaptations to changing conditions in the environment.

7. The capacity to evolve, based ultimately on variations in traits that arise through mutations in DNA.

Figure 1.7 shows a few representatives of the five kingdoms. When looking at this figure, look also at Table 1.1, which summarizes the main characteristics of life that have been described so far in this chapter. *Every living organism in all five kingdoms displays these characteristics.*

An Evolutionary View of Diversity

If organisms are so much alike in so many ways, what could possibly account for their diversity? In biology, a key explanation is called evolution by means of natural selection.

c

e

f

By way of example, suppose a DNA mutation gives rise to a different form of a trait in one member of a population—say, black moth wings instead of white. Suppose black wings prove to be adaptive in concealing the moth from predators. Because the black-winged moth has an advantage over a white moth right next to it on a soot-covered tree trunk, it lives to reproduce. So do its black-winged offspring—and so do *their* offspring.

The variant form of this trait is now popping up with greater frequency. In time it may even become the more common form, so that what was once a population of mostly white-winged moths now consists mostly of dark-winged moths. **Evolution** is taking place—the char-

acter of the population is changing through successive generations.

Long ago, Charles Darwin used pigeons to explain how evolution might occur. Domesticated pigeons show great variation in their traits. Darwin pointed out that pigeon breeders who wish to promote certain traits, such as black tail feathers with curly edges, will "select" individual pigeons having the most black and the most curl in their tail feathers. By permitting only those birds to mate, they will foster the desired traits and eliminate others from their captive population.

Thus Darwin used *artificial* selection as a model for natural selection.

Figure 1.8 A few examples of the more than 300 varieties of domesticated pigeons. Such forms have been derived, by selective breeding, from the wild rock dove (**a**).

Figure 1.8 shows a few of the variations in size, feather color, and other traits of pigeons that resulted from *artificial* selection. In later chapters, we will look at the mechanisms and consequences of *natural* selection. For now, these are the points to remember:

1. Members of a population vary in form, function, and behavior, and much of this variation is heritable.

2. Some forms of heritable traits are more adaptive than others; they improve chances of surviving and reproducing. Thus individuals with adaptive traits tend to make up more of the reproductive base in each new generation.

3. Natural selection is simply a measure of the difference in survival and reproduction that has occurred among individuals that differ from one another in one or more traits.

4. Any population *evolves* when some forms of traits increase in frequency and others decrease or disappear over the generations. In this manner, variations have accumulated in different lines of organisms. Life's diversity is the sum total of those variations.

THE NATURE OF BIOLOGICAL INQUIRY

On Scientific Methods

Species evolve in myriad directions, like branches growing in many directions on a single tree of life. Today we say this with confidence, but it was not always so. Awareness of evolution developed over centuries, as naturalists and travelers collected specimens of living and extinct organisms, then asked questions about the similarities and differences among them. Darwin and others proposed only tentative answers to those questions. Evidence supporting some of their answers came much later, after generations of scientists devised a staggering number of ingenious ways to test them.

If you have not had much exposure to the way scientists think and the methods they use to track down answers, you might believe that "doing science" is a mysterious ritual. There they are, exquisitely trained persons pondering terribly complex problems and doing experiments late into the night. Becoming a practicing scientist generally does require special training—but thinking scientifically does not.

For example, any reasonably alert person might wonder about some of the organisms illustrated in this chapter. Why does the silkworm moth (*Hyalophora cecropia*) have such distinct, boldly patterned wings? Why don't all trees grow as tall as redwoods? Why do zebras have striped coats? Why don't dung beetles simply burrow under a dung pat instead of rolling a ball of the stuff great distances before putting it underground?

There is no such thing as a single "scientific method" of investigating such questions. However, the following list is a good starting point for understanding how a scientist might proceed with such an investigation:

1. Identify a problem or ask a question about some aspect of the natural world.

2. Develop one or more **hypotheses**, or educated guesses, about what the solution or answer might be. This might involve sorting through what has been learned already about related phenomena.

3. Think about what predictably will occur or will be observed if the hypothesis is correct. This is sometimes called the "if-then" process. (*If* gravity pulls objects toward the earth, *then* it should be possible to observe apples falling down, not up, from a tree.)

4. Devise ways to *test* the accuracy of predictions drawn from the hypothesis. This typically involves making observations, developing models, and performing experiments.

5. If the tests do not turn out as expected, check to see what might have gone wrong. (Maybe a substance being tested was tainted, maybe a dial was set incorrectly or a relevant factor overlooked. Or maybe the hypothesis just isn't a good one.)

6. Repeat or devise new tests—the more the better. Hypotheses that have been supported by many different tests are more likely to be correct.

7. Objectively report the results from tests and the conclusions drawn from them.

Using this list as a guide, let's return to the question posed earlier about the silkworm moth's wing pattern. Is it a mating "flag"? If so, it would help the male and female moths identify each other, rather than dallying with members of the wrong species and producing defective offspring. Or does the pattern camouflage the moths from bird predators? Maybe it blends with plants that the moths rest on during the day. (Moths fly at night, not during daylight hours.)

These actually are two plausible explanations, based on what scientists have already learned about mate discrimination and camouflaging among insects in general. Returning to still another question, do dung beetles roll away their prize ball so other dung beetles won't get it? Or do they bury the ball in distant concealed places where predators are less likely to find the beetle larvae that will grow inside it?

In nearly all cases, questions about what causes a natural phenomenon have more than one possible answer. In science, *alternative hypotheses are the rule, not the exception.*

Testing Alternative Hypotheses

Identifying which of two or more hypotheses may be correct depends on tests. The trick here is to use each hypothesis as a guide for producing testable predictions. *If* the moth wing pattern functions in mate discrimination (the hypothesis), *then* it follows logically that mating should occur only at times of day when moths can see each other's wing patterns (the prediction).

Scientists do not use "prediction" as fortunetellers do, to "look into the future." They use it as a statement of what you should be able to observe in nature, if you were to go looking for it. Start from the observation that moths mate at night. If their wing pattern helps potential mates identify each other, then there won't be any moths mating on moonless nights. They won't be able to see the patterns.

Suppose you test this prediction by stealthily watching moths on a moonless night. To your surprise, you see moths mating just as often in total darkness as in moonlight. Here is evidence of a mistaken prediction—and, by extension, a mistaken hypothesis.

Whether test results support or undermine your hypothesis, you should repeat the test several times. Doing so will provide insight into the reliability of your findings. It's safe to say that most respectable conclusions in science rest on numerous studies, carried out by people who tried to test many alternative hypotheses.

By testing a prediction, you test the underlying hypothesis.

The Role of Experiments

The preceding example shows how you might test a prediction by simple observation. You also might be able to test it by **experiments**. These are tests in which nature is manipulated as a way to reveal its secrets. Generally, scientists try to design experiments in such a way that the results will clearly show that a hypothesis is mistaken. Why? It is often easy to *disprove* a hypothesis but almost impossible to *prove* one. (An infinite number of experiments would have to be performed to show that it holds true under all possible conditions.)

Suppose you come up with another hypothesis about the wing pattern of *H. cecropia*. You propose that individuals with *altered* color patterns should have difficulty securing a mate. To test this, you sit out night after night, waiting for a peculiarly patterned moth to fly by. None does, so you decide to do an experiment. You *paint* the wings of a group of moths.

In a fair test of your prediction, you do more than put the group of painted moths in a cage with unaltered ones to observe the outcome. You also have a **control group**, which is used to evaluate possible side effects of the manipulation of the experimental group. Ideally, members of a control group should be *identical* to those of an experimental group in every respect—*except* for the key factor, or **variable**, that is under study. The variable here is wing color pattern. You also make sure that the number of individuals in both groups is large enough to give you more confidence that the experimental outcome will not be due to chance alone. Two or three moths will tell you nothing, as you will discover later in the book, in a section on probability (pages 169 and 170).

How can you be sure there are no other variables between the two groups that might influence the outcome of an experiment? After all, maybe paint fumes are as repulsive to a potential mate as the painted pattern. Maybe you rough up the moths when you handle them and somehow make them less desirable than those in the control group. Maybe the paint weighs enough to change the flutter frequency of the wings.

And so you decide that the control group also must be painted, using the same kind of paint, the same kind of brushes, and the same amount of handling. But for this group, you *duplicate* the natural wing color pattern as you paint. Thus your experimental group and control group are identical except for the variable under study. If only those moths with altered wing patterns turn out to be unlucky in love, then your control group will help substantiate your hypothesis.

Experiments are tests in which nature is manipulated. They require careful design of a set of controls to evaluate possible side effects of the manipulation.

About the Word "Theory"

More than a century ago, Darwin unveiled his ideas about the evolution of species and ushered in one of the most dramatic of all scientific revolutions. The core of his thinking—that life's diversity is the outcome of evolution by natural selection—became popularly known as "the theory of evolution." We will be looking closely at Darwin's scientific work in a later chapter, but here let's focus on the word *theory*.

In science, a **theory** is a related set of hypotheses which, taken together, form a broad-ranging explanation about some fundamental aspect of the natural world. A scientific theory differs from a scientific hypothesis in its *breadth of application.* Darwin's theory fits this description—it is a big, encompassing "Aha!" explanation that, in a few intellectual strokes, makes sense of a huge number of observable phenomena. Think about it. His

theory explains what has caused most of the diversity among many millions of different living things!

There are many other major theories in biology. One explains what causes all offspring to resemble their parents, no matter what the species. Another explains what caused the mass extinction of the dinosaurs about 65 million years ago. We will be looking at these and other theories throughout the book. None is bigger in scope than Darwin's, but all do a good job of attempting to make sense of the natural world.

Like hypotheses, theories are accepted or rejected on the basis of tests. For example, several competing theories about evolution were pushed vigorously in Darwin's time, but since then they have been tested and essentially rejected. After thousands of different tests, Darwin's theory still stands, with only some modification. Today, most biologists accept the modified theory as correct—but they still keep their eyes open for new evidence that might call it into question.

Scientists admire tested theories for good reason. A tested theory serves as a general frame of reference for additional research, and research is what scientists do. If Darwin's theory is correct, then the diverse attributes of living things exist because, at least in the past, they helped individuals leave descendants. Therefore, when biologists look at a moth's wings, an immediate question comes to mind—"I wonder how that wing color helps that moth leave descendants." When using Darwin's theory as a guide to developing plausible hypotheses, they will probably focus on possible answers that relate directly or indirectly to reproductive success—as we saw in our earlier example.

A scientific theory is an explanation about the cause or causes of a broad range of related phenomena. Like hypotheses, theories are open to tests, revision, and tentative acceptance or rejection.

Uncertainty in Science

Isn't anything ever "for sure" in science? Are there no comfortable, final conclusions? In an ultimate sense, no. Scientists must be content with *relative* certainty about whether an idea is correct or not. When a theory or hypothesis withstands exhaustive testing by many independent researchers, that "relative certainty" can be very great. Even so, there is always a chance that one of the tests has hidden flaws, which would invalidate the results. That is why scientific papers include a section on the methods employed for the tests they describe. This allows other scientists to check the procedures used, even to the point of duplicating the research.

Knowing that others will scrutinize your ideas and the methods used to test them has a wonderful effect. It forces you to try to remain objective—even if you believe fiercely in what you propose.

In short, individual scientists must keep asking themselves: "Are there tests or observations that will show my ideas to be incorrect?" They are expected to put aside pride or bias by testing their ideas, even in ways that might prove them wrong. Even if an individual scientist doesn't (or won't) do this, *others will*—for science proceeds as a community that is both cooperative and competitive. Ideas are shared, with the understanding that it is just as important to expose errors as it is to applaud insights.

The fact that scientists can and do change their mind when presented with new evidence is a *strength* of their profession, not a weakness.

The Limits of Science

The call for objectivity strengthens the theories that do emerge from scientific studies. Yet it also puts limits on the kinds of studies that can be carried out. Beyond the realm of scientific analysis, certain events remain unexplained. Why do we exist, for what purpose? Why does any one of us have to die at a particular moment and not another?

Answers to such questions are *subjective*. This means they come from within, as an outcome of all the experiences and mental connections that shape our consciousness. Because individuals differ so enormously in this regard, subjective answers do not readily lend themselves to scientific analysis.

This is not to say that subjective answers are without value. No human society can function without a shared commitment to standards for making judgments, however subjective those judgments might be. Moral, aesthetic, economic, and philosophical standards vary from one society to the next. But all guide their members in deciding what is important and good, and what is not. All attempt to give meaning to what we do.

Every so often, scientists stir up controversy when they explain part of the world that was previously considered beyond natural explanation—that is, belonging to the "supernatural." This is sometimes true when moral codes are interwoven with religious narratives, which grew out of observations by ancestors. Exploring some longstanding view of the world from a scientific perspective may be misinterpreted as questioning morality, even though the two are not remotely synonymous.

For example, centuries ago Nicolaus Copernicus studied the movements of planets and stated that the earth circles the sun. Today the statement seems obvious. Back then, it was heresy. The prevailing belief was that the Creator had made the earth (and, by extension, humankind) the immovable center of the universe! Not long afterward a respected professor, Galileo Galilei, studied the Copernican model of the solar system. He thought it was a good one and said so. He was forced to retract his statement publicly, on his knees, and to put the earth back as the fixed center of things. (Word has it that when he stood up he muttered, "But it moves nevertheless.")

Today, as then, society has its sets of standards. Today, as then, those standards may be called into question when a new, natural explanation runs counter to supernatural belief. When this happens it doesn't mean that scientists as a group are less moral, less lawful, less sensitive, or less caring than any other group. Their work, however, is guided by one additional standard: *The external world, not internal conviction, must be the testing ground for scientific beliefs.*

Systematic observations, hypotheses, predictions, tests—in all these ways, science differs from systems of belief that are based on faith, force, authority, or simple consensus.

SUMMARY

1. All organisms are alike in the following characteristics:
    a. Their structure, organization, and interactions arise from the basic properties of matter and energy.
    b. They rely on metabolic and homeostatic processes.
    c. They have the capacity for growth, development, and reproduction.
    d. Their heritable instructions are encoded in DNA.

2. There are many millions of different kinds of organisms. Each distinct kind of organism is called a species. Distinct species resembling one another more than they resemble other species are grouped into the same genus, and so on with increasingly inclusive groupings into family, order, class, phylum (or division), and kingdom.

3. Diversity among organisms arises through mutations that introduce changes in the DNA. These changes lead to heritable variation in the form, functioning, or behavior of individual offspring.

4. Individuals in a population vary in their heritable traits, and the variations influence their ability to survive and reproduce. Under prevailing conditions, certain varieties of a given trait may be more adaptive than others. They will be "selected" and others eliminated

through successive generations. The changing frequencies of different traits change the character of the population over time; it evolves. These points are central to the principle of evolution by natural selection.

5. There are many scientific methods of gathering information. These are key terms associated with scientific inquiry:

a. Theory: An explanation of a broad range of related phenomena. An example is the theory of how processes of natural selection bring about evolution.

b. Hypothesis: A possible explanation of a specific phenomenon. Sometimes called an educated guess.

c. Prediction: A claim about what an observer can expect to see in nature *if* a particular theory or hypothesis is correct.

d. Test: The attempt to secure actual observations in order to determine whether they match the predicted or expected observations.

e. Conclusion: A statement about whether and to what extent a particular theory or hypothesis can be accepted or rejected, based on the tests of predictions derived from it.

6. The external world, not internal conviction, is the testing ground for scientific theories.

Review Questions

1. Why is it difficult to give a simple definition of life? (For this and subsequent chapters, *italic numbers* following review questions indicate the pages on which the answers may be found.) 2

2. What does *adaptive* mean? Give some examples of environmental conditions to which plants and animals must be adapted. 9

3. Study Figure 1.3. Then, on your own, arrange and define the levels of biological organization. What concept ties this organization to the history of life, from the time of origin to the present? 5

4. In what fundamental ways are all organisms alike? 9–10

5. What is metabolic activity? 4–5

6. What are the "instructions" contained in DNA? What is a mutation? Why are most mutations likely to be harmful? 3–4

7. Outline the one-way flow of energy and the cycling of materials through the biosphere. 6

8. What does evolution mean? 11

9. Witnesses in a court of law are asked to "swear to tell the truth, the whole truth, and nothing but the truth." What are some of the problems inherent in the question? Can you think of a better alternative?

10. Design a test to support or refute the following hypothesis: The body fat in rabbits appears yellow in certain mutant individuals—but only when those mutants also eat leafy plants containing a yellow pigment molecule called xanthophyll.

Self-Quiz *(Answers in Appendix IV)*

1. The complex patterns of structural organization characteristic of life are based on instructions contained in _____ .

2. Directly or indirectly, all living organisms depend on one another for _____ .

3. _____ is the ability of organisms to extract and transform energy from the environment and use it during maintenance, growth, and reproduction.

4. _____ means maintaining the body's internal operating conditions within a tolerable range even when environmental conditions change.

5. Diverse structural, functional, and behavioral traits are considered to be _____ to changing conditions in the environment.

6. The capacity to evolve is based on variations in traits, which originally arise through _____ .

7. Organisms show _____ , the ability to transmit instructions for heritable traits from parents to offspring.

8. That each of us has great-great-great-great-grandmothers and grandfathers is an example of a unique property of life known as _____ .

    a. metabolism   c. reproduction
    b. homeostasis   d. organization

9. A scientific approach to explaining various aspects of the natural world includes all of the following except _____ .

    a. hypothesis   c. faith and simple consensus
    b. testing   d. systematic observations

10. A related set of hypotheses that collectively explain some aspect of the natural world is a scientific _____ .

    a. prediction   d. authority
    b. test   e. observation
    c. theory

Selected Key Terms

adaptive trait 9	experiment 13	photosynthesis 5
aerobic respiration 5	Fungi 9	Plantae 9
Animalia 9	genus 9	population 4
biosphere 4	homeostasis 7	prediction 13
cell 4	hypothesis 13	Protista 9
community 4	inheritance 8	reproduction 8
control group 14	metabolism 4	species 9
DNA 3	Monera 9	test 13
ecosystem 4	multicelled	theory 14
energy 3	organism 4	variable 14
evolution 11	mutation 8	

FACING PAGE: *Living cells of a green alga (Elodea), with their chemical factories called chloroplasts.*

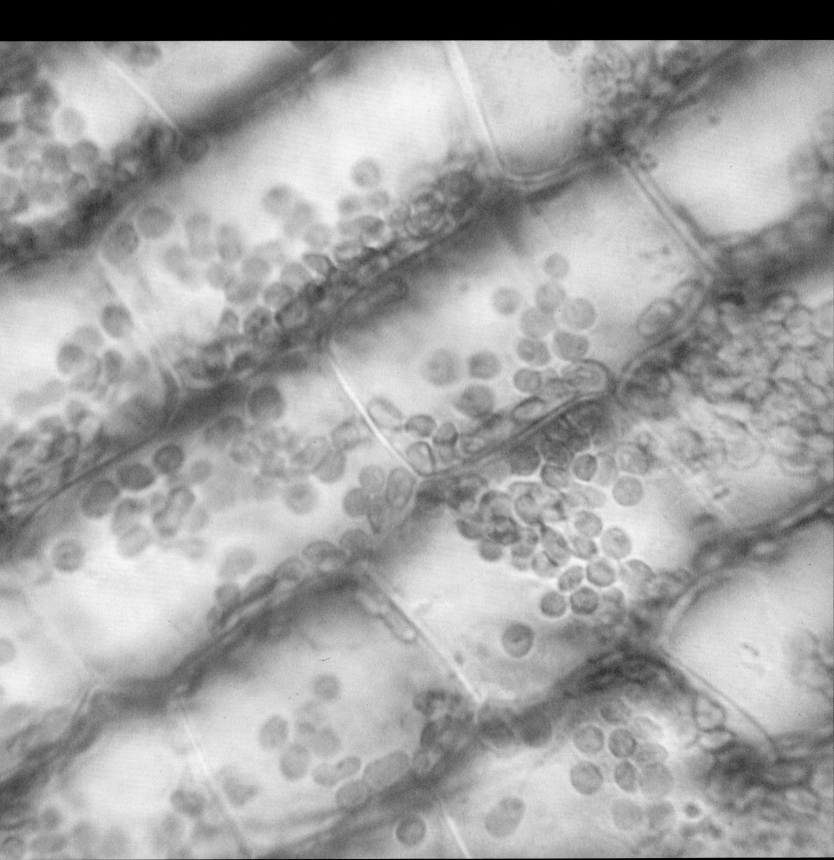

2 CHEMICAL FOUNDATIONS FOR CELLS

The Chemistry In and Around You

As you read this page, thousands of chemical reactions are proceeding inside you in ways that keep your body running smoothly. Whether it is daytime or the middle of the night, streams of sunlight are reaching half of the earth's surface. And countless plants are converting energy contained in the sun's rays into forms that can be used for assembling the carbohydrates and other building blocks of roots, stems, and leaves. Your very life depends on breathing in the oxygen released into the atmosphere during chemical reactions such as these, the first of which occurred more than 2.5 billion years ago.

In the past two centuries—a mere blip of evolutionary time—we have managed to discover what chemical substances are made of, how they can be transformed into different substances, and what it takes to accomplish the transformations. Some of the products of these discoveries are synthetic fabrics, fertilizers, vaccines, antibiotics, and the plastic components of refrigerators, computers, television sets, jet planes, and cars.

Our chemical "magic" brings us great benefits *and* monumental problems. For example, fertilizer applications and other agricultural practices maintain food supplies for the 5.4 billion people on earth. Without them, much of the human population would starve to death. But weeds don't understand that the fertilizers are for crop plants, and plant-eating insects don't understand that the crops are for us, not them. Each year they gobble up or ruin about 45 percent of what we grow. In 1945 we began battling back with synthetic compounds that kill weeds as well as insects, worms, rodents, and other animals. In 1988 alone, Americans managed to spread more than a billion pounds of herbicides and pesticides through homes, gardens, offices, industries, and farmlands (Figure 2.1).

Among the insect killers we find the carbamates, organophosphates (including malathion), and halogenated compounds (including chlordane). Most are neurotoxins, meaning they block vital communication signals between nerve cells. Some remain active for days, others for weeks or years. Unfortunately, they kill many of nature's insect eaters, including dragonflies and birds, as well as the targeted pests. Over time, the pests build up resistance to the pesticides, for reasons that will become apparent in later chapters. And only 10 percent of the insecticides being sold today have been assessed for potential health hazards.

Consider that *we* inhale pesticides, ingest them with food, or absorb them through skin. After entering the body, many pesticides can cause headaches, rashes, asthma, and bronchitis in susceptible people and can increase their vulnerability to chronic infections. Susceptibility depends on genetic makeup, overall health, nutritional habits, and concurrent exposure to other

Figure 2.1 Cropduster with its rain of pesticides.

toxic substances. Exposure to certain pesticides can trigger hives, pain in the joints, and other moderate allergic reactions in about 11 million Americans. They can trigger severe, life-threatening immune reactions in another 5 million.

Maintaining our food supplies, industries, and health depends on chemistry—and so do our chances of reducing harmful side effects of its application. You owe it to yourself and others to gain greater understanding of chemical substances and their interactions. By demystifying the "magic" of chemistry, you will be better equipped to assess the benefits and risks of its application to the world of life.

K E Y   C O N C E P T S

1. Atoms give up, acquire, or share their electrons with other atoms in specific ways. These interactions are the basis for the structural organization and activities of all living things.

2. Chemical bonds are unions between the electron structures of different atoms. The most common bonds in biological molecules are ionic bonds, hydrogen bonds, and covalent bonds.

3. Life depends on the properties of water, including its temperature-stabilizing effects, cohesiveness, and capacity to dissolve many substances.

4. Even though cells continuously produce and use hydrogen ions (H^+) during chemical reactions, they have the means to maintain the H^+ concentration within narrow limits.

ORGANIZATION OF MATTER

"Matter" is anything that occupies space and has mass. It includes the solids, liquids, and gases around you and within your body. All of these forms of matter are made of one or more **elements**, or fundamental substances that cannot be broken down to a different substance by ordinary chemical means. About ninety-two elements occur naturally on earth. It takes only four kinds—hydrogen, carbon, nitrogen, and oxygen—to make up most of the human body.

Some elements are vital for normal body functioning, even though they are present in what might seem to be insignificant amounts. Collectively, these so-called **trace elements** represent less than 0.01 percent of all the atoms in any organism. Copper is an example. Carefully dry out and analyze the tissues of a maple tree and you will find they are only about 0.006 percent copper. But if the tree has a copper deficiency, its leaf buds will die, yellow or dead spots will form on its leaves, and its overall growth will be stunted.

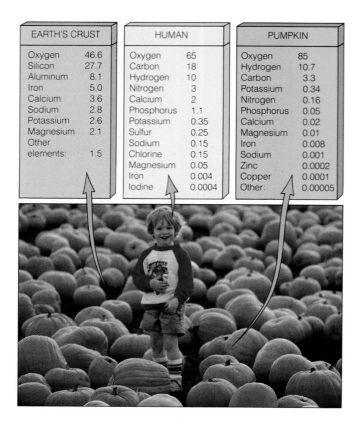

EARTH'S CRUST		HUMAN		PUMPKIN	
Oxygen	46.6	Oxygen	65	Oxygen	85
Silicon	27.7	Carbon	18	Hydrogen	10.7
Aluminum	8.1	Hydrogen	10	Carbon	3.3
Iron	5.0	Nitrogen	3	Potassium	0.34
Calcium	3.6	Calcium	2	Nitrogen	0.16
Sodium	2.8	Phosphorus	1.1	Phosphorus	0.05
Potassium	2.6	Potassium	0.35	Calcium	0.02
Magnesium	2.1	Sulfur	0.25	Magnesium	0.01
Other		Sodium	0.15	Iron	0.008
elements:	1.5	Chlorine	0.15	Sodium	0.001
		Magnesium	0.05	Zinc	0.0002
		Iron	0.004	Copper	0.0001
		Iodine	0.0004	Other:	0.00005

Figure 2.2 Comparison of the proportions of different elements in the earth's crust, the human body, and a pumpkin as percentages of the total weight of each.

Table 2.1	Atomic Number and Mass Number of Elements Common in Living Things		
Element	Symbol	Atomic Number	Most Common Mass Number
Hydrogen	H	1	1
Carbon	C	6	12
Nitrogen	N	7	14
Oxygen	O	8	16
Sodium	Na	11	23
Magnesium	Mg	12	24
Phosphorus	P	15	31
Sulfur	S	16	32
Chlorine	Cl	17	35
Potassium	K	19	39
Calcium	Ca	20	40
Iron	Fe	26	56
Iodine	I	53	127

Take a look at Figure 2.2, which gives an example of how organisms differ in the proportions of their constituent elements, relative to nonliving material.

By international agreement, a one- or two-letter chemical symbol stands for each element, regardless of the element's name in different languages. What we call *nitrogen* is called *azoto* in Italian and *stickstoff* in German. But the symbol for this element is always N. Similarly, the symbol for the element sodium is always Na (from the Latin *natrium*). Table 2.1 lists the chemical symbols of elements that are common in living things.

In substances called **compounds**, two or more elements are combined in fixed proportions. Water is a compound; it is always 11.9 percent hydrogen and 88.1 percent oxygen, no matter where you find it.

The Structure of Atoms

Look at some water in a glass, then imagine your eyes probing ever deeper into its underlying structure. First you would discover molecules composed of hydrogen and oxygen. By definition, a **molecule** is a unit of two or more atoms (of the same or different elements) bonded together. For any compound, it is the smallest unit that maintains the compound's elemental composition. In turn, each kind of **atom** is the smallest unit of matter that is unique to a particular element. Below the level of atoms, you always find the same types of particles. The particles are protons, neutrons, and electrons, the universal building blocks of atoms.

Protons and neutrons make up the atom's core region, or nucleus. Protons have a positive charge (p^+); neutrons are electrically neutral. Electrons, which have a negative charge (e^-), are attracted to the nucleus. They move rapidly around it and occupy most of the atom's volume. An atom has just as many protons as electrons, so it has no *net* charge, overall.

The number of protons in the nucleus, called the **atomic number**, differs for each element. The hydrogen atom has only one proton; its atomic number is 1. The carbon atom has six protons; its atomic number is 6. Table 2.1 lists other examples. The total number of protons *and* neutrons in the nucleus is the **mass number**. For example, the mass number of a carbon atom with six protons and six neutrons is 12. (The relative masses of atoms are also called atomic weights. This term is not precise—mass is not quite the same thing as weight— but its use continues.)

As you will see, knowing an atom's atomic number and mass number tells us something about how it will interact (if at all) with other atoms. Knowing those values gives us an idea of whether that atom can give up, acquire, or share its electrons with other atoms. *Such electron activity is the basis for the flow of materials and energy through the living world.*

Dating Fossils, Tracking Chemicals, and Saving Lives — Some Uses of Radioisotopes

In the winter of 1896, the physicist Henri Becquerel tucked a heavily wrapped rock of uranium into a desk drawer, on top of an unexposed photographic plate. A few days later, he opened the drawer and discovered a faint image of the rock on the plate—apparently caused by energy emitted from the rock. One of his coworkers, Marie Curie, gave the name "radioactivity" to the phenomenon.

As we now know, radioisotopes are unstable atoms, with too many protons or neutrons. The instability causes them to capture or emit electrons or some other particle. This spontaneous process, called radioactive decay, continues until the original isotope has changed to a new, stable isotope, one that is not radioactive.

Radioactive Dating

Each type of radioisotope has a characteristic number of protons and neutrons, and it decays spontaneously at a characteristic rate into a different isotope. The *half-life* is the time it takes for half the nuclei in any given amount of a radioactive element to decay into another isotope. The half-life cannot be modified by temperature, pressure, chemical reactions, or any other environmental factor. That is why radioactive dating is such a reliable method of determining the age of rock layers in the earth—hence the age of fossils they may contain. To determine a rock's age, we can compare the amount of one of its radioisotopes with the amount of the decay product for that isotope.

(a) Fossilized frond of a tree fern, one of many species that lived more than 250 million years ago. (b) Fossilized sycamore leaf that dropped 50 million years ago.

For example, ^{40}potassium has a half-life of 1.3 billion years and decays to ^{40}argon, a stable isotope. The age of anything that contains ^{40}potassium can be determined by measuring the ratio of ^{40}argon to ^{40}potassium. In this way, researchers have dated fossils that are millions, even billions, of years old (Figures *a* and *b*). Radioactive dating

Main Radioisotopes Used in Dating

Radioisotope (unstable)		Stable Product	Half-Life (years)	Useful Range (years)
^{87}rubidium	→	^{87}strontium	49 billion	100 million
^{232}thorium	→	^{208}lead	14 billion	200 million
^{238}uranium	→	^{206}lead	4.5 billion	100 million
^{40}potassium	→	^{40}argon	1.3 billion	100 million
^{235}uranium	→	^{207}lead	704 million	100,000
^{14}carbon	→	^{14}nitrogen	5,730	0–60,000

a

b

Isotopes

Although all atoms of an element have the same number of protons, they may vary slightly in how many neutrons they have. Atoms having the same atomic number but a different mass number are **isotopes**. Thus "a carbon atom" might be carbon 12 (containing six protons, six neutrons), carbon 13 (six protons, seven neutrons), or carbon 14 (six protons, eight neutrons). These can be written as ^{12}C, ^{13}C, and ^{14}C. All isotopes of an element have the same number of electrons, so they interact with other atoms the same way. Accordingly, cells can use any carbon isotope for a metabolic reaction.

You have probably heard of radioactive isotopes, or **radioisotopes**. They are unstable isotopes that tend to break apart (decay) into more stable atoms. The *Commentary* describes some of the ways in which radioisotopes are used in research, in medicine, and in establishing the age of fossil-containing rocks.

All atoms of an element have the same number of electrons and protons, but they can vary slightly in the number of neutrons. The variant forms are isotopes.

with ^{238}uranium, which has a half-life of 4.5 billion years, indicates that the earth formed 4.6 billion years ago. The list above Figure *a* shows the useful ranges of the main radioisotopes used in dating methods.

Tracking Chemicals

Emissions from radioisotopes can be detected by a scintillation counter and other devices. This means that isotopes can be used as *tracers*. They can be used to identify the pathways or destination of a substance that has been introduced into a cell, the human body, an ecosystem, or some other "system."

For example, because all isotopes of an element have the same number of electrons, they all interact with other atoms in the same way. Accordingly, cells can use isotopes of carbon for a given metabolic reaction. Carbon happens to be a key building block for photosynthesis. By putting

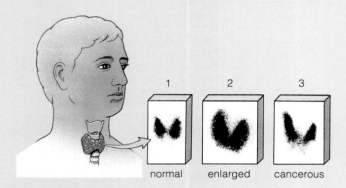

plant cells in a medium enriched in ^{14}carbon, researchers identified the steps by which plants take up carbon and incorporate it into newly forming carbohydrates. Tracers also are helping us increase crop production by providing insights into how plants use synthetic fertilizers and naturally occurring nutrients.

What about medical applications? As one example, the thyroid is the only gland in the body to take up iodine. A tiny amount of the radioisotope ^{123}iodine can be injected into a patient's bloodstream, then the thyroid can be scanned with a scintillation counter. This is called a radioisotope scan. Figure *c* shows examples of what these scans may reveal.

Saving Lives

In nuclear medicine, radioisotopes are used to diagnose and to treat diseases. Patients with irregular heartbeats use pacemakers, which are powered by energy emitted from ^{238}plutonium. (This otherwise dangerous radioisotope is sealed in a case to prevent its emissions from damaging body tissues.) With PET (positron-emission tomography), radioisotopes provide diagnostic information about abnormalities in the metabolic functions of specific tissues. The radioisotopes are incorporated in glucose or some other biological molecule, then they are injected into a patient, who is moved into a PET scanner. Cells in certain tissues absorb the glucose. The radioisotopes give off energy that can be used to produce a vivid image of the variations in metabolic activity among different cells (Figure *d*).

Finally, some cancer treatments make use of the fact that radioisotopes can damage or destroy living cells. In radiation therapy, localized cancers are deliberately bombarded with ^{226}radium or ^{60}cobalt.

c Scans of human thyroid glands after ^{123}iodine was injected into the bloodstream. The thyroid normally takes up iodine (including radioisotopes) and uses it in hormone production. (1) Uptake by a normal gland. (2) Enlarged gland of a patient with a thyroid disorder. (3) Cancerous thyroid gland.

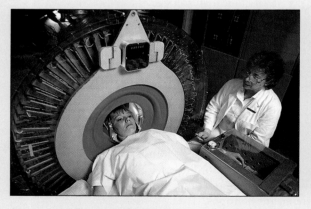
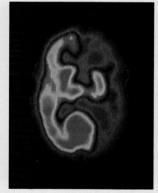

d Patient being moved into a PET scanner. The inset shows a vivid image of a brain scan of a child with a severe neurological disorder. The different colors signify differences in metabolic activity in one half of the brain; the other half shows no activity.

BONDS BETWEEN ATOMS

Let's turn now to the nature of reactions among atoms. In case you are not familiar with such reactions, take a moment to review Figure 2.3, which summarizes a few conventions used in describing them.

The Nature of Chemical Bonds

A **chemical bond** is a union between the electron structures of atoms. In other words, *it is an energy relationship.* Usually an atom gives up, gains, or shares one or more electrons with another atom. Some atoms do this rather easily, but others do not. Such differences in bonding behavior arise through differences in the number and arrangement of electrons in the atoms of each kind of element.

Picture three narcissistic actresses arriving at the Academy Awards wearing the same bright-red designer dress. Each has a compulsion to be in the limelight but dreads being photographed next to the others. Two might maneuver themselves *near* the center of attention while scooting away from each other to some extent; but by unspoken agreement, all three never, ever stay in the same place at the same time.

Electrons behave roughly the same way. They are attracted to the protons of a nucleus but repelled by other electrons that may be present. They spend as much time as possible near the nucleus and as far away from each other by moving in different *orbitals,* which are regions of space around the nucleus in which electrons are likely to be at any instant. Each orbital has enough room for two electrons, at most.

In all atoms, the orbital closest to a nucleus is ball-shaped (Figure 2.4). In a hydrogen atom, a single electron occupies that orbital. In a helium atom, two electrons occupy it. Electrons in the closest orbital to the nucleus are said to be at the *lowest energy level.*

Atoms larger than helium have two electrons in the first orbital. They also have other electrons that occupy different orbitals. On the average, those other electrons are farther away from the nucleus, and they are said to be at *higher energy levels.*

A simple although not quite accurate way to think about electron orbitals is to imagine them occupying the

Figure 2.4 Very simplified model of atomic structure, using a hydrogen atom and a helium atom as examples. The nucleus, a core region, consists of some number of protons and (except for hydrogen) neutrons. Electrons occur in orbitals around the nucleus.

Figure 2.3 Chemical bookkeeping.

Symbols for elements are used in writing *formulas,* which identify the composition of compounds. (For example, water has the formula H_2O. The subscript indicates two hydrogen atoms are present for every oxygen atom.) Symbols and formulas are used in *chemical equations:* representations of reactions among atoms and molecules.

In written chemical reactions, an arrow means "yields." Substances entering a reaction (*reactants*) are to the left of the arrow. Products of the reaction are to the right. For example, the overall process of photosynthesis is often written this way:

$$6CO_2 \quad + \quad 6H_2O \longrightarrow C_6H_{12}O_6 \quad + \quad 6O_2$$

| 6 carbons 12 oxygens | 12 hydrogens 6 oxygens | 6 carbons 12 hydrogens 6 oxygens | 12 oxygens |

Notice there are as many atoms of each element to the right of the arrow as there are to the left, even though they are combined in different forms. Atoms taking part in chemical reactions may be rearranged but they are never destroyed. The *law of conservation of mass* states that the total mass of all materials entering a reaction equals the total mass of all the products.

When thinking about cellular reactions, keep in mind that no atoms are lost, so the equations you use to represent them must be balanced in this manner.

Both the reactants and products can be expressed in moles. A "mole" is a certain number of atoms or molecules of any substance, just as "a dozen" can refer to any twelve cats, roses, and so forth. Its weight (in grams) equals the total atomic weight of the atoms that compose the substance.

For example, the atomic weight of carbon is 12. Hence one mole of carbon weighs 12 grams. A mole of oxygen (atomic weight 16) weighs 16 grams. Can you show why a mole of water (H_2O) weighs 18 grams, and why a mole of glucose ($C_6H_{12}O_6$) weighs 180 grams?

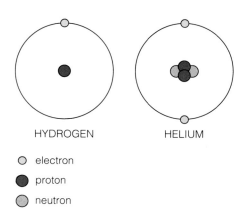

HYDROGEN   HELIUM

○ electron
● proton
○ neutron

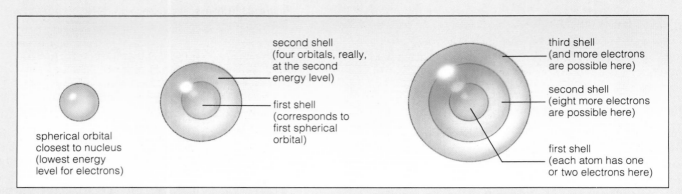

a The shell model of electron distribution in atoms.

second shell
(four orbitals, really,
at the second
energy level)

first shell
(corresponds to
first spherical
orbital)

spherical orbital
closest to nucleus
(lowest energy
level for electrons)

third shell
(and more electrons
are possible here)

second shell
(eight more electrons
are possible here)

first shell
(each atom has one
or two electrons here)

Figure 2.5 Arrangement of electrons in atoms. (**a**) In every atom, one or at most two electrons occupy a ball-shaped volume of space (an orbital) close to the nucleus. At this scale, the nucleus would be an invisible speck at the ball's center. This orbital is at the lowest energy level.

At the next (higher) energy level, there can be as many as eight more electrons—two in each of the four orbitals shown in (**b**). As you can see, the orbital shapes get tricky. For our purposes, we can ignore the shapes and simply think of the total number of orbitals at a given energy level as being somewhere inside a "shell" of the sort sketched in (**a**).

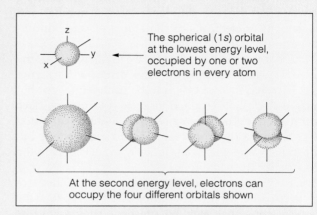

The spherical (1s) orbital at the lowest energy level, occupied by one or two electrons in every atom

At the second energy level, electrons can occupy the four different orbitals shown

b Shapes of electron orbitals that occur at the first and second energy levels, which correspond to the first and second "shells" in **a**.

Table 2.2	Electron Distribution for a Few Elements				
			Electron Distribution		
Element	Chemical Symbol	Atomic Number	First Shell	Second Shell	Third Shell
Hydrogen	H	1	1	—	—
Helium	He	2	2	—	—
Carbon	C	6	2	4	—
Nitrogen	N	7	2	5	—
Oxygen	O	8	2	6	—
Neon	Ne	10	2	8	—
Sodium	Na	11	2	8	1
Magnesium	Mg	12	2	8	2
Phosphorus	P	15	2	8	5
Sulfur	S	16	2	8	6
Chlorine	Cl	17	2	8	7

space inside hollow *shells* around the nucleus. The shell closest to the nucleus has one orbital, so it can hold no more than two electrons. The next shell can have as many as eight electrons, two in each of four orbitals. Successive shells can have still more electrons, as Figure 2.5 shows.

Hydrogen, the simplest atom, has one electron in its first (and only) shell. Notice in Figure 2.6 that sodium, with eleven electrons, has a lone electron in its outermost shell. In other words, electrons only *partly fill* the highest occupied shell of either atom. *Such atoms tend to react with other atoms.* As you can see from Table 2.2, atoms that tend to enter into reactions include not only hydrogen but also carbon, nitrogen, and oxygen. These atoms are the main building blocks of organisms.

An atom tends to react with other atoms when its outermost shell is only partly filled with electrons.

Ionic Bonding

Sometimes the balance between the protons and electrons of an atom is disturbed so much that one or more electrons are knocked out of the atom, pulled away from it, or added to it. When an atom loses or gains one or more electrons, it becomes positively or negatively charged. In this state, it is an **ion**.

When an atom loses or gains electrons, another atom of the right kind must be nearby to accept or donate the electrons. Since one loses and one gains electrons, *both* become ionized. Depending on the surroundings, the two ions can go their separate ways or stay together through the mutual attraction of their opposite charges. An association of two oppositely charged ions is an **ionic bond**. NaCl, the table salt we sprinkle on food, has ions of sodium (Na^+) and chloride (Cl^-) linked together this way. Figure 2.7 shows their arrangement.

An ion is an atom or molecule that has gained or lost one or more electrons, and so has acquired an overall positive or negative charge.

In an ionic bond, a positive and a negative ion are linked by the mutual attraction of opposite charges.

Covalent Bonding

Often, an attraction between two atoms is not quite enough for one atom to pull electrons completely away from the other. The atoms end up sharing electrons, in what is called a **covalent bond**. We can use a line to represent a single covalent bond between two atoms, as in H—H. If we want to focus on the number of electrons being shared, we can use a dot to represent each one:

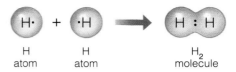

In a double covalent bond, two atoms share two pairs of electrons. An example is the O_2 molecule, or O=O. In a triple covalent bond (such as N≡N), two atoms share three pairs of electrons.

Covalent bonds may be nonpolar or polar. In a *nonpolar* covalent bond, both atoms exert the same pull on shared electrons. The word nonpolar implies there is no difference between the two "ends" (or poles) of the bond. An example is the H—H molecule. Its hydrogen atoms, with one proton each, attract the shared electrons equally.

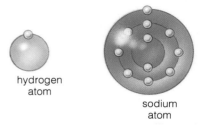

Figure 2.6 Distribution of electrons (yellow dots) for hydrogen and sodium atoms. Each atom has a lone electron (and room for more) in its outermost shell. Atoms having such partly filled shells tend to enter into reactions with other atoms.

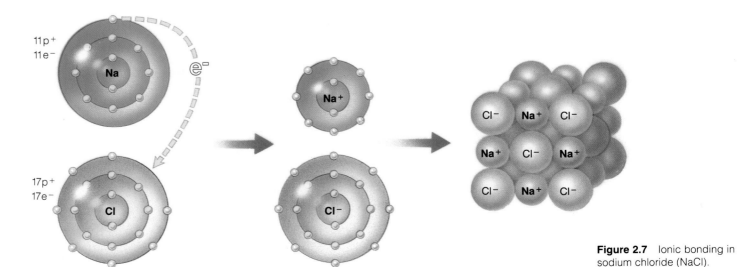

Figure 2.7 Ionic bonding in sodium chloride (NaCl).

In a *polar* covalent bond, atoms of different elements (which have different numbers of protons) do not exert the same pull on electrons. The more attractive atom ends up with a slight negative charge (it is "electronegative"). But this is balanced out by the other atom, which ends up with a slight positive charge. In other words, the two atoms together have no *net* charge, but the *distribution* of charge within the bond is asymmetric.

A water molecule (H—O—H) has polar covalent bonds. Its electrons are less attracted to the hydrogens than to the oxygen (which has more protons).

In a covalent bond, atoms share electrons.

If electrons are shared equally, the bond is nonpolar. If they are not shared equally, the bond is polar (electronegative at one end and electropositive at the other).

Covalent bonds provide strong structural links between atoms in the carbohydrates, lipids, proteins, and other building blocks of cell architecture. In DNA, a double-stranded molecule, they link atoms together in each strand.

Hydrogen Bonding

In a **hydrogen bond**, an atom of a molecule interacts weakly with a neighboring hydrogen atom that is already taking part in a polar covalent bond. (The hydrogen, which has a slight positive charge, is attracted to the slight negative charge of the other atom.) Hydrogen bonds can form between two different molecules, as shown in Figure 2.8. They also can form between two different regions of the same molecule where it twists back on itself.

Hydrogen bonds are common in large biological molecules. For example, many occur between the two strands of a DNA molecule, along the lines shown in Figure 2.8b. Individually, those bonds are easily broken, but collectively they help stabilize DNA structure. In later chapters, we will look at the energy it takes to break these and other bonds. Here, we will turn next to the hydrogen bonds between water molecules. These bonds impart some structure to liquid water—and they are responsible for many of water's life-sustaining properties.

In a hydrogen bond, an atom or molecule interacts weakly with a neighboring hydrogen atom that is already taking part in a polar covalent bond.

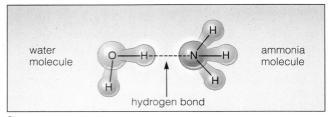

a

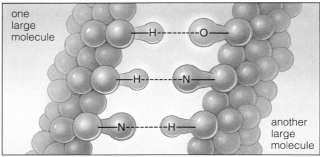

b

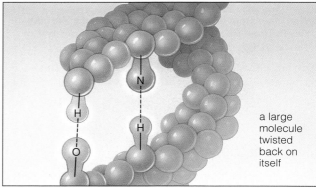

c

Figure 2.8 Some examples of hydrogen bonds. These are easier to break than covalent bonds. But where they occur in profusion, their collective action helps give water and other substances some of their characteristic properties. And collectively they help stabilize the shape of large molecules.

PROPERTIES OF WATER

Of all the planets in the solar system, only the earth is bathed in liquid water (Figure 2.9). Water covers about three-fourths of its surface. Life apparently originated in water. Many kinds of organisms still live in it, and the ones that don't carry an abundance of water around with them, in cells and in tissue spaces. The water molecules inside and outside cells are absolutely crucial for life. They are required for many important reactions, and they have major roles in the shape and internal molecular organization of cells.

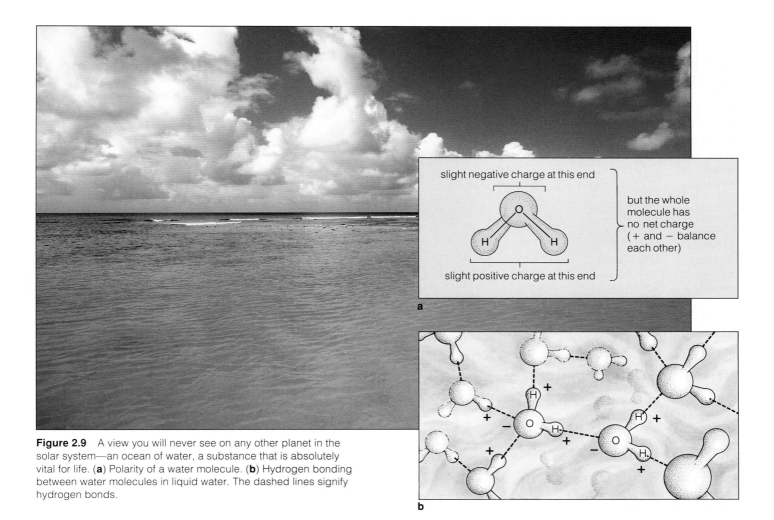

slight negative charge at this end

but the whole molecule has no net charge (+ and − balance each other)

slight positive charge at this end

a

b

Figure 2.9 A view you will never see on any other planet in the solar system—an ocean of water, a substance that is absolutely vital for life. (**a**) Polarity of a water molecule. (**b**) Hydrogen bonding between water molecules in liquid water. The dashed lines signify hydrogen bonds.

Polarity of the Water Molecule

Even though a water molecule carries no net charge, it interacts weakly with many other substances. How? Remember that the water molecule carries *partial* charges. As a result of its electron arrangements and bond angles, the whole molecule is slightly electronegative at the oxygen "end." And it is slightly electropositive at the "end" where the two hydrogens are positioned (Figure 2.9b).

Their polarity allows water molecules to interact with one another. Figure 2.9c shows how their hydrogen atoms form weak hydrogen bonds with oxygen atoms of neighboring molecules in liquid water. The same type of interaction is possible between water and many different polar substances. Polar substances are **hydrophilic** (water-loving). This means they are attracted to one end or the other of a water molecule and may form weak hydrogen bonds with it.

By contrast, nonpolar substances are **hydrophobic** (water-dreading). This means they tend to be repelled by water. Shake a bottle containing water and salad oil (a hydrophobic substance), then put it on a counter. The

oil and water hold little attraction for each other. Gradually, hydrogen bonds reunite the water molecules (they replace bonds that were broken when you shook the bottle). As they do, the oil molecules are pushed aside and forced to cluster in droplets or in a film at the water's surface.

In a hydrophilic interaction, polar substances form weak hydrogen bonds with water.

In a hydrophobic interaction, nonpolar groups cluster together in water. The clustering is not a true bond. The surrounding water molecules simply hydrogen-bond with one another and push out the nonpolar groups.

Hydrogen bonds and hydrophobic interactions underlie three properties of water that are biologically important. These are its resistance to changes in temperature, its internal cohesion, and its capacity to dissolve many substances.

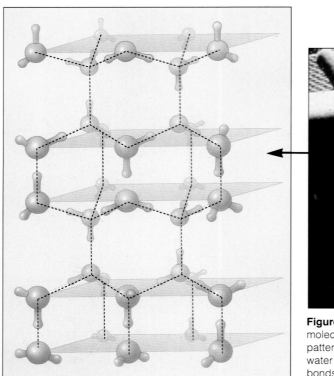

Figure 2.10 Hydrogen bonding between water molecules in ice. Below 0°C, each water molecule becomes locked by four hydrogen bonds into a crystal lattice. In this bonding pattern, molecules are spaced farther apart than in liquid water at room temperature. When water is liquid, constant molecular motion usually prevents the maximum number of hydrogen bonds from forming.

Temperature-Stabilizing Effects

Have you ever plunged into a solar-heated swimming pool well after sundown during a hot summer month? Even if the evening air has cooled off considerably, the water is still comfortably warm. Why doesn't the water temperature drop as fast as that of the surrounding air?

The *temperature* of a substance is a measure of how fast its molecules are moving. When water becomes heated, its molecules cannot move faster until hydrogen bonds among them are broken. It takes considerable heat to break those countless bonds and keep them from re-forming among molecules in liquid water. That is why it takes longer to boil a pot of water than it does to boil, say, alcohol or most other liquids.

When you heat the water to its boiling point, of course, its molecules have absorbed enough energy to separate from one another and you start to see steam (water vapor) rising from the pot. This is an example of **evaporation**, the process by which water molecules escape from a fluid surface and enter the surrounding air. Keep in mind that each water molecule in the liquid has *absorbed* some of the excess energy. When those molecules break free and depart in large numbers, they carry that energy away and the surface temperature of the water is lowered.

You cool off on hot, dry days through evaporative water loss from water that sweat glands secrete to your skin surface. (Sweat is 99 percent water.) You can lose quite a bit of water, given that you have more than 2½ million sweat glands distributed over nearly all of your body parts. Oasis plants of hot, dry deserts don't have sweat glands, but they too rely on evaporative water loss. Few plants elsewhere can do this—they would be severely stressed by such ongoing water losses. But the springs of oases provide constant replacements.

Water also resists freezing. At room temperature and down to about 0°C, hydrogen bonds between its molecules constantly break and form again, permitting some freedom of movement and keeping the water liquid. Below 0°C, however, the molecules become locked in the extended bonding pattern of ice. Because of this pattern, which is shown in Figure 2.10, ice is less dense than liquid water and is able to float on it. During winter freezes, sheets of ice typically form on surfaces of ponds, lakes, and streams. They act like a blanket that holds in the water's heat.

Cohesive Properties

Swimming in a pool on a hot summer night can be deliciously refreshing, as long as you don't mind the large night-flying bugs that accidentally hit the water's surface and float about on it. Like other substances with cohesive properties, liquid water resists rupturing when stretched

a

b

c

Figure 2.11 Evidence of water's cohesive properties. (**a, b**) A kingfisher plunging into water after a fish dinner. Notice how the water molecules stay put as a continuous sheet at the stream-air interface in the first photograph. Notice how they form many droplets when the kingfisher forces them away from the surface. The water's appearance in both cases results from hydrogen bonding among water molecules. The hydrogen bonds resist breaking apart when put under tension. (**c**) A water strider "walking" on water.

(placed under tension). Hydrogen bonds impart cohesion to water.

Where air and water meet at the pool surface, hydrogen bonds exert a constant inward pull on the uppermost water molecules. Their collective action imparts a high surface tension—insects have a tough time penetrating it, as do leaves and other bits of debris that land on the water (Figure 2.11).

In large land-dwelling plants, cohesion helps pull up water through narrow cellular pipelines that extend from the roots through stems and up to leaves. Hydrogen bonds hold the water molecules together in narrow columns along the route. Water evaporates from the leaf cells—and more water molecules are "pulled" into leaf cells as replacements (Chapter 28). The cohesion is strong enough to pull water even to the top of mature redwoods and other impressively tall trees.

Solvent Properties

Because of the polar nature of its molecules, water is an excellent solvent for ions and polar molecules. A *solvent* is any fluid in which one or more substances can be dissolved. The dissolved substances themselves are called *solutes*. What does "dissolved" actually mean? Consider what happens when you pour some table salt into a glass of water. The salt crystals eventually separate into Na^+ and Cl^- ions. Water molecules cluster around each

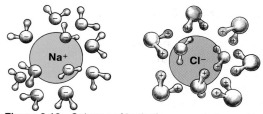

Figure 2.12 Spheres of hydration around charged ions.

positively charged ion with their "negative" ends pointing toward it. They also cluster around each negatively charged ion with their "positive" ends pointing toward it (Figure 2.12).

These clusterings are "spheres of hydration." They shield charged ions and keep them from interacting, so the ions can remain dispersed in water. *A substance is "dissolved" in water when spheres of hydration form around its individual ions or molecules.* This happens to solutes in cells, in the sap of maple trees, in your blood, and in the body fluids of all other organisms.

Cell structure and function depend on three properties of water: its internal cohesion, temperature-stabilizing effects, and capacity to dissolve many substances.

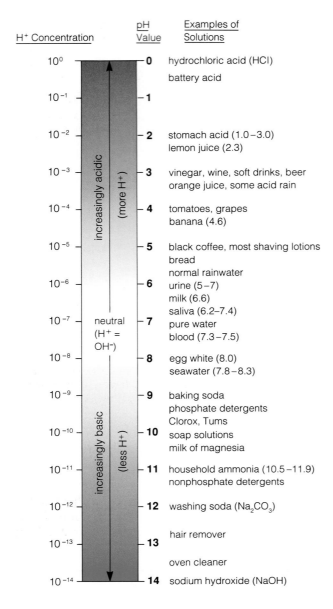

H+ Concentration	pH Value	Examples of Solutions
10^0	0	hydrochloric acid (HCl)
		battery acid
10^{-1}	1	
10^{-2}	2	stomach acid (1.0–3.0)
		lemon juice (2.3)
10^{-3}	3	vinegar, wine, soft drinks, beer
		orange juice, some acid rain
10^{-4}	4	tomatoes, grapes
		banana (4.6)
10^{-5}	5	black coffee, most shaving lotions
		bread
		normal rainwater
10^{-6}	6	urine (5–7)
		milk (6.6)
		saliva (6.2–7.4)
10^{-7}	7	pure water
		blood (7.3–7.5)
10^{-8}	8	egg white (8.0)
		seawater (7.8–8.3)
10^{-9}	9	baking soda
		phosphate detergents
		Clorox, Tums
10^{-10}	10	soap solutions
		milk of magnesia
10^{-11}	11	household ammonia (10.5–11.9)
		nonphosphate detergents
10^{-12}	12	washing soda (Na_2CO_3)
10^{-13}	13	hair remover
		oven cleaner
10^{-14}	14	sodium hydroxide (NaOH)

(increasingly acidic (more H+) from top; neutral (H+ = OH−) at 7; increasingly basic (less H+) toward bottom)

Figure 2.13 The pH scale, in which a liter of fluid is assigned a number according to the number of hydrogen ions present. The most useful part of the scale ranges from 0 (most acidic) to 14 (most basic), with 7 representing neutrality.

A change of only 1 on the pH scale means a tenfold change in hydrogen ion concentration. For example, the gastric juice in your stomach is ten times more acidic than vinegar, and vinegar is ten times more acidic than tomatoes.

ACIDS, BASES, AND SALTS

Acids and Bases

Some substances release one or more protons when they dissolve in water. Such "naked" protons are also known as **hydrogen ions** (H^+). A substance that releases H^+ in water is an **acid**.

Think about what happens when you inhale the fragrance of some fried chicken, then chew and swallow it, thereby sending the chicken on its way to the fluid contained in your saclike stomach. Cells in the stomach lining are stimulated into secreting hydrochloric acid (HCl), which separates into H^+ and Cl^-. The stomach fluid becomes more acidic, and a good thing, too. The increased acidity switches on enzymes that can digest the chicken proteins. It also helps kill bacteria that may have lurked in or on the chicken. Of course, if you stuff yourself with too much fried food, you may end up with a truly "acid stomach"—and then you might reach for an antacid tablet.

Milk of magnesia is one kind of antacid. It contains magnesium hydroxide [$Mg(OH)_2$]. This substance can separate into a magnesium ion (Mg^{++}) and a **hydroxide ion** (OH^-). The hydroxide ion may then *combine with* one of those excess hydrogen ions in the potent fluid inside your stomach and so help settle things down. Any substance that releases OH^- in water is a **base**.

The pH Scale

The **pH scale** is used to measure the concentration of free (unbound) hydrogen ions in different solutions. As Figure 2.13 shows, the scale ranges from 0 (most acidic) to 14 (most basic). The midpoint, 7, represents a neutral solution in which H^+ and OH^- concentrations are the same.

Pure water, with a pH of 7, is a neutral solution. Acidic solutions, such as hydrochloric acid and lemon juice, have a pH below 7. (That is, they have more H^+ than OH^- ions.) The opposite is true of alkaline solutions, which have a pH above 7. *The greater the H^+ concentration, the lower the pH value.* For example, a tenfold increase in the H^+ concentration of a fluid corresponds to a decrease by one unit of the pH scale.

Buffers

The life of each cell in each organism depends on ongoing chemical reactions—which happen to be extremely sensitive to even slight shifts in pH. Various mechanisms control the pH inside cells so that it hovers close to a neutral 7. Even though chemical reactions are con-

Figure 2.14 Sulfur dioxide emissions from a coal-burning power plant. Special camera filters revealed these otherwise invisible emissions. Together with other airborne pollutants, sulfur dioxides dissolve in atmospheric water to form acidic solutions. They are a major component of acid rain.

tinually using up and producing hydrogen ions, the cellular pH normally will not swing abruptly.

Control mechanisms also maintain the pH of blood and tissue fluids that bathe all living cells in your body. The pH of this "extracellular fluid" is a little more alkaline than the fluid inside cells; it generally ranges between 7.35 and 7.45. Later in the book, we will consider how the lungs and kidneys help control the body's overall acid-to-base balance. Here, simply keep in mind that the balancing acts depend largely on buffer molecules. A **buffer** is any molecule that can combine with hydrogen ions, release them, or both, and so help stabilize pH.

Different kinds of buffers can sponge up or release hydrogen ions when conditions dictate. Bicarbonate (HCO_3^-) in blood is a good example. It can combine with hydrogen ions to form carbonic acid (H_2CO_3). In doing so, it decreases the acidity of blood:

$$H_2CO_3 \rightleftharpoons HCO_3^- + H^+$$

carbonic acid   bicarbonate

Conversely, when the H^+ concentration declines, bicarbonate can release hydrogen ions and help increase blood's acidity. The buffering action normally helps stabilize the pH of blood. This is important, because if the pH value shoots above or drops below the optimal range, death may follow.

So far, we have been talking about pH values inside cells and multicelled organisms. The pH of the outside environment is often much higher or lower. Cells of sphagnum mosses grow in peat bogs, where the pH is 3.2 to 4.6 (highly acidic). Some roundworms live in places where the pH is 3.4. River water ranges between 6.8 and 8.6. Industrial wastes are sometimes

so acidic they affect the pH of rain (Figure 2.14 and Chapter 48).

Buffer molecules help maintain cellular pH near 7. They also help maintain the pH of extracellular fluid within the range of about 7.35 to 7.45. Environmental pH may be notably above or below that range.

Dissolved Salts

We have mentioned salts in passing but have not yet defined them. A **salt** is an ionic compound, formed when an acid reacts with a base. Sodium chloride can form this way:

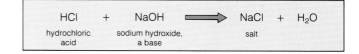

$$HCl + NaOH \longrightarrow NaCl + H_2O$$

hydrochloric acid   sodium hydroxide, a base   salt

Many salts dissolve into ions that play vital roles in cells. For example, ions of potassium (K^+) and sodium take part in the "messages" traveling through the nervous system. Calcium ions (Ca^{++}) take part in cell movements, cell division, nerve cell function, muscle contraction, and blood clotting.

Salts are ionic compounds that form when an acid reacts with a base, and that usually dissociate into positively and negatively charged ions in water.

WATER AND BIOLOGICAL ORGANIZATION

The organization of almost all large biological molecules is influenced by their interaction with water. Proteins, one of the topics of the next chapter, are a prime example. Depending on a variety of factors, the protein surface can be positively or negatively charged overall. The charged regions and polar groups attract water molecules. They also attract ions—which in turn attract water. The result is an electrically charged "cushion" of ions and water around the protein's surface:

Through such interactions with water and ions, the protein remains dispersed in cellular fluid rather than randomly settling against some cell structure. Why is it important to prevent settling? Many chemical reactions are played out on specific molecular regions of proteins—*they are the stages for life-sustaining tasks performed inside cells.*

Many more examples could be given of interactions between the polar water molecule and other substances characteristic of the cellular world. For now, the point to keep in mind is this:

The properties of water profoundly influence the organization and behavior of substances that make up cells as well as the cellular environment.

SUMMARY

1. Matter is composed of elements, the atoms of which are composed of protons, neutrons, and electrons. An atom has a *net* electric charge of zero. An ion is an atom or compound that has gained or lost one or more electrons and so has acquired an overall positive or negative charge.

2. All atoms of an element have the same number of protons but they can vary slightly in the number of neutrons (they are isotopes). Radioisotopes decay spontaneously into atoms of different types.

3. Electrons occupy orbitals in shells around the nucleus. An orbital can hold no more than two electrons. Atoms of hydrogen, carbon, nitrogen, and oxygen (the main elements of biological molecules) have an unfilled orbital in their outermost shell and tend to form bonds with other elements.

4. The following chemical bonds are common in cells:
   a. Ionic bond: A positive ion and a negative ion remain together by the mutual attraction of opposite charges.
   b. Nonpolar covalent bond: Atoms share one or more electrons equally; there is no difference in charge between the two poles of the bond.
   c. Polar covalent bond: Atoms share one or more electrons unequally, causing a slight difference in charge between the two poles of the bond.
   d. Hydrogen bond: An atom of a molecule interacts weakly with a hydrogen atom that is already taking part in a polar covalent bond.

5. Acids release hydrogen ions (H^+) in solution; bases combine with hydrogen ions. Hydrogen ions are the same thing as free (unbound) protons.

6. Cellular pH is commonly kept close to neutral, meaning the H^+ and OH^- concentrations are nearly equal. Reactions between an acid and a base produce salts—ionic compounds with vital roles in cell functions.

7. A water molecule shows polarity. Due to its electron arrangements, one end carries a partial negative charge and the other, a partial positive charge.
   a. Other polar molecules are attracted to water (they are hydrophilic).
   b. Nonpolar molecules are repelled by water (they are hydrophobic).

8. The properties of water influence the organization and behavior of substances that make up cells and cellular environments.
   a. Hydrogen bonding between its individual molecules gives water its temperature-stabilizing and cohesive properties.
   b. The polarity of the water molecules allows ions and polar molecules to dissolve readily in water. It also influences the shapes of large molecules in cells through hydrophobic and hydrophilic interactions.

Review Questions

1. Define element, atom, molecule, and compound. What are the six main elements (and their symbols) in most organisms? *19–20*

2. Define an atom, an ion, and an isotope. *20, 21, 25*

3. Explain the differences among covalent, ionic, and hydrogen bonds. *25, 26*

4. What is the difference between a hydrophilic and a hydrophobic interaction? Is a film of oil on water an outcome of bonding between the molecules making up the oil? *27*

5. Define an acid, a base, and a salt. On a pH scale from 0 to 14, what is the acid range? Why are buffers important in cells? *30–31*

6. What type of bond is associated with the temperature-stabilizing, cohesive, and solvent properties of water? Is that bond also important in hydrophobic interactions? *27–29*

Self-Quiz *(Answers in Appendix IV)*

1. Atoms are constructed of protons, neutrons, and _____ .

2. An _____ has a net charge of zero; an _____ has gained or lost one or more electrons, and so has become negatively or positively charged.
   a. ion; ion   c. atom; atom
   b. ion; atom   d. atom; ion

3. Interactions between atoms as they give up, acquire, or share _____ help determine the organization and activities of living things.

4. _____ are atoms of the same element that vary only in the number of neutrons they possess.

5. The main chemical elements found in biological molecules are:
   a. hydrogen, sulfur, nitrogen, oxygen
   b. phosphorus, hydrogen, carbon, oxygen
   c. carbon, oxygen, hydrogen, nitrogen
   d. carbon, oxygen, nitrogen, sulfur

6. Orbitals within shells around the nucleus of an atom can each hold no more than _____ electrons.
   a. one   c. three
   b. two   d. four

7. Electrons are shared unequally in a(n) _____ bond.
   a. nonpolar covalent   c. hydrogen
   b. ionic   d. polar covalent

8. Polar substances are _____; nonpolar substances are _____ .
   a. hydrophilic; also hydrophilic
   b. hydrophilic; hydrophobic
   c. hydrophobic; also hydrophobic
   d. hydrophobic; hydrophilic

9. Which characterizes the internal pH of most cells?
   a. high concentration of H^+
   b. nearly equal concentration of H^+ and OH^-
   c. high concentration of OH^-
   d. both b and c are correct

10. A(n) _____ can combine with hydrogen ions or release them in response to changes in cellular pH.
   a. acid   c. base
   b. salt   d. buffer

11. Match these chemistry concepts appropriately:
   _____ water molecule   a. close to neutral
             polarity   b. temperature-stabilizing
   _____ common bonds   and cohesive properties
             in biological   c. permits ions and polar
             molecules   molecules to dissolve
   _____ cellular pH   more easily
   _____ hydrogen bonds   d. produced by reaction
             between water   between acid and base
             molecules   e. ionic, covalent, and
   _____ salt   hydrogen

Selected Key Terms

acid *30*	hydrogen ion (H^+) *30*	molecule *20*
atom *20*	hydrophilic	nonpolar covalent
atomic number *20*	substance *27*	bond *25*
base *30*	hydrophobic	pH scale *30*
buffer *31*	substance *27*	polar covalent
chemical bond *23*	hydroxide ion	bond *26*
compound *20*	(OH^-) *30*	radioisotope *21*
element *19*	ion *25*	salt *31*
evaporation *28*	ionic bond *25*	solute *29*
hydrogen bond *26*	mass number *20*	solvent *29*

Readings

Lehninger, A. 1982. *Principles of Biochemistry.* New York: Worth. Classic reference book in the field.

Miller, G. T. 1991. *Chemistry: A Contemporary Approach.* Third edition. Belmont, California: Wadsworth. Simple introduction to basic chemical concepts and their application to everyday life.

3 CARBON COMPOUNDS IN CELLS

Ancient Carbon Treasures

More than 400 million years ago, immense geologic forces were changing the contours of the earth's surface. Vast seas drained away as the seafloor rose slowly beneath them, and they left behind sediments loaded with the nutrient-rich remains of untold generations of marine organisms. Think of it! For perhaps a billion years, simple green plants had been confined to the seas—yet now they were at the threshold of a sunlit, nutrient-rich, uncrowded world. Some pioneer species made the most of the promising new environment. Within a mere 50 million years, their descendants included ferns, broadleafed shrubs, and huge trees, some more than twelve stories tall (Figure 3.1).

Between 360 million and 280 million years ago, however, sea levels swung dramatically, and the continents were submerged and drained no less than fifty times. When the seas moved out, swamp forests with large, scaly-barked trees gradually carpeted the wet lowlands. When the seas moved in, the forests were submerged again. The organic mess left behind eventually became transformed into the vast coal deposits of Britain and of the Appalachian Mountains of North America.

Coal is 55 to 95 percent carbon; the rest is mostly hydrogen, oxygen, nitrogen, and sulfur. It is a legacy of photosynthesis in those ancient swamp forests. Countless numbers of plants converted the energy of sunlight into chemical energy, which became stored in glucose and other carbon-containing compounds, and then into plant tissues. When the seas made their incursions, the tissue remains were submerged and became buried in sediments that protected them from decay. Gradually, the sediments compressed the saturated, undecayed remains into what we now call *peat*. As more sediments accumulated, the increased heat and pressure squeezed out some of the hydrogen, sulfur, and other elements from the peat. And so the peat became even more compact, with a higher percentage of carbon—it became coal, one of the "fossil fuels" we use as energy resources.

It took a fantastic amount of photosynthesis, burial, and compaction to form each major seam of coal in the earth. It has taken humans only a few centuries to deplete much of the known coal deposits. Often you will hear about our annual "production rates" for coal or some other fossil fuel. But how much do we really produce each year? None. We simply *extract* it from the earth. Given the millions of years required for its formation, coal is, for all intents and purposes, a nonrenewable resource.

With this bit of history as background, we turn to the substances that hold onto energy as it flows through the living world. Those substances are the so-called

a

Figure 3.1 (**a**) Reconstruction of the kinds of plants that formed the vast swamp forests of the Carboniferous Period, about 50 million years ago. Burial and compaction of the remains of those forests produced the world's deposits of coal, a carbon-rich fossil fuel (**b**).

biological molecules, which range from complex carbohydrates to proteins, lipids, and nucleic acids. Carbon atoms, linked together, form the structural backbone of every one of them. Those molecules are the foundation for the structure and function of every cell, every organism; they were the foundation for every chunk of coal. Today your body uses them as building materials, as "worker" molecules such as enzymes, and as storehouses of energy that can drive all of our activities—from eating a leaf of lettuce to adding coal to the furnace fire.

b

KEY CONCEPTS

1. Carbon atoms linked together into chains and rings serve as the skeletons of organic compounds—skeletons to which hydrogen, oxygen, and other atoms are attached.

2. Cells are able to assemble simple sugars, fatty acids, amino acids, and nucleotides. Those four families of small organic molecules serve as energy sources or as building blocks for the large molecules characteristic of life—the complex carbohydrates, lipids, proteins, and nucleic acids.

3. Carbohydrates include glucose and other simple sugars. They also include polysaccharides, which may consist of hundreds or thousands of simple sugars linked into straight or branched chains. What we call fats are only one of several types of lipids. Carbohydrates and lipids serve as energy reserves and building blocks.

4. Some proteins have structural roles, but others have functional roles. One class of proteins, the enzymes, makes metabolic reactions proceed much, much faster than they would on their own. The nucleic acids called DNA and RNA are the basis of inheritance and cell reproduction.

PROPERTIES OF CARBON COMPOUNDS

By far, the three most abundant elements in living things are oxygen, hydrogen, and carbon. Much of the oxygen and hydrogen is linked together in the form of water. But significant amounts of those two elements also are linked to carbon, the most important structural element in the body.

Carbon's central role in the molecules of life arises from its bonding properties. *Each carbon atom can form as many as four covalent bonds with other carbon atoms as well as with other elements.* In cells, carbon atoms that are linked in chains or rings form the backbones, or skeletons, for diverse compounds. The backbones occur in strandlike, globular, and sheetlike molecules, some of which contain

Figure 3.2 Carbon compounds. There is a Tinkertoy quality to carbon compounds, in that a single carbon atom can be the start of truly diverse molecules assembled from "straight-stick" covalent bonds. Start with a hydrocarbon, which consists only of carbon and hydrogen. Methane (CH_4) is the simplest hydrocarbon. If you were to strip one hydrogen from methane, the result would be a methyl group, which occurs in fats, oils, and waxes:

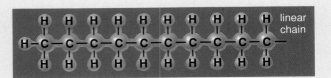

To such a linear chain, you could add branches:

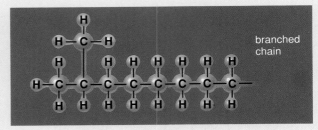

Now imagine that two methane molecules are each stripped of a hydrogen atom and bonded together. If the resulting structure were to lose a hydrogen atom, you would end up with an ethyl group:

You might even have the chains coiled back on themselves into rings, which may be represented in any of these ways:

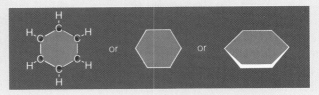

You could go on building a continuous chain, with all the carbon atoms arranged in a line:

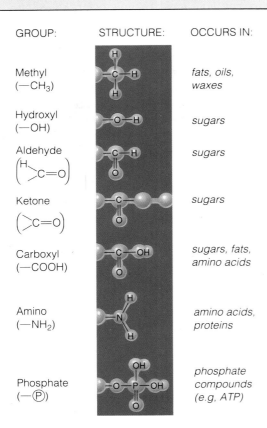

GROUP:	STRUCTURE:	OCCURS IN:
Methyl ($-CH_3$)		fats, oils, waxes
Hydroxyl ($-OH$)		sugars
Aldehyde		sugars
Ketone		sugars
Carboxyl ($-COOH$)		sugars, fats, amino acids
Amino ($-NH_2$)		amino acids, proteins
Phosphate		phosphate compounds (e.g, ATP)

Figure 3.3 Major functional groups that confer distinctive properties upon carbon compounds.

thousands, even millions, of atoms. Figure 3.2 shows some of the carbon bonding arrangements.

The carbon compounds assembled in cells are *organic* molecules. The term distinguishes them from the simple *inorganic* compounds, such as water and carbon dioxide, which have no carbon chains or rings.

Families of Small Organic Molecules

Compounds having no more than twenty or so carbon atoms are considered to be small organic molecules. The four main families of small organic molecules found in cells are *simple sugars, fatty acids, amino acids, and nucleotides*. Usually the compounds are present in cellular fluid. Cells use them as energy sources and as building blocks for large molecules, or "macromolecules." The main macromolecules are *polysaccharides* (one of three classes of carbohydrates), *lipids*, *proteins*, and *nucleic acids*. We will survey their characteristics shortly.

Functional Groups

The structure and behavior of organic compounds depend on the properties of their **functional groups,** which are atoms covalently bonded to the carbon backbone. For example, butter and other fats have "methyl groups" (—CH$_3$). Nonpolar covalent bonds link the hydrogen atoms to carbon in a methyl group (Figure 3.3). Water cannot form hydrogen bonds with non-polar groups—and that is why butter does not dissolve in water. Neither does wax, for the same reason. Alcohols, which include sugars, have "hydroxyl groups" (—OH) attached to the carbon backbone. Water *can* form hydrogen bonds at hydroxyl groups—and that is why sugars dissolve in water. Figure 3.3 shows the main functional groups that characterize organic compounds.

Because its atoms share electrons equally, the carbon backbone of an organic compound is a stable structure that does not break down easily. By comparison, functional groups are much more prone to take part in chemical reactions. They can influence the electron arrangements of neighboring atoms and so affect the structure and reactivity of the molecule as a whole.

The characteristic behavior of organic compounds depends largely on the type of functional groups that are attached to their carbon backbone.

Condensation and Hydrolysis

Small organic compounds are linked together into macromolecules with the help of enzymes. *Enzymes* are a special class of proteins that speed up reactions between specific substances. We will study these remarkable proteins in later chapters. For now, it is enough to know that when enzymes go to work on a target molecule, they usually are recognizing specific functional groups and bringing about specific changes in the structure of those groups.

Think about a **condensation reaction**, which results in the covalent linkage of small molecules and, often, the formation of water. Enzyme action causes an H atom to be split away from one molecule and an —OH group to be split away from another. A covalent bond forms between the two molecules at the exposed sites (Figure 3.4a). The parts that were split off are now free ions (H$^+$ and OH$^-$), and they may combine to form a water molecule.

Repeated condensation reactions can produce a polymer. A *polymer* is a molecule composed of anywhere from three to millions of relatively small subunits, which may or may not be identical. The individual units incorporated in polymers are often called *monomers*.

Now think about **hydrolysis**. As you can see from Figure 3.4b, this process is like condensation in reverse.

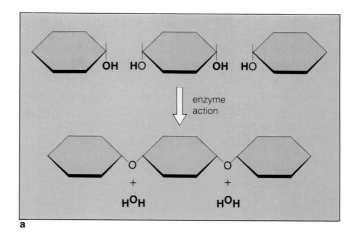

Figure 3.4 (**a**) Condensation. In this generalized example, three monomers become covalently bonded to form a larger molecule, and two water molecules are released during the reaction. (**b**) Hydrolysis. In this example, two covalent bonds of a molecule are split, and the H$^+$ and OH$^-$ ions derived from water molecules become attached to the molecular fragments.

During hydrolysis, enzyme action splits a molecule into two or more parts by breaking covalent bonds. At the same time, H$^+$ and OH$^-$ derived from a water molecule are attached to the exposed bonding sites. Hydrolysis reactions that break apart starch and other polymers are common in cells. The released subunits can be used as building blocks or energy sources.

Condensation is the covalent linkage of small molecules in a reaction that can also involve the formation of water.

Hydrolysis is the cleavage of a molecule into two or more parts by reaction with water.

CARBOHYDRATES

A **carbohydrate** is a simple sugar or a polymer assembled from a number of sugar units. Carbohydrates are probably the most abundant molecules in the world of life. All cells use them as transportable packets of quick energy, storage forms of energy, and structural materials. We recognize three classes of carbohydrates: the monosaccharides, oligosaccharides, and polysaccharides.

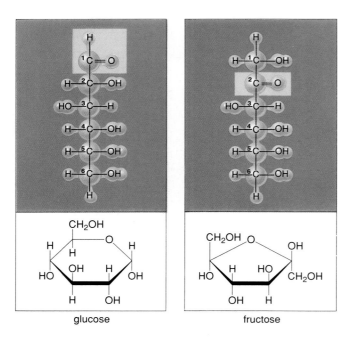

Figure 3.5 Straight-chain and ring forms of glucose and fructose. (For reference purposes, sometimes the carbon atoms of sugars are numbered in sequence, starting at the end of the molecule closest to the aldehyde or ketone group.)

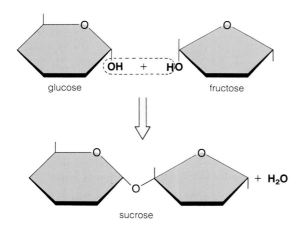

Figure 3.6 Condensation of two monosaccharides (glucose and fructose) into a disaccharide (sucrose).

Monosaccharides

The simplest type of carbohydrate is the sugar monomer, or **monosaccharide**. ("Saccharide" comes from a Greek word meaning sugar.) Sugar monomers are soluble in water, and most are sweet-tasting. As Figure 3.3 shows, they have two or more —OH groups and either an aldehyde or a ketone group. The most common ones have a backbone of five or six carbon atoms and tend to form ring structures when dissolved in cellular fluid.

Ribose and deoxyribose, which occur in RNA and DNA respectively, are five-carbon sugars. Figure 3.5 shows the structure of glucose, a six-carbon sugar ($C_6H_{12}O_6$). You will be encountering glucose repeatedly in this book. It is the main energy source for most organisms. And it is the precursor, or "parent" molecule, of many other compounds. Sucrose, fructose, and other specialized sugar molecules are derived from glucose. Large carbohydrates, including starch, are assembled from many glucose units.

You also will be encountering three other important compounds derived from sugar monomers. Glycerol (a sugar acid) is a component of fats. Vitamin C (a sugar acid) is essential in human nutrition. Glucose-6-phosphate (a sugar phosphate) is a premier entrant into major reaction pathways, including aerobic respiration.

Oligosaccharides

An **oligosaccharide** is a short chain of two or more sugar monomers. Sucrose, lactose, and maltose are examples. Each belongs to the subclass called *disaccharides*, meaning they each consist of two covalently joined monomers. A sucrose molecule is simply a glucose monomer joined to a fructose monomer (Figure 3.6). Sucrose is probably the most plentiful sugar in nature. It is the form in which carbohydrates are transported to and from different parts of leafy plants. We make table sugar by extracting and crystallizing sucrose from plants such as sugar cane. Lactose (a glucose and a galactose unit) is present in milk. Maltose (two glucose units) is present in germinating seeds. Oligosaccharides with three or more sugar monomers are usually attached as short side chains to proteins and other large molecules. Some of these side chains have roles in cell membrane function and in immunity, which are topics of later chapters.

Polysaccharides

A **polysaccharide** is a straight or branched chain of hundreds or thousands of sugar monomers, of the same or different kinds. The most common polysaccharides—starch, cellulose, glycogen, and chitin—are all polymers of glucose.

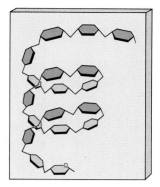

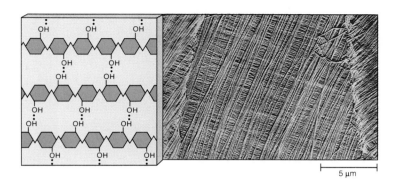

Figure 3.7 Oxygen bridges between the glucose subunits of amylose, a form of starch. The boxed inset depicts the coiling of an amylose molecule, which is stabilized by hydrogen bonds.

Figure 3.8 Structure of cellulose, which is composed of glucose subunits. Neighboring cellulose molecules hydrogen-bond together at —OH groups to form a fine strand. Such strands may twist together and then coil up as cellulose threads, of the sort shown in this micrograph of the cell wall of *Cladophora*, a green alga.

Starch is a storage form for sugar, and it can be readily hydrolyzed when cells require sugar units for energy or for building programs. Cellulose, a fiberlike structural material in the cell walls of plants, is tough and insoluble in water. It has been likened to the steel rods in reinforced concrete; it can withstand enormous weight and stress.

Given that both starch and cellulose consist of glucose units, why do they have such different properties? The answer is that adjacent glucose units are linked together in a different way in each substance. In starch, the linkages allow chains of glucose to twist into a coil, with many —OH groups facing outward (Figure 3.7). In cellulose, the chains stretch out side by side, and they hydrogen-bond to one another at —OH groups (Figure 3.8). The bonds stabilize the chains in tight bundles that resist breakdown. Only termites and a few other organisms have enzymes that can digest cellulose.

Glycogen, a highly branched polysaccharide, is the animal's equivalent of starch. Liver and muscle tissues are notable for their stores of glycogen. When blood sugar levels fall, liver cells break down glycogen and the glucose units are released to the blood. Similarly, when muscle cells are being given a workout, they tap into their glycogen supplies. The glycogen gives the cell quick access to energy. Glucose units can be released simultaneously from the ends of glycogen's numerous branches, a few of which are shown in Figure 3.9.

Many animals and fungi have cells that secrete chitin, a polysaccharide with nitrogen atoms attached to

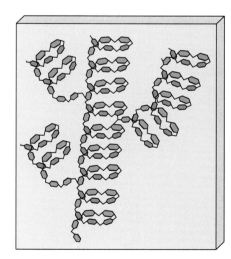

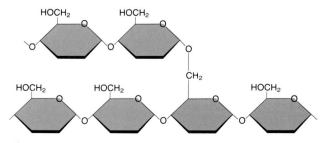

Figure 3.9 Branched structure of glycogen, a form in which sugars are stored in some animal tissues.

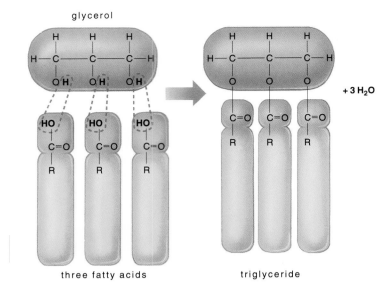

glycerol

three fatty acids   triglyceride

+3 H₂O

Figure 3.11 Condensation of fatty acids into a triglyceride. Here, the R signifies the "rest" of the carbon chain in each fatty acid molecule.

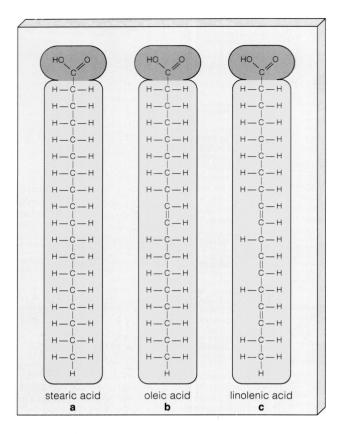

stearic acid   oleic acid   linolenic acid
     a   **b**   **c**

Figure 3.10 (*Above*) Structural formulas for representative fatty acids. (**a**) Stearic acid is fully saturated. (**b**) Oleic acid, with its double bond in the carbon backbone, is unsaturated. (**c**) Linolenic acid, with three double bonds, is one of the "polyunsaturated" fatty acids.

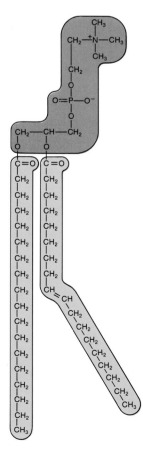

Figure 3.12 Structural formula of a typical phospholipid found in animal cell membranes. The hydrophilic head is shown in orange, and the hydrophobic tails in gold.

the glucose backbone. Chitin is the main structural material in external skeletons and other hard body parts of many insects and crustaceans, including crabs. Chitin is the reason fresh mushrooms, which contain a great deal of water, are firm as well as soft to the touch. Chitin is the main structural material in the cell walls of most fungal species, not only the mushroom-producing ones.

LIPIDS

Lipids are greasy or oily compounds that show little tendency to dissolve in water, but they dissolve in nonpolar solvents such as ether. Like polysaccharides and proteins, lipids can be broken down by hydrolysis reactions. Some lipids function in energy storage. Others are structural materials in membranes, coatings, and other cell structures. Here we will focus on two types: lipids with and without fatty acid components.

Lipids With Fatty Acids

A "fatty acid" is a long, water-insoluble chain of mostly carbon and hydrogen, with a —COOH group at one end (Figure 3.10). When a fatty acid is part of a more complex lipid molecule, it is usually stretched out like a flexible tail. Let's look briefly at three common lipids having fatty acid tails: the glycerides, phospholipids, and waxes.

Glycerides. The substances commonly called fats and oils are composed of glycerides. A **glyceride** molecule has one to three fatty acid tails attached to a backbone of

Figure 3.13 Penguins of the Antarctic, one of several types of animals that have a very thick, insulative layer of tryglycerides under the skin. Penguins also have a large, pear-shaped oil gland where their tail joins the body. These birds use their face and bill to spread the oil over their feathers. The oily coating keeps the feathers watertight and dry. And a good thing, too. Penguins may spend half a year in the open ocean without going ashore. They would become waterlogged and die in a few hours without their oil coating.

glycerol (Figure 3.11). Monoglycerides have one tail, diglycerides have two, and triglycerides have three. Glycerides are abundant lipids and a rich source of stored energy. Gram for gram, triglycerides yield more than twice as much energy as carbohydrates. Some animals, including seals and penguins, have a very thick layer of triglycerides beneath the skin (Figure 3.13). The layer serves as insulation against the near-freezing temperatures of their surroundings.

Saturated fats, including butter and lard, tend to be solids at room temperature. "Saturated" means all the carbon atoms in the fatty acid tails are joined by single C—C bonds and as many hydrogen atoms as possible are linked to them. Figure 3.10a shows this bonding scheme. The tails of adjacent molecules snuggle together in parallel array.

Unsaturated fats, or oils, tend to be liquid at room temperature. In this case, one or more double bonds occur between the carbon atoms in the fatty acid tails (Figure 3.10b and c). Oils are liquid because the double bonds create kinks that disrupt packing between tails.

Some amount of unsaturated fats is important in nutrition. In one study, immature rats were placed on a fat-free diet. The rats did not grow normally, their hair fell off, their skin turned scaly, and they died young. The abnormal conditions never developed in other rats that were fed small amounts of linoleic acid and linolenic acid, two unsaturated fats required in the diets of rats (and humans).

Phospholipids. In contrast to triglycerides, the phospholipids seldom serve as energy storage molecules. Rather, they are the main structural material of cell membranes. A **phospholipid** has two fatty acid tails attached to a glycerol backbone. It also has a hydrophilic "head," composed of a small polar group and a phosphate group, that dissolves in water (Figure 3.12). As you will see in Chapter 5, cell membranes have two layers of lipids

Figure 3.14 Beads of water on the waxy cuticle of cherries.

pressed against each other. One layer has its molecular heads dissolved in the fluid inside the cell, the other layer has its heads dissolved in the fluid surroundings, and all the tails are sandwiched between them.

Waxes. The lipids called **waxes** have long-chain fatty acids linked to long-chain alcohols or to carbon rings. Wax secretions form coatings that help keep the skin and hair of various animals protected, lubricated, and pliable. Waxes secreted from special glands in waterfowl and other birds help make feathers water-repellant (Figure 3.13). In many plants, waxes are embedded in a matrix of lipid polymers (cutin or suberin). For example, waxes and cutin form the *cuticle,* a covering on the surface of aboveground plant parts that helps restrict water loss. A waxy cuticle gives cherries, apples, and many other fruits a shiny appearance (Figure 3.14).

Lipids Without Fatty Acids

Lipids that have no fatty acid tails are less abundant than the ones described so far, but many have roles in membrane structure and in controls over metabolism.

Cholesterol Invasions of Your Arteries

Arteriosclerosis is a condition in which the blood vessels called arteries thicken and lose elasticity. Conditions worsen with *atherosclerosis,* the buildup of lipid deposits in the wall of arteries and the subsequent shrinking of arterial diameter. The lipid culprits are saturated fats and cholesterol.

Your body requires certain amounts of cholesterol, but the liver normally manufactures enough to meet the body's demands. Many of the foods you eat also contain cholesterol, and it ends up circulating in the blood along with cholesterol produced by the liver. If you habitually eat too much cholesterol-rich food, you may end up with high levels of cholesterol in the blood. If you have a certain heritable (genetic) disorder, the same thing might happen no matter what kinds of food you eat.

Cholesterol does not float freely in the blood. When it leaves the liver, it is bound to *low-density lipoproteins* (LDLs), which the liver also produces. The LDLs in turn bind to receptor molecules on cells in different parts of the body. So bound, they can be engulfed and their cholesterol cargo used for a range of cell activities. Other molecules, called *high-density lipoproteins* (HDLs), cart cholesterol back to the liver, where it can be metabolized.

Cells of some people with high blood cholesterol do not seem to have enough LDL receptors, so they cannot remove as much LDL from the blood. LDL levels increase, and so does the risk of atherosclerosis. LDLs, with their cholesterol cargo, seem to have a penchant for infiltrating arterial walls. Abnormal cells and cell products multiply at the infiltration sites, then calcium salts and a fibrous net form over the whole mass. The result is an atherosclerotic plaque. You can see such plaques in Figure *a.* Blood clots may form where they occur and narrow or block the arteries, leading to a heart attack (Chapter 38).

In contrast, HDLs seem to attract cholesterol out of arterial walls and transport it to the liver. Atherosclerosis is uncommon in rats, which have mostly HDLs. It is common in humans, who in general have mostly LDLs.

People who want to reduce their LDL levels are usually advised to restrict their intake of saturated fats. At the same time, they are encouraged to add fish oils and unsaturated vegetable oils (such as olive oil) to their diet. Some research suggests that HDL levels are higher in people who exercise regularly. They also appear to be higher in people who do not smoke cigarettes or other forms of tobacco, and who drink little or no alcohol.

a Plaques in arteries of a heart patient.

Some are long, water-insoluble chains. Others, including the steroids, have ring structures. All **steroids** have the same backbone of four carbon rings, but they vary in the number, position, and type of functional groups attached to it:

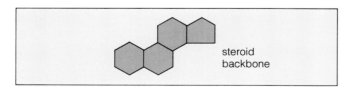

steroid backbone

You have probably heard of cholesterol. This steroid is an important component of animal cell membranes. It also is used in the synthesis of vitamin D, which is necessary for the proper development of bones and teeth. In excess amounts, however, cholesterol contributes to atherosclerosis. This is a disorder in which lipids become deposited in the walls of blood vessels (see *Commentary*). Plant tissues have no cholesterol, but they do contain steroids called phytosterols.

Many *hormones* are steroids. Among them are testosterone and estrogens, two major kinds of sex hormones. Hormones help regulate the body's growth, development, and reproduction, as well as its everyday functions. Bodybuilders and athletes sometimes use certain hormonelike steroids to increase their muscle mass. Unfortunately, use of those substances also can result in pronounced behavioral disorders and other abnormalities (Chapter 36).

PROTEINS

Proteins are polymers of amino acids. An **amino acid** is a small organic molecule having a hydrogen atom, an amino group, a carboxyl group, and one or more atoms called its R group. All four parts are covalently bonded to the same carbon atom. Under cellular conditions, the amino and the carboxyl parts are ionized (charged), as shown here:

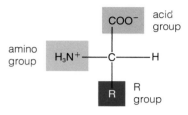

Proteins are assembled from only twenty different kinds of amino acids. (Figure 3.15 shows some of them.) Yet proteins are the most diverse of all biological molecules. Among their ranks are the enzymes, which make specific reactions proceed faster than they would on their own. Many molecules concerned with cell movements and transport of cell substances are proteins. So are many hormones and substances called antibodies, which help defend the body against disease-causing agents. Still other proteins are structural materials, the stuff of muscles, bone and cartilage, hoof and claw.

Protein Structure

Primary Structure. Cells build proteins by stringing amino acids together, one after the other, with *peptide bonds*. These are simply covalent bonds that form between the amino group of one amino acid and the carboxyl group of another, as shown in Figure 3.16. Three or more amino acids linked together are called a **polypeptide chain**.

Different sequences of amino acids occur in different proteins. For example, the two polypeptide chains making up the protein insulin always have the amino acid

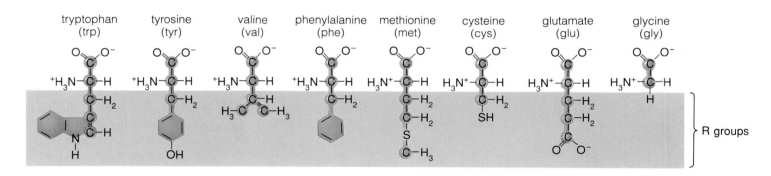

Figure 3.15 Structural formulas for eight of the twenty common amino acids. The R groups are highlighted by the green-shaded box.

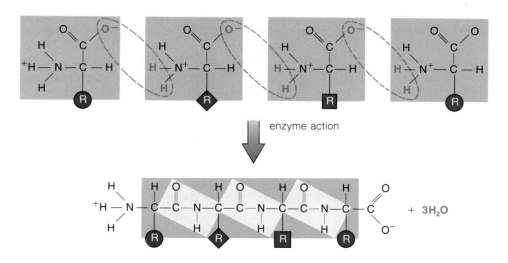

Figure 3.16 Condensation of a polypeptide chain from four amino acids.

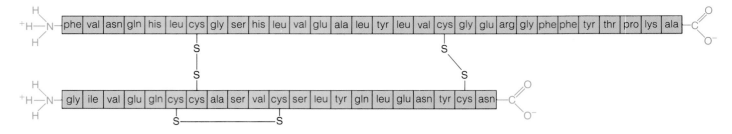

Figure 3.17 Linear sequence of amino acids in cattle insulin, as determined by Frederick Sanger in 1953. This protein is composed of two polypeptide chains, linked by disulfide bridges (—S—S—).

Figure 3.18 Hydrogen bonds (dotted lines) in a polypeptide chain. Such bonds can give rise to a coiled chain (**a**) or sheetlike array of chains (**b**).

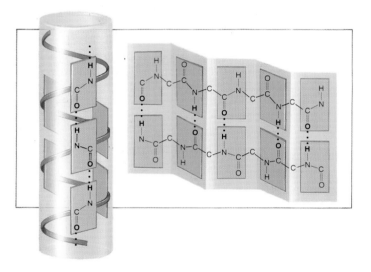

sequence shown in Figure 3.17. The specific sequence of amino acids in a polypeptide chain is the protein's *primary structure*.

Three-Dimensional Structure. The sequence of amino acids influences the shape that a protein can assume, what its function will be, and how it will interact with other substances. It does so in two major ways. First, oxygen and other atoms of specific amino acids in the sequence affect the patterns of hydrogen bonding between different amino acids along the chain. Second, R groups in the sequence interact and determine how the chain bends and twists into its three-dimensional shape.

In Figure 3.16, notice the peptide groups, which are indicated by the tan-shaded squares. Because of the way electrons are shared in each square, the atoms linked by the red bonds tend to be positioned rigidly in the same plane (the square). Only the atoms outside the squares have some freedom in how they become oriented. These

bonding patterns impose limits on the protein structures possible. In many cases, hydrogen bonds between every third amino acid hold the chain in a helical coil (Figure 3.18a). The structure of hemoglobin, an oxygen-carrying protein, includes coils like this. In other cases, the chain is extended, and hydrogen bonds hold two or more chains side by side in a sheetlike structure (Figure 3.18b). This bonding pattern occurs in silk proteins. The term *secondary structure* refers to the helical or extended pattern brought about by hydrogen bonds at regular intervals along a polypeptide chain.

Most helically coiled chains become further folded into a characteristic shape when one R group interacts with another R group some distance away, with the polypeptide backbone itself, or with other substances present in the cell. The term *tertiary structure* refers to the folding that arises through interactions among R groups of a polypeptide chain. Figure 3.19a shows one of the diverse shapes achieved through such interactions. Some proteins have *quaternary structure*, meaning they incorporate two or more polypeptide chains.

Proteins can be fiberlike, globular, or some combination of the two. Hemoglobin, shown in Figure 3.19b, has a globular shape, overall. Keratin, the structural material of hair and fur, is fibrous (Figure 3.20). So is collagen, the most common animal protein. Skin, bone, tendons, cartilage, blood vessels, heart valves, corneas—these and other components of the animal body depend on the strength inherent in collagen.

The amino acid sequence of a polypeptide chain dictates the final three-dimensional structure of a protein—and that structure dictates how the protein will interact with other cell substances.

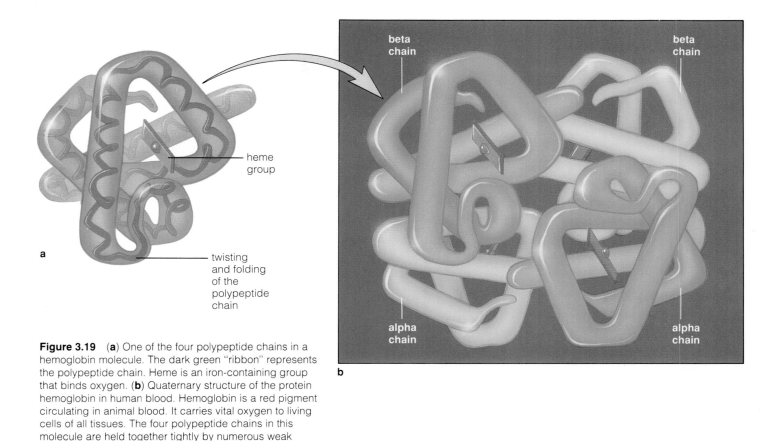

Figure 3.19 (**a**) One of the four polypeptide chains in a hemoglobin molecule. The dark green "ribbon" represents the polypeptide chain. Heme is an iron-containing group that binds oxygen. (**b**) Quaternary structure of the protein hemoglobin in human blood. Hemoglobin is a red pigment circulating in animal blood. It carries vital oxygen to living cells of all tissues. The four polypeptide chains in this molecule are held together tightly by numerous weak bonds.

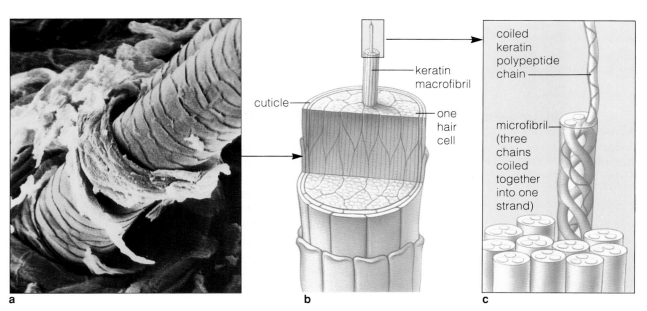

Figure 3.20 Structure of hair. Polypeptide chains of the protein keratin are synthesized inside hair cells, which are derived from epidermal cells of the skin. The chains become organized into fine fibers (microfibrils), which become bundled together into larger, cablelike fibers (macrofibrils) that eventually fill the cells. The dead, flattened cells form a tubelike cuticle around the developing hair shaft.

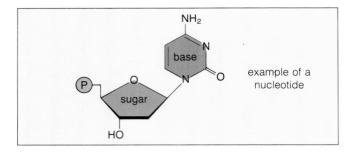

example of a
nucleotide

Figure 3.21 Generalized structure of a nucleotide.

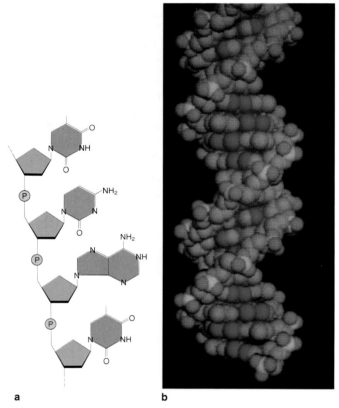

a
b

Figure 3.22 (**a**) Examples of bonds between nucleotides in a nucleic acid molecule. (**b**) Model of a segment of DNA, a molecule that is central to maintaining and reproducing the cell. The nucleotide bases are shown in blue.

Lipoproteins and Glycoproteins

Many proteins are normally combined with other types of molecules. **Lipoproteins**, for example, have both lipid and protein components. Lipoproteins circulate in the blood, where they transport fats and cholesterol (see *Commentary*). Most **glycoproteins** are proteins to which oligosaccharides are covalently bonded. Some of the oligosaccharides are linear chains, others are branched. Glycoproteins make up nearly all of the proteins on the outer surface of animal cells, most of the protein products secreted from cells, and most of the proteins found in blood.

Protein Denaturation

The hydrogen bonds and other interactions that hold a protein in its normal, three-dimensional shape are relatively weak. They also are sensitive to pH and temperature. When the H^+ concentration shifts or temperatures increase above 60°C, those interactions can be disrupted. **Denaturation** refers to the loss of a molecule's three-dimensional shape through disruption of the weak bonds responsible for it. When protein molecules have undergone denaturation, their polypeptide chains have unwound or have changed shape, so the proteins are no longer functional.

Think about the "egg white" of an uncooked chicken egg, which is a concentrated solution of the protein albumin. When you cook an egg, heat does not affect the strong covalent bonds of albumin's primary structure—but it destroys the weaker bonds responsible for its secondary and tertiary structure. Although denaturation can be reversed for some kinds of proteins when normal conditions are restored, albumin isn't one of them. There is no way to uncook a cooked egg.

NUCLEOTIDES AND NUCLEIC ACIDS

The small organic compounds called nucleotides are central to life. Each **nucleotide** contains three components: a five-carbon sugar (ribose or deoxyribose), a nitrogen-containing base that has either a single-ring or double-ring structure, and a phosphate group. The three components of a nucleotide are hooked together as shown in Figure 3.21.

Three kinds of nucleotides or nucleotide-based molecules are the adenosine phosphates, the nucleotide coenzymes, and the nucleic acids. We will explore the structure and function of these molecules in later chapters.

Adenosine phosphates are relatively small molecules that function as chemical messengers within and between cells, and as energy carriers. Cyclic adenosine monophosphate (cAMP) is a chemical messenger. The nucleotide called adenosine triphosphate (ATP) serves as a carrier of chemical energy from one reaction site to another in cells.

Nucleotide coenzymes transport hydrogen atoms and electrons from one reaction site to another in cells. The hydrogens and electrons are necessary in metabolism. Nicotinamide adenine dinucleotide (NAD^+) and flavin adenine dinucleotide (FAD) are two of the major coenzymes.

Nucleic acids are large, single or double strands of nucleotides. Adjacent nucleotides in the strand are connected by phosphate bridges, and their bases stick out to the side (Figure 3.22a). The sequence in which the four kinds of bases follow one another in the strand varies among nucleic acids.

Ribonucleic acids (RNAs) and deoxyribonucleic acids (DNAs) are built according to the plan just outlined. RNA is a single nucleotide strand, most often. DNA is usually a double-stranded molecule that twists helically, like a spiral staircase (Figure 3.22b). The two strands are held together by hydrogen bonds that form between them. You will be reading more about these molecules in chapters to come. For now, it is enough to know the following: (1) Genetic instructions are encoded in the sequence of bases in DNA, and (2) RNA molecules function in the processes by which genetic instructions are used to build proteins.

SUMMARY

1. Structurally, carbon atoms are the starting point for the large organic compounds in cells—the polysaccharides, lipids, proteins, and nucleic acids. Each carbon atom can form up to four covalent bonds with other atoms. Organic compounds are those with a backbone of carbon atoms covalently linked into a chain or ring structure.

2. Cells assemble larger organic compounds from simple sugars, fatty acids, amino acids, and nucleotides. By definition, all of these are small organic molecules that include no more than twenty or so carbon atoms.

3. The structure and behavior of organic compounds depend on the properties of functional groups, which are atoms or clusters of atoms bonded covalently to the carbon backbone. Unlike the relatively stable carbon backbone, the functional groups tend to enter into chemical reactions.

4. Organic compounds are put together and split apart by enzyme-mediated reactions. In condensation reactions, small molecules become covalently linked and water forms. In hydrolysis, a water molecule is used in a reaction that splits a molecule into two or more parts.

5. Table 3.1 on the next page summarizes the main categories of biological molecules that have been described in this chapter. Included in this table are the most common classes of molecules within each category. We will have occasion to return to their nature and roles in diverse life processes.

Review Questions

1. Four main families of small organic molecules are used in cells for the assembly of carbohydrates, lipids, proteins, and nucleic acids (the large biological molecules). What are they? *36*

2. Identify which of the following is the carbohydrate, fatty acid, amino acid, and polypeptide: *38, 40, 43–44*
   a. $^+NH_3$—CHR—COO^- c. (glycine)$_{20}$
   b. $C_6H_{12}O_6$ d. $CH_3(CH_2)_{16}COOH$

3. Is this statement true or false? Enzymes are proteins, but not all proteins are enzymes. *43*

4. Describe the four levels of protein structure. How do the side groups of a protein molecule influence its interactions with other substances? Give an example of what happens when the bonds holding a protein together are disrupted. *43–44, 46*

5. Distinguish between the following:
   a. monosaccharide, polysaccharide *38*
   b. peptide bond, polypeptide *43*
   c. glycerol, fatty acid *40–41*
   d. nucleotide, nucleic acid *46–47*

Self-Quiz   *(Answers in Appendix IV)*

1. The backbone of organic compounds is formed by the chemical bonding of _____ atoms into chains and rings.

Table 3.1 Summary of the Main Carbon Compounds in Living Things

Category	Main Subcategories	Some Examples and Their Functions	
CARBOHYDRATES *contain an aldehyde or a ketone group, and one or more hydroxyl groups*	**Monosaccharides**	Glucose	Structural roles, energy source
	Oligosaccharides	Sucrose (a disaccharide)	Form of sugar transported in plants
	Polysaccharides	Starch Cellulose	Energy storage Structural roles
LIPIDS *are largely hydrocarbon, generally do not dissolve in water but dissolve in nonpolar substances*	**Lipids with fatty acids:** *Glycerides*: one, two, or three fatty acid tails attached to glycerol backbone	Fats (e.g., butter) Oils (e.g., corn oil)	Energy storage
	Phospholipids: phosphate group, another polar group, and (often) two fatty acids attached to glycerol backbone	Phosphatidylcholine	Key component of cell membranes
	Waxes: long-chain fatty acid tails attached to alcohol	Waxes in cutin	Water retention by plants
	Lipids with no fatty acids: *Steroids*: four carbon rings; the number, position, and type of functional groups vary	Cholesterol	Component of animal cell membranes; can be rearranged into other steroids (e.g., vitamin D, sex hormones)
PROTEINS *are polypeptides (up to several thousand amino acids, covalently linked)*	**Fibrous proteins:** Individual polypeptide chains, often linked into tough, water-insoluble molecules	Keratin Collagen	Structural element of hair, nails Structural element of bones and cartilage
	Globular proteins: One or more polypeptide chains folded and linked into globular shapes; many roles in cell activities	Enzymes Hemoglobin Insulin Antibodies	Increase in rates of reactions Oxygen transport Control of glucose metabolism Tissue defense
NUCLEOTIDES *are units (or chains) having a five-carbon sugar, phosphate, and a nitrogen-containing base*	**Adenosine phosphates**	ATP	Energy carrier
	Nucleotide coenzymes	NAD^+, $NADP^+$	Transport of protons (H^+) and electrons from one reaction site to another
	Nucleic acids Chains of thousands to millions of nucleotides	DNA, RNAs	Storage, transmission, translation of genetic information

2. Each carbon atom can form up to _____ bonds with other atoms.
   a. four
   b. six
   c. eight
   d. sixteen

3. The four types of large organic molecules characteristic of life are the _____, _____, _____, and _____.

4. All of the following *except* _____ belong to the four families of small organic molecules that serve as building blocks for large biological molecules or as energy sources.
   a. fatty acids
   b. simple sugars
   c. lipids
   d. nucleotides
   e. amino acids

5. Which of the following would *not* be included in the family of carbohydrates?
   a. glucose molecules
   b. simple sugars
   c. fats
   d. polysaccharides

6. Increasing the rate of metabolic reactions is the role of functional proteins known as _____.
   a. DNA
   b. amino acids
   c. fatty acids
   d. enzymes

7. Nucleic acids, the basis of inheritance and of cell reproduction, include _____.
   a. polysaccharides
   b. DNA and RNA
   c. proteins
   d. simple sugars

8. Which of the following best describes the role of functional groups?
   a. assembling large organic compounds from smaller ones
   b. determining the structure and behavior of organic compounds
   c. splitting molecules into two or more parts
   d. speeding up metabolic reactions

9. In _____ reactions, small molecules become covalently linked, and water can also form.
   a. symbiotic
   b. hydrolysis
   c. condensation
   d. ionic

10. Match each type of molecule with the correct description.
    _____ long chain of amino acids   a. carbohydrate
    _____ energy carrier   b. phospholipid
    _____ glycerol, fatty acids, phosphate   c. protein
    _____ long chain of nucleotides   d. DNA
    _____ one or more sugar monomers   e. ATP

Selected Key Terms

adenosine phosphate *47*
amino acid *43*
carbohydrate *38*
condensation reaction *37*
denaturation *46*
enzyme *37*
fatty acid *36*
functional group *36*
glyceride *40*

glycoprotein *46*
hydrolysis *37*
inorganic molecule *36*
lipid *40*
lipoprotein *46*
monomer *37*
monosaccharide *38*
nucleic acid *47*
nucleotide *46*
nucleotide coenzyme *47*
oligosaccharide *38*

organic molecule *36*
peptide bond *43*
phospholipid *41*
polymer *37*
polypeptide chain *43*
polysaccharide *38*
protein *43*
saturated fat *41*
steroid *42*
unsaturated fat *41*
wax *41*

Readings

Hegstrom, R., and D. Kondepudi. January 1990. "The Handedness of the Universe." *Scientific American* 262 (1): 108–115.

Lehninger, A. 1982. *Principles of Biochemistry.* New York: Worth.

Scientific American. "The Molecules of Life." October 1985. This entire issue is devoted to articles on current insights into DNA, proteins, and other biological molecules. Excellent illustrations.

CELL STRUCTURE AND FUNCTION: AN OVERVIEW

Pastures of the Seas

Drifting through the surface waters of the world ocean are vast populations of living, single cells, busily engaged in photosynthesis. You can't see them without a microscope—a row of 7 million cells of one species would be less than a quarter-inch long—yet are they abundant! In some parts of the world, a cup of seawater may hold 24 million cells of one species, and that doesn't even include all the cells of *other* species.

Many of the photosynthetic drifters are protistans, such as the exquisitely shelled diatoms shown in Figure 4.1a. Others are bacteria, and still others are single-celled members of the plant kingdom. Together they are the pastures of the seas, grazed upon by microscopic animals that in turn are food for squids, fishes, and other predators. The pastures "bloom" in spring, when the waters have become warmer and enriched with nutrients, churned up from the deep by winter currents. Then, populations burgeon as their cellular members divide again and again. Some of those populations double in size not once, not twice, but *seven times* in a single day.

Biologists have known about these drifting populations for more than a century. They gave them the not-quite-accurate name "phytoplankton" (*phyto-* for plant, *plankton* for drifting). But no one suspected that the number of cells and their distribution were truly mind-boggling—until satellites started sending back photographs from space. For example, satellite images of the surface waters of the North Atlantic revealed springtime blooms that stretch from North Carolina to Spain. Those huge blooms also extend downward, several hundred feet beneath the ocean's surface!

Collectively, those single microscopic cells have enormous impact on the world's climate. They use about half of the carbon dioxide that is released each year during fossil fuel burning and other human activities. If they did not do this, atmospheric concentrations of carbon dioxide would be building up more rapidly to levels that may warm the planet, by way of the greenhouse effect described in Chapter 46. Warming can lead to changes in sea level and patterns of rainfall that can flood coastal areas and cause global shifts in the areas suitable for food production.

All cells of the phytoplankton have the same dependency on sunlight, carbon dioxide, and traces of phosphorus, nitrogen, and other nutrients. They all are sensitive to changes in the chemical composition of seawater. Even so, we daily dump industrial wastes, fertilizers and other chemicals of agricultural runoff, and raw sewage into the world ocean. How do the drifting cells tolerate that noxious chemical brew? How much more *can* they tolerate?

With these questions we start thinking about the way cells are put together and how they respond to changing conditions in their world. As our example of the photosynthetic drifters makes clear, tiny cells have significance beyond their capacity to survive and keep busy. Viewed collectively, cells work together in keeping you and all other kinds of multicelled organisms alive—and untold numbers work in ways that shape the physical character of the biosphere.

Figure 4.1 (**a**) Representative diatoms—members of the floating pastures of the seas, the marine phytoplankton. (**b, c**) Satellite images that reveal the concentration of chlorophyll at the ocean surface. (Chlorophyll, a type of light-trapping pigment, occurs in nearly all photosynthetic organisms.) In these images, red indicates high chlorophyll concentrations, purple indicates low. During winter (**b**), phytoplankton are abundant in a few coastal areas only. During the spring (**c**), the pastures spread across the entire North Atlantic.

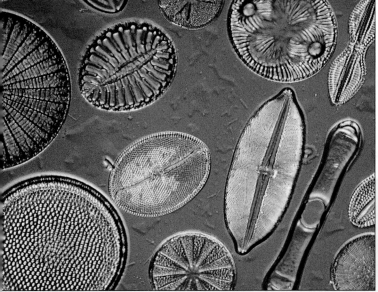

a

10 μm

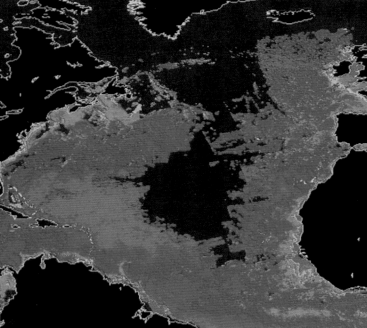

b *(Above)* Winter   c *(Below)* Spring

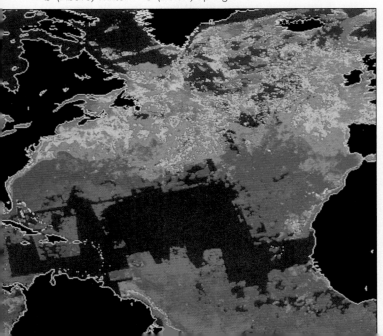

KEY CONCEPTS

1. Cells are the smallest units that still retain the characteristics of life, including complex organization, metabolic activity, and reproductive behavior.

2. All cells have a plasma membrane, and it surrounds an inner region of cytoplasm. The plasma membrane keeps events within the cell separate from the surrounding environment, so that the events proceed in organized, controlled ways.

3. Eukaryotic cells have a nucleus and other organelles (membrane-bound compartments) within the cytoplasm. The membranes of these organelles separate different chemical reactions in the space of the cytoplasm and so allow the reactions to proceed in orderly fashion. Prokaryotic cells (bacteria) do not have comparable organelles.

GENERALIZED PICTURE OF THE CELL

Emergence of the Cell Theory

Early in the seventeenth century, Galileo Galilei arranged two glass lenses in a cylinder. With this instrument he happened to look at an insect, and afterward he described the stunning geometric patterns of its tiny eyes. Thus Galileo, who was not a biologist, was the first to record a biological observation made through a microscope. The study of the cellular basis of life was about to begin. First in Italy, then in France and England, biologists set out to explore a world whose existence had not even been suspected.

At mid-century Robert Hooke, "Curator of Instruments" for the Royal Society of England, was at the forefront of the studies. When Hooke first turned one of his microscopes to a thinly sliced piece of cork from a mature tree, he observed tiny, empty compartments. He gave them the Latin name *cellulae* (meaning small rooms); hence the origin of the biological term "cell." They were actually walls of dead cells, which is what cork is made of. But Hooke did not think of them as being dead because he

a   b   c

Figure 4.2 Microscopy then and now. (**a**) Robert Hooke's compound microscope and his drawing of dead cork cells. (**b**) Anton van Leeuwenhoek, microscope in hand. (**c**) An electron microscope in a modern laboratory.

did not know that cells could be alive. He also noted that cells in other plant materials contained "juices." Figure 4.2 shows Hooke's microscope.

Given the simplicity of their instruments, it is amazing that the pioneers in microscopy saw as much as they did. Antony van Leeuwenhoek, a shopkeeper, had great skill in constructing lenses and possibly the keenest vision. He even observed a single bacterium—a type of organism so small it would not be seen again for another two centuries! Yet this was mostly an age of exploration, not of interpretation. Once the limits of their simple instruments had been reached, the early microscopists had to give up interest in cell structure without being able to explain what they had seen.

Then, in the 1820s, improvements in lens design brought cells into sharper focus. Robert Brown, a botanist, observed a spherelike structure in every plant cell he examined. He called the structure a "nucleus." By 1839, the zoologist Theodor Schwann reported the presence of cells in animal tissues. He began working with Matthias Schleiden, a botanist who had concluded that all plant tissues are composed of cells and that the nucleus is somehow paramount in a cell's development. Both investigators proposed that each cell develops as an independent unit, even though its life is influenced by the organism as a whole. Schwann distilled the mean-

ing of the new observations into what became known as the first two generalizations of the **cell theory:**

All organisms are composed of one or more cells.

The cell is the basic *living* unit of organization for all organisms.

Yet a question remained: Where do cells come from? A decade later the physiologist Rudolf Virchow completed studies of growth and reproduction (their division into two cells). He reached the following conclusion, which became the third generalization of the cell theory:

All cells arise from preexisting cells.

Not only was a cell viewed as the smallest living unit, the continuity of life was now seen to arise directly from the growth and division of single cells. Within each tiny cell, events were going on that had profound implications for all levels of biological organization!

Basic Aspects of Cell Structure and Function

Cells vary in size, shape, and complexity. However, they are alike in a few basic features. They all have a *plasma membrane* surrounding an inner region, the *cytoplasm*. Their hereditary material, DNA, is not scattered haphazardly through the cell interior. Bacterial cells have DNA physically concentrated in a part of the cytoplasm designated the nucleoid. All other cell types have DNA organized in a *nucleus:*

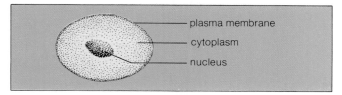
- plasma membrane
- cytoplasm
- nucleus

These features, which are the structural basis of all cellular events, can be defined in this way:

1. Plasma membrane. This is the cell's outermost membrane, composed of two lipid layers in which proteins are embedded. It separates internal events from the environment so that they can proceed in organized, controlled ways. The membrane does not totally isolate the interior; many substances move across it. It also has receptors for outside information that can alter cell behavior.

2. Nucleus. This membrane-bound compartment contains hereditary instructions (DNA) and other molecules that function in how the instructions are read, modified, and dispersed.

3. Cytoplasm. The cytoplasm is everything enclosed by the plasma membrane, except for the nucleus. In all but bacterial cells, it has compartments in which specific metabolic reactions occur. It includes particles and filaments bathed in a semifluid substance. The filaments form a skeleton that imparts shape and permits movement.

Cell Size and Cell Shape

Can any cell be seen with the unaided human eye? There are a few, including the "yolks" of bird eggs, cells in the red part of a watermelon, and the fish eggs we call caviar. Generally, however, cells are too small to be observed without microscopes; they are measured in micrometers. A micrometer is only *one-millionth* of a meter long. Your red blood cells are about 6 to 8 micrometers across—and a string of about 2,000 would only be as long as your thumbnail is wide! As Figures 4.3 through 4.5 indicate, light microscopes reveal details down to about 0.2 micrometer. Electron microscopes are used for details smaller than this.

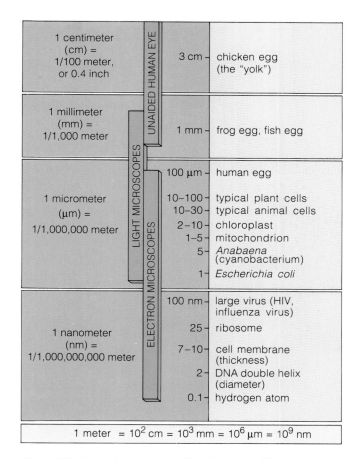

Figure 4.3 Units of measure used in microscopy. The micrometer is used in describing whole cells or large cell structures. The nanometer is used in describing cell ultrastructures and large organic molecules.

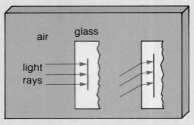

air   glass

light rays

a *Refraction of light rays (The angle of entry and the molecular structure of the glass determine how much the rays will bend)*

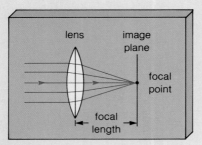

lens   image plane

focal point

focal length

b *Focusing of light rays*

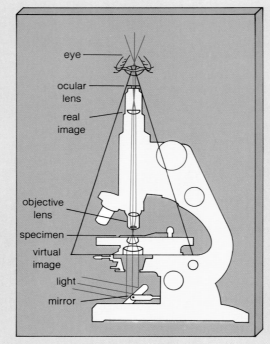

eye

ocular lens

real image

objective lens

specimen

virtual image

light

mirror

c *Compound light microscope*

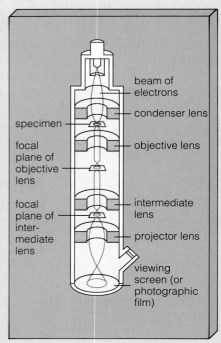

beam of electrons

condenser lens

specimen

objective lens

focal plane of objective lens

intermediate lens

focal plane of intermediate lens

projector lens

viewing screen (or photographic film)

d *Transmission electron microscope*

Figure 4.4 Microscopes—gateways to the cell.

Light Microscopes. (**a**) Light microscopy relies on the bending, or refraction, of light rays. Light rays pass straight through the center of a curved lens. The farther they are from the center, the more they bend. (**b, c**) The *compound light microscope* is a two-lens system. All rays coming from the object being viewed are channeled through the system of lenses to a single place where they can be seen by the human eye.

If you wish to observe a *living* cell, it must be small or thin enough for light to pass through. Also, structures inside cells can be seen only if they differ in color and contrast from their surroundings—but most are nearly colorless and optically uniform in density. Specimens can be stained (exposed to dyes that react with some cell structures but not others), but staining usually alters the structures and kills the cells. Finally, dead cells begin to break down at once, so they must be preserved (fixed) before staining. Most observations have been made of dead, fixed, or stained cells. Largely transparent living cells can be observed through the *phase-contrast microscope*. Here, small differences in the way different structures refract light are converted to larger variations in brightness.

No matter how good a glass or quartz lens system may be, when magnification exceeds 2,000× (when the image diameter is 2,000 times as large as the object's diameter), cell structures appear large but are not clearer. By analogy, when you hold a magnifying glass close to a newspaper photograph, you see only black dots. You cannot see a detail as small as or smaller than a dot; the dot would cover it up. In microscopy, something like dot size intervenes to limit *resolution* (the property that determines whether small objects close together can be seen as separate things). That limiting factor is the physical size of wavelengths of visible light.

Light comes in different wavelengths, or colors. The red wavelengths are about 750 nanometers and violet wavelengths are about 400 nanometers; all other colors fall in between. If an object is smaller than about one-half the wavelength, light rays passing by it will overlap so much that the object won't be visible. The best light microscopes resolve detail only to about 200 nanometers.

Transmission Electron Microscopes. Electrons are usually thought of as particles, but they also behave like waves. The wavelengths of electrons used in electron microscopes are about 0.005 nanometer—about 100,000 times shorter than those of visible light! Ordinary lenses cannot be used to focus such accelerated streams of electrons, because glass scatters them. But each electron carries an electric charge, which responds to magnetic force. A magnetic field can divert electrons along defined paths and channel them to a focal point. Magnetic lenses are used in *transmission electron microscopes* (**d**).

Electrons must travel in a vacuum, otherwise they would be randomly scattered by molecules in the air. Cells can't live in a

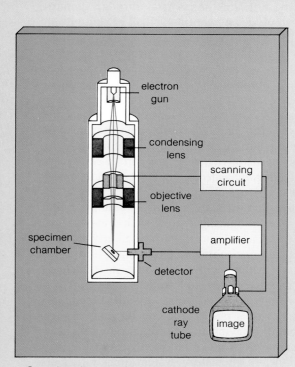

electron
gun

condensing
lens

scanning
circuit

objective
lens

specimen
chamber

amplifier

detector

cathode
ray
tube

image

e *Scanning electron microscope*

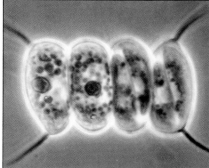

As on other micrographs in this book, the short bar provides a reference for size. Each micrometer (µm) is only 1/1,000,000 of a meter.

a Light micrograph (phase-contrast).

b Light micrograph (Nomarski process).

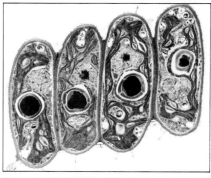

c Transmission electron micrograph, thin section.

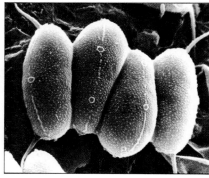

d Scanning electron micrograph.

vacuum, so living cells cannot be observed at this higher magnification. In addition, specimens must be sliced extremely thin so that electron scattering corresponds to the density of different structures. (The more dense the structure, the greater the scattering and the darker the area in the final image formed.) Specimen fixation is crucial. Fine cell structures are the first to fall apart when cells die, and artifacts (structures that do not really exist in cells) may result. Because most cell materials are somewhat transparent to electrons, they must be stained with heavy metal "dyes," which can create more artifacts.

With *high-voltage electron microscopes*, electrons can be made ten times more energetic than with the standard electron microscope. With the energy boost, intact cells several micrometers thick can be penetrated. The image produced is something like an x-ray plate and reveals the three-dimensional internal organization of cells (see, for example, Figure 4.21).

Scanning Electron Microscopes. (e) With a *scanning electron microscope*, a narrow electron beam is played back and forth across a specimen's surface, which has been coated with a thin metal layer. Electron energy triggers the emission of secondary electrons in the metal. Equipment similar to a television camera detects the emission patterns, and an image is formed. Scanning electron microscopy does not approach the high resolution of transmission instruments. However, its images have fantastic depth.

Figure 4.5 Comparison of how different types of microscopes reveal cellular details. The specimen is the green alga *Scenedesmus;* the magnification is the same in all cases. (**a**) Phase-contrast and (**b**) Nomarski techniques create optical contrasts without staining the cells; both have enhanced the value of light microscopes. (**c**) Details of a cell's internal structure show up best with transmission electron microscopy. (**d**) Scanning electron microscopy provides a three-dimensional view of surface features.

Why are most cells so small? There is a physical constraint on increases in cell size, called the **surface-to-volume ratio**. Simply put, *as a cell expands in diameter, its volume increases more rapidly than its surface area.* Figure 4.6 illustrates the relationship: Volume increases with the cube of the diameter, but surface area increases only with the square.

For example, suppose we figure out a way to make a round cell grow four times in diameter. Its cytoplasmic volume increases (4 × 4 × 4), or sixty-four times. But the surface area of its plasma membrane increases by only sixteen times (4 × 4). To survive, cells must constantly exchange materials with their surroundings. The greater the volume of cytoplasm, the more plasma membrane is required to handle the increased traffic. Unfortunately for our rotund cell, each unit of plasma membrane must now serve four times as much cytoplasm as before! Past a certain point, the inward flow of nutrients and outward flow of wastes (some toxic) is not fast enough, and we have a dead cell on our hands.

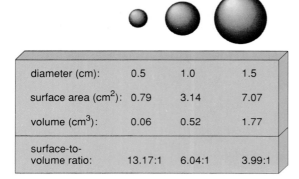

diameter (cm):	0.5	1.0	1.5
surface area (cm^2):	0.79	3.14	7.07
volume (cm^3):	0.06	0.52	1.77
surface-to-volume ratio:	13.17:1	6.04:1	3.99:1

Figure 4.6 Relationship between the surface area and volume when a sphere is enlarged. Notice that as the diameter increases, the volume increases more rapidly than the surface area.

A very large, round cell would have the added problem of moving nutrients and wastes through its large volume of cytoplasm. By contrast, small or skinny cells don't have this problem. The random motion of molecules easily distributes substances through their small or stretched-out volume of cytoplasm. This molecular motion, called diffusion, is an important topic of the next chapter.

So you can see why most cells are small—or long and thin, or have outfoldings and infoldings that increase their membrane surface relative to the volume of cytoplasm. *The smaller or more stretched out or frilly-surfaced the cell, the more efficiently materials can cross its plasma membrane and diffuse through the cytoplasm.*

Surface-to-volume constraints also influence multicelled body plans. Some algae grow as delicate strands. Their cells are attached end to end, and each interacts directly with the environment. Other algae and a few protistans are sheetlike, with all cells at or near the body surface. In massive plants and animals, transport systems move materials to and from the millions, billions, even trillions of cells packed together in their tissues. That is the point of having an incessantly pumping heart and an elaborate network of blood vessels inside your body. This efficient circulatory system quickly delivers materials from the environment to the doorstep of all living cells and sweeps away wastes. Its "highways" cut through the volume of tissue and so shrink the distance that would otherwise have to be traversed by diffusion.

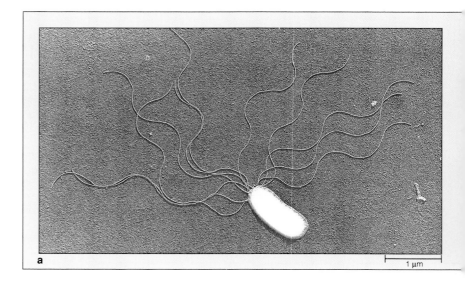

a   1 μm

PROKARYOTIC CELLS—
THE BACTERIA

Let's now turn to the characteristics of specific cell types, starting with bacteria. Bacteria are the smallest cells and, in structural terms, the simplest to think about. For most bacteria, a rigid or semirigid cell wall surrounds the plasma membrane. The wall, formed by secretions from the bacterium, supports the cell and imparts shape to it (Figure 4.7). Beneath the wall, the plasma membrane controls the movement of substances into and out of the cytoplasm. Bacterial cells have only a small volume of cytoplasm, although many ribosomes are dispersed through it.

A **ribosome** consists of two molecular subunits, each composed of RNA and protein molecules. In all cells, not just bacteria, ribosomes serve as workbenches for making proteins. At the ribosomal surface, enzymes speed the construction of polypeptide chains. Each new protein consists of one or more of those chains (page 44).

Bacterial cells are said to be prokaryotic because they do not have a nucleus. The DNA is concentrated in an irregularly shaped region of cytoplasm called the *nucleoid*. The word *prokaryotic* means "before the nucleus," and it implies that some forms of bacteria existed on earth before the evolution of cells having a nucleus. Chapters 19 and 22 provide closer looks at the evolution, structure, and functioning of bacteria. Here our focus will be on the nucleated cells, the eukaryotes.

EUKARYOTIC CELLS

Function of Organelles

Outside the realm of bacteria, all cells of all organisms—from the diatoms of phytoplankton to polar bears and peach trees and puffball mushrooms—are eukaryotic. Only eukaryotic cells have a profusion of organelles. **Organelles** are membranous sacs, envelopes, and other compartmented portions of the cytoplasm. The most conspicuous organelle is the nucleus. (Hence the name *eukaryotic*, which means "true nucleus.")

No chemical apparatus in the world can match the eukaryotic cell for the sheer number of chemical reactions that proceed in so small a space. Many of the reactions are incompatible. For example, a starch molecule can be put together by some reactions and taken apart by others—but a cell would gain nothing if the reactions proceeded at the same time on the same molecule. Yet reactions proceed smoothly in the cytoplasm, and they do so largely for these reasons:

Organelles physically separate chemical reactions (many of which are incompatible) into different regions of the cytoplasm.

Organelles separate different reactions in time, as when molecules are produced in one organelle, then used later in other reaction sequences.

Figure 4.7 Bacterial body plans. (**a**) Surface view of *Pseudomonas marginalis*, which is equipped with bacterial flagella. Like other types of flagella, these structures are used in propelling the cell through its environment. (**b**) Sketch and transmission electron micrograph of *Escherichia coli*.

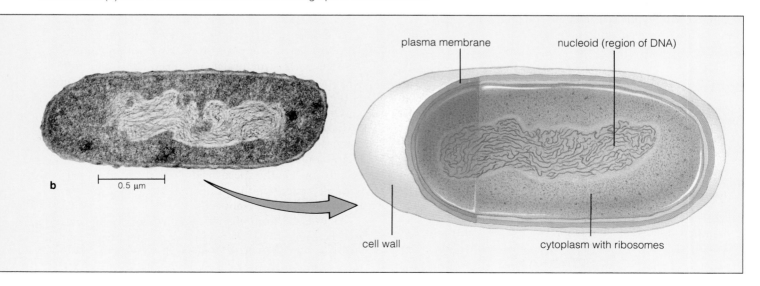

plasma membrane   nucleoid (region of DNA)

b   0.5 µm

cell wall   cytoplasm with ribosomes

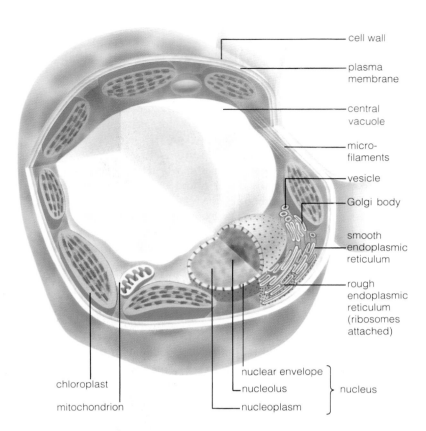

cell wall

plasma
membrane

central
vacuole

micro-
filaments

vesicle

Golgi body

smooth
endoplasmic
reticulum

rough
endoplasmic
reticulum
(ribosomes
attached)

chloroplast

mitochondrion

nuclear envelope
nucleolus
} nucleus
nucleoplasm

Figure 4.8 Generalized sketch of a plant cell, showing the types of organelles that may be present.

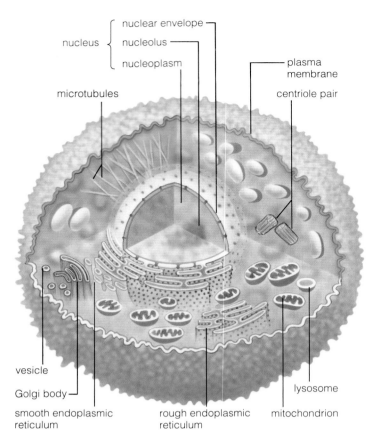

nucleus {
nuclear envelope
nucleolus
nucleoplasm
}

microtubules

plasma
membrane

centriole pair

vesicle

Golgi body

smooth endoplasmic
reticulum

rough endoplasmic
reticulum

lysosome

mitochondrion

Figure 4.9 Generalized sketch of an animal cell, showing the types of organelles that may be present.

Typical Components of Eukaryotic Cells

In general, eukaryotic cells contain the following organelles, each with specific functions:

nucleus	*physical isolation and organization of DNA*
endoplasmic reticulum	*modification of polypeptide chains into mature proteins; lipid synthesis*
Golgi bodies	*further modification, sorting, and shipping of proteins and lipids for secretion or for use in cell*
lysosomes	*digestion (breakdown) within the cell*
transport vesicles	*transport of a variety of materials to and from organelles and plasma membrane*
mitochondria	*ATP formation*

Besides the organelles just listed, eukaryotic cells have many thousands of *ribosomes*, either "free" in the cytoplasm or attached to certain membranes. They also have a *cytoskeleton*, an internal network of protein filaments. The cytoskeleton imparts shape to a cell and keeps its internal parts structurally organized. It also underlies movements of cell structures and organelles—and often of the entire cell through the environment.

Only photosynthetic cells of plants contain *chloroplasts*. Inside these organelles, sunlight energy is converted to forms that are used in the synthesis of biological molecules. One or more *central vacuoles* often occupy most of the space inside fungal and plant cells. A *cell wall* surrounds the plasma membrane of many protistan, fungal, and plant cells.

Figures 4.8 through 4.11 show where organelles and structures might be located in a typical plant and animal cell. Keep in mind that calling a cell "typical" is like calling a squid or cactus a "typical" animal or plant. Mindboggling variations exist on the basic plan.

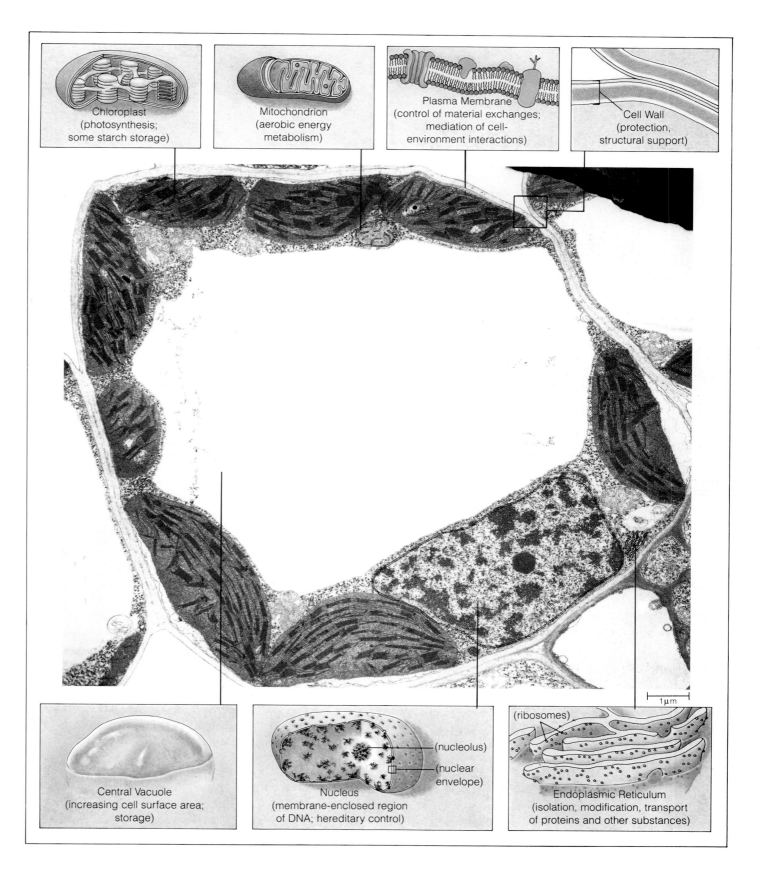

Chloroplast
(photosynthesis;
some starch storage)

Mitochondrion
(aerobic energy
metabolism)

Plasma Membrane
(control of material exchanges;
mediation of cell-
environment interactions)

Cell Wall
(protection,
structural support)

1 μm

(ribosomes)

Central Vacuole
(increasing cell surface area;
storage)

Nucleus
(membrane-enclosed region
of DNA; hereditary control)

(nucleolus)

(nuclear
envelope)

Endoplasmic Reticulum
(isolation, modification, transport
of proteins and other substances)

Figure 4.10 Transmission electron micrograph of a plant cell from a blade of Timothy grass, cross-section.

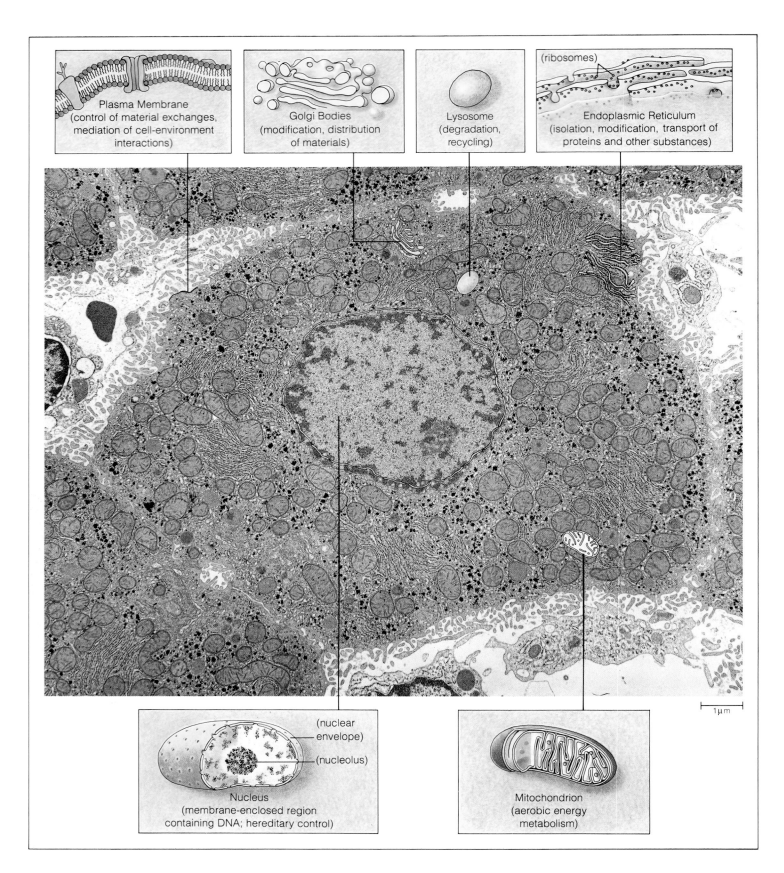

Plasma Membrane
(control of material exchanges, mediation of cell-environment interactions)

Golgi Bodies
(modification, distribution of materials)

Lysosome
(degradation, recycling)

(ribosomes)

Endoplasmic Reticulum
(isolation, modification, transport of proteins and other substances)

1 μm

(nuclear envelope)

(nucleolus)

Nucleus
(membrane-enclosed region containing DNA; hereditary control)

Mitochondrion
(aerobic energy metabolism)

Figure 4.11 Transmission electron micrograph of an animal cell from a rat liver, cross-section.

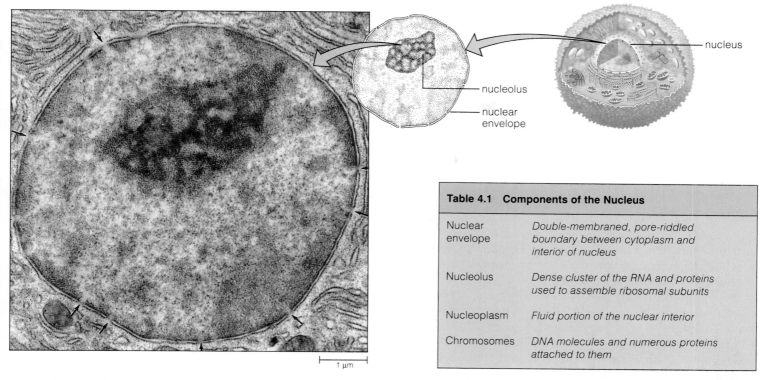

Figure 4.12 Transmission electron micrograph of the nucleus from a pancreatic cell. Arrows point to pores in the nuclear envelope.

Table 4.1	Components of the Nucleus
Nuclear envelope	*Double-membraned, pore-riddled boundary between cytoplasm and interior of nucleus*
Nucleolus	*Dense cluster of the RNA and proteins used to assemble ribosomal subunits*
Nucleoplasm	*Fluid portion of the nuclear interior*
Chromosomes	*DNA molecules and numerous proteins attached to them*

THE NUCLEUS

There would be no cells whatsoever without carbohydrates, lipids, proteins, and nucleic acids. It takes a special class of proteins called enzymes to build and use those molecules. Thus, *cell structure and function begin with proteins—and instructions for building the proteins themselves are contained in DNA.*

A membrane-bound compartment, the **nucleus**, isolates the DNA in eukaryotic cells. Figure 4.12 shows its structure. Table 4.1 lists its components.

The nucleus serves two functions. First, it helps control access to the instructions contained in DNA. Second, the nucleus keeps the DNA separate from all the substances and metabolic machinery in the cytoplasm. This makes it easier to package up the DNA when the time comes for a cell to divide. The DNA can be sorted efficiently into parcels, one for each new cell that forms.

Nuclear Envelope

Imagine a golf ball sheathed in a double layer of Saran Wrap. The outermost part of the nucleus, the **nuclear envelope**, is something like that; it has a two-membrane structure. (As with cell membranes in general, each of the two membranes of the nuclear envelope is actually two layers of lipid molecules, studded with a variety of

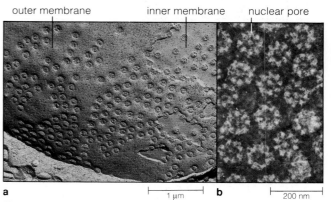

Figure 4.13 (**a**) Electron micrograph of a freeze-fractured nucleus (page 78). The two membranes that form the nuclear envelope are positioned like sheets, one atop the other. Each membrane has a lipid bilayer structure, as described on page 76. (**b**) Closer look at the pores that cross both membranes.

proteins.) Ribosomes dot the side of the membrane facing the cytoplasm.

As Figures 4.12 and 4.13 indicate, pores occur at regular intervals over the entire nuclear envelope. Observations show that the pores serve as passageways for the controlled exchange of specific substances between the nucleus and cytoplasm.

Nucleolus

As eukaryotic cells grow, two or more dense masses of irregular size and shape appear in the nucleus. Each mass is a **nucleolus** (plural, nucleoli). Nucleoli are sites where the protein and RNA subunits of ribosomes are assembled. The subunits are then shipped out of the nucleus, into the cytoplasm, where they come together as intact ribosomes. Figure 4.12 shows a nucleolus in a nondividing cell.

Chromosomes

Eukaryotic DNA is threadlike, with a great number of proteins attached to it like beads on a string. Some of the proteins are enzymes. Many others form a scaffold that organizes the DNA during cell division.

Before a cell divides, its DNA molecules are duplicated (both new cells get all the DNA instructions this way). Then the duplicated molecules fold and twist into condensed structures, proteins and all. Early microscopists could see only the condensed structures, and they called them **chromosomes** ("colored bodies"). Today, we call DNA and its proteins a chromosome regardless of whether it is in threadlike or condensed form.

The point is this: *Chromosomes do not always look the same during the life of a cell.* We will consider different aspects of "the chromosome" in chapters to come, and it will help to keep this point—and the following sketch—in mind:

| unduplicated, uncondensed chromosome (a DNA double helix + proteins) | duplicated uncondensed chromosome (two DNA double helices + proteins) | duplicated, condensed chromosome |

THE CYTOMEMBRANE SYSTEM

The polypeptide chains of proteins are assembled in the cytoplasm. What happens to the newly formed chains? Many are dissolved in the cytoplasm, and others enter the cytomembrane system. The **cytomembrane system** includes the *endoplasmic reticulum, Golgi bodies, lysosomes,*

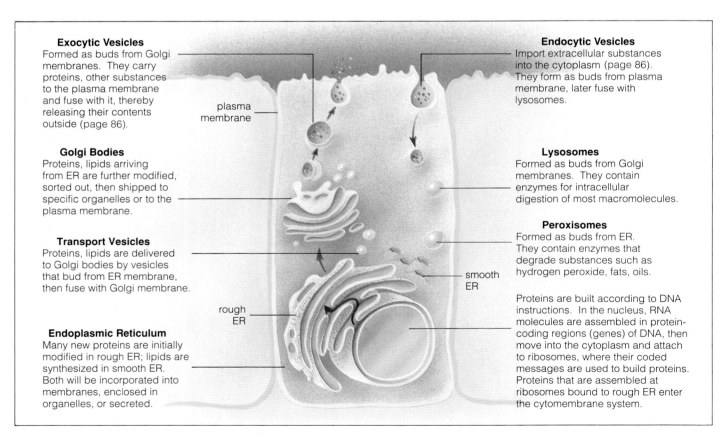

Exocytic Vesicles
Formed as buds from Golgi membranes. They carry proteins, other substances to the plasma membrane and fuse with it, thereby releasing their contents outside (page 86).

Golgi Bodies
Proteins, lipids arriving from ER are further modified, sorted out, then shipped to specific organelles or to the plasma membrane.

Transport Vesicles
Proteins, lipids are delivered to Golgi bodies by vesicles that bud from ER membrane, then fuse with Golgi membrane.

Endoplasmic Reticulum
Many new proteins are initially modified in rough ER; lipids are synthesized in smooth ER. Both will be incorporated into membranes, enclosed in organelles, or secreted.

plasma membrane

rough ER

smooth ER

Endocytic Vesicles
Import extracellular substances into the cytoplasm (page 86). They form as buds from plasma membrane, later fuse with lysosomes.

Lysosomes
Formed as buds from Golgi membranes. They contain enzymes for intracellular digestion of most macromolecules.

Peroxisomes
Formed as buds from ER. They contain enzymes that degrade substances such as hydrogen peroxide, fats, oils.

Proteins are built according to DNA instructions. In the nucleus, RNA molecules are assembled in protein-coding regions (genes) of DNA, then move into the cytoplasm and attach to ribosomes, where their coded messages are used to build proteins. Proteins that are assembled at ribosomes bound to rough ER enter the cytomembrane system.

Figure 4.14 Cytomembrane system. Endoplasmic reticulum, transport vesicles, Golgi bodies, and endocytic vesicles are components of the secretory pathway of this system (upward-directed arrows).

and a variety of *vesicles*. As Figures 4.14 and 4.15 indicate, many proteins as well as lipids take on their final form and are distributed by way of this system.

Endoplasmic Reticulum and Ribosomes

Endoplasmic reticulum, or ER, is a membrane with rough and smooth regions, owing largely to the presence or absence of ribosomes on the surface facing the cytoplasm. In animal cells, the ER begins at the nuclear envelope and curves through the cytoplasm.

Rough ER has many ribosomes attached. Often it is arranged as stacked, flattened sacs of the sort shown in Figure 4.16a. Polypeptide chains are assembled on the ribosomes. Only the newly forming chains that have a "signal" enter the sacs (the rest join the cytoplasmic pool of proteins). The signal is a sequence of about fifteen to twenty specific amino acids. As the chains pass through the rough ER, enzymes attach oligosaccharides to them. They are destined for membranes, for secretion outside the cell, or for delivery to other organelles.

Many kinds of cells specialize in secreting proteins, and rough ER is notably abundant in them. In your own body, for example, some cells of the pancreas produce and secrete proteins (enzymes) that end up in your small intestine, where they function in digestion. Proteins also are produced in quantities by immature egg cells that grow rapidly in size (frog and other amphibian eggs are like this).

Smooth ER is free of ribosomes and curves through the cytoplasm like connecting pipes (Figure 4.16b). It is the main site of lipid synthesis in many cells. Smooth ER is highly developed in seeds and in animal cells that secrete steroid hormones. Drugs and some harmful

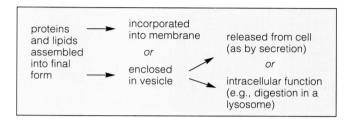

Figure 4.15 Destination of proteins and lipids that take on their final form in the cytomembrane system.

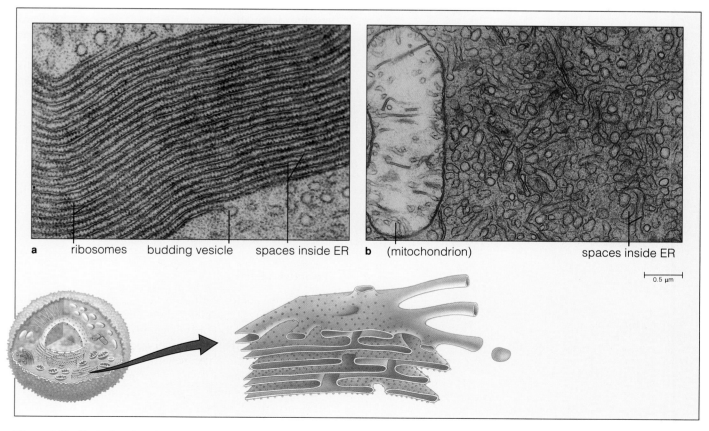

a   ribosomes   budding vesicle   spaces inside ER   b   (mitochondrion)   spaces inside ER

0.5 μm

Figure 4.16 Endoplasmic reticulum. (**a**) Rough ER, showing how the membrane surface facing the cytoplasm is studded with ribosomes. (**b**) Smooth ER, in cross-section.

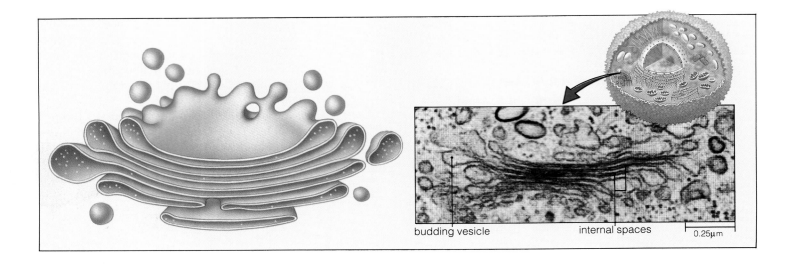

budding vesicle   internal spaces   0.25μm

Figure 4.17 Electron micrograph and sketch of a Golgi body.

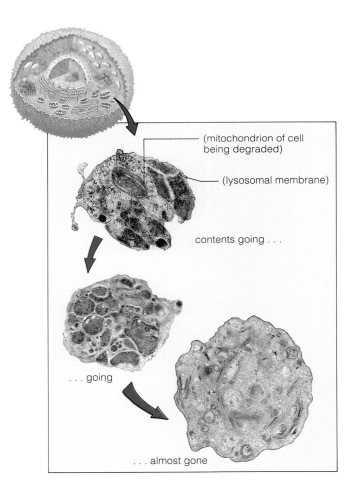

(mitochondrion of cell being degraded)

(lysosomal membrane)

contents going . . .

. . . going

. . . almost gone

Figure 4.18 Digestion of organelles from a destroyed cell, as seen in a lysosome.

by-products of metabolism are inactivated in the smooth ER of liver cells. One type of smooth ER (the sarcoplasmic reticulum) occurs in skeletal muscle; it stores and releases calcium ions that play a role in muscle contraction.

Golgi Bodies

In **Golgi bodies**, many proteins and lipids undergo final modification, then they are sorted out and packaged for specific destinations. Here, polysaccharides that were attached to each protein in the ER become trimmed and embellished in specific ways. The final results allow components of the Golgi membrane to "recognize" differences among many products and to form vesicles with special mailing tags around them.

All eukaryotic cells have one or more Golgi bodies. In outward appearance, a Golgi body resembles a stack of pancakes—usually eight or less, and usually curled at the edges (Figure 4.17). Each "pancake" is a flattened compartment. The topmost pancakes bulge at the edges, then the bulges break away as vesicles. Some secretory cells concentrate and store products in such vesicles until the cell is signaled to release them. Figure 4.14 shows where the Golgi bodies fit in the secretory pathway.

Assorted Vesicles

In animals, some of the vesicles budding from Golgi bodies becomes **lysosomes**, the main organelles of digestion inside the cell. Forty or so enzymes in these membrane bags can break down every polysaccharide, nucleic acid, and protein, along with some lipids. (Comparable enzymes are in central vacuoles of plant cells.) Lysosomes fuse with vesicles that carry a variety of substances or damaged cell parts to be degraded (Figure 4.18). They also can destroy bacteria and foreign particles.

Some vesicles that bud from the ER are called *peroxisomes*. They contain enzymes that use oxygen to break down fatty acids and amino acids. Hydrogen peroxide, a potentially harmful substance, is a product of the reactions. Another enzyme converts the hydrogen peroxide to water and oxygen or uses it to break down alcohol. If you drink alcohol, nearly half of it is degraded in peroxisomes of your liver and kidney cells.

Another type of vesicle, the *glyoxysome*, is abundant in lipid-rich seeds, such as those of peanut plants. Its enzymes help convert stored fats and oils to the sugars necessary for rapid, early growth of the plant.

MITOCHONDRIA

As mentioned earlier in the book, the energy of ATP drives nearly all cell activities. Eukaryotic cells with high demands for ATP rely heavily on the **mitochondrion** (plural, mitochondria). This organelle specializes in liberating the energy stored in sugars and using it to form *many* ATP molecules. Mitochondria extract far more energy from sugars and fats than can be done by any other means, and they do so with the help of oxygen. When you breathe in, you are taking in oxygen for your mitochondria.

Some cell types have only a sprinkling of mitochondria. The cells with the greatest demands for ATP may have more than a thousand. Not surprisingly, mitochondria are especially profuse in muscle cells, parts of nerve cells, and cells that specialize in absorbing or secreting substances.

In size and certain biochemical features, mitochondria resemble bacteria. As described on page 361, they may have originated from ancient bacteria that had been engulfed by predatory cells. In brief, by this scenario, the bacteria managed to escape digestion. They even thrived and reproduced in the cytoplasm of the predatory cell and *its* descendants. As they became permanent, protected residents, they lost many structures and functions necessary for independent life. In time the stripped-down bacterial descendants became mitochondria.

In shape, mitochondria can resemble balls, lumpy potatoes, tubes, or threads. But their shapes change. Depending on chemical conditions in the cell, mitochondria grow and branch out, even fuse with one another and divide in two. Each mitochondrion has an outer membrane facing the cytoplasm and an inner membrane, usually with many deep, inward folds called cristae (Figure 4.19). The double-membrane system creates two compartments. Hydrogen ions move from one compartment to the other in ways that cause ATP to form. This important process is described more fully in Chapter 8.

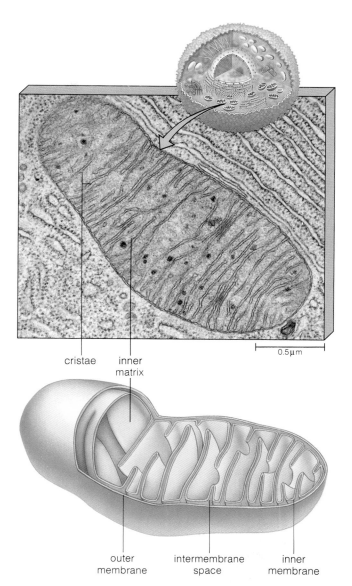

cristae   inner matrix

0.5µm

outer membrane   intermembrane space   inner membrane

Figure 4.19 Micrograph and generalized sketch of a mitochondrion.

SPECIALIZED PLANT ORGANELLES

Chloroplasts and Other Plastids

Most plant cells contain one or more "plastids," a category of organelles specialized for photosynthesis and storage. Three kinds are common:

chloroplasts   *with photosynthetic pigments and starch-storing capacity*

chromoplasts   *with pigments that often may function in pollination and seed dispersal by visually attracting animals*

amyloplasts   *with starch-storing capacity; no pigments*

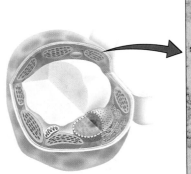

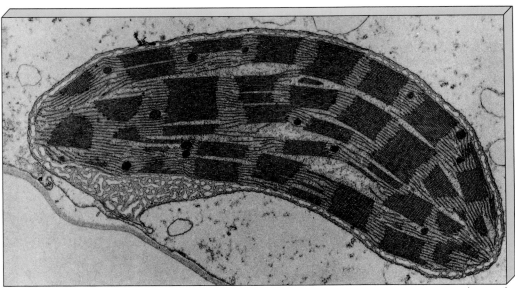

Figure 4.20 Micrograph and generalized sketch of a chloroplast. A semifluid substance (the stroma) surrounds an elaborate system of membrane compartments. Commonly, many of the compartments are organized as stacks of flattened disks (grana, singular, granum). Like mitochondria, chloroplasts have a double-membrane envelope.

chloroplast envelope
outer boundary membrane
intermembrane space
inner boundary membrane
granum
photosynthetic membranes
stroma

0.5µm

The **chloroplast** functions in photosynthesis. It has a double-membrane envelope that surrounds a semifluid substance, *stroma*.

Within the chloroplast's stroma is an elaborate system of flattened membrane compartments. These interconnect with one another, and it is common to see many disk-shaped compartments organized into stacks (Figure 4.20). Each stack is a *granum* (plural, grana). Pigments, enzymes, and other molecules of the membrane system trap sunlight energy and have roles in ATP formation.

In the stroma, enzymes speed the assembly of sugars, starch, and other products of photosynthesis. Clusters of new starch molecules ("starch grains") are often stored briefly inside the chloroplast.

Chloroplasts often are oval or disk-shaped and may be green, yellow-green, or golden-brown. Their color depends on the kinds and amounts of light-absorbing pigment molecules in their membranes. Chlorophyll, a green pigment, is an example. Chlorophyll is present in all chloroplasts, but it may be masked by other pigments, as it is in brown algae.

Chromoplasts store red or brown pigments that give flower petals, fruits, and some roots (such as carrots) their characteristic colors. The colorless amyloplasts, which occur in plant parts exposed to little (if any) sunlight, are often storage sites for starch grains. They are abundant in cells of stems, potato tubers, and many seeds.

Central Vacuoles

Mature, living plant cells often have a large, fluid-filled **central vacuole** (Figure 4.10). This organelle usually occupies 50 to 90 percent of the cell interior, so there is only a narrow zone of cytoplasm between the vacuole and the plasma membrane.

A central vacuole can store amino acids, sugars, ions, and toxic wastes. It also increases cell surface area. During growth, cell walls enlarge under the force of water pressure that builds up inside the vacuole. The cell also enlarges, but its cytoplasm is "stretched out" between the vacuole and wall. The improved surface-to-volume ratio enhances mineral absorption.

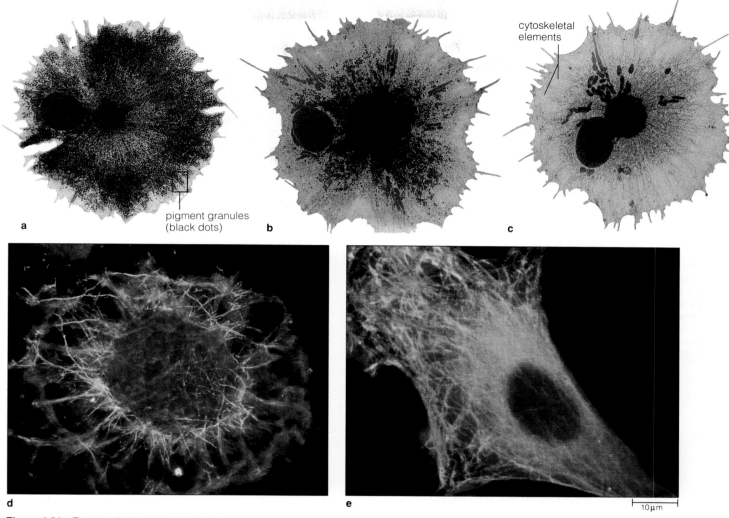

a

pigment granules
(black dots)

b

c

cytoskeletal
elements

d

e

10μm

Figure 4.21 The cytoskeleton—the basis of a cell's shape, internal organization, and capacity for motion.

(**a-c**) High-voltage electron micrographs of a pigment-containing cell from a squirrelfish. Amphibian skin and fish scales often have pigment-containing cells that collectively cause color changes. By changing color, the animal often can blend better with its surroundings and so escape the attention of predators.

The color darkens when, in response to signals from the nervous and endocrine systems, pigment granules are moved to the periphery of these cells (**a**). The color lightens when granules become condensed near the center of the cells (**b**). Granules are moved rapidly along abundant tracks of microtubules and other cytoskeletal elements (**c**).

(**d**) Cytoskeleton of a plant cell (African blood lily), made visible with *fluorescence microscopy*. By this process, cells take up molecules that bind only to certain types of proteins—*after* those molecules have been labeled with fluorescent dyes. The glow from the bound molecules marks the location of different proteins. Here, the green filaments are microtubules, composed of the protein tubulin. They are outside the nucleus, in which chromosomes (stained purple) are clustered.

(**e**) Cytoskeleton of a fibroblast, a cell that gives rise to certain connective tissues in animals. Fluorescence microscopy reveals the location of three different proteins. In this composite of three images, actin (blue) and vinculin (red) are associated with microfilaments. Tubulin (green) is associated with microtubules.

THE CYTOSKELETON

Components of the Cytoskeleton

Each cell type has a characteristic shape and internal organization made possible by its own tiny **cytoskeleton**. This interconnected system of bundled fibers, slender threads, and lattices extends from the nucleus all the way to the plasma membrane (Figure 4.21).

Some parts of the cytoskeleton are *transient;* they appear and disappear at different times in the life of a cell. Before a cell divides, for instance, some fibers assemble into a "spindle" structure that attaches to chromosomes and moves them about, then they disassemble when the task is done. But many parts of the cytoskeleton are *permanent*. These include the filaments in skeletal muscle cells, which are the basis of contraction. Other

Figure 4.22 Three major classes of protein fibers making up the cytoskeleton of eukaryotic cells. None occurs in prokaryotic cells. Microtubules consist of globular protein subunits (tubulins) linked in parallel rows. The ones involved in cell movements are typically assembled and disassembled within minutes.

Intermediate filaments are unique to specialized types of animal cells. For example, some help form "spot welds" that help hold adjacent cells together in certain tissues. The intermediate filaments in different cell types are composed of different protein subunits.

The protein actin is the key subunit of microfilaments. Actin is present in all eukaryotic cells. Often it is the most abundant cytoplasmic protein. In contractile cells, it functions in association with thick fibers composed of myosin. Myosin is another type of microfilament (Chapter 36).

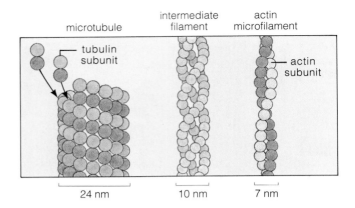

Figure 4.23 (**a**) Scanning electron micrograph showing the hairlike cilia on the surface of *Paramecium*. (**b**) The 9 + 2 array of microtubules in a cilium or flagellum.

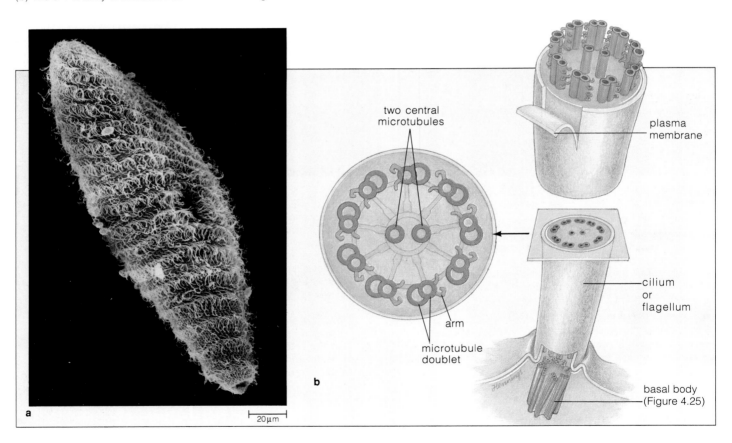

examples are flagella and cilia, two kinds of motile structures that will be described shortly.

Microtubules and *intermediate filaments* are the main cytoskeletal elements. *Microfilaments* occur in some specialized animal cells. All three types are assembled from protein subunits (Figure 4.22).

Flagella and Cilia

Microtubular structures propel many free-living eukaryotic cells through their surroundings. These structures are **flagella** (singular, flagellum). Certain protistans have one or more flagella. Dinoflagellates, a large contingent in marine phytoplankton, have a pair of them. Human sperm cells have one flagellum that makes up most of their length.

Cilia (singular, cilium), a similar kind of microtubular structure, typically are arrayed at the cell surface (Figure 4.23). Many free-living cells use cilia for propulsion. Cells with fixed positions in some tissues use cilia for stirring up their surroundings. For example, airborne bacteria and other particles are drummed out of your lungs when many thousands of cells lining the air tubes beat their cilia in coordination.

Cilia are shorter and more numerous than flagella, but both have the same organization. Nine pairs of microtubules ring two central microtubules, in what is called a *9 + 2 array*. An extension of the plasma membrane surrounds the array. Figure 4.24 describes how interactions between microtubules cause the cilium or flagellum to bend, this being the basis of propulsion.

MTOCs and Centrioles

As each new cell develops, microtubules influence what its shape will be. We know, for example, that cellulose strands maintain the shape of plant cell walls. But a temporary "scaffold" of microtubules laid down earlier by those cells during their growth guides the placement of cellulose deposits in the newly forming walls. So the question becomes this: *What organizes the microtubules?*

The organization and orientation of microtubules depend on the number, type, and location of small collections of proteins and other substances in the cytoplasm. Each such collection is a **microtubule organizing center** (MTOC). In most animal cells, a prominent MTOC near the nucleus also includes a pair of centrioles. **Centrioles** are small cylinders composed of triplet microtubules (Figure 4.25). While DNA is being duplicated before cell division, centrioles also are duplicated; a new one grows at right angles to the parent structure.

Centrioles play a vital role in ciliated or flagellated cells. When such cells are first forming, a centriole moves away from the nucleus and through the cytoplasm, then becomes positioned near the plasma membrane. There,

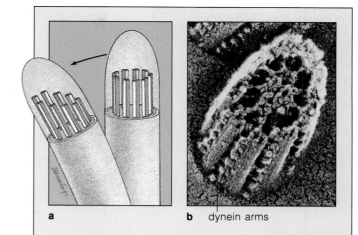

a   b   dynein arms

Figure 4.24 Movement of cilia and flagella. (**a**) In an unbent cilium or flagellum, all microtubule doublets extend the same distance into the tip. With bending, the doublets on the outside of the arc are displaced farthest from the tip. This relationship shows that microtubule doublets slide over each other, rather than contracting, when the cilium or flagellum bends.

(**b**) Many clawlike "arms" extend at regular intervals along the length of each microtubule doublet in the outer ring. The arms are composed of dynein, an ATP-hydrolyzing enzyme. All arms protrude in the same direction—toward the next doublet in the ring. When dynein binds and splits an ATP molecule, the angle of the arm changes with respect to the doublet in front of it. The arm bends and is strongly attracted to the doublet in front of it. On contact, the arm "unbends" with great force and causes its neighbor to move. The dynein releases its hold, after which it can grab another ATP molecule and attach to a new binding site on the doublet.

In other words, dynein arms on one doublet swing back and forth like tiny oars, displacing the neighboring doublet with each oarlike arc. The displacement produces the doublet sliding responsible for bending the flagellum or cilium.

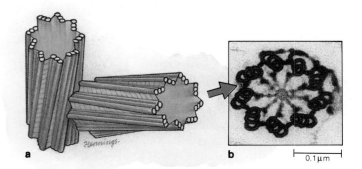

a   b   ⊢——⊣ 0.1 μm

Figure 4.25 (**a**) A pair of centrioles, which occur near the nucleus of many cells. Centrioles apparently help organize the cytoskeleton. In many species they become basal bodies, which give rise to the microtubular core of cilia and flagella. (**b**) Electron micrograph of a basal body, thin section, from a protistan (*Saccinobacculus*).

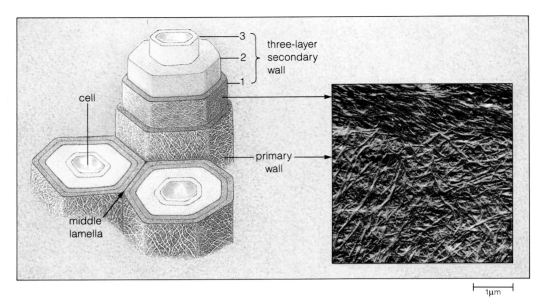

Figure 4.26 Primary and secondary walls of plant cells as they would appear in partial cross-section. The micrograph shows primary and secondary walls of a plant cell. The strands are largely cellulose.

it gives rise to the microtubules that form the core structure of a cilium or flagellum (Figure 4.23). After giving rise to the microtubules, the centrioles remain attached to the motile structure as a **basal body**.

At one time, centrioles also were thought to direct cell division. Yet division proceeds as usual in flowering plants, conifers, and other species that have no centrioles. It also proceeded even when the centrioles near the nucleus of a fertilized mouse egg were experimentally removed. However, those centrioles may govern the *plane* of cell division.

During development, each cell must divide at a prescribed angle relative to the other cells. The successive division planes influence the shape of the developing embryo and the adult form. Intriguingly, a mouse embryo did result from repeated divisions of the centriole-deprived cells. But the divisions followed a disorganized pattern, so the embryo was deformed!

CELL SURFACE SPECIALIZATIONS

Cell Walls

For many cells, surface deposits outside the plasma membrane form coats, capsules, sheaths, or walls. **Cell walls** occur among bacteria, protistans, fungi, and plants. Animal cells do not produce walls, although some secrete products to the surface layer of tissues in which those cells occur.

Most cell walls have carbohydrate frameworks. They generally provide support and resist mechanical pressure, as when they keep plant cells from stretching too much while the cells expand with incoming water during growth. Even the most solid-looking walls have microscopic spaces that make them porous, so water and solutes can move to and from the plasma membrane.

In new plant cells, cellulose strands are bundled together and added to a developing *primary cell wall*. After the main growth phase, many types of plant cells also deposit materials to form an inner, rigid *secondary cell wall*, which often consists of several layers (Figure 4.26). Cutin, suberin, and waxes commonly are embedded in cell walls at or near the plant's surface. They play a protective role and help reduce water loss.

Extracellular Matrix and Cell Junctions

In multicelled organisms, each living cell must interact with its physical surroundings, but it also must interact with its cellular neighbors. At the surface of organs, body cavities, or the body itself, cells must link tightly together so that the interior of the organism (or organ) is not indiscriminately exposed to the outside world. In all tissues, cells of the same type must recognize one another and physically stick together. Finally, in tissues where cells must act in coordinated fashion, the cells must share channels to exchange signals, nutrients, or both. Heart muscle cells are an example.

Extracellular Matrix. In animals, a meshwork of macromolecules called the *extracellular matrix* holds the cells of many tissues together. The shape of the matrix influences how cells of a given tissue will divide and what their shape will be. Its composition influences cell metabolism. Components of the matrix often include collagen and other fibrous proteins, glycoproteins, and specialized polysaccharides that form a jellylike or watery "ground substance." Nutrients, hormones, and other molecules readily diffuse from cell to cell through the ground substance. Much of the body weight of vertebrates consists of extracellular matrix material, as in bone.

In multicelled plants, adjacent cells are cemented together at their primary walls (Figure 4.26). The cementing material (the middle lamella) has an abundance of pectin compounds. Home cooks use some of those pectins to "bind" jams and jellies.

Cell Junctions in Animals. Cell-to-cell junctions are common in animal tissues. They are illustrated in Chapter 31, but here we can mention the three most common types. *Tight junctions* occur between cells of epithelial tissues, which line the body's outer surface, inner cavities, and organs. They form tight seals that keep molecules from freely crossing the epithelium and entering deeper tissues. Such seals keep stomach acids from leaking into other tissues, for example.

Adhering junctions are like spot welds at the plasma membranes of two adjacent cells. They help hold cells together in tissues that are subject to stretching, such as epithelium of the skin, heart, and stomach.

At *gap junctions*, small, open channels directly link the cytoplasm of adjacent cells. In heart muscle and smooth muscle, gap junctions provide rapid communication between cells. In liver and other tissues, they allow small molecules and ions to pass directly from one cell to the next.

Cell Junctions in Plants. In land plants, living cells are linked wall-to-wall, not membrane-to-membrane. However, channels called *plasmodesmata* (singular, plasmo-

desma) extend across adjacent walls and connect the cytoplasm of neighboring cells. There can be 1,000 to 100,000 plasmodesmata penetrating the walls of a cell. The total number affects the rate at which nutrients and other substances are transported between adjacent cells.

We will be returning to such cell-to-cell interactions. For now, the point to remember is this: *In multicelled organisms, coordinated cell activities depend on specialized forms of linkage and communication between cells.*

SUMMARY

1. The cell theory has three main points. First, all living things are made of one or more cells. Second, each cell is the basic living unit. It either can exist independently or has the potential to do so. Third, a new cell arises only from cells that already exist.

2. At the minimum, all cells have a nucleus (or nucleoid, in the case of bacteria), cytoplasm, and a plasma membrane.

3. The plasma membrane separates internal cellular events from the environment so that they can proceed in organized ways. It also acts as a boundary for exchanges between the cell's interior and its surroundings.

4. Eukaryotic cells have a variety of organelles—membranous compartments concerned with acquiring and using energy, building molecules, and tearing down molecules in controlled, specialized ways.

5. Organelle membranes separate different chemical reactions in the space of the cytoplasm and so allow them to proceed in orderly fashion.

6. Table 4.2 on the next page summarizes the organelles and other structures found in both prokaryotic and eukaryotic cells.

Review Questions

1. State the three principles of the cell theory. *52*

2. All cells share three features: a nucleus (or nucleoid), cytoplasm, and a plasma membrane. Describe the functions of each. *53*

3. Why is it highly improbable that you will ever encounter a predatory two-ton living cell on the sidewalk? *56*

4. Suppose you want to observe details of the surface of an insect's compound eye. Would you benefit most from a compound light

microscope, transmission electron microscope, or scanning electron microscope? *54–55*

5. Are all cells microscopic? Is the micrometer used in describing whole cells or extremely small cell structures? Is the nanometer used in describing whole cells or cell ultrastructure? *53*

6. Eukaryotic cells generally contain these organelles: nucleus, endoplasmic reticulum, Golgi bodies, and mitochondria. Describe the function of each. *61–65*

7. What are the components of the cytomembrane system? Sketch their general arrangement, from the nuclear envelope to the plasma

Table 4.2 Summary of Typical Components of Prokaryotic and Eukaryotic Cells

Cell Component	Function	Prokaryotic	Eukaryotic			
		Moneran	Protistan	Fungus	Plant	Animal
Cell wall	Protection, structural support	✓*	✓*	✓	✓	none
Plasma membrane	Regulation of substances moving into and out of cell	✓	✓	✓	✓	✓
Nucleus	Physical isolation and organization of DNA	none	✓	✓	✓	✓
DNA	Encoding of hereditary information	✓	✓	✓	✓	✓
RNA	Transcription, translation of DNA messages into specific proteins	✓	✓	✓	✓	✓
Nucleolus	Assembly of ribosomal subunits	none	✓	✓	✓	✓
Ribosome	Protein synthesis	✓	✓	✓	✓	✓
Endoplasmic reticulum	Modification of many proteins into mature form; lipid synthesis	none	✓	✓	✓	✓
Golgi body	Final modification of proteins, lipids; sorting and packaging them for shipment inside cell or for export	none	✓	✓	✓	✓
Lysosome	Intracellular digestion	none	✓	✓*	✓*	✓
Mitochondrion	ATP formation	**	✓	✓	✓	✓
Photosynthetic pigment	Light-energy conversion	✓*	✓*	none	✓	none
Chloroplast	Photosynthesis, some starch storage	none	✓*	none	✓	none
Central vacuole	Increasing cell surface area, storage	none	none	✓*	✓	none
Cytoskeleton	Cell shape, internal organization, basis of cellular motion	none	✓*	✓*	✓*	✓
Complex flagellum, cilium	Movement	none	✓*	✓*	✓*	✓

*Known to occur in at least some groups.
**Aerobic reactions do occur in many groups, but mitochondria are not involved.

membrane, and describe the role of each in the flow of materials between these two boundary layers. *62–65*

8. Lysosomes dismantle and dispose of malfunctioning organelles and foreign particles. Can you describe how? *64*

9. Describe the structure and function of chloroplasts and mitochondria. Mention the ways in which they are similar. *65–66*

10. Is this statement true or false? All chloroplasts are plastids, but not all plastids are chloroplasts. *65*

11. What are the functions of the central vacuole in mature, living plant cells? *66*

12. What is a cytoskeleton? How do you suppose it might aid in cell functioning? *67–69*

13. Are all components of the cytoskeleton permanent? *67*

14. What gives rise to the microtubular array of cilia and flagella? Distinguish between a centriole and a basal body. *69–70*

15. Cell walls occur among which organisms: bacteria, protistans, plants, fungi, or animals? Are cell walls solid or porous? *70*

16. In plants, is a secondary cell wall deposited inside or outside the surface of the primary cell wall? Do all plant cells have secondary walls? *70*

17. What are some functions of the extracellular matrix in animal tissues? *70–71*

18. In multicelled organisms, coordinated interactions depend on linkages and communication between adjacent cells. What types of junctions occur between adjacent animal cells? Plant cells? *71*

19. With a sheet of paper, cover the Table 4.2 column entitled Function. Can you now name the primary functions of the cell structures listed in this table?

20. Having done the preceding exercise, can you now write a paragraph describing the differences between prokaryotic and eukaryotic cells?

Self-Quiz *(Answers in Appendix IV)*

1. _____ are the smallest units that have complex organization, show metabolic activity, reproduce, and exhibit other characteristics of life.

2. The plasma membrane _____
   a. surrounds an inner region of cytoplasm
   b. separates the nucleus from the cytoplasm
   c. separates internal cell events from the environment
   d. acts as a nucleus in prokaryotic cells
   e. only a and c are correct

3. Which is *not* a key point of the cell theory?
   a. all living things are made of one or more cells
   b. the cell is the basic living unit
   c. no cell can exist unless at least one other cell is present
   d. new cells arise only from cells that already exist

4. Unlike eukaryotic cells, prokaryotic cells _____
   a. do not have a plasma membrane
   b. have RNA, not DNA
   c. do not have a nucleus
   d. all of the above

5. Organelles _____
   a. are membrane-bound compartments
   b. are typical of eukaryotic cells, not prokaryotic cells
   c. separate chemical reactions in time and space
   d. all of the above are functions of organelles

6. Plant cells but not animal cells have _____.
   a. mitochondria   c. ribosomes
   b. a plasma membrane   d. a cell wall

7. Eukaryotic DNA is contained within the _____.
   a. central vacuole   d. Golgi body
   b. nucleus   e. b and d are correct
   c. lysosome

8. The cytomembrane system does *not* include:
   a. ER   d. plastids
   b. transport vesicles   e. all of the above are parts
   c. Golgi bodies   of the system

9. The _____ is responsible for cell shape, internal structural organization, and cell movement.

10. Match each organelle with its correct function.
    _C_ protein synthesis   a. mitochondrion
    _H_ movement   b. chloroplast
    _G_ intracellular digestion   c. ribosome
    _d e_ modification of new proteins   d. smooth ER
    _D_ lipid synthesis   e. rough ER
    _B_ photosynthesis   f. nucleolus
    _A_ ATP formation   g. lysosome
    _F_ ribosome subunit assembly   h. flagellum

Selected Key Terms

basal body *70*
cell *52*
cell theory *52*
cell wall *70*
central vacuole *66*
centriole *69*
chloroplast *65*
chromosome *62*
cilium *69*
cytomembrane system *62*
cytoplasm *53*
cytoskeleton *67*
endoplasmic reticulum (ER) *63*
eukaryotic cell *57*
flagellum *69*
Golgi body *64*
intermediate filament *71*

lysosome *64*
microfilament *71*
microtubule *71*
microtubule organizing center (MTOC) *69*
mitochondrion *58, 65*
nuclear envelope *61*
nucleoid *57*
nucleolus *62*
nucleus *53, 61*
organelle *57*
plasma membrane *53*
plastid *65*
prokaryotic cell *57*
ribosome *57*
surface-to-volume ratio *56*

Readings

Alberts, B., et al. 1989. *Molecular Biology of the Cell.* Second edition. New York: Garland.

Bloom, W., and D. Fawcett. 1986. *A Textbook of Histology.* Eleventh edition. Philadelphia: Saunders. Outstanding reference book on cell structure.

deDuve, C. 1985. *A Guided Tour of the Living Cell.* New York: Freeman. Beautifully illustrated introduction to the cell; two short volumes.

Kessel, R., and C. Shih. 1974. *Scanning Electron Microscopy in Biology.* New York: Springer-Verlag. Stunning micrographs.

Rothman, J. September 1985. "The Compartmental Organization of the Golgi Apparatus." *Scientific American* 253(3):74–89. Describes the three specialized compartments of Golgi bodies.

Weber, K., and M. Osborn. October 1985. "The Molecules of the Cell Matrix." *Scientific American* 253(4):110–120. Summarizes techniques used to study the cytoskeleton.

Weibe, H. 1978. "The Significance of Plant Vacuoles." *Bioscience* 28: 327–331.

5 MEMBRANE STRUCTURE AND FUNCTION

Water, Water Everywhere

Not one living thing on earth can survive without water—the stuff that bathes cells inside and out, donates its molecules to metabolic reactions, and dissolves vital ions and keeps them from settling in a heap. But "water" in one place may not be the same as "water" in another. Seawater is much saltier than lake water (many more ions are dissolved in it), and some seas are saltier than others. And not all places on earth have a continuous supply of the stuff. Water obviously is available year-round in seas and lakes, but it comes and goes in brief, seasonal pulses in deserts and it stays frozen in the far north except for slushy surface thaws during the brief summer months.

If all goes well, each organism holds on to enough water and dissolved ions—not too little, not too much—to maintain the structure and metabolic activities of its living cells. Both single-celled and multicelled organisms have mechanisms by which they conserve the volume and ionic composition of the fluids inside them.

But who is to say that life consistently goes well? Think about a hard-shelled goose barnacle drifting offshore on a log, more or less at the mercy of ocean currents. For better or worse, the log is its home. A stalk at one end of this marine organism is attached to the wood, and featherlike appendages at the other end extend into the water, collecting microscopic bits of food (Figure 5.1a). The salty fluid inside the living cells that make up the barnacle body is in balance with the surrounding seawater. But suppose the log drifts by chance into the dilute waters from a melting glacier (Figure 5.1b). The balance is upset, and the goose barnacle has no way to deal with this. Its cells become horrifically stressed, and it is likely that the barnacle may die.

The same thing can happen to burrowing worms and other soft-bodied organisms that live between the high and low tide marks along a rocky shore. During an unpredictably heavy storm, when the seawater becomes dilute with runoff from the land, the balance between

Figure 5.1 Goose barnacles, which live attached to logs and other floating objects in the seas. Their featherlike appendages extend from the shell and comb the water for edibles. The cells of these animals are vulnerable to changes in salt concentration. If their log home were to drift into glacial meltwaters—which are low in salinity—the animals would die. In the photograph to the right, the dark blue water has a high salt concentration; the lighter blue water is flowing out from the glacier in the background.

a

the fluid in their cells and the surrounding water is similarly upset. At such times, the resulting deaths in the intertidal zone reach catastrophic proportions.

With this example we begin to see the cell for what it is: a tiny, organized bit of life in a world that is, by comparison, unorganized and sometimes harsh. *No matter what goes on outside, the cell must bring in certain substances, release or keep out others, and conduct its internal activities with great precision.* For this bit of life, the bastion against disorganization is the **plasma membrane**—a thin, seemingly flimsy surface layering of little more than lipids and proteins, dotted here and there with guarded passageways. Across this membrane, materials are exchanged between the cytoplasm and the surroundings. Then, within the cytoplasm of eukaryotic cells, exchanges are made across **internal cell membranes**, which form the compartments called organelles. This most fundamental of all cell structures, the membrane, will be our focus here.

b

KEY CONCEPTS

1. Cell membranes are composed mainly of phospholipids and proteins. The phospholipids form a double layer (bilayer) that gives the membrane its basic structure and serves as a barrier to water-soluble substances. The proteins carry out most membrane functions.

2. The plasma membrane helps control the types and amounts of substances moving into and out of cells. It also has built-in mechanisms for receiving chemical signals from the outside and for chemically recognizing other cells.

3. Internal cell membranes form compartments, which separate the many different metabolic reactions that proceed in the space of the cytoplasm.

4. Simple diffusion is the natural, unassisted movement of a solute from one region to another region where it is not as concentrated. Many small molecules having no net charge diffuse across cell membranes. Membrane transport proteins assist ions and many large molecules across.

5. Some transport proteins are open channels for water-soluble substances; some have molecular gates that open in controlled ways. Other transport proteins (carriers) bind solutes and shunt them across the membrane, either passively or actively (after receiving an energy boost).

FLUID MEMBRANES IN A LARGELY FLUID WORLD

The Lipid Bilayer

Fluid bathes the outer surface of living cells and fills most of their interior. You might think that the plasma membrane would have to be a rather solid structure, given that both sides of it are immersed in fluid, yet the plasma membrane is fluid, too! When you puncture a cell with a fine needle, the cell does not lose cytoplasm when the needle is withdrawn. Instead, the cell surface seems to flow over and seal the puncture. How can a fluid cell membrane remain distinct from fluid surroundings? The answer is found in the properties of lipid molecules.

Lipid molecules cluster spontaneously when surrounded by water. Consider the phospholipids, the most abundant type of lipid in cell membranes. A **phospholipid**, recall, has a hydrophilic (water-loving) head and two fatty acid tails, which are largely hydrophobic (water-dreading):

When many phospholipid molecules are immersed in water, hydrophobic interactions may force them together into two layers, with all the fatty acid tails sandwiched between the hydrophilic heads. This arrangement is called a **lipid bilayer**, and it is the structural basis of all cell membranes:

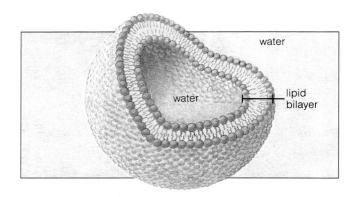

Because lipid bilayers minimize the number of hydrophobic groups exposed to water, the fatty acid tails do not have to spend a lot of energy fighting the water molecules, so to speak. Thus the reason a punctured plasma membrane tends to seal itself is that the puncture is energetically unfavorable (it leaves too many hydrophobic groups exposed).

Ordinarily, of course, few cells are ever punctured with fine needles. But the self-sealing behavior of their membrane lipids is useful for more than damage control. This behavior underlies the formation of vesicles, as described briefly in the preceding chapter. When vesicles bud off ER or Golgi membranes, for example, hydrophobic interactions with water molecules of the surrounding fluids push lipid molecules together and so close off the breakaway sites. The same thing happens when some of the vesicles fuse with the plasma membrane, and when pinched-off bits of the plasma membrane become vesicles that move into the cytoplasm. We will return to this membrane behavior later in the chapter.

Lipid molecules give a cell membrane its basic structure and its relative impermeability to water-soluble molecules.

Fluid Mosaic Model of Membrane Structure

Three types of lipids are common in cell membranes: phospholipids, glycolipids, and sterols, all of which are largely hydrophobic with a hydrophilic head. The phospholipids differ in their head regions. They also differ in the length and degree of saturation of their fatty acid tails. (Fully saturated fatty acids have no double bonds in the carbon backbone; unsaturated fatty acids have one or more.) Figure 5.2 shows phosphatidylcholine, a phospholipid that is common to cell membranes.

Glycolipids are similar in structure but they have one or more sugar monomers attached at the head end. The steroid cholesterol is abundant in animal cell membranes but nonexistent in plant cell membranes, which have phytosterols.

Within a lipid bilayer, individual lipid molecules show quite a bit of movement. They move sideways, they spin

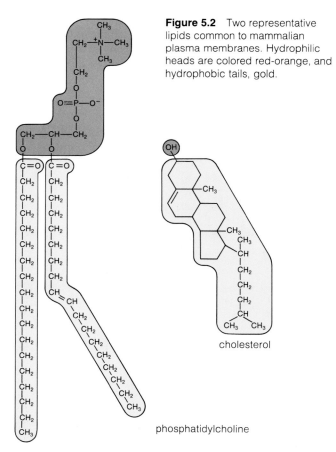

Figure 5.2 Two representative lipids common to mammalian plasma membranes. Hydrophilic heads are colored red-orange, and hydrophobic tails, gold.

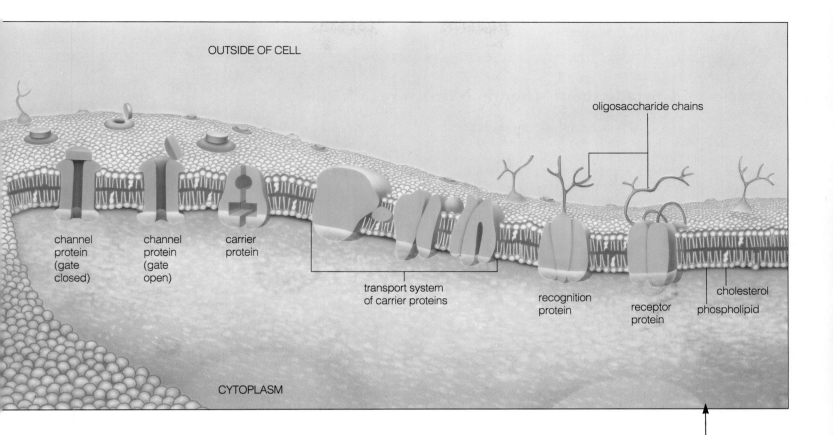

OUTSIDE OF CELL

channel protein (gate closed)

channel protein (gate open)

carrier protein

transport system of carrier proteins

oligosaccharide chains

recognition protein

receptor protein

cholesterol

phospholipid

CYTOPLASM

about their long axis, and their tails flex back and forth. The movements help keep adjacent lipids from packing into a solid layer. Packing is also disrupted by the presence of cholesterol and by lipids with short tails (which cannot interact as strongly as long-tailed neighbors do) or unsaturated tails (which tend to kink at the sites of double bonds).

So far, you may have the impression that a cell membrane consists only of lipids. Actually, a variety of proteins are nestled in the lipid bilayer and positioned at its two surfaces. In other words, *the membrane is a mosaic* of lipids and proteins. Most proteins of the plasma membrane are *glycoproteins*, meaning they have sugars covalently bonded to them. Usually the sugars are short-chain oligosaccharides. They always extend outward into the extracellular fluid, never into the cytoplasm. Although certain proteins are tethered to specific locations, many are not so restricted; their position can shift in the plane of the membrane. Taken together, the molecular movements and packing variations help account for the fact that a membrane behaves more like a *fluid* than a solid.

The lipids and proteins of one of the membrane's surfaces differ in number, kind, and arrangement from those of the other surface. In other words, the membrane has an *asymmetrical structure*. As you will see, this

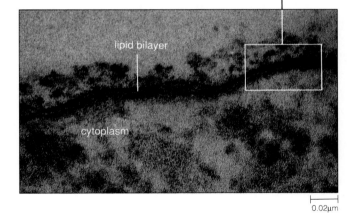

lipid bilayer

cytoplasm

0.02μm

Figure 5.3 Views of the plasma membrane, based on the fluid mosaic model of membrane structure. The micrographs show the plasma membrane of a eukaryotic cell, thin section.

asymmetry reflects differences in the tasks carried out at the two membrane surfaces.

Figure 5.3 shows a model of membrane structure. It is a recent version of a model put together in 1972 by S. J. Singer and G. Nicolson. Evidence favoring this model comes from many sources, including freeze-fracture microscopy, as described next in the *Doing Science* essay.

Discovering Details About Membrane Structure

Cells come in a spectacular array of shapes and sizes, and they perform a spectacular array of different tasks. Particularly in multicelled organisms, specialized cells engage in a kind of export-import business, selectively shipping off materials that they have manufactured for use by other cells and just as selectively receiving substances that neighboring or distant cells of other types have produced. The shipping and receiving functions are related to the structure of the plasma membrane.

If you wanted to deduce the structure of a plasma membrane, you would probably begin with attempts to identify its molecular components. Your first challenge is to secure a large sample of plasma membrane that has not been contaminated with internal cell membranes, such as those of the nucleus. Using red blood cells will simplify your task. Such cells are abundant, they are easy to collect—and structurally they are simple. As red blood cells develop, the nucleus becomes inactivated and is expelled from the cell body. At maturity, each cell is little more than a sack of hemoglobin, ribosomes, and a few other parts that will keep the cell functional for its life span of about 120 days.

By placing a sample of red blood cells in a test tube containing distilled water, you can separate the plasma membranes from the rest of the cell components. Red blood cells are hypotonic in such a solution. This means there are more solutes (and fewer water molecules) inside the cell body than outside. Because water can move across the plasma membrane, it tends to follow its concentration gradient and diffuse into the cell body (by osmosis).

Red blood cells have no mechanisms for actively taking in or expelling water, so they tend to swell when placed in hypotonic solutions. In time the cells in your sample burst and their contents spill out. This gives you a mixture of membranes, hemoglobin, and other cell parts in the water.

How do you separate the bits of membrane? The trick here is to place a tube containing a special solution of cell parts in a *centrifuge*, a device that spins test tubes at high speed. Each component of a cell has its own molecular composition, and this gives it a characteristic density. As the centrifuge spins at the appropriate speed, components with the greatest density will move toward the bottom of the tube. Other components will take up layered positions above them according to their relative densities.

If you have done the centrifugation properly, one layer in the solution inside the tube will contain only shreds of cell membrane. You can carefully draw off the layer and examine it microscopically to verify that your membrane sample is not contaminated.

Through standard chemical analysis, you find that the plasma membrane is composed of lipids and proteins. Today we know how those molecules are organized in a plasma membrane, but in the past there were two competing models of membrane structure. In the "protein coat" model, the membrane was composed of a bilayer of lipids,

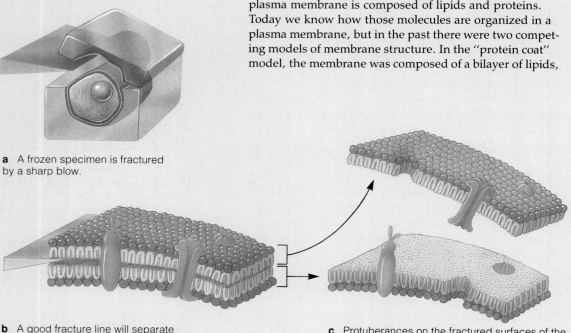

a A frozen specimen is fractured by a sharp blow.

b A good fracture line will separate the layers of the lipid bilayer.

c Protuberances on the fractured surfaces of the membrane are mostly membrane proteins.

coated on both surfaces with a layer of proteins. In the "fluid mosaic" model, proteins were largely embedded within the bilayer.

How would you test the protein coat model? One way would be to predict and then measure the amount of protein present in a given sample of membranes from red blood cells. First you calculate how much protein would be required to coat the inner and outer surface of a known number of cells. Then you separate a membrane sample into its lipid and protein fractions. By measuring the *observed* ratio of proteins to lipids, you can match the results against the ratio *predicted* on the basis of the model.

Such tests have been done on membrane samples, and they reveal that there is far too little protein to cover both surfaces of a lipid bilayer. The tests provide evidence against the protein coat model, at least for red blood cells.

What would be a direct test of the fluid mosaic model? You could employ freeze-fracturing and freeze-etching techniques. As Figures *a–e* show, these are special methods of preparing cells for electron microscopy. First a sample of cells is immersed in liquid nitrogen, an extremely cold fluid. The cells freeze instantly, after which they can be struck with a microscopically small chisel. A properly directed blow will fracture the cell in such a way that one layer of the lipid bilayer separates from the other. Those preparations can then be inspected under extremely high magnifications.

(If the *protein coat* model were correct, what do you suppose one such layer would reveal upon examination?)

What you actually see is not a perfectly smooth, pure lipid layer. Freeze-fractured cell membranes reveal many bumps and other irregularities in lipid layers. The bumps are the proteins that are incorporated whole, directly in the bilayer—just as the fluid mosaic model requires.

(**a**) Freeze-fracturing and freeze-etching. In the freeze-fracture step, specimens being prepared for electron microscopy are rapidly frozen, then fractured by a sharp blow from the edge of a fine blade. (**b, c**) Fractured membranes commonly split down the middle of the lipid bilayer. Typically, one inner surface is studded with particles and depressions, and the other is a complementary pattern of depressions and particles. The particles are membrane proteins.

(**d**) Sometimes specimens are freeze-etched: more ice is evaporated from the fracture face to expose the outer membrane surface. (**e**) In a process called metal shadowing, the fractured surface is coated with a layer of carbon and heavy metal such as platinum. The coating is thin enough to replicate details of the exposed specimen surface. The metal replica, not the specimen itself, is used for micrographs. (**f**) The micrograph shows part of a replica of a red blood cell, prepared by freeze-fracturing and freeze-etching.

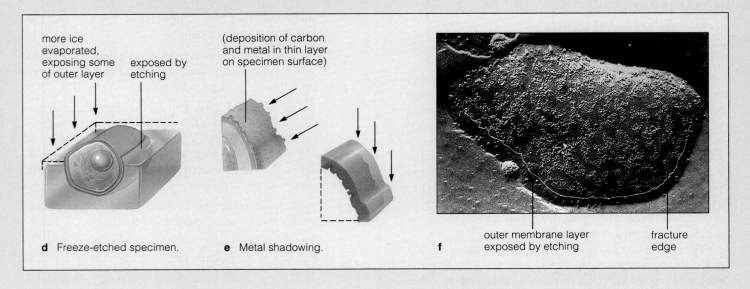

more ice evaporated, exposing some of outer layer   exposed by etching

(deposition of carbon and metal in thin layer on specimen surface)

d Freeze-etched specimen.   **e** Metal shadowing.   **f** outer membrane layer exposed by etching   fracture edge

Fluid mosaic model of membrane structure: **A cell membrane is an asymmetrical mosaic of lipids *and* proteins. The membrane shows fluid behavior because of movements and packing variations among its lipids and proteins.**

Keep in mind that the fluid mosaic model is only a starting point for discussing membrane structure and function. Membranes vary not only in their composition but also in their fluidity. Consider one effect of cholesterol, an important lipid in animal cell membranes. Cholesterol has a rigid ring structure (page 42). When its rings interact with the tails of neighboring phospholipids, they immobilize parts of their neighbors' tails and so make the membrane less fluid at that particular spot. Yet at high concentration, cholesterol actually helps keep the membrane fluid by preventing saturated fatty acid tails from packing together.

Why is maintaining membrane fluidity so important? Cell survival depends on it. For example, we know that a bacterial or yeast cell doesn't have many mechanisms for dealing with cold spells. When temperatures fall, membranes tend to become less fluid, and the functioning of their enzymes and other proteins suffers. In response to the cold, however, those cells rapidly synthesize lipids that have double bonds in their tails—and the infusion of more kinky lipids into the membrane helps keep it from stiffening up.

Functions of Membrane Proteins

Whereas the lipid bilayer serves as the overall structural framework for a cell membrane, proteins carry out most membrane functions. Three categories of proteins are crucial in this respect:

> Membrane transport proteins
> Recognition proteins
> Receptor proteins

As is the case for other biological molecules, each of these proteins has certain numbers and kinds of atoms linked to its carbon backbone. The specific arrangement and location of those atoms are the basis for chemical interactions with other substances. For example, when a substance weakly binds with some of those atoms, the binding might induce changes in the protein's shape. The physical changes can "bump" the bound substance from one side of the membrane to the other. This is a very simplified description of a membrane transport function, but you get the idea.

There are two classes of proteins with transport functions: channel proteins and carrier proteins. A **channel protein** serves as a pore through which ions or other water-soluble substances move from one side of the membrane to the other. Some channels are perpetually open. Others, being equipped with molecular "gates," open and close in controlled ways that permit or block the passage of specific ions. A **carrier protein** binds specific substances and changes its shape in ways that shunt the substances across the membrane. As you will see shortly, some carrier proteins do this rather passively. Others use energy as they actively pump substances in a specific direction.

A **recognition protein** is like a molecular fingerprint at the cell surface; it identifies the cell as being of a certain type. When a multicelled embryo is developing, recognition proteins help guide the ordering of cells into tissues. Later, they function in cell-to-cell interactions. The oligosaccharide chains of certain recognition proteins help form a "sugar coat" at the outer surface of the plasma membrane. Some proteins of the coat promote adhesion between cells. Others function in cell-to-cell recognition and in coordinating cell behavior within tissues.

A **receptor protein** is like a switch that turns on or off when particular substances bind to it. Receptor proteins have binding sites for hormones or other substances that can trigger alterations in a cell's metabolism or behavior. For example, the binding of the hormone somatotropin to a receptor turns on enzymes that crank up the machinery for cell growth and division. Different cell types have different combinations of receptors. In vertebrates, some receptors are widespread among the body's cells. Others are restricted to only a few cell types. Most commonly, the receptors are arrayed at the plasma membrane; some are located inside the cell.

To sum up, all cell membranes have the following characteristics in common:

1. Cell membranes are composed of lipids (especially phospholipids) and proteins.

2. In membranes, lipid molecules have their hydrophilic heads at the two outer faces of a bilayer and their hydrophobic tails sandwiched in between.

3. The lipid bilayer gives cell membranes their overall *structure* and serves as a barrier between two solutions. (The plasma membrane separates the fluids inside and outside the cell. The membranes of organelles or other cellular compartments separate different solutions within the space of the cytoplasm.)

4. Proteins embedded in the bilayer or positioned at its surfaces carry out most membrane *functions*.

5. Cell membranes show fluid behavior as a result of movements and packing variations among their lipid and protein components.

DIFFUSION

With the preceding overview in mind, we are almost ready to explore the details of membrane function. First, however, let's consider the natural, unassisted movements of water and solutes. Many membrane properties involve mechanisms that work with or against these movements.

Gradients Defined

"Concentration" refers to the number of molecules (or ions) of a substance in a given volume of fluid. In the absence of other forces, molecules of a given type move down their **concentration gradient**. They tend to move from a region of greater concentration to a region where they are less concentrated. They are driven to do so because they are constantly colliding with one another millions of times a second. Random collisions do send the molecules back and forth, but the *net* movement is outward from the region of greater concentration. Similarly, for any defined volume, gradients can also exist between two regions that differ in pressure, temperature, or net electric charge.

When your heart contracts, it generates fluid pressure that is greatest in the first artery leaving the heart and lowest in the last vein leading back into it. When you slip while skiing and find yourself sitting in a snowbank, heat energy flows down a thermal gradient and is transferred away from your body to the snow. Every moment of your life, you depend on electrical gradients that are associated largely with differences in the concentration of sodium ions inside and outside the plasma membrane of your nerve cells. When sodium ions are permitted to flow down their concentration gradient, they create an electric current across the membrane. Such currents are the "messages" that travel through your nervous system. As you will see throughout this book, gradients are central to a variety of dynamic processes.

Simple Diffusion

The random movement of like molecules or ions down a concentration gradient is called **simple diffusion**. Simple diffusion accounts for the movements of many small molecules across cell membranes.

The direction in which a substance diffuses depends on its own concentration gradient, not on any others. In other words, each substance diffuses *independently* of other substances that may be present. Suppose you put a few drops of red food coloring into one end of a pan filled with water. At first all of the dye molecules remain at that end, but many molecules start careening toward the other end. Even though collisions are also sending

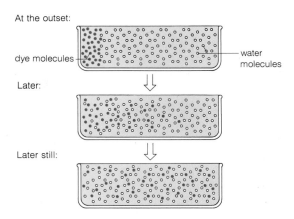

Figure 5.4 Diagram of the diffusion of dye molecules in one direction and water molecules in the opposite direction in a pan of water.

some dye molecules back, *more* molecules are leaving the concentrated region, and the net movement is down the gradient. For the same reason, water moves in the opposite direction, to the region where water molecules are less concentrated in the pan (Figure 5.4).

Even when dye molecules and water molecules are dispersed evenly in the pan, collisions still occur at random. But now there are no more gradients, hence there is no *net* movement in either direction. The molecules are said to be at dynamic equilibrium.

The *rate* of diffusion depends on several factors. For example, diffusion proceeds faster when the concentration gradient is steep. It also proceeds faster at higher temperatures, because heat energy causes molecules to move more rapidly (hence to collide more frequently). Diffusion rates are also affected by molecular size. Other factors being equal, smaller molecules move faster than large ones do.

Bulk Flow

Diffusion rates are often enhanced by **bulk flow**. This is the tendency of all the different substances in a fluid to move together in the same direction in response to a pressure gradient.

For example, in large multicelled plants and animals, nutrients and other substances do not diffuse slowly across large tissue masses to reach the plasma membrane of individual cells. They move rapidly by bulk flow through transport tubes, which occur in the respiratory and circulatory systems of animals and in the vascular systems of plants. In effect, bulk flow "shrinks" the diffusion distances for molecules that must move between the environment and interior cells. It also shrinks the diffusion distance between two body regions, as between leaves and roots.

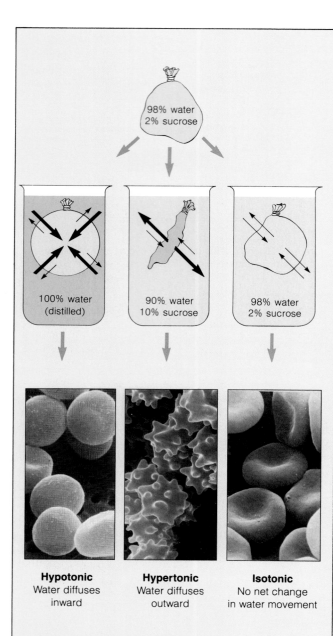

Hypotonic
Water diffuses inward

Hypertonic
Water diffuses outward

Isotonic
No net change in water movement

Figure 5.5 Effects of osmosis in different environments. The sketches show why it is important for cells to be matched to solute levels in their environment. (In each sketched container, arrow width represents the relative amount of water movement.)

The micrographs correspond to the sketches. They show the kinds of shapes that might be seen in human red blood cells placed in *hypotonic* solutions (influx of water into the cell), *hypertonic* solutions (outward flow of water from the cell), and *isotonic* solutions (internal and external solute concentrations are matched, no net movement of water).

Red blood cells have no special mechanisms for actively taking in or expelling water molecules. Hence they would swell or shrivel up, if solute levels in their environment were to change.

OSMOSIS

Osmosis Defined

A membrane is "differentially permeable" when some substances but not others can pass through it. **Osmosis** is the movement of water across any differentially permeable membrane in response to solute concentration gradients, a pressure gradient, or both. (A *solute* is any dissolved substance.)

Imagine you have a plastic bag that acts like a membrane, in that it is permeable to water molecules but impermeable to larger molecules. You fill the bag with water containing a small amount of table sugar, then you put it in a container of distilled water. "Distilled" means the water is nearly free of solutes. Inside the bag, the concentration of water molecules per unit volume is less than it is outside—so water moves into the bag (Figure 5.5). However, only the water is moving across the membrane—the sugar molecules are too large to do so. Thus, the sugar inside the bag cannot follow *its* concentration gradient and move out. Soon the bag swells with water, and eventually it may spring a leak or burst.

Suppose you had immersed the bag in water having more dissolved sugar than did the solution inside the bag. The net movement of water would have been outward, and the bag would have shriveled.

Finally, suppose you had used distilled water inside and outside the bag. Because the solute concentrations would be the same, there would be no water concentration gradient—and there would be no net movement of water in either direction.

Osmosis is the passive movement of water across a differentially permeable membrane in response to solute concentration gradients, a pressure gradient, or both.

Tonicity

Osmotic movements across cell membranes are affected by **tonicity**—that is, the relative concentrations of solutes in two fluids. In this case, we are talking about the extracellular fluid and the fluid portion of cytoplasm. When solute concentrations are equal in both fluids, or *isotonic*, there is no net osmotic movement of water in either direction. When the solute concentrations are not equal, one fluid is *hypotonic* (has less solutes) and the other is *hypertonic* (has more solutes). Water molecules tend to move from a hypotonic fluid to a hypertonic one.

If cells did not have mechanisms for adjusting to differences in tonicity, they would shrivel or burst like the plastic bag described above. For example, many solutes are dissolved in your bloodstream and in the cytoplasm

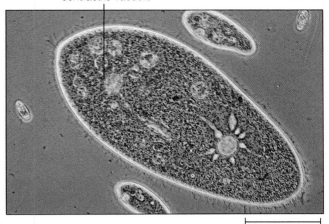

contractile vacuole

10 μm

Figure 5.6 Osmosis and *Paramecium*, a single-celled protozoan that lives in fresh water (a hypotonic environment). Because the cell interior is hypertonic relative to the surroundings, water tends to move into the cell by osmosis. If the influx were left unchecked, the cell would become bloated and the plasma membrane would rupture. However, the excess is expelled by an energy-requiring transport process that is based on a specialized organelle called the contractile vacuole. Tubelike extensions of this organelle extend through the cytoplasm. Water enters these extensions and collects in a central vacuolar space. When filled, the vacuole contracts and the water is forced into a small pore that empties to the outside.

of red blood cells. But suppose you immerse a red blood cell in a hypotonic solution. That cell contains many large organic molecules, which cannot cross the plasma membrane. Although those molecules cannot move out, water can move in. Internal pressure builds up, but red blood cells have no mechanism for disposing of such a large volume of excess water. The cell continues to swell until the membrane ruptures. Then the cell undergoes *lysis*—it becomes grossly "leaky"—and is destroyed.

If the red blood cell had been placed in a hypertonic solution, water would have moved out of the cytoplasm and the cell would have shriveled. This particular type of cell maintains its volume only when solute concentrations stay much the same on both sides of the plasma membrane. Other cell types can live in hypotonic or hypertonic environments, as Figure 5.6 illustrates.

Water Potential

The soil in which most land plants grow is hypotonic relative to the cells in those plants. Thus water tends to move by osmosis into the plant. When individual cells absorb the water, pressure increases against the cell wall, which is fairly rigid. This internal fluid pressure on a cell wall is called **turgor pressure**.

Pressure will build up inside any walled cell when water moves inward by osmosis, but water will also be squeezed back out when turgor pressure is great enough

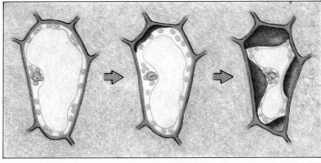

a   b   c

Figure 5.7 Effects of osmosis on plant cells. Most land plants grow in hypotonic soil, so water tends to move into their cells by osmosis. The cells have fairly rigid walls outside the plasma membrane; when water is absorbed, pressure increases against the wall. *Turgor pressure* refers to the internal pressure on a cell wall resulting from the inward osmotic movement of water.

If turgor pressure becomes high enough to counter the effects of cytoplasmic solutes, water also will be squeezed back out. When the *outward* flow equals the *inward* flow, turgor pressure is constant, cell walls cannot collapse and the soft plant parts stay erect. When soil solutes reach high concentrations, however, the net water movement is outward and wilting occurs.

(**a**) At the start of this experiment, 10 grams of salt (NaCl) in 60 milliliters of water are added to a pot containing tomato plants. (**b**) The plants start collapsing after about 5 minutes. (**c**) Wilting is severe in less than 30 minutes. The corresponding sketches show progressive plasmolysis (shrinking of cytoplasm away from the cell walls).

to counter the effects of internal solutes. Both forces have the potential to cause the directional movement of water. The sum of these two opposing forces is called the **water potential**.

Turgor pressure is constant when the outward flow of water from a plant cell equals the rate of inward diffusion. There is enough pressure to keep cell walls from collapsing, and the soft parts of the plant body are maintained in erect positions. When the environment becomes so dry that the movement of water into the plant dwindles or stops, water moves out of the cells and soft parts of the plant body wilt. Wilting also occurs when solutes reach high concentrations in the environment (Figure 5.7).

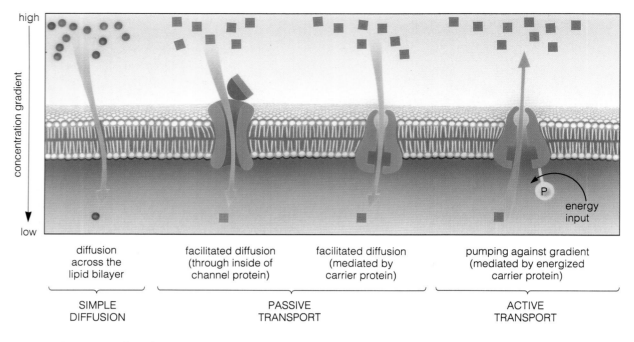

Figure 5.8 Overview of active and passive transport mechanisms.

MOVEMENT OF WATER AND SOLUTES ACROSS CELL MEMBRANES

The Available Routes

We are now ready to consider how substances move into and out of cells, with or without assistance from membrane proteins. Only small, electrically neutral molecules readily diffuse across the bilayer. Examples are oxygen, carbon dioxide, and ethanol, all of which have no net charge:

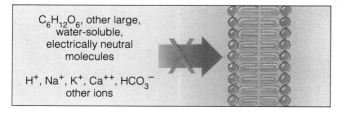

In contrast, glucose and other large, water-soluble molecules that are electrically neutral almost never diffuse freely across the lipid bilayer. Neither do positively or negatively charged ions, no matter how small they are:

Different types of proteins move these substances across the membrane by active and passive transport mechanisms. In **active transport**, the protein becomes activated into moving a solute against its concentration gradient. In **passive transport**, the protein does *not* require an energy boost. The solute simply moves through the protein's interior, following its concentration gradient.

Figure 5.8 provides an overview of how substances move across cell membranes. Through a combination of simple diffusion, passive transport, and active transport, cells or organelles are supplied with numerous raw materials and they are rid of wastes, at controlled rates. These mechanisms help maintain pH and volume inside the cell or organelle within functional ranges.

Facilitated Diffusion

A membrane protein with passive transport functions is highly selective about which solutes it will assist across the membrane. A protein that transports glucose, for instance, will not transport amino acids. And the protein *only* helps a particular solute move in the direction that simple diffusion would take it (down its concentration gradient). That is why this transport mechanism is called **facilitated diffusion**.

Membrane proteins involved in facilitated diffusion may have an interior space that can be opened up to one side of the membrane at a time (Figure 5.9). Probably a specific array of hydrophilic groups project into that

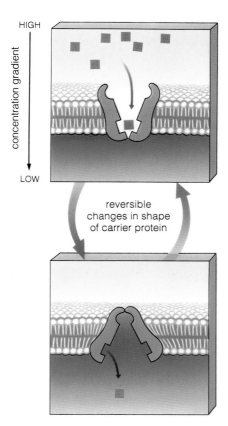

space. When water-soluble molecules bind with the groups, the binding seems to trigger a change in the protein shape. This change permits the solute to move through the hydrophilic interior. While the solute makes its passage, the protein closes in behind it and returns to its original shape.

Active Transport

A membrane protein that *actively* moves specific solutes into and out of cells undergoes changes in shape, somewhat like the changes induced in the proteins responsible for facilitated diffusion. In this case, however, an energy boost leads to a *series* of changes, which cause the protein to pump solutes across the membrane (Figure 5.10). Most often, ATP donates energy to the protein.

One active transport system, the *sodium-potassium pump,* helps maintain high concentrations of potassium and low concentrations of sodium inside the cell. Another, the *calcium pump,* helps keep calcium concentrations at least a thousand times lower inside the

Figure 5.9 One model of facilitated diffusion across a cell membrane. The carrier protein shown here can exist in two configurational states. In one state, the binding sites for a solute are exposed to the fluid outside the lipid bilayer of the membrane. In the other state, the same binding sites are exposed to cytoplasmic fluid. The transition from one state to the other is reversible and depends on the direction of the solute concentration gradient across the membrane. The changes in shape are induced when the solute is bound in place.

Figure 5.10 (*Below*) Simplified picture of an active transport system in animal cell membranes. In this example, transport of one kind of solute across the membrane is coupled with transport of another kind in the opposite direction. The transport protein receives an energy boost from ATP and so undergoes changes in its shape that are necessary for the transport process. The net result of active transport is that a solute is moved across the membrane against its concentration gradient, an event that is coupled to the expenditure of energy.

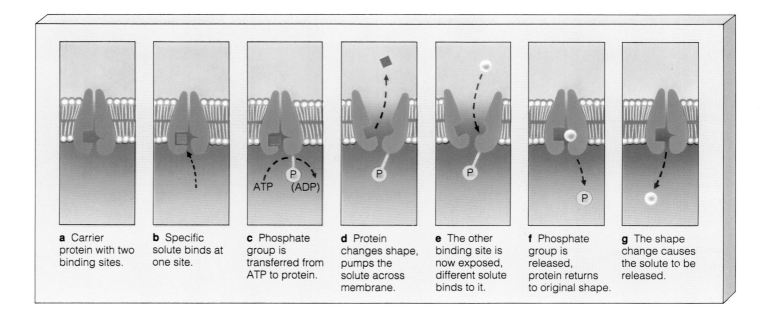

a Carrier protein with two binding sites.

b Specific solute binds at one site.

c Phosphate group is transferred from ATP to protein.

d Protein changes shape, pumps the solute across membrane.

e The other binding site is now exposed, different solute binds to it.

f Phosphate group is released, protein returns to original shape.

g The shape change causes the solute to be released.

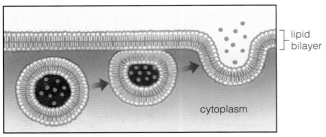

a Exocytosis

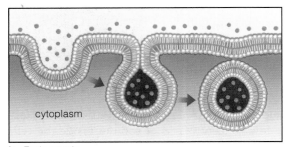

b Endocytosis

Figure 5.11 (**a**) Fusion of a vesicle with the plasma membrane during exocytosis. (**b**) Formation of a vesicle during endocytosis.

cell than outside. The features that all active transport systems hold in common can be summarized this way:

1. In **active transport,** small ions, small charged molecules, and large molecules are pumped across a cell membrane, against their concentration gradient.

2. Certain proteins spanning the lipid bilayer are the active transport systems. They are highly selective about which solutes they will bind and transport.

3. The proteins act when a specific solute is bound in place and when they receive an energy boost, as from ATP.

Exocytosis and Endocytosis

Exocytosis and endocytosis are processes by which small regions of plasma membrane or organelle membranes pinch off and form transport vesicles around substances (Figure 5.11). In **exocytosis**, vesicles in the cytoplasm travel to the plasma membrane, fuse with it, and have their contents released to the cell's surroundings. Vesicles derived from Golgi bodies and released from secretory cells are an example; these are illustrated in Figure 4.14. In **endocytosis**, a region of the plasma membrane encloses particles at or near the cell surface, then pinches off to form a vesicle that moves into the cytoplasm.

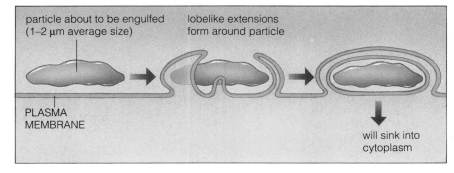

a Phagocytosis

b Pinocytosis

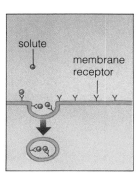

c Receptor-mediated endocytosis

Figure 5.12 Mechanisms of endocytosis.

The amoeba relies on endocytosis. This single-celled organism is phagocytic—a "cell eater." When it encounters a chemically "tasty" particle or cell, one or two lobes form at its surface. The lobelike extensions curve back and form a compartment around the particle, and this compartment becomes an endocytic vesicle (Figure 5.12). Many endocytic vesicles fuse with lysosomes. As mentioned on page 64, lysosomes are filled with digestive enzymes. Following fusion, the contents of the endocytic vesicle are digested. Phagocytic white blood cells rely on endocytosis when they destroy harmful agents such as bacteria.

Endocytosis also transports liquid droplets into animal cells. (Sometimes this process is called pinocytosis, which means "cell drinking.") A depression forms at the surface of the plasma membrane and dimples inward around extracellular fluid. An endocytic vesicle forms and moves inside the cytoplasm, where it fuses with lysosomes.

In *receptor-mediated endocytosis*, specific molecules are brought into the cell through involvement of specialized regions of the plasma membrane that form coated pits. Each pit, or shallow depression, is coated on its cytoplasmic side with a dense lattice of proteins (Figure 5.13). The pit appears to be lined with surface receptors that are specific (in the example shown here) for lipoprotein particles. When lipoproteins are bound to the receptors, the pit sinks into the cytoplasm and forms an endocytic vesicle.

Figure 5.14 shows the possible destinations of vesicles that form by way of exocytosis and endocytosis.

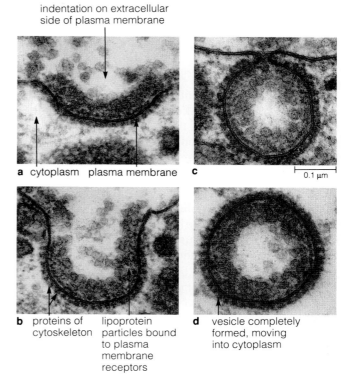

a cytoplasm plasma membrane   **c**   0.1 μm

b proteins of   lipoprotein
cytoskeleton   particles bound
to plasma
membrane
receptors

d vesicle completely
formed, moving
into cytoplasm

indentation on extracellular side of plasma membrane

Figure 5.13 A closer look at receptor-mediated endocytosis and the formation of a transport vesicle. The electron micrographs show part of the plasma membrane of an immature egg from a hen. The shallow indentation in (**a**) is an example of a *coated pit.* On the cytoplasmic side of the pit, proteins (clathrin) form a dense lattice. On the extracellular side of the membrane, the pit is thought to be lined with surface receptors that are specific for lipoprotein particles (**b**). In (**c**), the pit has deepened and has become rounded. In (**d**), the coated vesicle is fully formed, with lipoproteins (and protein receptors) enclosed within it.

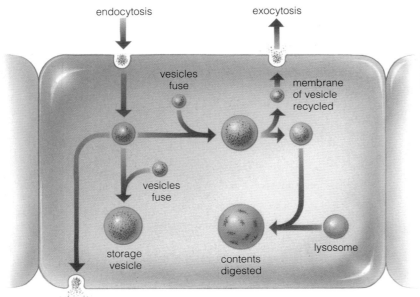

endocytosis   exocytosis

vesicles fuse

membrane of vesicle recycled

vesicles fuse

storage vesicle

contents digested

lysosome

substance released at opposing cell surface

Figure 5.14 Possible fates of substances in vesicles formed by endocytosis. Compare Figure 4.14, which shows the formation of vesicles in the cytomembrane system.

SUMMARY

Membrane Structure

1. Living cells are bathed in fluid of one sort or another, and their interior is also fluid. The plasma membrane is the boundary between the external fluid world and the cytoplasm. Internal cell membranes are boundaries between different cytoplasmic regions.

2. The lipid bilayer is the basic structure of all cell membranes. It has two layers of lipids (phospholipids especially), with the hydrophobic tails of the molecules sandwiched in between the hydrophilic heads. The membrane is rather fluid because of packing variations and rapid movements among the individual molecules.

3. Most membrane functions are carried out by diverse proteins embedded in the bilayer or weakly bonded to one of its surfaces.

Membrane Functions

1. Three classes of proteins are crucial for membrane functions. These are membrane transport proteins, recognition proteins, and receptor proteins. The following points summarize their functions.

2. *Control of substances moving into and out of cells.* Transport proteins passively or actively move specific solutes across the lipid bilayer portion of the plasma membrane. The selective movements of solutes affect metabolism, cell volume and cellular pH, nutrient stockpiling, and the removal of harmful substances.

3. *Compartmentalization of internal cellular space.* Membranes within the cytoplasm form compartments (organelles) in which specialized activities occur. The nucleus, chloroplasts, and the inner and outer mitochondrial membranes are examples.

4. *Signal reception.* Some receptor proteins bind signaling molecules such as hormones, and when they do, they trigger alterations in cell behavior or metabolism.

5. *Cell-to-cell recognition.* In multicelled animals, recognition proteins help cells of like type identify and adhere to one another during tissue formation and, at later stages of development, during tissue interactions.

6. *Transport of macromolecules and particles.* Through exocytosis and endocytosis, cells eject or take in large molecules or particles across the plasma membrane.

Movement of Water and Solutes Across Membranes

1. Different types of solutes move across the membrane, in some cases with and in some cases against the solute concentration gradient.

2. Simple diffusion is a natural, unassisted movement of solutes from a region of high concentration to a region of lower concentration.

3. Osmosis is the movement of water across a differentially permeable membrane in response to solute concentration gradients, a pressure gradient, or both. When conditions are isotonic (equal concentrations of solutes across the membrane), there is no osmotic movement in either direction. Water tends to move from hypotonic fluids (with lower concentrations of solutes) to hypertonic fluids (with higher concentrations of solutes).

4. Channel and carrier proteins carry out membrane transport functions. Open or gated channel proteins serve as pores through which water-soluble substances cross the membrane. Carrier proteins bind solutes and so shunt them across. Some carrier proteins do this passively, others require an energy boost to pump solutes across.

5. The facilitated diffusion of a solute through channel proteins or carrier proteins is called passive transport. The solute moves passively through the protein interior, in the direction that its concentration gradient takes it.

6. The pumping of solutes across membranes by energized carrier proteins is called active transport.

Review Questions

1. Describe the fluid mosaic model of plasma membranes. What makes the membrane fluid? What parts constitute the mosaic? *76–80*

2. List the structural features that all cell membranes have in common. *80*

3. Describe some functions of membrane proteins. *80*

4. How does diffusion work? *81*

5. What is osmosis, and what causes it? *82*

6. Explain the difference between active and passive transport mechanisms. *84*

7. What types of substances can readily diffuse across the lipid bilayer of a cell membrane? What types of substances must be actively or passively transported across? *84*

8. Can you explain the difference between exocytosis and endocytosis? *86–87*

1. Substances move into and out of the cell by crossing the _____ .

2. Cell membranes consist mainly of a _____ .
   a. carbohydrate bilayer and proteins
   b. protein bilayer and phospholipids
   c. phospholipid bilayer and proteins
   d. nucleic acid bilayer and proteins

3. Most membrane functions are carried out by _____ associated with the bilayer.
   a. proteins   c. nucleic acids
   b. phospholipids   d. hormones

4. When a cell is placed in a hypotonic solution, _____ .
   a. water tends to move into the cell
   b. water tends to move out of the cell
   c. water does not move into or out of the cell
   d. exocytosis will have to occur

5. _____ cannot diffuse across the lipid bilayer of the plasma membrane.
   a. ethanol   c. oxygen
   b. ions   d. carbon dioxide

6. Three classes of proteins essential for membrane functions are _____ , _____ , and _____ .

7. _____ are membrane proteins that bind hormones and other substances that can trigger changes in cell behavior or metabolism.
   a. gated channel protein   c. carrier protein
   b. receptor protein   d. recognition protein

8. The passive movement of a substance through channel proteins as it follows its concentration gradient across a cell membrane is an example of _____ .
   a. osmosis   c. diffusion
   b. active transport d. facilitated diffusion

9. _____ help cells of like type identify and adhere to one another during tissue development.
   a. open channel proteins   c. recognition proteins
   b. energized carrier proteins d. receptor proteins

10. Match each membrane structure/function with the appropriate description.
   _____ substance expelled from cell when vesicle fuses with plasma membrane
   _____ water moves across membrane due to concentration or pressure gradients
   _____ natural, unassisted movement of solutes down a concentration gradient
   _____ facilitated diffusion of solutes via channel or carrier proteins
   _____ substances move into cell by invagination of plasma membrane
   _____ carry out membrane transport functions
   _____ energized carrier protein pumps solutes across membrane

   a. passive transport
   b. channel and carrier proteins
   c. active transport
   d. simple diffusion
   e. endocytosis
   f. osmosis
   g. exocytosis

Selected Key Terms

active transport *84*
bulk flow *81*
calcium pump *85*
carrier protein *80*
channel protein *80*
concentration gradient *81*
endocytosis *86*
exocytosis *86*
facilitated diffusion *84*
fluid mosaic model *80*
hypertonic *82*
hypotonic *82*
internal cell membrane *75*
isotonic *82*
lipid bilayer *76*
lysis *83*
osmosis *82*
passive transport *84*
phospholipid *76*
plasma membrane *75*
recognition protein *80*
receptor protein *80*
receptor-mediated endocytosis *87*
simple diffusion *81*
sodium-potassium pump *85*
solute *82*
tonicity *82*
turgor pressure *83*
water potential *83*

Readings

Alberts, B. et al. 1989. *Molecular Biology of the Cell*. Second edition. New York: Garland.

Bretscher, M. October 1985. "The Molecules of the Cell Membrane." *Scientific American* 253(4):100–108. Fairly recent description of the structure and function of the plasma membrane.

Dautry-Varsat, A., and H. Lodish. May 1984. "How Receptors Bring Proteins and Particles Into Cells." *Scientific American* 250(5): 52–58. Describes receptor-mediated endocytosis.

Singer, S., and G. Nicolson. 1972. "The Fluid Mosaic Model of the Structure of Cell Membranes." *Science* 175:720–731.

Unwin, N., and R. Henderson. February 1984. "The Structure of Proteins in Biological Membranes." *Scientific American* 250(2): 78–94.

6 GROUND RULES OF METABOLISM

The Old Man of the Woods

Deep in a Michigan forest, on a moist and rotting trunk of a fallen tree, a quiet celebration of life is about to unfold. The huge trunk is jarring in the midst of such luxuriant greenery, an unsettling reminder of the death that comes to all organisms with intricately specialized cells. Yet through a patch of decaying bark a mushroom erupts. Soon it has grown several inches tall. It is dirty brown and ugly to boot, with a cap that is shaggy, cracked, and lumpy with gray-black scales (Figure 6.1). Mushroom hunters love it. Rather than using its scientific name, *Strobilomyces floccopus*, they commonly call it the "old man of the woods" and welcome its presence at the dinner table.

Suppose you come across an old man of the woods and decide to dissect it gently. You see almost immediately that it consists of a highly ordered array of parts. It has a stalk, which had grown tall and strong enough to support the cap some distance above the tree trunk. Inside the cap, hundreds of tiny tubes point downward. Even tinier reproductive bodies called spores have been growing inside the tubes and still are sequestered there.

If you had left the mushroom on the trunk, those spores would have been released from the tubes, and air currents would have dispersed them to good (or bad) places for germination.

As you probe the rotting wood where the mushroom had been growing, you realize that the old man of the woods is more than a stalk and a cap. You have yanked the mushroom from the rest of the fungal body—slender filaments threading every which way through its substrate. Collectively the filaments represent the mycelium, the part of the fungus devoted to securing food energy from the surroundings. Like other fungi, the old man of the woods produces and secretes enzymes to the outside of its body. Those enzymes digest large organic molecules in the tree's once-living tissues. The digested molecular bits are small enough for fungal cells to absorb—and to use as energy sources and raw materials.

Thus in death, the fallen tree had become the basis of new life. *The energy and materials derived from its organic molecules had helped another organism grow, maintain its highly organized state, make spores, and so serve as a bridge to a new generation of life.*

You might have focused your attention on a living tree in the forest, a robin, a squirrel, an earthworm, or any other organism besides a fungus. You would have arrived at the same understanding of the metabolic connections between energy and the living state. Metabolism—the controlled acquisition and use of energy in the synthesis and breakdown of organic compounds—happens only in living organisms. It depends on specific enzymes and organized arrays of cell parts, the construction of which depends ultimately on instructions encoded in DNA.

This chapter is our starting point for exploring the nature of energy and how it can be used to do cellular work. It will serve as our background for considering specific mechanisms by which organisms acquire and release the energy that keeps them alive—the central topics of the next two chapters.

Figure 6.1 Reproductive structures of the old man of the woods, assembled with energy and materials that had been stored previously in another organism in the web of life.

KEY CONCEPTS

1. Cells have the controlled capacity to trap and use energy for stockpiling, breaking apart, building, and eliminating substances in ways that contribute to survival and reproduction. This capacity is called metabolism.

2. The complex organization characteristic of life is maintained by a steady input of energy. To stay alive, cells must replace the energy that they inevitably lose as heat by metabolic reactions. Directly or indirectly, the sun is the source of energy replacements for nearly all organisms.

3. In biosynthetic pathways, organic molecules are assembled and energy becomes stored in them. In degradative pathways, organic molecules are broken apart and energy is released.

4. Enzymes (which speed reaction rates) take part in nearly all metabolic pathways. So does ATP, an organic compound that transfers energy from one reaction site to another in the cell.

By definition, **metabolism** is the controlled capacity to acquire and use energy for stockpiling, breaking apart, building, and eliminating substances in ways that contribute to survival and reproduction.

You see outward signs of metabolic activity whenever you look at a living cell through a microscope. Most obviously, the cell is moving about. Through its movements, it is identifying and taking in raw materials suspended in the water droplet on the slide. To power those tiny movements, the cell is extracting energy from food molecules stored away earlier. Even as you watch it, the cell is using energy and materials as it builds and maintains its membranes, its stores of chemical compounds, its DNA, its pools of enzymes. It is alive, it is growing, it may divide in two. Multiply this activity by *65 trillion cells* and you have an inkling of the metabolic activity going on in your own body as you sit quietly, observing that single cell!

ENERGY AND LIFE

Energy is a capacity to do work. You use energy when you run or sleep or think about something bad, dangerous, or pleasant. In all cases, some cells of your muscles, brain, and other body parts are being put to work. Each cell is using energy, as when it is contracting or when it is producing and secreting hormone molecules that can alter your behavior.

You (or the cell) cannot create your own energy from scratch, however. You must get it from someplace else. That is the message of the **first law of thermodynamics:**

The total amount of energy in the universe remains constant. More energy cannot be created; existing energy cannot be destroyed. It can only be converted from one form to another.

Consider what this law means. The universe has only so much energy, distributed in a variety of forms. One form can be converted to another, as when corn plants absorb sunlight energy and convert it to the chemical energy of starch. By eating corn, you can extract and convert its energy to other forms, such as mechanical energy for your movements. Energy conversions of all kinds are notable for this reason: Whenever one takes place, a little energy escapes to the surroundings as heat. (Your body steadily gives off about as much heat as a

100-watt light bulb because of ongoing conversions in your cells.) As Figure 6.2 suggests, however, none of the energy vanishes. It just ends up someplace else.

This brings us to another important concept, the *quality* of energy available. Energy concentrated in a starch molecule is high quality, since it lends itself to conversions. Heat energy spread out in the atmosphere is low quality since, for all practical purposes, it can't be gathered up and converted to other forms. Heat can be transferred from a hot object to a cooler one, but it cannot on its own be transferred in the opposite direction.

The amount of low-quality energy is increasing in the grand scheme of things. That is the point of the **second law of thermodynamics:**

The spontaneous direction of energy flow is from high-quality to low-quality forms. With each conversion, some energy is randomly dispersed in a form that is not as readily available to do work.

Without energy to maintain it, any organized system tends to become disorganized over time. As Figure 6.2 indicates, "system" means all the matter in a specified region—a plant, a strand of DNA, a galaxy, and so on. **Entropy** is a measure of the degree of randomness or disorder of these and all other systems. The second law says that, in any process that happens spontaneously, the total entropy of the system and its surroundings must increase.

Think about the Egyptian pyramids—originally organized, presently crumbling, and many thousands of years from now, dust. The *ultimate* destination of the pyramids and everything else in the universe is a state of maximum entropy. Why? Through the eons, high-quality energy that can be tapped to maintain order in systems ultimately will be converted to energy of the lowest quality—perhaps uniformly dispersed heat—and how will that ever be gathered up to do work? Billions of years from now, energy conversions as we know them may never happen again.

Can it be that life is one glorious pocket of resistance to the rather depressing flow toward complete entropy? After all, every time a new organism grows, atoms become linked together into precise arrays and energy becomes more concentrated and organized, not less so! We see order everywhere, in patterned butterfly wings, in the petaled symmetry of flowers, in the structure of DNA. Yet a simple example will show that the second law does indeed apply to life on earth.

Think of each living cell as a speck of order in a universe that is tending toward disorder. The cell stays that

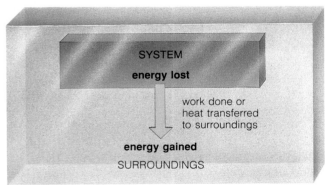

net energy change = 0

Figure 6.2 The nature of energy. According to the first law of thermodynamics, the total energy content of any system and its surroundings remains constant. "System" means all matter within a specific region, such as a plant, a DNA molecule, or a galaxy. The "surroundings" are everything in the universe *except* the system.

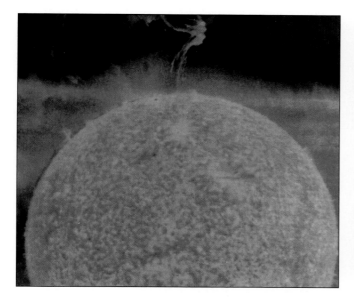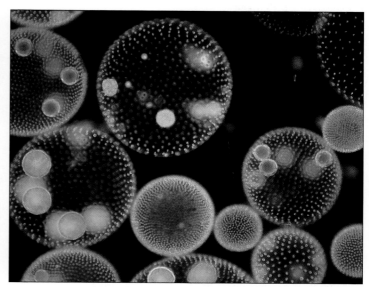

Figure 6.3 All events large and small, from the birth of stars to the death of a microorganism, are governed by laws of thermodynamics. Shown here, eruptions on the sun's surface and, to the right, *Volvox*—each sphere a colony of microscopically small single cells able to capture sunlight energy that indirectly drives their life processes.

way as long as it continually taps into energy from an outside source. The primary source of energy for life on earth is the sun—which is steadily losing energy (Figure 6.3). Plants capture some sunlight energy, and they convert it to other forms of energy such as that of glucose and other organic molecules. Some of the stored energy gets transferred to organisms that feed, directly or indirectly, on plants. At every energy transfer along the way, however, some energy is lost, usually as heat. This adds more heat to the universal pool.

Overall, then, energy is still flowing in one direction. *The world of life maintains a high degree of organization only because it is being resupplied with energy lost from someplace else.*

There is a steady flow of sunlight energy into the interconnected web of life, and this compensates for the steady flow of energy leaving it.

THE NATURE OF METABOLISM

Energy Changes in Metabolic Reactions

The next two chapters focus mainly on the chemical reactions of photosynthesis and aerobic respiration, the main pathways of energy flow in the world of life. Those reac-

tions will make more sense if you keep the following three concepts in mind:

1. The substances present at the end of a reaction (the products) may have *less* or *more* energy than did the starting substances (the reactants).

2. Most reactions are reversible. Besides proceeding in the forward direction (from reactants to end products), they can proceed in the reverse direction (from end products back to reactants).

3. A reversible reaction tends to approach equilibrium, a state in which it proceeds at about the same rate in both directions.

Energies of Cellular Substances. When we speak of the energy of a cellular substance, we really are speaking of the usable energy that is released when the substance is broken down to simpler materials. Chemical energies are often measured in terms of kilocalories per mole. A *kilocalorie* is the same thing as a thousand calories—the amount of energy needed to heat 1,000 grams of water from 14.5°C to 15.5°C at standard pressure. Because energy can be converted from one form to another, bond energies can be expressed in kilocalories even though the energy is used for something other than heating water.

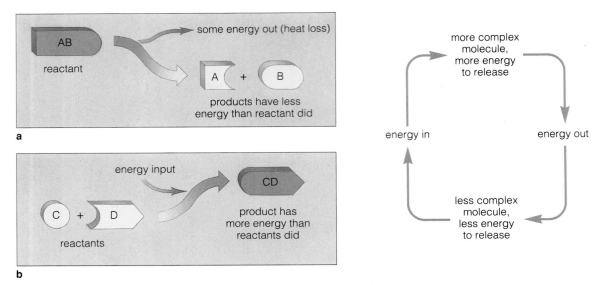

Figure 6.4 Energy changes in metabolic reactions. (**a**) In exergonic reactions, products have less energy than the reactants did. Some energy released by the reactions is harnessed to do cellular work. The breakdown of glucose into carbon dioxide and water is an example (page 120). (**b**) In endergonic reactions, products have more energy than the reactants did. An energy input drives these reactions. An example is photosynthesis, in which sunlight energy drives the linkage of carbon dioxide and water into sugars and other compounds of higher energy content.

Energy Losses and Gains. Why is it important to think about chemical energies? Just as you might use firewood as a source of energy, so do cells use glucose and other molecules. You light a match to wood; cells use ATP like a "match" to start breaking the covalent bonds of glucose. The bonds occur between carbon, hydrogen, and oxygen atoms of glucose ($C_6H_{12}O_6$). When a glucose molecule is fully broken down, all that is left are six molecules each of carbon dioxide (CO_2) and water (H_2O). These are small molecules, with new covalent bonds among their atoms. The new bonds are stronger than the covalent bonds between the carbon atoms in glucose. Thus carbon dioxide and water are more stable compounds than glucose. Said another way, the overall breakdown reactions *release* energy—as much as 686 kilocalories per mole of glucose—just as a fire releases heat energy.

Glucose breakdown is a good example of a reaction in which the products end up with less energy than the reactants had. Reactions that show a net *loss* in energy are said to be *exergonic,* meaning "energy out" (Figure 6.4a). Your body runs on a good deal of energy released during exergonic reactions. On the average, the breakdown of food molecules in your body releases about 2,000 to 2,800 kilocalories each day.

Conversely, many other reactions simply won't proceed on their own without an energy boost. The energy-requiring reactions in which starch and other large molecules are assembled from smaller, energy-poor ones are examples of this. Reactions that show a net *gain* in energy are said to be *endergonic,* meaning "energy in" (Figure 6.4b).

Reversible Reactions. Like most chemical reactions, the ones that proceed in living cells are "reversible." In other words, once product molecules have formed, they can be converted back to reactant molecules.

To understand how this works, recall that living cells are bathed in fluid and are largely fluid themselves. Molecules in this fluid move about constantly and bump into each other, and most often they react spontaneously in regions where they are most concentrated. The more concentrated they are, the more their random movements put them on collision courses. When they collide, one molecule might cause the other to split apart or change its shape, or one might join up with the other.

As long as the concentration of reactant molecules is high enough, a chemical reaction proceeds in the forward direction to product molecules. But when the concentration of product builds up, some number of product molecules will revert to reactants. Reversible reactions are indicated by arrows running in the "forward" and "reverse" directions (Figure 6.5).

Dynamic Equilibrium. Unless other events in the cell (or body) keep it from doing so, a reaction that is reversible approaches **dynamic equilibrium**. Then, the forward and reverse reactions proceed at equal rates (Figure 6.6). There is no *net* change in the concentrations of reac-

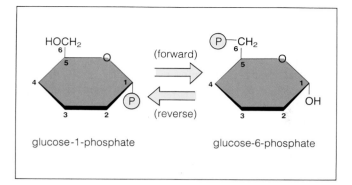

Figure 6.5 A reversible reaction. In most cells, phosphate can become attached to a glucose molecule in several different ways. With high concentrations of glucose-1-phosphate, the reaction shown here tends to run in the forward direction; with high concentrations of glucose-6-phosphate, it runs in reverse. (The "1" and "6" identify the particular carbon atom of the glucose ring to which phosphate is attached.)

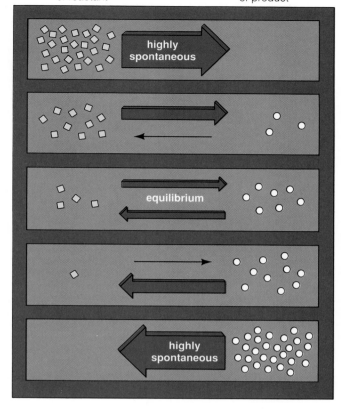

Figure 6.6 Chemical equilibrium. With high concentrations of reactant molecules, reactions generally proceed most strongly in the forward direction. With high concentrations of product molecules, they proceed most strongly in reverse. At chemical equilibrium, the *rates* of the forward and reverse reactions are equal.

tants and products even though molecules are still being converted from one form to the other. It's like a party with as many people wandering in as wandering out of two adjoining rooms. The total number in each room stays the same—say, thirty in one and ten in the other—even though the mix of people in each room continually changes.

How many reactant molecules and how many product molecules will there be at equilibrium? That depends on the substances involved. *Every reaction has its own characteristic ratio of products to reactants at equilibrium.*

In one reaction, glucose-1-phosphate is rearranged into glucose-6-phosphate. As Figure 6.5 shows, these are simply glucose molecules with a phosphate group attached to one of their six carbon atoms. Suppose we allow the reaction to proceed when the concentrations of both substances are the same. At this point the forward reaction occurs 19 times faster than the reverse reaction does. Said another way, the forward reaction is producing more molecules in the same amount of time. Only when there finally are 19 molecules of glucose-6-phosphate for every molecule of glucose-1-phosphate will the forward and reverse reactions proceed at the same rate. For this example, then, the characteristic ratio of products to reactants at equilibrium is 19:1.

You may be thinking that which way a reaction proceeds is a remote issue, of concern only to laboratory chemists. As you will now see, this is really a central issue in our study of what it takes to be alive.

Metabolic Pathways

Cells never stop building up and tearing down substances until they are dead. This incessant activity requires control over the directions in which different reactions proceed. Why? Cells use only so many molecules of different substances at a given time, and they have only so much internal space to hold any excess. If cells were to produce more than they could use, store, or secrete, the excess might cause problems.

Think about what happens to a baby born with a mutation that can give rise to the genetic disorder phenylketonuria (PKU). In this disorder, a series of reactions is blocked and a substance (phenylalanine) reaches high concentrations in the baby's body. The excess enters into reactions that produce phenylketones, and the accumulation of those substances leads to severe mental retardation. The symptoms can be prevented if affected individuals are detected soon enough. They are immediately placed on a diet that makes it difficult for phenylalanine to accumulate in the body.

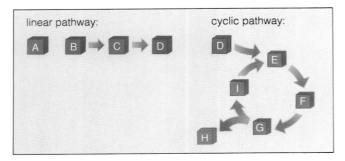

Figure 6.7 Linear and cyclic metabolic pathways.

Normally, cells maintain, increase, and decrease the concentrations of substances by coordinating a variety of metabolic pathways. A **metabolic pathway** is an orderly sequence of reactions, the steps of which are quickened with the help of specific enzymes. Most sequences are linear; some are circular (Figure 6.7). Branches often link different pathways, with products of one pathway serving as reactants for others.

Overall, the main metabolic pathways are either degradative or biosynthetic. In **degradative pathways**, carbohydrates, lipids, and proteins are broken down in stepwise reactions that lead to products of lower energy. Often, some of the energy released is used to do cellular work. In **biosynthetic pathways**, small molecules are assembled into larger molecules such as polysaccharides, proteins, and nucleic acids.

The participants in metabolic pathways can be defined in the following way:

reactants	*substances able to enter into a reaction; also called substrates or precursors*
intermediates	*compounds formed between the start and the end of a metabolic pathway*
enzymes	*proteins that catalyze (speed up) reactions*
cofactors	*small molecules and metal ions that help enzymes or that carry atoms or electrons from one reaction site to another*
energy carriers	*mainly ATP, which readily donates energy to diverse reactions*
end products	*the substances present at the conclusion of a metabolic pathway*

Let's take a closer look at some of these substances and at their vital roles in metabolism.

ENZYMES

A cup of sugar left undisturbed for twenty years changes very little. But when you eat sugar, it rapidly undergoes chemical change. Enzymes secreted by some of your cells account for the difference in the rate of change. **Enzymes** are molecules with enormous catalytic power, meaning that they greatly enhance the rate at which specific reactions approach equilibrium. The vast majority are proteins, although a few forms of RNA also have catalytic properties (page 301).

Characteristics of Enzymes and Their Substrates

Enzymes have four characteristics in common. *First,* enzymes do not make anything happen that would not eventually happen on its own. They just make it happen faster—at least a million times faster, usually. *Second,* enzyme molecules are not permanently altered or used up in the reactions they catalyze; they can be used over and over again. *Third,* each type of enzyme is highly selective about its substrates.

Substrates are specific molecules that an enzyme can chemically recognize, bind briefly to itself, and modify in a specific way. Thrombin, an enzyme involved in blood clotting, helps break a specific peptide bond in a specific protein. It chemically recognizes a bond between two amino acids (arginine and glycine) and speeds its breakdown:

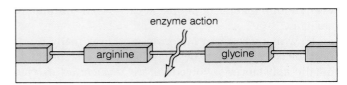

Fourth, an enzyme can recognize both the reactants and the products of a given reaction as its substrates. In other words, the same enzyme can catalyze a reaction in both the forward and reverse directions.

Enzyme Structure and Function

Substrates interact with an enzyme at one or more regions called active sites. An **active site** is a surface region where an enzyme molecule is folded in the shape of a crevice and where a particular reaction is catalyzed. Figure 6.8 shows an example.

As long ago as 1890, Emil Fischer thought the shape of some region of the enzyme's surface must match a complementary region on its substrate, like a lock precisely matching its key. Today we know the match is not so rigid. In Daniel Koshland's **induced-fit model**, an active site almost *but not quite* matches its substrate when

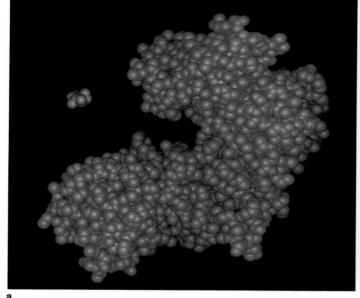

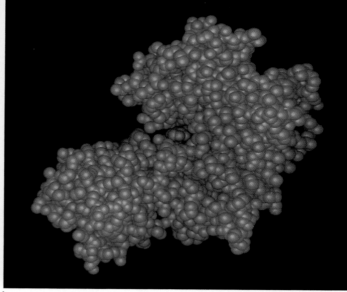

a   b

Figure 6.8 Model of the induced fit between the enzyme hexokinase (blue) and its bound substrate (a glucose molecule, shown in red).

(**a**) The cleft into which the glucose is heading is the enzyme's active site. (**b**) In this enzyme-substrate complex, notice how the enzyme shape is altered temporarily: the upper and lower parts now close in around the substrate.

first making contact with it. The enzyme-substrate interaction is enough to strain certain bonds within the substrate molecule. The strained bonds are easier to break, and this paves the way for the formation of the new bonding arrangements characteristic of the product molecules.

Figure 6.9 is a simple picture of how an enzyme catalyzes a reaction between two substrate molecules. When the molecules fit most precisely into the active site, they are in an activated condition called the "transition state." Now the reaction between them proceeds spontaneously, just as a boulder pushed up and over the crest of a hill rolls down on its own.

In the cellular world, reactants typically do not enter the transition state without an energy "push." Simply put, the reactant molecules must collide with some minimum amount of energy. For any given reaction, the minimum amount of energy needed to bring all the reactant molecules to the transition state is called the **activation energy**. That amount is like a hill over which the molecules must be pushed (Figure 6.10).

An enzyme increases the rate of a given reaction by *lowering* the required activation energy. How? Among other things, weak but extensive bonding at the active site puts the enzyme's substrates in orientations that promote reaction. In contrast, reactants colliding on their own do so from random directions, so mutually attractive chemical groups may not make contact and reaction may not occur.

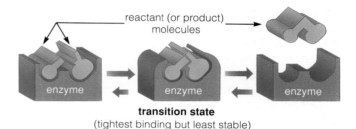

transition state
(tightest binding but least stable)

Figure 6.9 Induced-fit model of enzyme-substrate interactions. Only when the substrate is bound in place is the enzyme's active site complementary to it. The most precise fit occurs during a transition state that precedes the reaction. An enzyme-substrate complex is short-lived, partly because only weak bonds hold it together.

Figure 6.10 Energy hill diagram showing the effect of enzyme action. An enzyme greatly enhances the rate at which a reaction proceeds because it lowers the required activation energy (E_a). In other words, not as much collision energy is needed to boost reactant molecules to the crest of the energy hill (transition state).

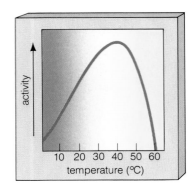

Figure 6.11 Effect of increases in temperature on enzyme activity.

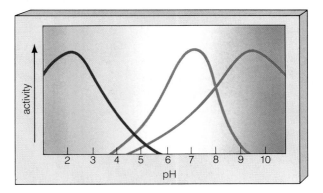

Figure 6.12 Visible effects of environmental temperature on enzyme activity. In Siamese cats, the fur on the ears and paws contains more dark-brown pigment (melanin) than the rest of the body. A heat-sensitive enzyme controlling melanin production is less active in warmer body regions, and this results in lighter fur in those regions.

Figure 6.13 Effect of pH on enzyme activity. The brown line charts the activity of an enzyme that is fully functional in neutral solutions. The red line charts the activity of one that is functional in basic solutions; the purple line, in acidic solutions.

An enzyme enhances the rate of a given reaction by lowering the activation energy required for that reaction.

Effects of Temperature and pH on Enzymes

Each type of enzyme functions best within a certain temperature range (Figures 6.11 and 6.12). When the temperature becomes too high, reaction rates decrease sharply. The increased thermal energy disrupts weak bonds holding the enzyme in its three-dimensional shape, and denaturation occurs (page 46). Because the active site becomes altered, substrates cannot bind to it.

Brief exposure to temperatures that are much higher than an organism typically encounters can destroy enzymes and adversely affect metabolism. This happens during fevers that are high enough to be extremely dangerous. Thus, when the internal body temperature of a human reaches 44°C (112°F), death generally follows.

Also, most enzymes function best within a limited range of pH, which commonly is near pH 7. Higher or lower pH values generally disrupt the enzyme's three-dimensional shape and its function (Figure 6.13). Pepsin is one of the exceptions; this enzyme functions in the extremely acidic fluid of the stomach.

Enzymes function only within limited ranges of temperature and pH.

Control of Enzyme Activity

Through controls over enzymes of different pathways, cells direct the flow of nutrients, wastes, and other substances in suitable ways. Some of the controls govern the number of enzyme molecules available. For example, certain mechanisms work to accelerate or slow down the production of enzymes. Others stimulate or inhibit the activity of enzymes already formed.

When you eat too much sugar, for example, enzymes in your liver cells act on the excess, converting it first to glucose and then to glycogen or fat. When your body uses up glucose and needs more, enzymes break down glycogen to release its glucose subunits. In this case, a hormone called glucagon acts as a control over enzyme activity. It stimulates the key enzyme in the pathway by which glycogen is degraded, and it inhibits the enzyme that catalyzes glycogen formation.

Inhibitors, which can bind with enzymes and interfere with their function, are one type of control mechanism. A certain inhibitor molecule acts on trypsin, a

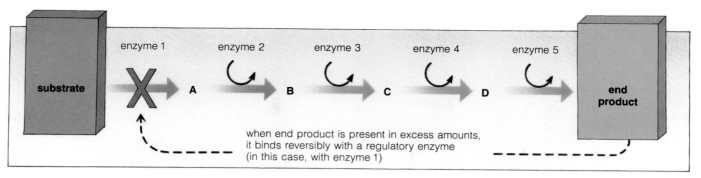

Figure 6.14 Example of feedback inhibition of a metabolic pathway. Here, the end product binds to the first enzyme in the pathway leading to its formation. When the product concentration drops, fewer molecules are around to inhibit the regulatory enzyme in the pathway, so production can rise again. Feedback inhibition allows concentrations of substances to be adjusted quickly to the cell's requirements.

protein-digesting enzyme. Cells in the pancreas produce and secrete trypsin into the small intestine. Prior to secretion, trypsin is kept isolated in vesicles, in inactive form. Any molecules that do escape packaging are shut down by the inhibitor. Without this safeguard, trypsin could be unleashed against the proteins of tissues and blood vessels in the pancreas. In *acute pancreatitis*, this enzyme and others are activated prematurely, sometimes with fatal results.

Some enzymes are governed by **allosteric control**. Besides the active site, "allosteric" enzymes have control sites where specific substances can bind and alter enzyme activity. For example, when a cell produces tryptophan molecules faster than it uses them, the unused molecules bind to and inhibit an allosteric enzyme. The enzyme happens to catalyze a step necessary in the synthesis of tryptophan.

Control of tryptophan production is a form of **feedback inhibition**. The output of the process works in a way that inhibits further output. Figure 6.14 illustrates this control mechanism.

Enzymes act only on specific substrates, and controls over their activity are central to the directed flow of substrates into, through, and out of the cell.

COFACTORS

Some enzymes speed up the transfer of electrons, atoms, or functional groups from one substrate to another. Many must be assisted by nonprotein components called **cofactors,** either to help catalyze the reaction or to serve fleetingly as the transfer agents.

Cofactors include some large organic molecules that function as *coenzymes*. Among them are **NAD$^+$** (nicotinamide adenine dinucleotide) and **FAD** (flavin adenine dinucleotide). Both types have roles in the breakdown of glucose and other carbohydrates. When enzymes dismantle a glucose molecule, unbound protons (H$^+$) and electrons become available. These are picked up by NAD$^+$, FAD, or both and turned over to other reaction sites. When carrying their cargo, the two coenzymes are abbreviated NADH and FADH$_2$ respectively.

NADP$^+$ (nicotinamide adenine dinucleotide phosphate) is a coenzyme with a central role in photosynthesis. It also serves as a link between some of the main degradative and biosynthetic pathways (for example, the ones by which fatty acids are assembled). Discerning enzymes recognize the phosphate "tag" on NADP$^+$ molecules and allow them to transfer protons and electrons to specific reaction sites. When carrying this cargo, the coenzyme is often abbreviated NADPH.

Some *metal ions* also serve as cofactors. They include ferrous iron (Fe^{++}), which is a component of cytochrome molecules. The cytochromes are carrier proteins bound in cell membranes, such as the membranes of chloroplasts and mitochondria.

ELECTRON TRANSFERS IN METABOLIC PATHWAYS

If you were to throw some glucose into a woodfire, its atoms would quickly let go of one another and combine with oxygen in the atmosphere, forming CO$_2$ and H$_2$O. The energy that had been stored in glucose would be lost as heat. Cells make better use of a glucose molecule; they do not "burn" it all at once and so waste its stored energy. Instead, atoms are plucked away from glucose in controlled steps, so that *intermediate* molecules form

Figure 6.15 (**a**) When hydrogen and oxygen are made to react (say, by an electric spark), energy is released as heat. (**b**) In cells, the same type of reaction is made to occur in many small steps that allow much of the released energy to be harnessed in usable form. These "steps" are electron transfers, often between molecules that operate together as an electron transport system.

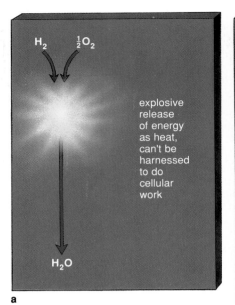

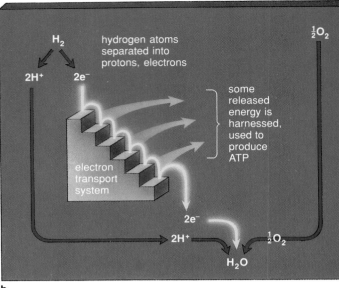

a

b

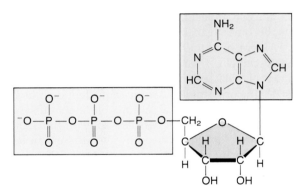

Figure 6.16 Structural formula for adenosine triphosphate, or ATP. The triphosphate group is shaded in gold, the sugar ribose in pink, and the adenine portion in blue.

along the route from $C_6H_{12}O_6$ to the CO_2 and H_2O. At each step, a specific enzyme lowers the activation energy for an intermediate. And at each step, only *some* energy is released.

Remember that a chemical bond is a union of the electron structures of two atoms. Breaking a chemical bond puts electrons up for grabs. In chloroplasts and mitochondria, such liberated electrons are sent through **electron transport systems**. The systems consist of enzymes and cofactors, bound in a cell membrane, that transfer electrons in a highly organized sequence. One molecule "donates" electrons, and the next in line "accepts" them. Each time a donor gives up electrons, it is said to be "oxidized." Each time an acceptor acquires

electrons, it is "reduced." *Oxidation-reduction* merely means an electron transfer.

Electron transfers can occur after an atom (or a molecule) absorbs enough energy to boost one or more of its electrons to a shell farther from its nucleus (page 24). An excited electron quickly returns to the lowest energy level available to it—and it gives off energy when it does this. *Electron transport systems "intercept" excited electrons and make use of the energy they release.*

By analogy, think of an electron transport system as a staircase (Figure 6.15). The electrons at the top of the staircase have the most energy. They drop down the staircase, one step at a time (they are transferred from one electron carrier to another). With each drop, some energy being released can be harnessed to do work—for example, to make hydrogen ions move in ways that establish pH and electric gradients across membranes. Such gradients, as you will discover, are central in ATP formation.

Electron transfers may seem rather abstract. The *Commentary* makes the idea more concrete by describing unusual but striking effects of electron transfers.

ATP: THE UNIVERSAL ENERGY CARRIER

Structure and Function of ATP

Sunlight is the primary energy source for the web of life. Before the sun's energy can be used in cell activities, it must be transformed into the chemical energy of **ATP** (adenosine triphosphate). Similarly, cells cannot directly

Commentary

You Light Up My Life—Visible Effects of Electron Transfers

Travel the ocean at night and you may see evidence of comb jellies near the water's surface. These relatives of jellyfishes give good displays of *bioluminescence*—which simply means luminescent flashes in body tissues (Figure *a*). Such flashes also occur in several groups of bacteria, fungi, fishes, and insects, including fireflies. The insects use flashes as signals that may attract a mate.

Highly fluorescent substances called luciferins can be prodded into giving up electrons. Once they have been released, the electrons drop to a lower energy level—and light is emitted when they do. Enzymes called luciferases catalyze the reactions.

Several years ago, the biochemist Keith Wood "stole" the soft, yellowish light that flashes in fireflies. Wood and his coworkers isolated the firefly gene that contains the instructions for building luciferase molecules. Just recently

they stole more lights from the Jamaican click beetle, otherwise known as the kittyboo. The kittyboo has four luciferase genes. One codes for green flashes, another for greenish-yellow, and the others for yellow and orange. Biochemists have been inserting the luciferase genes into a variety of organisms, including *Escherichia coli*. Figure *b* shows four bacterial cells, glowing with the borrowed light.

The ability to transfer luciferase genes into *E. coli* may prove useful in studying the mechanisms by which particular genes are turned on and off. Imagine a light bulb that couldn't be seen when you turned on the switch. What biochemists are doing is effectively substituting a light bulb for a gene. They are interested in the switches, not the bulb. When light flashes in a modified *E. coli* cell, they have observable evidence that a *switch* controlling a gene has been activated.

row of combs   mouth

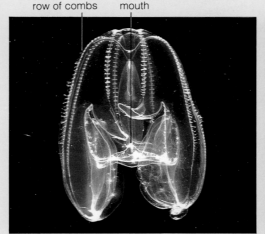

a Bioluminescence in a comb jelly

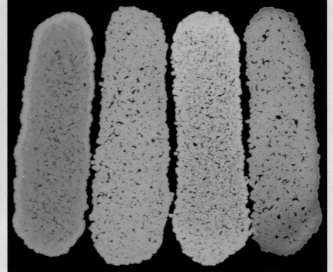

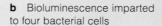
b Bioluminescence imparted to four bacterial cells

use the energy that is released during the breakdown of carbohydrates or other large organic molecules. They must first use the stored energy to produce ATP.

How does ATP act as an energy carrier from one reaction to another? As Figure 6.16 shows, ATP is composed of adenine (a nitrogen-containing compound), ribose (a five-carbon sugar), and a triphosphate (three phosphate groups), all linked together by covalent bonds.

Under cellular conditions, one of the phosphates can be split off easily by hydrolysis, and a good deal of usable energy is released when this happens. Said another way, ATP is an unstable molecule that can readily break down to more stable products.

Many hundreds of different enzymes can couple ATP hydrolysis to many hundreds of different metabolic reactions. Energy released during ATP hydrolysis provides energy for biosynthesis, active transport across cell membranes, and molecular displacements, such as those underlying muscle contraction. In fact, ATP molecules are like the coins of a nation—they are a common currency of energy in all cells.

ATP directly or indirectly delivers energy to or picks up energy from almost all metabolic pathways.

The ATP/ADP Cycle

In the **ATP/ADP cycle**, an energy input drives the linkage of **ADP** (adenosine diphosphate) and a phosphate group (or inorganic phosphate) into ATP. Then the ATP donates a phosphate group elsewhere and reverts back to ADP:

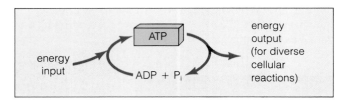

Adding phosphate to a molecule is called **phosphorylation**. What is important about it? When a molecule becomes phosphorylated by ATP, its store of energy generally increases *and it becomes primed to enter a specific reaction.*

With the ATP/ADP cycle, cells have a renewable means of conserving energy and transferring it to specific reactions. The ATP turnover is breathtaking. Even if you were bedridden for twenty-four hours, your cells would turn over approximately 40 kilograms (88 pounds) of ATP molecules simply for routine maintenance!

SUMMARY

1. Cells acquire and use energy to synthesize, accumulate, break down, and rid themselves of substances in controlled ways. These activities, which sustain cell growth, maintenance, and reproduction, are called metabolism.

2. Cellular use of energy conforms to two laws of thermodynamics. According to the first law, energy can be converted from one form to another but the total amount in the universe never changes. According to the second law, with each energy conversion, some energy is dispersed in a form that is not as readily available to do work.

3. Cells lose energy during metabolic reactions, but in nearly all cases, they replace it with energy derived directly or indirectly from the sun.

4. A metabolic pathway is a stepwise sequence of reactions in a cell. In *biosynthetic* pathways, organic molecules are assembled and energy becomes stored in them. In *degradative* pathways, organic molecules are broken apart and energy is released.

5. The following substances take part in metabolic reactions:

   a. Reactants or substrates: the substances that enter a reaction.

   b. Enzymes: proteins that serve as catalysts (they speed up reactions). Enzymes do not change the expected outcome of a reaction. They only change the rate at which the reaction proceeds.

   c. Cofactors: coenzymes (including NAD^+) and metal ions that help catalyze reactions or carry functional groups stripped from substrates.

   d. Energy carriers: mainly ATP, which readily donates energy to other molecules. Most biosynthetic pathways run directly or indirectly on ATP energy.

   e. End products: the substances formed at the end of a metabolic pathway.

6. If left undisturbed, most metabolic reactions approach a state of dynamic equilibrium. Then, there is no further net change in the concentrations of reactants and products.

7. Enzymes do not change what the concentration ratio of reactant to product molecules will be at equilibrium. They only increase the rate at which a reaction approaches equilibrium (by lowering the required activation energy).

8. During many electron transfers (oxidation-reduction reactions), energy is released that can be used to do work—for example, to make ATP.

Review Questions

1. State the first and second laws of thermodynamics. Which law deals with the *quality* of available energy, and which deals with the *quantity?* Give some examples of high-quality energy. *92–93*

2. Does the living state violate the second law of thermodynamics? In other words, how does the world of living things maintain a high degree of organization, even though there is a universal trend toward disorganization? *92–93*

3. In metabolic reactions, does equilibrium imply equal concentrations of reactants and products? Can you think of a cellular event that might keep a reaction from approaching equilibrium? *94*

4. Describe an enzyme and its role in metabolic reactions. *96–98*

5. What are the three molecular components of ATP? What is the function of ATP, and why is phosphorylation of a molecule by ATP so important? *101–102*

6. What is an oxidation-reduction reaction? What is its function in cells? *100*

Self-Quiz *(Answers in Appendix IV)*

1. A cell's capacity to acquire and use energy for building and breaking apart molecules is called _____ .

2. Two laws of _____ govern how cells acquire, convert, and transfer energy during metabolic reactions.

3. The ultimate source of energy for nearly all organisms on earth is _____ .
   a. food   c. the sun
   b. water   d. ATP

4. Which of the following statements is *not* true of dynamic equilibrium?
   a. The concentration of product is equal to the concentration of reactant.
   b. The rate of forward reaction is equal to the rate of reverse reaction.
   c. There is no further change in the concentrations of reactant and product.
   d. It is unchanged by enzyme activity.

5. In biosynthetic pathways, _____ .
   a. organic molecules are simply broken apart
   b. energy is not required for the reactions
   c. organic molecules are assembled
   d. energy is stored in organic molecules
   e. both c and d are correct

6. Which of the following is *not* true? Metabolic pathways _____ .
   a. occur in stepwise series of chemical reactions
   b. are speeded up by enzyme activity
   c. may degrade or assemble molecules
   d. overcome the second law of thermodynamics

7. Enzymes _____ .
   a. enhance reaction rates   c. act on specific substrates
   b. are affected by pH   d. all of the above are correct

8. All electron transport systems involve _____ and _____
   a. enzymes, cofactors   c. chloroplasts, mitochondria
   b. electron transfers,   d. both a and b are correct
   released energy

9. The main energy carriers in cells are _____ .
   a. NAD^+ molecules   c. ATP molecules
   b. cofactors   d. enzymes

10. Match each substance with its correct description.
    _____ a coenzyme or metal ion   a. reactant
    _____ substance formed at end of a   b. enzyme
    metabolic pathway   c. cofactor
    _____ mainly ATP   d. energy carrier
    _____ substance entering a reaction   e. end product
    _____ protein that catalyzes a
    reaction

Selected Key Terms

Readings

Atkins, P. 1984. *The Second Law.* New York: Freeman.

Doolittle, R. 1985. "Proteins." *Scientific American* 253(4):88–99.

Fenn, J. 1982. *Engines, Energy, and Entropy.* New York: Freeman. Deceptively simple introduction to thermodynamics; good analogies. Paperback.

Fersht, A. 1985. *Enzyme Structure and Mechanism.* Second edition. New York: Freeman.

7 ENERGY-ACQUIRING PATHWAYS

Sunlight, Rain, and Cellular Work

Just before dawn in the Midwest the air is dry and motionless. The heat that has scorched the land for weeks still rises from the earth and hangs in the air of a new day. There are no clouds in sight. There is no promise of rain. For hundreds of miles, crops stretch out, withered or dead. All the marvels of modern agriculture can't save them now. In the absence of one vital resource—water—life in each cell of those many thousands of plants has ceased.

In Los Angeles, a student reading the morning newspaper complains that the Midwest drought will probably cause a hike in food prices. In Washington, D.C., economists calculate the crop failures in terms of decreased tonnage available for domestic consumption and export; government officials brood about what that

means to the nation's balance of payments. In Ethiopia, a child with bloated belly and spindly legs waits passively for death. Even if food donations were to reach her now, it would be too late. Deprived of food resources too long, the cells of her body will never grow normally again.

You are about to explore the ways in which cells trap and use energy. At first the cellular pathways may seem to be far removed from the world of your interests. *Yet the food molecules on which you and nearly all other organisms depend cannot be built or used without those pathways and the raw materials—including water—required for their operation.*

We will return repeatedly to this point in later chapters, when we address such concerns as human

Figure 7.1 Links between photosynthesis and aerobic respiration—the main energy-acquiring and energy-releasing pathways in the world of life.

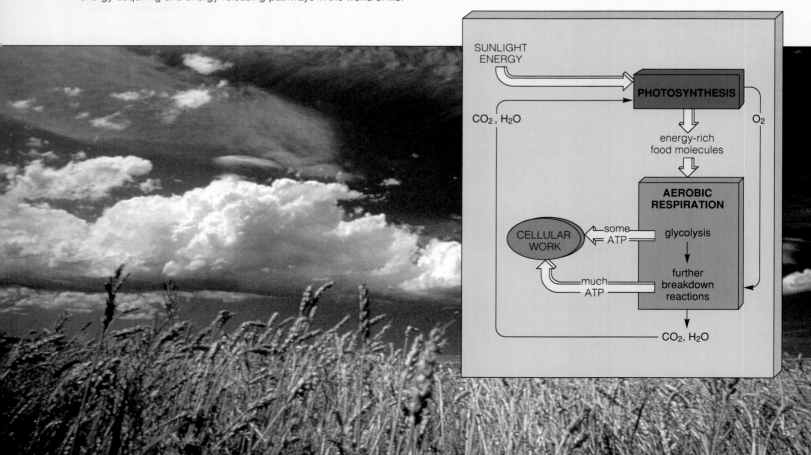

nutrition, human population growth and environmental limits on agriculture, genetic engineering of new crop plants, and the effects of pollution on food production. Here, our point of departure is the *source* of food, which isn't a farm or a supermarket or a refrigerator. What we call "food" was put together somewhere in the world by living cells from organic compounds. Given that the basic structure of every organic compound is a carbon backbone, the questions become these:

1. *Where does the carbon come from in the first place?*

2. *Where does the energy come from to drive the linkage of carbon and other atoms into organic compounds?*

3. *How does the energy inherent in those compounds become available to do cellular work?*

The answers vary, depending on whether you are talking about autotrophic or heterotrophic organisms.

Autotrophs obtain carbon and energy from the physical environment. They are "self-nourishing" (which is what autotroph means). Their carbon source is carbon dioxide (CO_2), a gaseous substance all around us in the air and dissolved in water. Only the *photosynthetic* autotrophs can get energy from sunlight. Plants, some protistans, and some bacteria fall in this category. A few kinds of bacteria are *chemosynthetic* autotrophs; they get energy by stripping electrons from sulfur or some other inorganic substance.

Heterotrophs are not self-nourishing. They feed on autotrophs, each other, and organic wastes. They must get carbon and energy from organic compounds *already built* by autotrophs. Animals, fungi, many protistans, and most bacteria are heterotrophs.

It follows, from the above, that carbon and energy enter the web of life primarily by **photosynthesis**. Energy stored in organic compounds as a result of photosynthesis can be released by several different pathways, all of which begin with the same breakdown reactions (*glycolysis*). Of these, the pathway called **aerobic respiration** releases the most energy. Figure 7.1 shows the links between photosynthesis and aerobic respiration—the focus of this chapter and the next.

1. Organic compounds, with their carbon backbones, are the key building blocks and energy stores for life. Plants and other photosynthetic autotrophs assemble their own organic compounds. They use carbon dioxide from the air as a source for the carbon, and they trap sunlight energy to drive the synthesis reactions.

2. Photosynthesis is the main biosynthetic pathway by which carbon and energy enter the web of life. Animals and other heterotrophs must obtain their carbon and energy from organic compounds already built by the autotrophs.

3. In plants, photosynthesis proceeds at clusters of pigment molecules and electron transport systems present in the membranes of organelles called chloroplasts.

4. During the light-dependent reactions, pigment molecules absorb light energy and are compelled to give up electrons to transport systems. Movement of electrons through the systems sets up electrochemical gradients across membranes, and these gradients drive the formation of ATP and NADPH.

5. During the actual synthesis reactions, which proceed independently of sunlight, ATP provides energy to drive the joining of carbon and oxygen (from carbon dioxide) with hydrogen and electrons (from NADPH). The reactions produce sugar phosphates, which are used to form sucrose, starch, and other end products of photosynthesis.

PHOTOSYNTHESIS

Simplified Picture of Photosynthesis

Photosynthesis proceeds in two stages, each with its own set of reactions. In the *light-dependent* reactions, sunlight energy is absorbed and converted to chemical energy, which is transferred to ATP and NADPH. (Here you may wish to refer to pages 99 and 100.) In the *light-independent* reactions, sugars and other compounds are assembled with the help of ATP and NADPH. Photosynthesis is typically summarized this way:

$$2H_2O + CO_2 \xrightarrow{\text{sunlight}} O_2 + (CH_2O) + H_2O$$

As you can see from this equation, hydrogen atoms obtained from water molecules end up in newly synthesized compounds that are based on some number of (CH_2O) units. For instance, for the reactions leading to a new glucose molecule $(C_6H_{12}O_6)$, you would have to multiply everything by six to get the six carbons, twelve hydrogens, and six oxygens in it:

$$12H_2O + 6CO_2 \xrightarrow{\text{sunlight}} 6O_2 + C_6H_{12}O_6 + 6H_2O$$

Here, glucose is shown as an end product in order to keep the chemical bookkeeping simple. The reactions

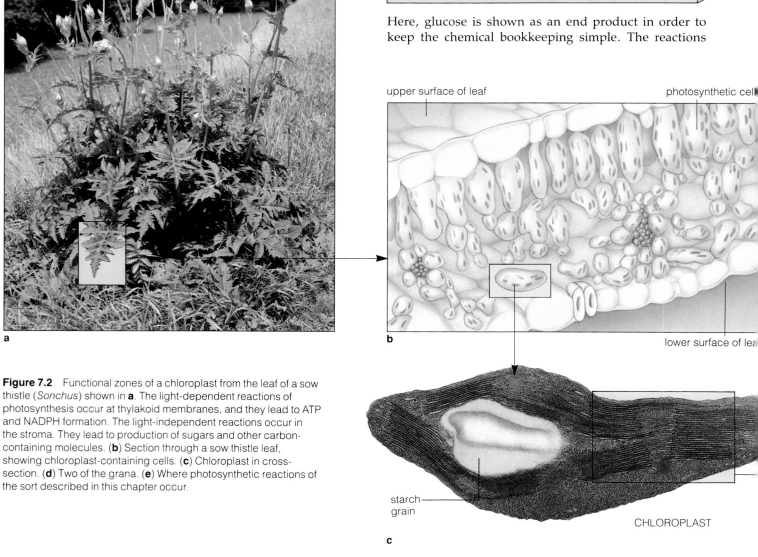

a

upper surface of leaf

photosynthetic cell

b

lower surface of leaf

Figure 7.2 Functional zones of a chloroplast from the leaf of a sow thistle (*Sonchus*) shown in **a**. The light-dependent reactions of photosynthesis occur at thylakoid membranes, and they lead to ATP and NADPH formation. The light-independent reactions occur in the stroma. They lead to production of sugars and other carbon-containing molecules. (**b**) Section through a sow thistle leaf, showing chloroplast-containing cells. (**c**) Chloroplast in cross-section. (**d**) Two of the grana. (**e**) Where photosynthetic reactions of the sort described in this chapter occur.

starch grain

CHLOROPLAST

c

don't really end with glucose, however. Newly formed glucose and other simple sugars are linked at once into sucrose, starch, and other carbohydrates—the true end products of photosynthesis.

Chloroplast Structure and Function

Let's take a look at how photosynthesis proceeds in the chloroplasts of leafy plants. As indicated earlier in Figure 4.20, this organelle has a double-membrane envelope around its semifluid interior, the *stroma*. An elaborate membrane system weaves through the stroma. It takes the form of flattened channels and disklike compartments organized into stacks, called *grana* (singular, granum). This is the **thylakoid membrane** system of many kinds of plants.

The spaces inside the thylakoid disks and channels connect with one another. They form a continuous compartment where hydrogen ions can be accumulated and used in ways that help produce ATP. The stroma is where enzymes speed the assembly of the actual products of photosynthesis, including starch molecules (Figure 7.2).

If you could line up 2,000 chloroplasts, one after another, the lineup would be no wider than a dime. Imagine all the chloroplasts in just one lettuce leaf, each a tiny factory for producing sugars and starch—and you get an idea of the magnitude of metabolic events required to feed you and all other organisms living together on this planet.

LIGHT-DEPENDENT REACTIONS

Three events unfold during the **light-dependent reactions**, the initial stage of photosynthesis. *First*, pigments absorb light energy and give up electrons. *Second*, electron and hydrogen transfers lead to ATP and NADPH formation. *Third*, the pigments that gave up electrons in the first place get electron replacements.

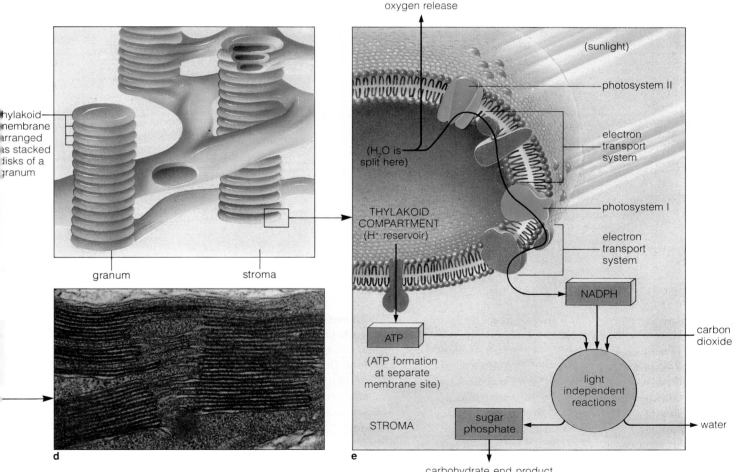

thylakoid membrane arranged as stacked disks of a granum

granum   stroma

d

oxygen release

(sunlight)

photosystem II

electron transport system

(H₂O is split here)

THYLAKOID COMPARTMENT (H⁺ reservoir)

photosystem I

electron transport system

NADPH

ATP

(ATP formation at separate membrane site)

carbon dioxide

light independent reactions

STROMA

sugar phosphate

water

e

carbohydrate end product (e.g., sucrose, starch, cellulose)

Figure 7.3 (**a**) Where wavelengths of visible light occur in the electromagnetic spectrum. Photosynthesis, vision, and other light-requiring processes typically use wavelengths ranging from about 400 to 750 nanometers. Shorter wavelengths (such as ultraviolet and x-rays) are so energetic they break bonds in organic compounds, so they can destroy cells. Longer wavelengths (such as infrared) are not energetic enough to drive the formation of NADPH during photosynthesis. NADPH is necessary for the biosynthetic programs of large, multicelled plants.

(**b**) Wavelengths absorbed by some of the photosynthetic pigments. Peaks in the ranges of absorption correspond to the measured amount of energy absorbed and used in photosynthesis. Colors used here correspond to the colors transmitted (or reflected) by each pigment type. (Thus chlorophylls absorb blue and red wavelengths best and transmit wavelengths in between.) Together, different photosynthetic pigments can absorb most of the energy available in the spectrum of visible light.

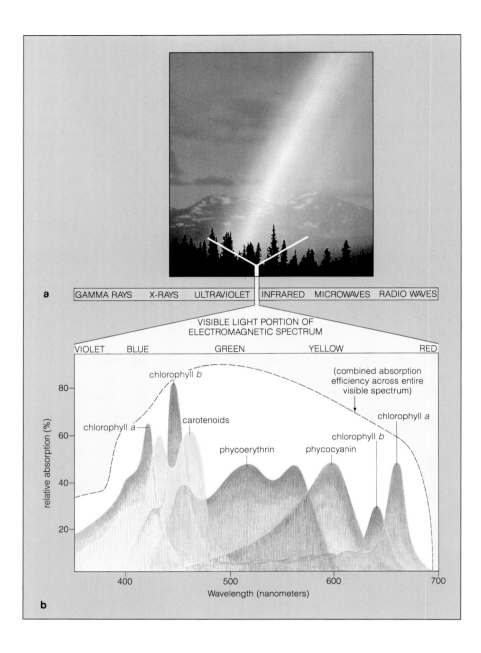

Light Absorption

Light-Trapping Pigments. The thylakoid membrane contains a variety of pigments, all of which are photon-absorbing molecules. A photon is a packet of light energy. Some photons have more energy than others, and the differences correspond to different wavelengths of light. The shorter the wavelength, the more energy. Most organisms use wavelengths ranging from about 400 to 750 nanometers for photosynthesis, vision, and other light-requiring processes. That is a very small part of the electromagnetic spectrum (Figure 7.3), but those are the wavelengths we perceive as colors of light.

Each type of pigment absorbs certain wavelengths and transmits or reflects the rest. Leaves appear green

because they have abundant *chlorophyll* pigments that absorb blue and red wavelengths but transmit green. Carotenoid pigments absorb violet and blue wavelengths but transmit yellow, orange, and red. Leaves also contain carotenoids and other pigments, but the more abundant chlorophylls usually mask their presence. Those pigments become visible in autumn, when the leaves stop producing chlorophyll (Figure 7.4).

Collectively, photosynthetic pigments can absorb most of the wavelength energy available in the spectrum of visible light. Figure 7.5 shows an early experiment designed to test which wavelengths are used most effectively by a green plant. In that green plant and others, the main pigments are chlorophyll *a*, chlorophyll *b*, and the carotenoids.

maple in
summer

maple in
autumn

Figure 7.4 Changes in leaf color in autumn. Chloroplasts of mature leaves contain chlorophylls, carotenoids (including the yellow carotenes and xanthophylls), and other pigments, each of which absorbs certain wavelengths of light. The intense green color of chloroplasts usually masks the presence of other pigments. In autumn, however, the gradual reduction in daylength and other factors trigger the breakdown of chlorophyll, and additional colors show through.

Also in autumn, water-soluble anthocyanins accumulate in the central vacuoles of leaf cells. These pigments appear red if plant fluids are slightly acidic, blue if basic (alkaline), and colors in between at intermediate levels of acidity.

The color of birch, aspen, and other tree species is always the same in autumn. The color of other species, including maple, ash, and sumac, varies around the country; it even varies from one leaf to the next, depending on the pigment combinations.

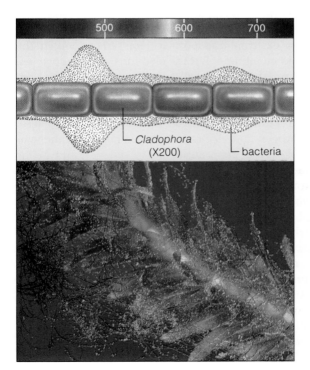

Cladophora
(X200)

bacteria

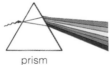

prism

Figure 7.5 T. Englemann's 1882 experiment, which revealed the most effective wavelengths for photosynthesis by *Cladophora,* a filamentous green alga.

Oxygen is a by-product of photosynthesis, as the photograph of the aquatic plant *Elodea* indicates. (The bubbles of oxygen sometimes are visible on sunny days.) Oxygen also is used by many organisms in the energy-releasing pathway called aerobic respiration. Englemann suspected that aerobic (oxygen-using) bacteria living in the same places as *Cladophora* would congregate in areas where oxygen was being produced.

Englemann used a crystal prism to cast a tiny spectrum of colors on a microscope slide. Then he positioned an algal filament to run parallel with the spectrum (see diagram). The bacteria did indeed cluster next to the filament where the most oxygen was being released. And those regions corresponded to colors (wavelengths) being absorbed most effectively—in this case, violet and red.

Photosystems. A thylakoid membrane has many thousands of **photosystems**, which are organized clusters of 200 to 300 light-trapping pigment molecules. The pigments are bound to membrane proteins. Over 90 percent of the pigments in a photosystem simply "harvest" sunlight. When they absorb photon energy, one of their electrons gets boosted to a higher energy level (page 23). The electron returns almost at once to a lower level, and the extra energy is released when it does:

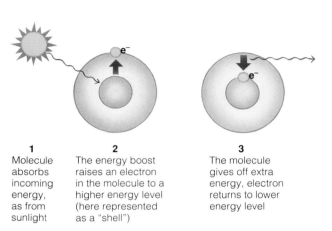

1
Molecule absorbs incoming energy, as from sunlight

2
The energy boost raises an electron in the molecule to a higher energy level (here represented as a "shell")

3
The molecule gives off extra energy, electron returns to lower energy level

The released energy hops from one molecule to another in the photosystem. With each hop, a little of that energy is usually lost (as heat). The energy remaining corresponds to longer and longer wavelengths, compared to the original photon energy. Only a few chlorophyll molecules in a photosystem can respond to the longest of those wavelengths. They act like a sink, or *trap,* for all the energy being harvested by all the other pigments. Those chlorophylls alone give up the electrons used in photosynthesis.

When energy flows into an energy trap, an electron is excited and is rapidly transferred to an acceptor molecule embedded in the thylakoid membrane. Thus, *the first event of photosynthesis is the light-activated transfer of an electron from a special chlorophyll molecule to an acceptor molecule.* It is over in less than a billionth of a second.

How ATP and NADPH Form in Chloroplasts

Electrons expelled from a chlorophyll molecule pass through one or two **electron transport systems** of the sort described on page 100. Each system is a series of molecules bound in the thylakoid membrane (Figure 7.2d). Each molecule accepts and then donates electrons to the next molecule in line. The electron flow sets up gradients across the membrane, and these drive the attachment of phosphate to ADP, so forming ATP. We call this *photo*phosphorylation, because the pathway depends on an earlier input of light energy. Such pathways are cyclic or noncyclic.

Cyclic Pathway. A special chlorophyll molecule, P700, dominates the type of pigment cluster called *photosystem I.* This molecule absorbs wavelengths of about 700 nanometers. In the simplest ATP-producing pathway, electrons "travel in a circle" from a P700 chlorophyll, through a transport system, then back to P700:

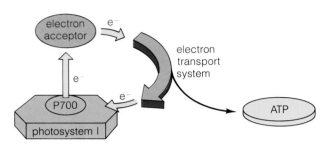

This pathway is called **cyclic photophosphorylation**. It is probably the oldest means of ATP production. Early photosynthetic autotrophs were no larger than existing bacteria, so their body-building programs could scarcely have been enormous. They could have used ATP alone to build organic compounds even though such reactions are rather inefficient. (ATP only carries energy to sites where organic compounds are built. The electrons and hydrogen atoms required must be obtained by other means.) However, energy from the cyclic pathway would not have sustained the evolution of larger photosynthesizers, including leafy plants.

Noncyclic Pathway. Today, leafy plants rely mostly on **noncyclic photophosphorylation**. In this ATP-producing pathway, electrons move through two photosystems and two electron transport systems embedded in the thylakoid membrane. But the electrons do not move in a circle through these membrane sites. At the end of the transport system that contains P700, they are picked up by a coenzyme, NADP$^+$, that delivers hydrogen as well as the electrons to a reaction site in the stroma. There, the hydrogen and electrons are used *directly* in the synthesis of organic compounds!

Figure 7.6 is a diagram of the sequence of events in the noncyclic pathway. The reactions begin at a photosystem distinguished by a chlorophyll molecule (P680) that absorbs wavelengths of about 680 nanometers. The photosystem having this type of molecule is designated *photosystem II.* When P680 absorbs light energy, it gives up an electron to an acceptor molecule. From there, the electron moves through a transport system—and then to chlorophyll P700 of photosystem I.

The excited electron has not yet returned to the lowest available energy level. When the P700 absorbs light energy, electrons are boosted even higher and passed to a second transport system. Transport systems, recall, are like steps on an energy staircase—and this boost places electrons at the top of a higher staircase. There is enough energy left at the bottom of this staircase to attach two electrons and a hydrogen ion (H^+) to $NADP^+$, the result being NADPH.

Thus, in the noncyclic pathway, electrons flow in one direction to NADPH. In the meantime, the P680 molecule that gives up electrons in the first place is getting replacements—from water. Inside the thylakoid compartment, water molecules are being split into oxygen, unbound protons (that is, hydrogen ions), and electrons. Photon energy indirectly drives this reaction sequence, which is called **photolysis** (Figure 7.6).

Oxygen atoms split from water molecules are by-products of the noncyclic pathway. Oxygen has been accumulating ever since this pathway evolved, more than 3.5 billion years ago. It profoundly changed the earth's atmosphere. And it made possible aerobic respiration, the most efficient pathway for extracting energy from organic compounds. The emergence of the noncyclic pathway ultimately allowed you and all other animals to be around today, breathing the oxygen that helps keep your cells alive.

With cyclic photophosphorylation, ATP alone forms.

With noncyclic photophosphorylation, ATP *and* NADPH form, and oxygen is released as a by-product.

After the noncyclic pathway evolved, oxygen accumulated in the atmosphere and made aerobic respiration possible.

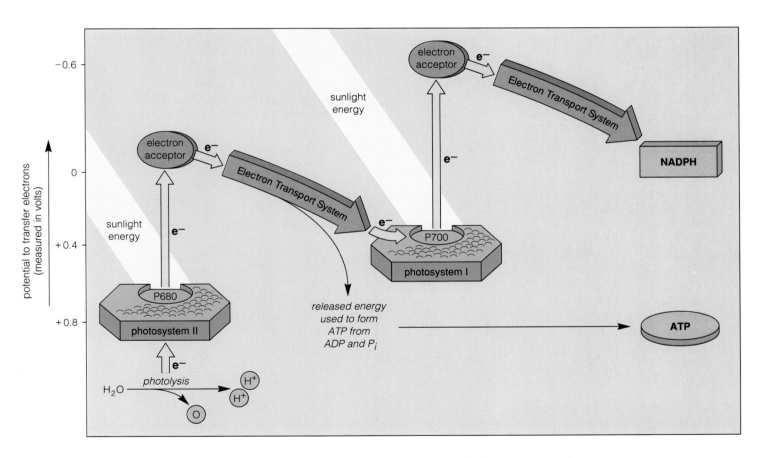

Figure 7.6 Noncyclic photophosphorylation, which yields NADPH as well as ATP. Electrons derived from the splitting of water molecules (photolysis) travel through two photosystems, which work together in boosting the electrons to an energy level high enough to lead to NADPH formation. Figure 7.7 provides a closer look at the mechanism by which ATP forms.

A Closer Look at ATP Formation

So far, you have seen what happens to the electrons and oxygen released when water molecules are split at the start of the noncyclic pathway. What happens to the hydrogen ions? They accumulate inside the thylakoid compartment of the chloroplast. Hydrogen ions also accumulate there when electron transport systems are operating. (This is true of both the cyclic and noncyclic pathways.) This accumulation sets up concentration and electric gradients across the thylakoid membrane—and the energy inherent in those gradients is tapped to form ATP.

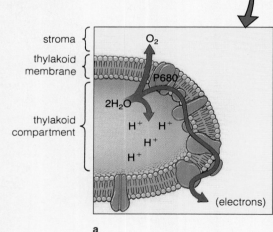

Figure 7.7 The chemiosmotic theory of ATP formation, as it occurs in chloroplasts of leafy plants.

thylakoid membrane system in chloroplast

Figure 7.7 shows how the combined force of the gradients propels hydrogen ions out of the thylakoid compartment and into the stroma. They flow through channel proteins that have enzyme activity. Through enzyme action, inorganic phosphate (P_i) becomes linked to ADP, and so ATP is formed.

The idea that the concentration and electric gradients across a membrane drive ATP formation is known as the **chemiosmotic theory**.

LIGHT-INDEPENDENT REACTIONS

The **light-independent reactions** are the "synthesis" part of photosynthesis. They are the reactions by which carbohydrates are put together in the stroma of chloroplasts. ATP provides energy and NADPH provides hydrogen and electrons for the reactions. The air around photosynthetic cells provides carbon and oxygen (in the form of carbon dioxide).

The reactions are called "light-independent" because they do not depend directly on sunlight. They *can* proceed as long as ATP and NADPH are available from the light-dependent reactions. But ATP and NADPH normally are produced during daylight, so the light-independent reactions occur mostly during the day.

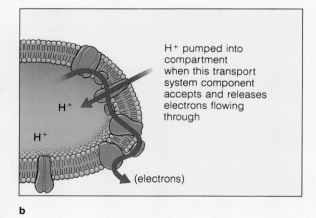

stroma
thylakoid membrane
thylakoid compartment

O_2

P680

$2H_2O$

H^+ H^+

H^+

H^+

(electrons)

a

H^+ pumped into compartment when this transport system component accepts and releases electrons flowing through

H^+

H^+

(electrons)

b

(**a**) The start of the noncyclic pathway of photosynthesis, water molecules are split into oxygen, "naked" protons (or hydrogen ions), and electrons. (The reaction sequence is called photolysis.)

The oxygen is released as an end product and the electrons are sent through transport systems. The hydrogen ions accumulate inside the thylakoid compartment of the chloroplast, as sketched above.

Hydrogen ions also accumulate in the compartment when electron transport systems are operating. (This is true of both the cyclic and noncyclic pathway.) When certain molecules of the transport system accept electrons, they also pick up hydrogen ions from the stroma and release them inside the compartment (**b**).

Through photolysis and electron transport, hydrogen ions become more concentrated in the thylakoid compartment than in the stroma. The lopsided distribution of those positively

Calvin-Benson Cycle

The heart of the light-independent reactions is the **Calvin-Benson cycle**, which is named after its discoverers, Melvin Calvin and Andrew Benson. During this cyclic pathway, carbon is captured from carbon dioxide, and a sugar phosphate forms in reactions that require ATP and NADPH. The cycle ends with regeneration of RuBP (ribulose bisphosphate)—a compound that is required to capture the carbon in the first place.

As Figure 7.8 shows, the reactions begin with **carbon dioxide fixation**. Carbon dioxide diffuses into leaves and is present in the spaces between photosynthetic cells. Those cells have enzymes that capture carbon dioxide and attach it to *RuBP*, which has a backbone of five carbon atoms. The attachment produces an unstable, six-carbon intermediate that splits at once into two molecules of *PGA* (phosphoglycerate).

Each PGA receives a phosphate group from ATP. The resulting intermediate receives hydrogen and electrons from NADPH to form *PGAL* (phosphoglyceraldehyde). It takes six carbon dioxide and six RuBP molecules to produce twelve PGAL. Most of the PGAL becomes rearranged into new RuBP molecules—which can be used to fix more carbon. But two PGAL are joined together to form a *sugar phosphate*. Such sugars have phosphate

groups attached that prime them for further reaction. Glucose-6-phosphate is an example. Its name simply means a phosphate group is attached to the sixth carbon atom of the glucose molecule (Figure 6.5).

The Calvin-Benson cycle yields enough RuBP to replace the ones used at the start of carbon dioxide fixation. The ADP, NADP$^+$, and phosphate leftovers are sent back to the light-dependent reaction sites, where they are converted once more to NADPH and ATP. The sugar phosphate formed in the cycle can serve as a building block for the plant's main carbohydrates, including sucrose, starch, and cellulose. Synthesis of those com-

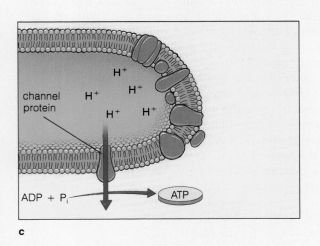

c

charged ions also creates a difference in electric charge across the thylakoid membrane. An electric gradient as well as a concentration gradient has been established.

(c) The combined force of the concentration and electric gradients propels hydrogen ions out of the compartment and into the stroma. The ions flow through channel proteins (called ATP synthases). The proteins span the membrane and have built-in enzyme machinery. The ion flow drives the enzyme machinery by which ADP combines with inorganic phosphate to form ATP.

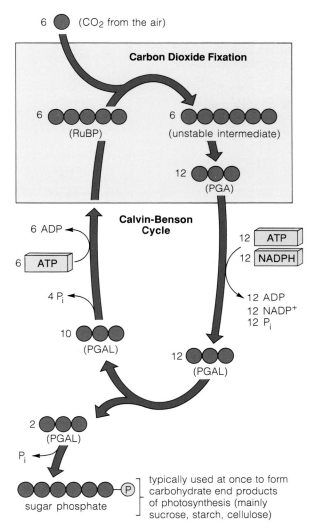

Figure 7.8 Summary of the light-independent reactions of photosynthesis. The carbon atoms of the different molecules are depicted in red. All of the intermediates have one or two phosphate groups; for simplicity we show only the one on the "end product" (sugar phosphate). The phosphate groups that have been detached from molecules during the reactions are designated P$_i$, meaning "inorganic phosphate."

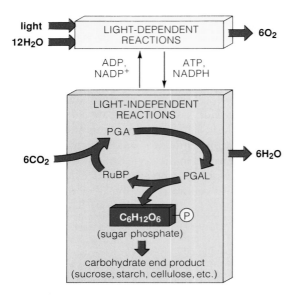

Figure 7.9 Summary of the main reactants, intermediates, and products of photosynthesis corresponding to the equation:

$$12H_2O + 6CO_2 \xrightarrow{sunlight} 6O_2 + C_6H_{12}O_6 + 6H_2O$$

pounds by different metabolic pathways marks the conclusion of the light-independent reactions.

Figure 7.9 summarizes the main reactants, intermediates, and products of photosynthesis, from start to finish. Beginning with the light-dependent reactions, photolysis of every two water molecules yields four electrons and one O_2 molecule. For every four electrons, three ATP and two NADPH molecules are produced. *Each turn* of the Calvin-Benson cycle requires three ATP and two NADPH molecules, as well as a CO_2 molecule. It takes six turns of the cycle to get the six CO_2 molecules required for each six-carbon sugar formed.

During the Calvin-Benson cycle, carbon is "captured" from carbon dioxide, a sugar phosphate forms in reactions requiring ATP and NADPH, and RuBP (needed to capture the carbon) is regenerated.

How Autotrophs Use Intermediates and Products of Photosynthesis

So here we are, with sugar phosphates formed in a microscopic speck of an organelle. Visualize millions of such specks in, say, a corn plant. Where do the sugar

phosphates go from here? Although a fraction of them are used at once as fuel to provide energy for cellular work, almost all are used as building blocks in the synthesis of sucrose, starch, and cellulose.

Of all the carbohydrates produced by photosynthesis, sucrose is the most easily transportable. Conducting tissues carry it from sites in the leaf to living cells in all parts of the corn plant body. Starch is the main storage form of carbohydrate in the leaves, stems, and roots.

What happens in other plants besides corn? Some of the sucrose produced in leaves of potato plants is converted and stored as starch in underground stem regions called tubers (the "potatoes"). In sugar beets, onions, and sugarcane, sucrose itself is the main storage form. Photosynthetic autotrophs also use intermediates and products of photosynthesis when they assemble lipids and amino acids. Indeed, some green algae use more than 90 percent of the carbon fixed when they construct proteins and lipids. These plants have a brief life cycle, and they live in places where sunlight and water are plentiful. They put most photosynthetic products into rapid growth and reproduction instead of diverting them to storage forms.

C4 Plants

Plant growth depends on photosynthesis, and photosynthesis depends on efficient carbon dioxide fixation. That mechanism in turn depends on how much carbon dioxide is concentrated around photosynthetic cells. Although there is plenty of carbon dioxide in the air, it is not always abundantly available to the cells *inside the leaves* of land plants.

Leaves have a waxy covering that retards moisture loss, and water escapes mainly through tiny passages (stomata) across the leaf's surface. On hot, dry days the stomata are closed and the plant conserves water. This means, however, that carbon dioxide can't enter the leaf. At the same time, oxygen builds up inside the leaf as a by-product of photosynthesis. The stage is set for a wasteful process called "photorespiration." In this process, oxygen instead of carbon dioxide becomes attached to the RuBP used in the Calvin-Benson cycle, with different results for the plant:

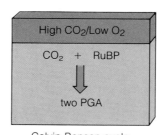

Calvin-Benson cycle predominates

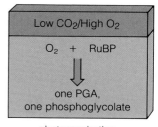

photorespiration predominates

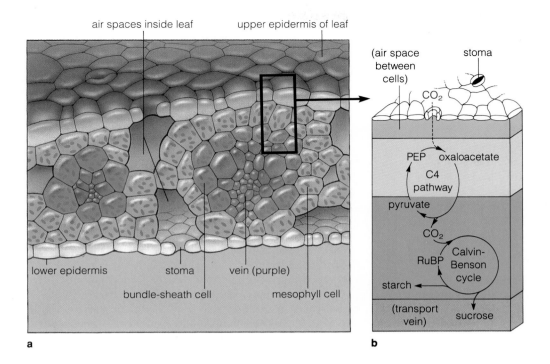

air spaces inside leaf   upper epidermis of leaf

lower epidermis   stoma   vein (purple)

bundle-sheath cell   mesophyll cell

a

(air space between cells)   stoma

CO_2

PEP   oxaloacetate

C4 pathway

pyruvate

CO_2

RuBP   Calvin-Benson cycle

starch

(transport vein)   sucrose

b

Figure 7.10 C4 pathway. (**a**) This is the internal structure of a leaf from corn (*Zea mays*), a typical C4 plant. Notice how the photosynthetic bundle-sheath cells (dark green) surround the veins and are in turn surrounded by photosynthetic mesophyll cells (lighter green). (**b**) C4 plants have a carbon-fixing system that *precedes* the Calvin-Benson cycle.

Formation of sugar phosphates depends on PGA. When photorespiration wins out, less PGA forms and the plant's capacity for growth suffers.

Many plants, including crabgrass, sugarcane, and corn, continue to fix carbon dioxide even when their stomata close on hot, dry days. In these plants, the first compound formed by carbon dioxide fixation is the four-carbon oxaloacetate. Hence their name, "C4" plants. (Three-carbon PGA is produced in "C3" plants.)

C4 plants fix carbon dioxide not once but twice, in two different types of photosynthetic cells. Carbon dioxide that has diffused into their leaves is fixed initially in *mesophyll cells*, forming oxaloacetate. This is a temporary fix, so to speak. The oxaloacetate is quickly transferred to *bundle-sheath cells*, which form a layer around every vein in the leaf. In those cells, the carbon dioxide is released, fixed again, and used to build carbohydrates (Figure 7.10). Thus carbon dioxide accumulates in the bundle-sheath cells, and its concentration can remain high even when the Calvin-Benson cycle is operating.

When the temperature climbs, stomata close and oxygen accumulates in the leaves of any type of plant. C3 plants photorespire even more than usual in hot weather, and their growth slows. By contrast, C4 plants do not photorespire as much, even in hot weather. Because of their special carbon-fixing system, the carbon dioxide concentration still increases in the bundle-sheath cells—so oxygen loses out in the competition for RuBP.

C4 species are abundant in regions with the highest temperatures during the growing season. For example, 80 percent of all native species in Florida are C4 plants—compared to 0 percent in Manitoba, Canada. Kentucky bluegrass and other C3 species have the advantage in regions where temperatures drop below 25°C; they are better adapted to cold. When we mix C3 and C4 species from different regions in our gardens, one or the other kind will have the advantage during at least some part of the year. That is why a lawn of Kentucky bluegrass can thrive during cool spring weather in San Diego, only to be overwhelmed during the hot summer months by a C4 plant—crabgrass.

CHEMOSYNTHESIS

Photosynthesis so dominates the energy-trapping pathways that sometimes it is easy to overlook other, less common routes. The *chemosynthetic autotrophs* obtain energy not from sunlight but rather from the oxidation of inorganic substances (ammonium ions, iron or sulfur compounds, and so on). That is their source of energy for building organic compounds.

For example, some bacteria that live in soil use ammonia (NH_3) molecules as an energy source, stripping them of protons and electrons. Nitrite ions (NO_2^-) and nitrate ions (NO_3^-) are the remnants of their activities. Compared with ammonium ions, nitrite and nitrate ions are readily washed out of the soil, so the action of the so-called nitrifying bacteria can lower soil fertility. When farmers are ready to plant their fields, they sometimes add chemicals to the soil that inhibit bacterial metabolism.

We will return later to the environmental effects of chemosynthetic autotrophs. In this unit, we turn next to pathways by which energy is released from carbohydrates and other biological molecules—the chemical legacy of autotrophs and, ultimately, of sunlight.

Commentary

Energy From Artificial Chloroplasts?

If innovative researchers are successful, motorists may one day fill their gas tanks with an abundant, nonpolluting hydrogen fuel. As motorists stand at the pump, they may also worry less about the greenhouse effect and global warming. When carried out on a large scale, the same technology that produces the fuel could gobble up carbon dioxide—a "greenhouse gas" that is accumulating to alarming levels in the atmosphere.

Such rewards might emerge from efforts to create artificial assemblies of the molecules required for photosynthesis. Such assemblies in principle will convert energy from sunlight into molecules such as methane and hydrogen. Methane is a type of natural gas, and hydrogen is a potential liquid fuel for use in automobiles or for generating electricity. Most importantly, sunlight is an abundant, "no-cost" energy source.

In attempting to devise a workable artificial system, scientists have focused on events in the photosystems of chloroplast membranes. In those natural systems, photons absorbed by chlorophyll molecules provide the energy to boost electrons to a higher energy level. An artificial system must incorporate natural or manmade molecules that can absorb light, donate electrons, and receive excited electrons. Such a system must also maintain the one-way flow of electrons. And it must keep the process going in a way that permits released energy to be captured in a useful chemical form—much as it is in the NADPH formed by the noncyclic pathway of photosynthesis.

In the past few years, researchers have put together molecular assemblies that actually mimic the light absorption and electron excitation in photosystems. For example, chemists at Arizona State University have synthesized a promising, five-part molecule of natural compounds, including some derived from chlorophyll and from electron transport systems. The artificial assembly maintains an energetic state long enough for useful products to be obtained. Another candidate is a chain of inorganic molecules that absorb light like chlorophyll. The chain is embedded in a matrix of a porous mineral (zeolite). The tiny pores of zeolite can be aligned to form molecule-sized tunnels, which can keep the photosynthetic apparatus physically oriented in the proper way.

While theoretical and technical problems remain, the promise of artificial photosynthesis is simply too great to be dismissed. We may soon learn the intricate details of chloroplast function. The human population—so inextricably dependent on green plants—will then owe chloroplasts even more.

SUMMARY

1. Plants and other photosynthetic autotrophs use sunlight (as an energy source) and carbon dioxide (as the carbon source) for building organic compounds. Animals and other heterotrophs must obtain carbon and energy from organic compounds already built by autotrophs.

2. Photosynthesis is the main biosynthetic pathway by which carbon and energy enter the web of life. It consists of two sets of reactions:

   a. The light-dependent reactions take place at the thylakoid membrane system of chloroplasts. These reactions produce ATP and NADPH (or ATP alone).

   b. The light-independent reactions take place in the stroma around the membrane system. They produce sugar phosphates that are used in building sucrose, starch, and other end products of photosynthesis.

3. These are the key points about the light-dependent reactions:

   a. Photosystems are clusters of photosynthetic pigments that are bound to proteins in the thylakoid membrane. Light absorption causes the transfer of electrons from a chlorophyll to an acceptor molecule, which donates them to a transport system in the membrane.

   b. Operation of electron transport systems causes H^+ to accumulate inside the thylakoid membrane system. This produces concentration and electric gradients that drive the formation of ATP.

   c. In the cyclic pathway, electrons travel in a circle, back to the photosystem that originally gave them up (photosystem I). This pathway yields ATP only.

   d. The noncyclic pathway also yields ATP, but the electrons end up in NADPH. In this pathway, electrons derived from the splitting of water molecules travel from photosystem II to a transport system, then to photosystem I and another transport system.

4. These are the key points about the light-independent reactions:

   a. The ATP produced in the light-dependent reactions provides energy and NADPH provides hydrogen atoms and electrons for the "synthesis" part of photosynthesis.

b. Sugar phosphates form by operation of the Calvin-Benson cycle. The cycle begins when carbon dioxide from the air is affixed to RuBP, making an unstable intermediate that splits into two PGA.

c. Each PGA receives a phosphate group from ATP. The resulting molecule receives H^+ and electrons from NADPH to form PGAL. Two of every twelve PGAL are used to produce a six-carbon sugar phosphate. The rest of the PGAL are rearranged to regenerate RuBP for the cycle.

5. Some plants living in hot climates have an additional carbon-fixing system, the C4 pathway. This pathway helps circumvent photorespiration, in which oxygen instead of carbon dioxide becomes affixed to RuBP. (Otherwise, photorespiration undoes much of what photosynthesis accomplishes.)

Review Questions

1. Define the difference between autotrophs and heterotrophs, and give examples of each. In what category do photosynthesizers fall? *105*

2. Summarize the photosynthesis reactions in words, then as an equation. Distinguish between the light-dependent and the light-independent stages of these reactions. *105–106*

3. A thylakoid compartment is a reservoir for which of the following substances: glucose, photosynthetic pigments, hydrogen ions, fatty acids? *107*

4. Sketch the reaction steps of noncyclic photophosphorylation, showing where the excited electrons eventually end up. Do the same for the cyclic pathway. *110–111*

5. Which of the following substances are *not* required for the light-independent reactions: ATP, NADPH, RuBP, chlorophyll, carotenoids, free oxygen, carbon dioxide, enzymes? *113*

6. Suppose a plant carrying out photosynthesis were exposed to carbon dioxide molecules that contain radioactively labeled carbon atoms ($^{14}CO_2$). In which of the following compounds will the labeled carbon first appear: NADPH, PGAL, pyruvate, PGA? *113*

7. How many CO_2 molecules must enter the Calvin-Benson cycle to produce one sugar phosphate molecule? Why? *114*

Self-Quiz *(Answers in Appendix IV)*

1. Molecules with backbones of _____ serve as the main building blocks of all organisms.

2. Photosynthetic autotrophs use _____ from the air as their carbon source and _____ as their energy source.

3. In plant cells, light-*dependent* reactions occur _____.
   a. in the cytoplasm
   b. at the plasma membrane
   c. in the stroma
   d. in the thylakoid membrane

4. In plant cells, light-*independent* reactions occur _____.
   a. in the cytoplasm
   b. at the plasma membrane
   c. in the stroma
   d. in the grana

5. In the light-dependent reactions, _____.
   a. carbon dioxide is incorporated into carbohydrates
   b. ATP and NADPH are formed
   c. carbon dioxide accepts electrons
   d. sugar phosphates are formed

6. When light is absorbed by a photosystem, _____.
   a. sugar phosphates are produced
   b. electrons are transferred to an acceptor molecule
   c. RuBP accepts electrons
   d. the light-dependent reactions are initiated
   e. both b and d are correct

7. The Calvin-Benson cycle begins when _____.
   a. light is available
   b. light is not available
   c. carbon dioxide is attached to RuBP
   d. electrons leave a photosystem

8. In the light-independent reactions, ATP furnishes phosphate groups to _____.
   a. RuBP
   b. $NADP^+$
   c. PGA
   d. PGAL

9. Match each event in photosynthesis with its correct description.
   _____ uses RuBP; produces PGA
   _____ uses ATP and NADPH
   _____ forms NADPH
   _____ produces ATP and NADPH
   _____ produces ATP only

   a. cyclic pathway
   b. noncyclic pathway
   c. carbon dioxide fixation
   d. formation of PGAL
   e. Transfer of H^+ and electrons to $NADP^+$

Selected Key Terms

autotroph *105*	electron transport	PGAL *113*
C4 plant *115*	system *110*	photolysis *111*
Calvin-Benson	granum *107*	photophosphorylation *110*
cycle *113*	heterotroph *105*	photosynthesis *105*
carbon dioxide	light-dependent	photosystem *110*
fixation *113*	reaction *107*	RuBP *113*
chemiosmotic	light-independent	stroma *107*
theory *112*	reaction *112*	sugar phosphate *113*
chlorophyll *108*	PGA *113*	thylakoid membrane *107*

Readings

Alberts, B. 1989. "Chloroplasts and Photosynthesis." In *Molecular Biology of the Cell*. Second edition. New York: Garland.

Moore, P. 1981. "The Varied Ways Plants Tap the Sun." *New Scientist* 12:394–397. Clear, simple introduction to the C4 plants.

Youvan, D., and B. Marrs. 1987. "Molecular Mechanisms of Photosynthesis." *Scientific American* 256:42–50.

8 ENERGY-RELEASING PATHWAYS

The Killers Are Coming!

"Killer" bees from South America buzzed across the border between Mexico and the United States in 1990 and are now working their way north. Alarming newspaper headlines accompanied their arrival, and for good reason. The bees are descended from aggressive African queen bees, and they can be terrifying.

In the 1950s, researchers had some queen bees shipped from Africa to Brazil for breeding experiments. Honeybees, you understand, are big business. Not only do they produce honey, they also are rented out to pollinate commercial orchards. (Put a screened cage around an orchard tree in bloom and less than 1 percent of the flowers will set fruit. Put a hive of honeybees in the same cage and 40 percent of the flowers will set fruit.) The honeybees in Brazil seemed sluggish, compared to their aggressive African relatives, so the researchers hoped to cross-breed the two types and come up with a mild-mannered but zippier pollinator.

One of the researchers accidentally released twenty-six of the African queens. This was bad enough. Then beekeepers got wind of the preliminary experimental results. After learning that the first few generations of offspring were jazzed-up but still nice honeybees, they imported hundreds of *additional* African queens and released them to mate with locals.

The genetic backlash came later. For some inexplicable reason, genes of the African bees became dominant in subsequent generations. Now there are "Africanized" honeybees with nasty tempers. When disturbed, they become astonishingly agitated. Whereas a mild-mannered honeybee might chase an animal fifty yards or so, a squadron of Africanized bees will chase it for a quarter of a mile. If they catch it, they can sting it to death.

Africanized bees do everything other bees do, but they do more of it faster. The developmental stages leading from egg to adult proceed more quickly. Adults fly more rapidly. (They beat their wings more than 200 times per second; by contrast, a butterfly wing flaps

about 4 times per second.) Adults produce more offspring in the same span of time. They die sooner. But while they are alive, they forage more vigorously than other bees do in the competition for nectar.

For the Africanized honeybee, doing things faster means having a nonstop supply of energy that can keep the metabolic fires burning. This bee's stomach can hold 30 milligrams of sugar-rich nectar—enough to keep the bee flying for 60 kilometers without running out of fuel. The cells of its flight muscles are packed with exceptionally large mitochondria—and mitochondria are truly efficient organelles in which energy originally stored in the sugar becomes converted to the energy of ATP. Only with mitochondria can the Africanized bee generate enough energy to sustain prolonged effort, whether this be day-long nectar gathering or chasing down intruders.

As you might deduce from all of this, Africanized bees probably will not survive in regions where winters are harsh and plants stop blooming for months at a time. Thus, although residents of California, Florida, and similar sunbelt states may have to keep an eye on the sky, it is not likely that Alaskans will be bothered by huge populations of killer bees.

In their reliance on a great deal of fuel and efficient metabolic furnaces, Africanized bees are not that different from you or any other highly active or rapidly growing organism. In their reliance on energy stored in sugars and other organic compounds, they are like *all* other organisms. Even though the energy-releasing pathways differ in some details from one organism to the next, each requires characteristic starting materials, then yields predictable products and by-products. And *all* of the energy-releasing pathways yield ATP. In fact, throughout the biosphere, there is startling similarity in the uses to which energy and raw materials are put. *At the biochemical level, there is undeniable unity among all forms of life.* We will return to this idea in the *Commentary* at the chapter's end.

Figure 8.1 A mild-mannered honeybee buzzing in for a landing on a flower, wings beating with energy provided by ATP. If this were one of its Africanized relatives, possibly you would not stay around to watch the landing.

1. Plants produce and use ATP during photosynthesis. Plants and all other organisms also produce ATP by degrading (breaking apart) organic compounds such as glucose. Aerobic respiration is the degradative pathway with the greatest energy yield.

2. Aerobic respiration proceeds through three stages. First, glucose is partially broken down to pyruvate with a net energy yield of two ATP. Second, the pyruvate is completely broken down to carbon dioxide and water. Electrons liberated during the reactions are delivered to a transport system. Third, energy is released as electrons are transferred through the system, and it drives the formation of thirty-six or more ATP for every glucose molecule. At the end of the system, free oxygen combines with the electrons and with hydrogen ions to form water.

3. Over evolutionary time, photosynthesis and aerobic respiration have become linked on a global scale. The oxygen by-products of photosynthesis serve as final electron acceptors for the aerobic pathway. And the carbon dioxide and water released in aerobic respiration are raw materials used in building organic compounds during photosynthesis:

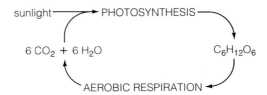

ATP-PRODUCING PATHWAYS

No organism stays alive without taking in energy. Plants get energy from the sun; animals get energy secondhand, thirdhand, and so on, by eating plants and one another. But no matter what the *source* of energy may be, organisms must convert it to a form of chemical energy that can drive metabolic reactions. In all cells, the "coins" most commonly minted and then spent on metabolic reactions are molecules of adenosine triphosphate, or **ATP**. Figure 6.16 shows the structure of ATP, which has three phosphate groups attached to the rest of the molecule.

Plants produce ATP during photosynthesis, as described in the preceding chapter. But plants and all other organisms also can produce ATP through degradative pathways that release chemical energy stored in carbohydrates, lipids, or proteins.

The main degradative pathway is **aerobic respiration**. The "aerobic" part of the name means that oxygen is the final acceptor of electrons stripped from glucose or some other molecule during energy-releasing reactions. With every breath you take, you are replenishing oxygen supplies for your busily respiring cells. That is why astronauts who have ventured to the moon wouldn't have lasted more than a little while if their spacesuits had not been plugged into oxygen tanks as they clambered about on the moon's surface. (Unlike the earth, the moon has no oxygen-rich atmosphere.) When you breathe out, you are ridding your body of the carbon dioxide and water leftovers of aerobic respiration (Figure 8.2).

Other degradative pathways are "anaerobic," in that something besides oxygen serves as the final electron acceptor for energy-releasing reactions. The most common anaerobic pathways are called **fermentation** and **anaerobic electron transport**. Cells in your body rely mostly on aerobic respiration, but they can use a fermentation pathway for short periods when oxygen supplies are low. Many microbes rely exclusively on anaerobic pathways. Some are indifferent to the presence or absence of oxygen; they include bacteria that are used in the production of yogurt. Other microbes are "strict anaerobes" that die if exposed to oxygen. Among them are the bacteria that cause botulism, tetanus, and some other serious diseases.

All energy-releasing pathways proceed through an orderly sequence of steps. As we examine the pathways in more detail, keep in mind that they do not proceed all by themselves. Enzymes catalyze each step, and the intermediate produced at a given step serves as a substrate for the next enzyme in the pathway.

AEROBIC RESPIRATION

Overview of the Reactions

Of all degradative pathways, aerobic respiration produces the most ATP for each glucose molecule being dismantled. Whereas fermentation has a net yield of two ATP, the aerobic route yields *thirty-six* ATP or more. If you were a microscopic bacterium, you would not require much ATP. Being large, complex, and highly active, you depend absolutely on the high ATP yield of aerobic respiration.

a

Using glucose as the starting material, the aerobic route is often summarized this way:

$$C_6H_{12}O_6 + 6O_2 \longrightarrow 6CO_2 + 6H_2O$$

GLUCOSE   CARBON DIOXIDE

The summary equation only tells us what the substances are at the start and the finish of the aerobic route. In between are *three stages* of reactions.

In the first stage, *glycolysis*, glucose is partially degraded (oxidized) to pyruvate. By the end of the second stage, which includes the *Krebs cycle*, glucose has been completely degraded to carbon dioxide and water. Neither stage produces much ATP. However, protons and electrons are stripped from intermediates during both stages, and these are delivered to a transport system. That system is used in the third stage of reactions, *electron transport phosphorylation*, which yields many ATP. Oxygen accepts the "spent" electrons from the transport system (Figure 8.3).

b

c

Figure 8.2 (**a**) Astronauts about to land in an anaerobic world—the Apollo 12 voyagers to the moon. (**b**) The view homeward, with the sun emerging from behind the earth—the only planet known to be enveloped in an oxygen-rich atmosphere. (**c**) A few humans and a number of trees. Like most kinds of organisms on earth, the trees as well as the women are using oxygen in metabolic reactions that sustain their activities.

Figure 8.3 Overview of aerobic respiration, the main energy-releasing pathway, with glucose as the starting material. In glycolysis, glucose is partially broken down to pyruvate. (Glycolysis is also the initial set of reactions in other pathways, including fermentation.) In the Krebs cycle, pyruvate is completely broken down to carbon dioxide. Coenzymes (NADH and FADH$_2$) accept protons and electrons being stripped from intermediates of the reactions and deliver them to an electron transport system. Oxygen accepts the electrons from the transport system. From start (glycolysis) to finish, the aerobic pathway typically has a net energy yield of thirty-six ATP.

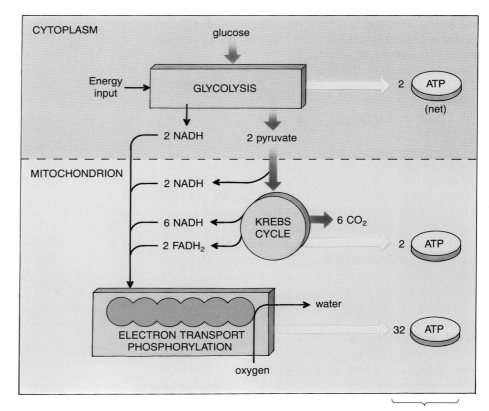

Figure 8.4 Glycolysis, showing glucose as the starting material. These breakdown reactions take place in the cytoplasm of all cells. They end with the formation of two pyruvate molecules and have a net energy yield of two ATP.

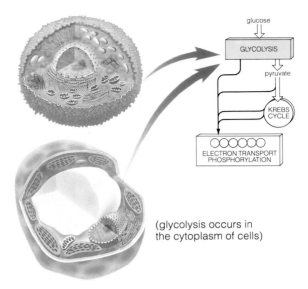

(glycolysis occurs in the cytoplasm of cells)

Glycolysis

Glycolysis is the partial breakdown of glucose or some other organic compound into two molecules of pyruvate. The breakdown occurs by way of a controlled sequence of reactions that take place in the cytoplasm of all cells. Enzymes lower the activation energy for the formation of each intermediate along the route, and energy is released when some of the intermediates are converted to different molecular forms.

Each glucose molecule has a backbone of six carbon atoms to which hydrogen atoms are attached (Figure 3.5). Here we show the backbone in simplified fashion:

The first steps of glycolysis are *energy-requiring;* they do not proceed without an energy input. The energy becomes available when two ATP molecules are subjected to hydrolysis (page 37). Through enzyme action, each ATP transfers a phosphate group to the six-carbon backbone. The backbone then splits apart to form two molecules of **PGAL** (phosphoglyceraldehyde), each with a three-carbon backbone. Formation of PGAL marks the start of the *energy-releasing* steps of glycolysis (Figure 8.4).

Through enzyme action, each PGAL now gives up one proton and two electrons to NAD$^+$, forming NADH. Recall that NAD$^+$ is a "reusable" coenzyme that functions at the active site of an enzyme. It accepts protons and electrons stripped from a substrate, then transfers them elsewhere and so becomes NAD$^+$ again. Meanwhile, the intermediate remaining at the site combines with inorganic phosphate (P_i). The result is a rather unstable molecule that readily gives up a phosphate group to ADP. In this manner, two ATP molecules form, one for each PGAL.

As you can see, *ATP has been formed by the direct, enzyme-mediated transfer of a phosphate group from a substrate to ADP.* The mechanism is called **substrate-level phosphorylation**. (This is different from "electron transport" phosphorylation, which requires oxygen and a transport system.)

Substrate-level phosphorylation occurs again during glycolysis. At a certain step, enzymes strip a proton and a hydroxide ion (OH$^-$) from each of two intermediate molecules. This leaves two molecules of PEP (phosphoenol pyruvate). PEP is rather unstable, and when it breaks down, the energy released is enough to drive the transfer of one of its phosphate groups to ADP. Because the phosphorylation occurs twice (once for each PGAL), two more ATP are formed. At this point, glucose has been broken down to two pyruvate molecules, each with a three-carbon backbone (Figure 8.4). Glycolysis is over.

In sum, the initial breakdown of glucose produces two NADH, four ATP (by substrate-level phosphorylation), and two pyruvate molecules. But remember, two ATP were invested at the start of glycolysis, so the *net* yield is only two ATP.

Glycolysis produces two NADH, two ATP (net), and two pyruvate molecules for each glucose molecule entering the reactions.

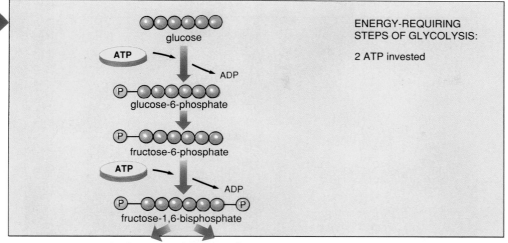

ENERGY-REQUIRING
STEPS OF GLYCOLYSIS:

2 ATP invested

glucose

ATP → ADP

glucose-6-phosphate

fructose-6-phosphate

ATP → ADP

fructose-1,6-bisphosphate

1. Glucose, a six-carbon molecule, gets an ATP-derived phosphate group attached to it at one end, then gets rearranged into fructose-6-phosphate.

2. Another ATP-derived phosphate group gets attached at the other end, forming fructose-1,6-bisphosphate.

3. This intermediate is split into PGAL and DHAP. Because each of those molecules is easily converted to the other, we can say that two PGAL have formed.

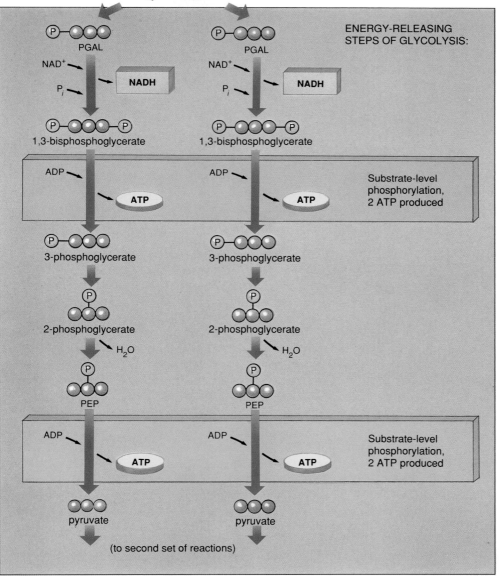

ENERGY-RELEASING
STEPS OF GLYCOLYSIS:

PGAL

NAD^+ → NADH
P_i →

1,3-bisphosphoglycerate

ADP → ATP

Substrate-level phosphorylation, 2 ATP produced

3-phosphoglycerate

2-phosphoglycerate

H_2O

PEP

ADP → ATP

Substrate-level phosphorylation, 2 ATP produced

pyruvate

(to second set of reactions)

4. Each PGAL gives up two electrons and one proton (H^+) to NAD^+, forming NADH. Each PGAL also combines with inorganic phosphate (P_i). Each of the two resulting intermediates donates a phosphate group to ADP, forming ATP.

5. With this formation of two ATP, the original energy investment of two ATP is paid off.

6. In the next two conversions, the intermediate gives up one proton and one hydroxide ion (OH^-) which combine to form water. The resulting intermediate is phosphoenol pyruvate (PEP).

7. The rather unstable PEP molecule readily gives up a phosphate group to ADP, forming ATP. Thus the net energy yield from glycolysis is two ATP for each glucose molecule entering the reactions.

8. Two molecules of pyruvate (the ionized form of pyruvic acid) are the end products of glycolysis.

NET ENERGY YIELD: 2 ATP

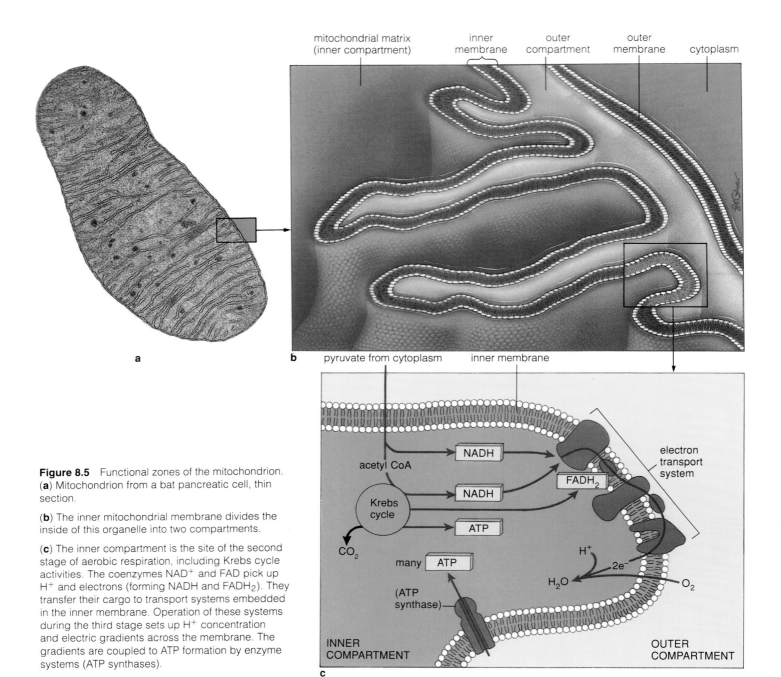

mitochondrial matrix (inner compartment)　inner membrane　outer compartment　outer membrane　cytoplasm

a

b

pyruvate from cytoplasm　inner membrane

NADH

acetyl CoA

NADH

FADH$_2$

Krebs cycle

ATP

electron transport system

CO$_2$

many ATP

H$^+$

H$_2$O　2e$^-$　O$_2$

(ATP synthase)

INNER COMPARTMENT

OUTER COMPARTMENT

c

Figure 8.5 Functional zones of the mitochondrion. (**a**) Mitochondrion from a bat pancreatic cell, thin section.

(**b**) The inner mitochondrial membrane divides the inside of this organelle into two compartments.

(**c**) The inner compartment is the site of the second stage of aerobic respiration, including Krebs cycle activities. The coenzymes NAD$^+$ and FAD pick up H$^+$ and electrons (forming NADH and FADH$_2$). They transfer their cargo to transport systems embedded in the inner membrane. Operation of these systems during the third stage sets up H$^+$ concentration and electric gradients across the membrane. The gradients are coupled to ATP formation by enzyme systems (ATP synthases).

Krebs Cycle

At the end of glycolysis, pyruvate molecules are still in the cytoplasm. They now may enter a *mitochondrion*, the only organelle where the second and third stages of the aerobic pathway can proceed. Figure 8.5 shows the structure of a representative mitochondrion.

The next series of reactions takes place in the inner compartment of the mitochondrion. There, the pyruvate molecules donate their carbon atoms to reactions that produce carbon dioxide molecules as by-products. Take a look at Figure 8.6. It shows that the second stage begins with conversion of each pyruvate to acetyl-CoA. The con-

version product becomes attached to oxaloacetate, the point of entry into a cyclic set of reactions called the **Krebs cycle**. (The cycle was named after Hans Krebs, who began working out its details in the 1930s.) Notice there is some symmetry here: Three carbon atoms enter the reactions (as the backbone of each pyruvate), and three carbon atoms leave (in three molecules of carbon dioxide).

The Krebs cycle itself serves three functions. First, many reaction steps are concerned with juggling the intermediate molecules back into the form of oxaloacetate. Cells have only so much oxaloacetate, and the cycle could not proceed over and over again if oxaloacetate

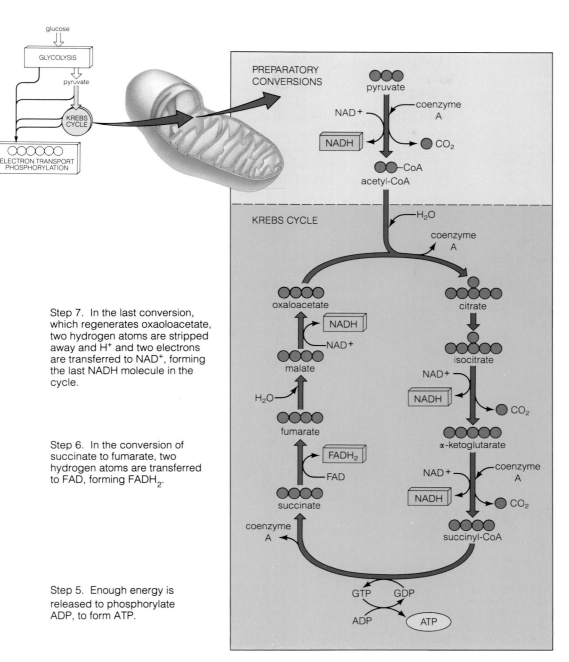

glucose

GLYCOLYSIS

pyruvate

KREBS CYCLE

ELECTRON TRANSPORT PHOSPHORYLATION

PREPARATORY CONVERSIONS

pyruvate

NAD^+ — coenzyme A

NADH — CO_2

CoA
acetyl-CoA

KREBS CYCLE

H_2O

coenzyme A

oxaloacetate

NADH
NAD^+

malate

H_2O

fumarate

FADH$_2$
FAD

succinate

coenzyme A

citrate

isocitrate

NAD^+
NADH
CO_2

α-ketoglutarate

NAD^+ — coenzyme A
NADH — CO_2

succinyl-CoA

GTP   GDP

ADP   ATP

Step 1. As three-carbon pyruvate enters the mitochondrion, enzymes split away its COO^- group, which departs as CO_2. Enzymes also transfer a proton (H^+) and two electrons to NAD^+, forming NADH. The two-carbon molecule remaining is linked to coenzyme A, forming the intermediate acetyl-CoA.

Step 2. The two-carbon molecule becomes attached to oxaloacetate, the point of entry into the Krebs cycle, to form the six-carbon citrate.

Step 3. Citrate is rearranged into the six-carbon isocitrate, which is stripped of two hydrogen atoms, with all but one H^+ of those atoms being transferred to NAD^+ to form NADH. It is also stripped of a COO^- group, this being the second CO_2 to depart.

Step 4. The resulting intermediate also gives up two hydrogen atoms and NADH forms. And it gives up a COO^- group, which is the last CO_2 to depart.

The rest of the reactions work to convert the intermediate remaining (succinyl-CoA) back to the oxaloacetate on which the Krebs cycle turns.

Step 7. In the last conversion, which regenerates oxaoloacetate, two hydrogen atoms are stripped away and H^+ and two electrons are transferred to NAD^+, forming the last NADH molecule in the cycle.

Step 6. In the conversion of succinate to fumarate, two hydrogen atoms are transferred to FAD, forming FADH$_2$.

Step 5. Enough energy is released to phosphorylate ADP, to form ATP.

were not regenerated. Second, one reaction step serves to produce ATP (by substrate-level phosphorylation). Third, some reaction steps liberate protons and electrons for transfer to the coenzymes NAD^+ and FAD, which will in turn transfer them to other reaction sites. When carrying their cargo, the coenzymes are abbreviated NADH and FADH$_2$.

It takes two preparatory reaction sequences and two turns of the Krebs cycle to use up the two pyruvate molecules. The two turns of the cycle add only two more ATP to the small yield from glycolysis. *But they also load many more coenzymes with protons and electrons that can be used in the third stage of the aerobic pathway.* The ATP

Figure 8.6 The Krebs cycle and the preparatory reactions preceding it. For *each* pyruvate molecule, 3 CO_2, 1 ATP, 4 NADH, and 1 FADH$_2$ are formed. But remember the steps shown occur *twice* for each glucose molecule broken down.

payoff from that final stage will be substantial indeed, as you might well induce from this chart:

Glycolysis:		2 NADH
Pyruvate conversion preceding Krebs cycle:		2 NADH
Krebs cycle:	2 FADH$_2$	6 NADH
Total electron carriers sent to third stage of aerobic pathway:	2 FADH$_2$ +	10 NADH

Electron Transport Phosphorylation

In the third stage of the aerobic pathway, the coenzymes NADH and FADH$_2$ give up protons and electrons to highly organized transport systems. Those systems con-

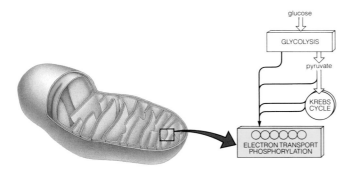

sist of enzymes and other proteins that are located in the inner membrane of the mitochondrion. Together, they work in ways that set up concentration and electric gradients across the membrane. And according to the **chemiosmotic theory**, those gradients drive the formation of ATP, in the following manner.

As Figures 8.5 and 8.7 show, the components of each transport system are arranged in series. Electrons are passed down the line, so to speak, and with each transfer they lose some energy. The energy released at certain transfers drives the pumping of unbound protons (H$^+$) out of the inner compartment of the mitochondrion. The pumping action produces H$^+$ concentration and electric gradients across the inner membrane. H$^+$ can flow across the membrane, down its combined electrochemical gradient. The flow occurs through ATP synthases, which are channel proteins that span the membrane and show enzyme activity. Here, enzyme action joins inorganic phosphate to ADP, forming ATP.

At the end of the electron transport system, free oxygen combines with the electrons and with hydrogen ions to form water. By removing electrons from the transport system, oxygen allows new electrons to flow through the system.

Operation of the transport system commonly leads to the formation of thirty-two ATP. When added to the ATP produced during the first and second stages, this brings the total net yield of the aerobic pathway to thirty-six ATP.

Figure 8.7 Electron transport phosphorylation. The reactions occur at transport systems and enzyme systems (ATP synthases) embedded in the inner mitochondrial membrane. The transport system consists of enzymes and other proteins (including cytochrome molecules) that operate one after the other, as shown in Figure 8.5.

The membrane itself creates two compartments. The reactions begin in the inner compartment, when NADH and FADH$_2$ give up H$^+$ and electrons to the transport system. Electrons are accepted and passed through the system, but the H$^+$ is left behind—in the outer compartment:

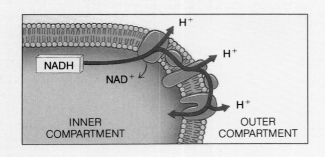

Soon there is a higher concentration of H$^+$ ions in the outer compartment than in the inner one. In other words, concentration and electric gradients now exist across the membrane. The H$^+$ ions follow the gradients and move back into the inner compartment. They do this by flowing through the ATP synthases that span the membrane. Energy associated with the flow drives the coupling of ADP and inorganic phosphate into ATP:

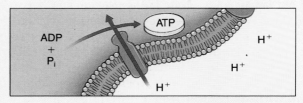

Do these events sound familiar? They should: ATP forms in much the same way in chloroplasts (Figure 7.7). The idea that concentration and electric gradients across a membrane drive ATP formation is called the chemiosmotic theory.

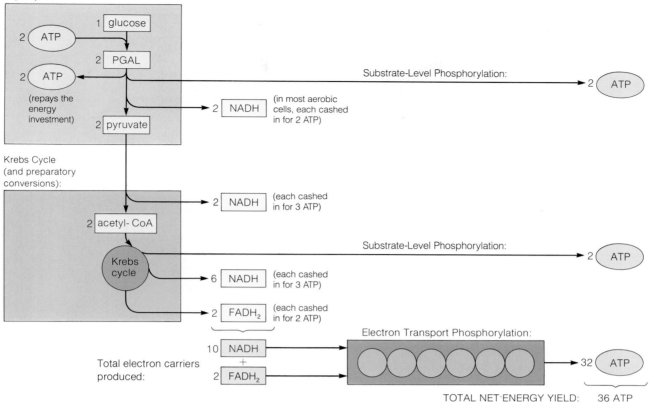

Figure 8.8 Summary of the energy harvest from one glucose molecule sent through the aerobic respiration pathway. Actual ATP yields vary, depending on cellular conditions and on the mechanism used to transfer energy from cytoplasmic NADH into the mitochondrion.

In aerobic respiration, glucose is completely broken down to carbon dioxide and water.

NAD⁺ and FAD accept unbound protons (H⁺) and electrons stripped from substrates of the reactions, then they deliver them to an electron transport system. Oxygen is the final acceptor of those electrons.

From glycolysis (in the cytoplasm) to the final reactions (in the mitochondrion), this pathway commonly yields thirty-six ATP for every glucose molecule.

Glucose Energy Yield

Figure 8.8 summarizes energy-yielding steps of aerobic respiration. Notice that the net yield from the complete breakdown of a glucose molecule commonly is thirty-six or thirty-eight ATP. The total varies, depending on what happens to the electrons being transferred by NADH molecules that have been produced in the cytoplasm (during glycolysis).

To understand what goes on, compare what happens to the electrons carried by an NADH or FADH₂ molecule formed *inside* the mitochondrion during the second-stage reactions. The NADH makes its delivery to the highest possible point of entry into a transport system. The delivery allows the pumping of enough H⁺ to produce three ATP molecules. The FADH₂ makes its delivery at a lower point of entry into the transport system. As a result, fewer H⁺ are pumped, and only two ATP are produced (Figure 8.5c).

It so happens that NADH formed in the cytoplasm makes its delivery *to* the mitochondrion, not *into* it. Once inside that organelle, its electrons might be transferred to an NAD⁺ *or* FAD molecule.

For example, consider what happens in liver, heart, and kidney cells. A shuttle built into the outer membrane accepts the electrons, then gives them up to NAD⁺ inside. Because NADH turns over electrons at the top of a transport system, three ATP form. Thus, in liver, heart, and kidney cells, the overall energy harvest is thirty-eight ATP.

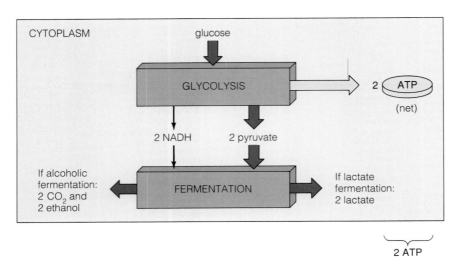

Figure 8.9 Overview of fermentation, a type of degradative pathway in which an intermediate or product of the reactions themselves serves as the final electron acceptor. The photograph shows the dustlike coating on grapes that contains yeasts, single-celled organisms that use a fermentation pathway.

More commonly, as in skeletal muscle and brain cells, a different shuttle accepts deliveries from cytoplasmic NADH. That shuttle donates electrons to FAD inside the mitochondrion. Because the electrons are delivered to a lower point of entry into the transport system, only two ATP can form. In such cells, the overall energy harvest from the complete breakdown of each glucose molecule is thirty-six ATP.

Remember that glucose is an unstable compound; its covalent bonds are much weaker than the covalent bonds of carbon dioxide and water. In fact, when glucose is broken down completely to carbon dioxide and water, about 686 kilocalories of energy are released. Of that, about 7.5 kilocalories become conserved for further use in each of the thirty-six ATP molecules typically formed. Thus the energy-conserving efficiency of this pathway is (36)(7.5)/(686), or 39 percent.

ANAEROBIC ROUTES

Our planet has its share of anaerobic environments. Among them are marshes, bogs, and the sediments and mud of lakes, rivers, and seas. Other anaerobic settings include the animal gut, canned foods, and wastewater treatment facilities (Chapter 48). Many bacterial species thrive in totally anaerobic environments. Other species thrive in environments that are well oxygenated some of the time but not all of the time. The aerobic pathway predominates under such conditions, then other pathways kick in when the oxygen concentration drops.

Let's look briefly at three common anaerobic pathways of glucose breakdown. Two are fermentation pathways (Figure 8.9); the third is called anaerobic electron transport. Unlike aerobic respiration, the fermentation pathways do not require an "outside" electron acceptor such as oxygen. The other pathway uses some molecule other than oxygen as an electron acceptor.

Alcoholic Fermentation

Like the aerobic pathway, the fermentation pathways begin with glycolysis: glucose is broken down to two pyruvate molecules, and two NADH molecules are formed. In **alcoholic fermentation**, the pyruvate molecules from glycolysis undergo rearrangements but are not used up completely. The NADH that forms during the reactions donates electrons to an intermediate, acetaldehyde, which thereupon becomes ethanol. Figure 8.10 shows the reaction sequence.

Yeasts, which are single-celled fungi, can produce ATP through alcoholic fermentation. This metabolic activity has widespread commercial applications. For example, bakers use cells of the yeast *Saccharomyces cerevisiae* to make bread dough rise (Figure 8.11). They mix the yeast with a small amount of sugar and blend the mixture into the dough. As the yeast converts the sugar to ethanol and carbon dioxide, the gaseous CO_2 expands and causes the dough to rise. Oven heat causes the gas to escape from the dough, leaving behind a yeasty-tasting, porous (and so light-textured) product. Baking powder also produces carbon dioxide, but it cannot impart the signature yeasty taste to baked goods.

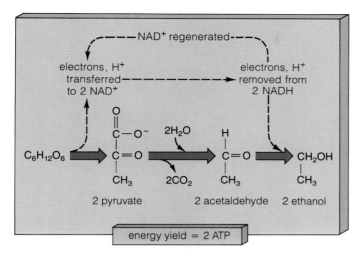

Figure 8.10 Alcoholic fermentation. The net ATP yield is from glycolysis, the first stage of the route.

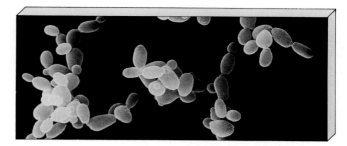

Figure 8.11 Budding cells of the yeast that makes bread dough rise.

Large-scale fermentation by yeasts also is the mainstay of the manufacture of beer and wine. Fruits are a natural home to wild yeasts, and the alcoholic results of their metabolic activities have been recognized since ancient times. Today, vintners still rely on the wild yeasts that are present on grapes (Figure 8.9), but they also add cultivated strains of *S. ellipsoideus* to the juice in fermentation vats. Unlike wild yeasts, which can tolerate an alcohol concentration of only 4 percent, cultivated strains can keep breaking down sugar molecules even when the alcohol concentration reaches 12 to 14 percent.

The fermentation products of wild yeasts still pack a punch. Robins get drunk in droves on naturally fermenting berries of *Pyracantha* shrubs. Where those shrubs are planted densely alongside freeways, the robins typically collide with cars and trees in a manner analogous to drunk drivers. Similarly, wild turkeys get a buzz on when they gobble up naturally fermenting apples in untended orchards.

Lactate Fermentation

In **lactate fermentation**, pyruvate from glycolysis accepts protons and electrons from NADH. The result is a three-carbon product, lactate. Figure 8.12 shows the reaction sequence. Sometimes lactate is called "lactic acid." However, it is more accurate to refer to the ionized form of the compound (lactate), which is far more common in cells.

One group of bacteria produces lactate exclusively as the fermentation product; milk or cream turned sour is a sign of their activity. Muscle cells may or may not use the lactate fermentation pathway, depending on the

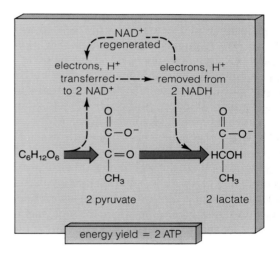

Figure 8.12 Lactate fermentation. The net ATP yield is from glycolysis, the first stage of the route.

demands being placed on the muscle to which they belong (page 633). Aerobic respiration is absolutely necessary for moderate, prolonged muscle action. But when demands are intense but brief—say, during a short race—lactate fermentation produces ATP quickly. This fermentation route cannot be used for long. It has such a low energy yield that the muscle cells would quickly exhaust their glycogen reserves. (Glycogen is the main storage form for glucose in animals.) Once such reserves are depleted, muscles fatigue quickly and lose their ability to contract.

Keep in mind that glucose is not completely degraded by either fermentation pathway, so considerable energy still remains in the products. No more ATP is produced, beyond the two molecules from glycolysis. *The final steps serve only to regenerate NAD⁺.* The low energy yield is quite enough for some single-celled anaerobic organisms. It can even help carry some otherwise "aerobic" cells through times of stress. But it is not enough to sustain the activities of large, multicelled organisms (this being one of the reasons you never will come across an anaerobic elephant).

Fermentation pathways have a net yield of two ATP (from glycolysis). The NAD⁺ necessary for glycolysis is regenerated during the reactions.

Anaerobic Electron Transport

Aerobic respiration and fermentation proceed in our own cells, and fermentation is of great commercial interest to us. But a variety of energy-releasing pathways exists in the natural world, especially within the bacterial kingdom. These pathways are vitally important to the cycling of nitrogen, sulfur, and other elements that are necessary components of biological molecules.

One such pathway is called **anaerobic electron transport**. Here, electrons stripped from various substrates are donated to transport systems bound in the bacterial plasma membrane. For example, sulfate-reducing bacteria produce ATP by stripping electrons from a variety of organic compounds and sending them through membrane transport systems. In some cases, the inorganic compound sulfate (SO_4^{--}) serves as the final electron acceptor and so is converted into sulfide (H_2S). Sulfate-reducing bacteria are often abundant in waterlogged soils and aquatic habitats that are rich in decomposed organic material. They also thrive on dissolved substances that move through the animal gut, an anaerobic setting.

Other kinds of bacteria in soils can produce ATP by stripping electrons from nitrate (NO_3^-), leaving nitrite (NO_2^-) as the end product. In Chapter 46, we will consider the role of these bacteria in the global cycling of nitrogen.

Anaerobic electron transport also is a significant energy-releasing pathway used by sulfate-reducing bacteria, nitrifying bacteria, and other species that live deep in the ocean, around hydrothermal vents. As you will see in Chapter 47, they form the food production base for unique communities.

ALTERNATIVE ENERGY SOURCES IN THE HUMAN BODY

Our cells require a steady supply of carbohydrates, lipids, and proteins for energy and raw materials. When we eat more carbohydrates than our cells are calling for, the excess can be stored as glycogen, most notably in liver and muscle cells. Excess fats are tucked away as glistening droplets in the cells of adipose tissue. Between meals, when demands for raw materials and energy exceed dietary intake, the body can draw upon a variety of stored organic compounds.

Of the foods we eat, carbohydrates are the main source of energy This is true of mammals generally. The carbohydrates are broken down into glucose and other monosaccharides, which are absorbed across the gut lining and then circulated to cells by the bloodstream. Our cells use free glucose for as long as it is available. Our brain cells essentially use nothing else, for they have virtually no stores of lipids for energy metabolism and only a limited amount of glycogen. During normal conditions, a mammal taps into its glycogen stores with great precision when the blood glucose level decreases slightly. A starving mammal will break into its fat reserves, thus "sparing" the glucose that is still available for the all-important brain. It will then start breaking down the body's proteins. However, after the blood glucose level falls precipitously, brain function may be so disrupted that death is likely to follow.

Figure 8.13 shows the points of entry into the aerobic pathway for complex carbohydrates, fats, and proteins. Chapter 37 describes the digestion of these compounds into their simpler components. Here we can simply out-

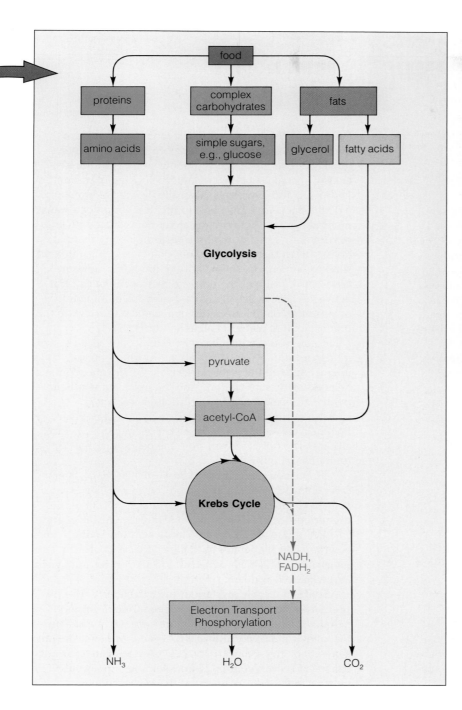

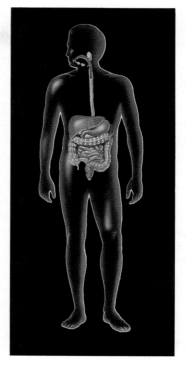

human digestive system, which breaks down food molecules into smaller molecules that can be taken up by individual cells

Figure 8.13 Points of entry into the aerobic pathway for complex carbohydrates, fats, and proteins after they have been reduced to their simpler components in the human digestive system.

line how fats and proteins can be used as alternative energy sources.

A fat molecule, recall, consists of a glycerol head and one, two, or three fatty acid tails. Once a fat molecule has been digested into its component parts, the glycerol can be converted to PGAL and inserted into the glycolytic reactions. The long carbon backbone of each fatty acid can be split many times into two-carbon fragments, these being easily converted into acetyl-CoA and picked up for the Krebs cycle. Whereas each glucose molecule has six carbon atoms, a fatty acid has many more—and its degradation produces much more ATP.

Following protein digestion, the amino acid subunits can have their amino group (NH_3) removed. The remnants can enter the Krebs cycle, where the hydrogens can be removed from the remaining carbon atoms and transferred to coenzymes. The amino groups are converted into ammonia, a waste product.

Perspective on Life

In this unit, you have read about pathways by which cells trap, store, and then release energy to drive their activities. Over evolutionary time, the main pathways—photosynthesis and aerobic respiration—have become interconnected on a grand scale.

When life began, little (if any) free oxygen was present in the earth's atmosphere. Most likely, early single-celled organisms produced ATP by pathways similar to glycolysis. And given the anaerobic conditions, fermentation routes must have predominated.

Photosynthetic organisms emerged more than 3 billion years ago, and they turned out to be a profound force in evolution. Oxygen, a by-product of their activities, began to accumulate in the atmosphere. Some photosynthesizers were opportunistic about the increasing oxygen levels. Perhaps through mutations in their metabolic machinery, they gained the capacity to use oxygen as an electron acceptor for degradative reactions—and in time some cells abandoned photosynthesis entirely. Among those cells were the forerunners of animals and other organisms able to survive with aerobic machinery alone.

With aerobic respiration, life became self-sustaining, for its final products—carbon dioxide and water—are precisely the materials used to build organic compounds in photosynthesis! Thus the flow of carbon, hydrogen, and oxygen through the metabolic pathways of living organisms came full circle:

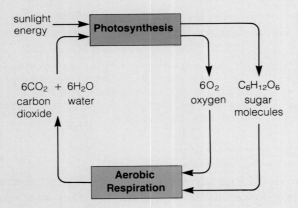

Perhaps one of the most difficult connections you are asked to perceive is the link between yourself—a living, intelligent being—and such remote-sounding things as energy, metabolic pathways, and the cycling of carbon, hydrogen, and oxygen. Is this really the stuff of humanity?

Think back, for a moment, to the description of a water molecule. A pair of hydrogen atoms competing with an oxygen atom for a share of the electron joining them doesn't exactly seem close to our daily lives. But from that simple competition, the polarity of the water molecule arises. As a result of the polarity, hydrogen bonds form between water molecules. And that is a beginning for the organization of lifeless matter that leads, ultimately, to the organization of matter in all living things.

For now you can imagine other kinds of molecules interspersed in water. Many are nonpolar and resist interaction with the water molecules. Others are polar and respond by dissolving in it. And the lipids among them (with water-soluble *and* water-insoluble regions) spontaneously assemble into a two-layered film. Such lipid bilayers are the basis for all cell membranes, hence all cells. The cell has been, from the beginning, the fundamental *living* unit.

With the boundary afforded by a cell membrane, chemical reactions can be contained and controlled. The essence of life *is* chemical control. This "control" is not some mysterious force. It is a chemical responsiveness to energy changes and to the kinds of molecules present in the environment. It operates by "telling" a class of protein molecules—enzymes—when and what to build, and when and what to tear down.

And it is not some mysterious force that creates the proteins themselves. DNA, the slender double strand of heredity, has the chemical structure—*the chemical message*—that allows molecule faithfully to reproduce molecule, one generation after the next. Those DNA strands tell many billions of cells in your body how countless molecules must be built and torn apart for their stored energy.

So yes, carbon, hydrogen, oxygen, and other organic molecules represent the stuff of you, and us, and all of life. But it takes more than molecules to complete the picture. You are alive because of the way molecules are organized and maintained by a constant flow of energy. It takes outside energy from sources such as the sun to drive their formation. Once molecules are assembled into cells, it takes outside energy derived from food, water, and air to sustain their organization. Plants, animals, fungi, protistans, and bacteria are part of a web of energy use and materials cycling that ties together all levels of biological organization. Should energy fail to reach any part of any level, life there will dwindle and cease.

For energy flows through time in only one direction—from forms rich in usable energy to forms having less usable stores of it. Only as long as sunlight flows into the web of life—and only as long as there are molecules to recombine, rearrange, and recycle—does life have the potential to continue in all its rich diversity.

In short, life is no more *and no less* than a marvelously complex system of prolonging order. Sustained by energy transfusions, it continues because of a capacity for self-reproduction—the handing down of hereditary instructions. With those instructions, energy and materials are organized, generation after generation. Even with the death of the individual, life is prolonged. With death, molecules are released and can be recycled once more, providing raw materials for new generations. In this flow of energy and cycling of material through time, each birth is affirmation of our ongoing capacity for organization, each death a renewal.

SUMMARY

1. ATP, the main energy carrier in cells, can be produced by photosynthesis. It also can be produced by aerobic respiration or fermentation, these being degradative pathways by which chemical energy is released from glucose (or some other organic compound).

2. Glycolysis, the partial breakdown of a glucose molecule, is the first stage of all the main degradative pathways. It takes place in the cytoplasm and requires an initial energy input from ATP. During glycolysis, two ATP (net), two NADH, and two pyruvate molecules are produced for each glucose molecule.

3. In aerobic respiration, oxygen is the final acceptor of electrons stripped from glucose. The pathway proceeds from glycolysis, through the Krebs cycle, and through electron transport phosphorylation. Its net energy yield is commonly thirty-six ATP.

4. In the second stage of aerobic respiration, pyruvate from glycolysis is converted to a form that can enter the Krebs cycle (a cyclic metabolic pathway). The conversion reactions and the cycle itself produce eight NADH, two $FADH_2$, and two ATP. In this stage, the glucose molecule is degraded completely to carbon dioxide and water.

5. The coenzymes NADH and $FADH_2$ (formed during glycolysis and the Krebs cycle) deliver electrons to a transport system embedded in the inner membrane of mitochondria. The third stage of the aerobic pathway (electron transport phosphorylation) depends on electron transfers through the system *and* on phosphorylation reactions that take place at channel proteins spanning the membrane.

6. Operation of the transport system sets up H^+ concentration and electric gradients across the membrane. H^+ moves down the gradients, through channel proteins. Energy associated with the flow of hydrogen ions drives the coupling of ADP and inorganic phosphate to form ATP. Oxygen accepts the "spent" electrons and combines with H^+ to form water.

7. Many microbes are not metabolically equipped to use oxygen as a final electron acceptor. They rely instead on alcoholic fermentation, lactate fermentation, or anaerobic electron transport. Also, some cells that normally use the aerobic pathway (such as skeletal muscle cells) rely on lactate fermentation when the demand for muscle action is brief but intense.

8. Compared with aerobic respiration, the anaerobic pathways have a small net yield (two ATP, from glycolysis), because glucose is not completely degraded. Following glycolysis, the remaining reactions serve to regenerate NAD^+.

Review Questions

1. ATP can be produced when carbohydrates are degraded. Define three types of energy-releasing pathways by which this occurs. Which yields the most ATP? *120*

2. Which energy-releasing pathways occur in the cytoplasm? In the mitochondrion of eukaryotes? *121, 122, 124*

3. Is the following statement true? Your muscle cells cannot function at all unless they are supplied with oxygen. *129*

4. Glycolysis is the first stage of all the main pathways by which glucose is degraded. Can you define those pathways in terms of the final electron acceptor for their reactions? If you include the two ATP molecules formed during glycolysis, what is the *net* energy yield from one glucose molecule for each pathway? *126–130*

5. In anaerobic routes of glucose breakdown, further conversions of pyruvate do not yield any more usable energy. What, then, is the advantage of the conversions? *130*

6. Describe the functions of the Krebs cycle. Describe the functions of electron transport phosphorylation. *120, 125, 126*

Self-Quiz *(Answers in Appendix IV)*

1. Plants as well as bacteria, protistans, fungi, and animals can produce ATP by degrading _____ .

2. Glucose can be degraded by way of two anaerobic pathways, called _____ and _____ , as well as by aerobic respiration.

3. In the first stage of aerobic respiration, glucose is partially broken down to _____ , which in the second stage is broken down completely to _____ and _____ .

4. ATP is best described as _____ .
   a. a high-energy phosphate compound
   b. a primary source of chemical energy
   c. being produced by plants, animals, bacteria, protistans, and fungi
   d. all of the above

5. Which of the following is *not* a product of glycolysis?
   a. two NADH
   b. two pyruvate
   c. two H_2O
   d. two ATP

6. Glycolysis occurs in which part of the cell?
   a. nucleus
   b. mitochondrion
   c. plasma membrane
   d. cytoplasm

7. The final acceptor of the electrons stripped from glucose during aerobic respiration is _____ .
   a. water
   b. hydrogen
   c. oxygen
   d. NADH

8. Electron transport systems for the aerobic reactions are located in the _____ .
   a. cytoplasm
   b. inner mitochondrial membrane
   c. outer mitochondrial membrane
   d. stroma

9. The flow of _____ through channel proteins in the inner mitochondrial membrane provides the energy to couple ADP and inorganic phosphate to form ATP.
   a. electrons
   b. hydrogen ions
   c. NADH
   d. $FADH_2$

10. Match each type of metabolic reaction with its function:
   _____ glycolysis
   _____ fermentation
   _____ Krebs cycle
   _____ electron transport phosphorylation

   a. produces ATP, NADH, and CO_2
   b. degrades glucose into two pyruvate
   c. regenerates NAD^+
   d. flow of H^+ through channel proteins that drives ATP formation

Selected Key Terms

aerobic respiration *120*	$FADH_2$ *125*
alcoholic fermentation *128*	glycolysis *122*
anaerobic electron transport *130*	Krebs cycle *124*
anaerobic pathway *120*	lactate fermentation *129*
ATP *120*	mitochondrion *124*
chemiosmotic theory *126*	NADH *125*
electron transport phosphorylation *126*	substrate-level phosphorylation *122*

Readings

Becker, W. 1986. *The World of the Cell.* Menlo Park, California: Benjamin/Cummings. Chapters 7 and 8 are a good place to start for further readings on anaerobic and aerobic metabolism.

Brock, T., B. Smith, and M. Madigan. 1988. *Biology of Microorganisms.* Fifth edition. Englewood Cliffs, New Jersey: Prentice-Hall. Clear descriptions of the energy-releasing pathways of microbes.

Lehninger, A. 1982. *Principles of Biochemistry.* New York: Worth. Clear, accessible introduction to metabolic pathways.

FACING PAGE: *Human sperm, one of which will penetrate this mature egg and so set the stage for the development of a new individual in the image of its parents.*

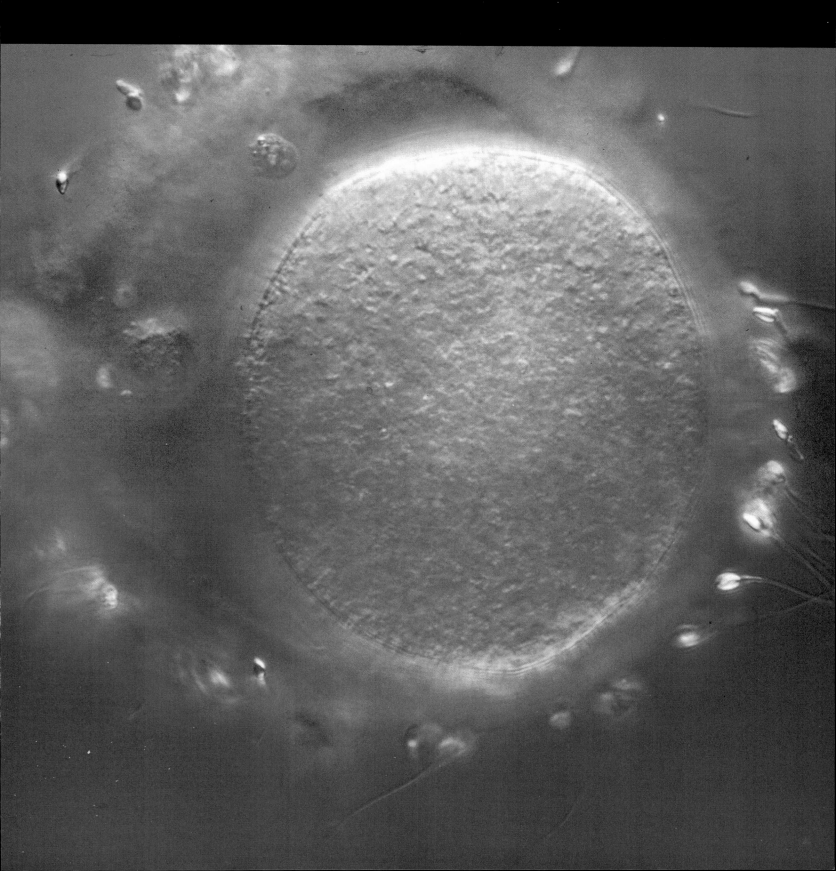

Silver in the Stream of Time

Five o'clock, and the first rays of the sun dance over the wild Alagnak River of the Alaskan tundra. This September morning, as every morning for several weeks, life is both ending and beginning in the clear, frigid waters. By the thousands, mature silver salmon have returned from the open ocean to spawn in their shallow native home. A female salmon rests briefly in a quiet eddy, then continues upstream a bit more (Figure 9.1). She is tinged with red, the color of spawners, and she is dying.

On this morning the female salmon pauses, then quickly hollows out a shallow "nest" in the gravel riverbed. Now scores of translucent pink eggs emerge from her body. Within moments a male salmon appears and sheds a cloud of sperm near the eggs. Trout and other voracious predators of the Alagnak will consume most of the eggs. But a few will survive, and following successful fertilization, they will give rise to a new generation.

The female lingers on for a few hours, but depleted of eggs and with vital organs failing, she soon dies. On

the riverbank, a bald eagle loses no time in consuming her carcass. Yet her remains speak of a remarkable journey. That female had herself started life as a pea-sized egg that had been fertilized in the Alagnak's gravel bed. Within three years, she had become a streamlined vertebrate, fashioned from billions of cells. And she had reached reproductive maturity. Early in her development, some of her cells had been set aside for reproduction, and in time they had given rise to eggs. On this morning, together with the sperm that eventually fertilized them, those eggs became part of an ongoing story of birth, cell divisions and growth, death, and rebirth for the silver salmon.

For you, as for the silver salmon and all other organisms, reproduction depends on the capacity of cells to divide. Starting with the fertilized egg in your mother's body, a single cell divided in two, then the two into four, and so on until billions of cells were growing, developing in specialized ways, and dividing at different times to produce your genetically prescribed body parts. Today your body is composed of approximately 65

trillion cells. Cell divisions are still proceeding in many parts of it. Every five days, for example, cell divisions manage to replace the lining of your small intestine. Even now, in males who are reading this page, divisions are probably proceeding that will give rise to mature sperm cells—part of the reproductive bridge to the next generation.

Understanding the nature of cell division—and, ultimately, how new individuals are put together in the image of their parents—begins with answers to three questions. *First*, what structures and substances are necessary for inheritance? *Second*, how are they divided and distributed into daughter cells? *Third*, what are the division mechanisms themselves? We will require more than one chapter to consider the answers (and best guesses) about cell reproduction and other mechanisms of inheritance. However, the points made in the first half of this chapter can help you keep the overall picture in focus.

Figure 9.1   The last of one generation and the first of the next in the Alagnak River of Alaska.

KEY CONCEPTS

1. Each cell of a new generation will not grow or function properly unless it receives the necessary hereditary information, in the form of parental DNA, and a portion of the cytoplasm from the parental cell. In eukaryotes, DNA is parceled out to daughter cells by mitosis or meiosis, both of which are nuclear division mechanisms.

2. A chromosome is a DNA molecule with certain proteins attached to it. The cells of each species have a characteristic number of chromosomes. "Diploid" cells contain two of each type of chromosome characteristic of the species.

3. Mitosis maintains the number of chromosomes from one cell generation to the next. Thus each daughter cell formed by the division of the nucleus and cytoplasm of a diploid parental cell will be diploid also.

4. Mitosis proceeds through four continuous stages: prophase, metaphase, anaphase, and telophase. Actual cytoplasmic division (cytokinesis) occurs toward the end of the nuclear division or at some point afterward.

5. Mitosis is the basis of asexual reproduction of single-celled eukaryotes as well as the growth of multicelled eukaryotes. Meiosis is the basis of gamete formation, hence of sexual reproduction.

DIVIDING CELLS: THE BRIDGE BETWEEN GENERATIONS

Overview of Division Mechanisms

In biology, **reproduction** means producing a new generation of cells or multicelled individuals. Reproduction is part of a *life cycle,* a recurring frame of events in which individuals grow, develop, maintain themselves, and reproduce according to instructions encoded in DNA, which they inherit from their parents. Reproduction begins with the division of single cells. And the ground rule for cell division is this: *Each cell of a new generation must receive hereditary information (encoded in parental DNA) and enough cytoplasmic machinery to start up its own operation.*

Recall that DNA contains instructions for making proteins. Some proteins serve as structural materials; many serve as enzymes with roles in the synthesis of carbohydrates, lipids, and other building blocks of the cell. Thus, unless new cells receive the necessary instructions for making proteins, they will not grow or function properly.

Also, the cytoplasm of the parental cell already has operating machinery—enzymes, organelles, and so on. When a daughter cell inherits what looks merely like a blob of cytoplasm, it really is getting "start-up" machinery for its operation, until it has time to use its inherited DNA for growing and developing on its own.

In multicelled plants and animals, cell division typically begins with mitosis or meiosis and ends with cytokinesis. **Mitosis** and **meiosis** are *nuclear* division mechanisms—the means by which DNA instructions are sorted out and distributed into new nuclei for the forthcoming daughter cells. The actual splitting of a parental cell into two daughter cells occurs by way of **cytokinesis**, or *cytoplasmic* division.

In multicelled organisms, mitosis is the basis for growth through repeated divisions of the body's cells, which are called *somatic* cells. Mitosis also is the basis for asexual reproduction in many plants, animals, and other organisms.

In contrast, meiosis occurs only in germ cells, a cell lineage set aside for sexual reproduction. By definition, *sexual reproduction* is a process that begins with meiosis, proceeds through the formation of gametes (sperm and eggs), and ends at fertilization. At fertilization, a sperm nucleus and egg nucleus fuse together in the zygote, the first cell of the new individual.

This chapter focuses on mitosis, and the chapter to follow, on meiosis. As you will see, the two division mechanisms have much in common but they differ in their end result. Both are limited to eukaryotes. The prokaryotes (bacteria) use a different division mechanism, as listed in Table 9.1 and described on page 355.

Some Key Points About Chromosome Structure

Before you track the distribution of DNA into daughter cells, reflect for a moment on its structural organization. Many proteins are attached to eukaryotic DNA, and they generally are equal in mass to the DNA itself. Together, the DNA and proteins form a structure called the **chromosome**.

Between divisions, a chromosome is stretched out in threadlike form. It is still threadlike when it is duplicated prior to cell division, and the two threads remain attached for a while as **sister chromatids** of the chromosome. Each "thread," of course, is a DNA double helix with its associated proteins, which we show here as a simplified version of a current model:

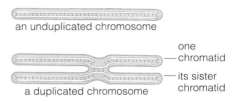

an unduplicated chromosome

one chromatid
its sister chromatid
a duplicated chromosome

Notice the small region of the chromosome where the DNA appears to be constricted. This is the **centromere**, a region having attachment sites for microtubules that will help move the chromosome during nuclear division:

centromere (constricted region)

Keep in mind that the location of the centromere varies from one type of chromosome to the next. Also keep in mind that this model is highly simplified. The DNA double helix in each chromatid is actually two molecular

Table 9.1 Cell Division Mechanisms	
Mechanisms	Used By
Mitosis, cytokinesis	*Single-celled eukaryotes (for asexual reproduction)*
	Multicelled eukaryotes (for bodily growth; also for asexual reproduction in some species)
Meiosis, cytokinesis	*Eukaryotes (basis of gamete formation and sexual reproduction)*
Prokaryotic fission	*Bacterial cells*

Figure 9.2 (**a**) Photograph of the 46 chromosomes from a diploid cell of a human male. All are in the duplicated state. (**b**) By cutting apart and arranging the chromosomes according to length and centromere location, we see that there are two sets of 23 chromosomes, with all the chromosomes in one set having a partner, or homologue, in the other. The partners don't pair at all during mitosis, but they pair up with each other during meiosis.

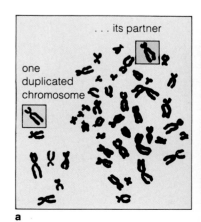

... its partner

one duplicated chromosome

a

strands twisted together repeatedly like a spiral staircase (that's what the "double helix" means), and it is much longer than can be shown here.

Mitosis, Meiosis, and the Chromosome Number

All individuals of the same species have the same number of chromosomes in their somatic cells. For example, there are 46 chromosomes in your somatic cells, 48 in a gorilla's, and 14 in a garden pea's (Table 9.2). When a gorilla or pea plant grows, mitosis assures that all the new cells added to its body will end up with the parental number of chromosomes. *With mitosis, the chromosome number is maintained, division after division:*

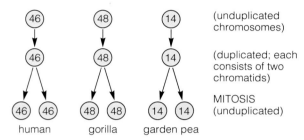

(unduplicated chromosomes)

(duplicated; each consists of two chromatids)

MITOSIS (unduplicated)

To get an initial sense of the difference between mitosis and meiosis, take a look at Figure 9.2, which shows all the chromosomes in a human somatic cell. Notice how the chromosomes can be lined up as pairs in two rows. (One row is coded pink and the other, green.) Except for sex chromosomes, designated X and Y, the members of each pair have the same length and centromere location, and their hereditary instructions deal with the same traits. The two members of each pair are **homologous chromosomes**.

Homologues don't interact at all during mitosis, but they pair with each other during meiosis. Although the X and Y chromosomes differ in most respects, they pair during meiosis so we still call them homologues.

During meiosis, the chromosome number is reduced by half for forthcoming gametes. And not just any half—each gamete ends up with *one of each pair* of homologous chromosomes. (It doesn't matter which of the two it gets.) Reducing the chromosome number requires two nuclear divisions:

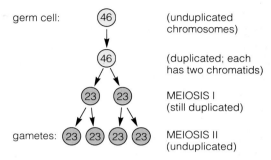

germ cell:

(unduplicated chromosomes)

(duplicated; each has two chromatids)

MEIOSIS I (still duplicated)

gametes:

MEIOSIS II (unduplicated)

Gametes end up with a **haploid** number of chromosomes, which we can take to mean as "half" of the parental number. When a sperm nucleus and an egg nucleus fuse at fertilization, the **diploid** number is restored. "Diploid" means having two chromosomes of each type in the somatic cells of sexually reproducing species.

Mitosis is a type of nuclear division that *maintains* the parental number of chromosomes for forthcoming cells. It is the basis for bodily growth and, in some cases, asexual reproduction of eukaryotes.

Meiosis is a type of nuclear division that *reduces* the parental chromosome number by half—to the haploid number. It occurs only in sexual reproduction.

Table 9.2	Number of Chromosomes in the Somatic Cells of Some Eukaryotes*	
Mosquito, *Culex pipiens*		6
Fruit fly, *Drosophila melanogaster*		8
Garden pea, *Pisum sativum*		14
Corn, *Zea mays*		20
Lily, *Lilium*		24
Yellow pine, *Pinus ponderosa*		24
Frog, *Rana pipiens*		26
Earthworm, *Lumbricus terrestris*		36
Rhesus monkey, *Macaca mulatta*		42
Human, *Homo sapiens*		46
Chimpanzee, *Pan troglodytes*		48
Gorilla, *Gorilla gorilla*		48
Potato, *Solanum tuberosum*		48
Amoeba, *Amoeba*		50
Horse, *Equus caballus*		64
Horsetail, *Equisetum*		216
Adder's tongue fern, *Ophioglossum reticulatum*		1,000+

*These examples are a sampling only. Chromosome number for most species falls between 10 and 50.

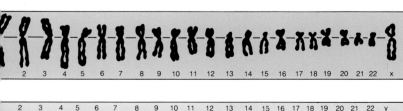

b

MITOSIS AND THE CELL CYCLE

Mitosis is a very small part of the **cell cycle**, which includes events that extend from the time a cell forms until its own division is completed. Normally, a cell destined to enter mitosis spends about 95 percent of the cell cycle in **interphase**, when it increases its mass, approximately doubles the number of its cytoplasmic components, and finally duplicates its DNA.

As Figure 9.3 and the following list indicate, "interphase" actually consists of three phases of activities:

Mitosis	M	*nuclear division, commonly followed by cytokinesis*
Interphase	G_1	*a "gap" (interval) before DNA replication*
	S	*"synthesis" (replication) of DNA and associated proteins*
	G_2	*a second "gap" after DNA replication, before mitosis*

The "gaps" were so named before biologists knew what was going on—specifically, that new cell components are synthesized during G_1 and assembled for distribution to daughter nuclei during G_2.

To get an idea of the challenges facing a cell that is destined to divide, imagine that one of your liver cells is blown up as large as a basketball. If you stretched out its diploid number of chromosomes end to end, the lineup would be about 20 miles long. Before that cell divides, another 20 miles' worth of chromosomes must be copied.

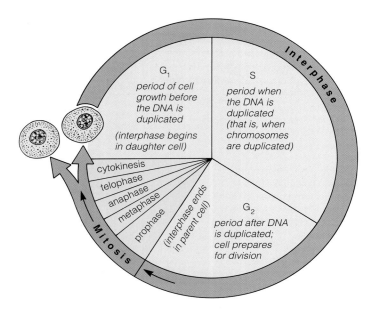

Figure 9.3 Eukaryotic cell cycle. This drawing has been generalized; the length of different stages varies greatly from one type of cell to the next.

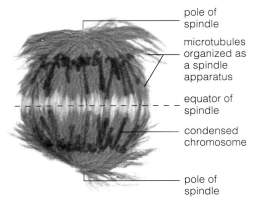

Figure 9.4 Mitosis in a plant cell (*Haemanthus*). The chromosomes are stained blue, and the microtubules that move them about are stained red.

pole of spindle

microtubules organized as a spindle apparatus

equator of spindle

condensed chromosome

pole of spindle

Before its nucleus finally divides, the 40 miles must be sorted into two subsets—each with one of each type of chromosome. Before its cytoplasm divides, the subsets must be packaged so that each daughter cell ends up with 20 miles of the proper chromosomes. All of this goes on in a space the size of a basketball. And all in a few hours.

The duration of the cell cycle varies widely, but it is fairly consistent for all cells of a given type. For example, all of your brain cells are arrested at interphase and never will divide again. Cells of a newly forming sea urchin may double in number every two hours. To be sure, adverse environmental conditions may disrupt the cycle, as when cells atypically become arrested in the G_1 phase. (For example, this happens among amoebas and other protistans when they are deprived of a vital nutrient.) Even so, if a cell progresses past a certain point in G_1, the cycle normally will be completed regardless of outside conditions. Because the cycle is so predictable for each cell type in each kind of organism, you might suspect that cells have built-in controls over the cycle's duration. As you will read on page 232, this is indeed the case.

Good health depends on the successful completion of cell cycles. Every error in DNA duplication or misshipment of chromosomes may lead to problems. Even the timing and regulation of cell division must be precisely controlled. If the cell cycle stops in growing tissues, the tissues die. If the cycle proceeds uncontrollably in mature tissues, cancer follows. A landmark case of unchecked cell divisions is described in the *Commentary* at this chapter's end.

STAGES OF MITOSIS

When a cell makes the transition from interphase to mitosis, it stops constructing new cell parts. Profound changes now proceed one after the other, through four stages. The sequential stages of mitosis are called *prophase, metaphase, anaphase,* and *telophase.*

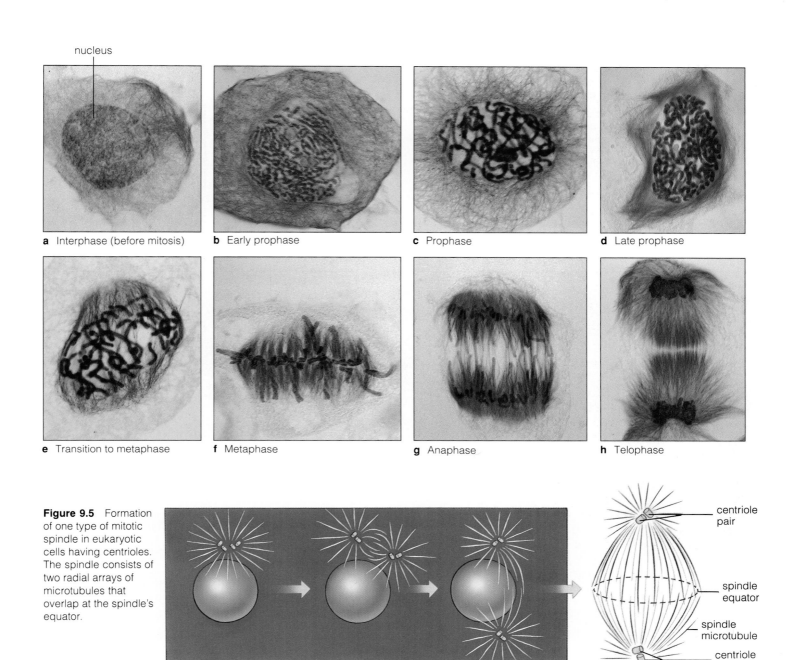

a Interphase (before mitosis) **b** Early prophase **c** Prophase **d** Late prophase

e Transition to metaphase **f** Metaphase **g** Anaphase **h** Telophase

Figure 9.5 Formation of one type of mitotic spindle in eukaryotic cells having centrioles. The spindle consists of two radial arrays of microtubules that overlap at the spindle's equator.

centriole pair

spindle equator

spindle microtubule

centriole pair

The Microtubular Spindle

Before considering the details of mitosis, take a look at Figure 9.4, which will give you an idea of the extent of chromosome movements through the different stages. The chromosomes do not move about on their own. They are moved by a **spindle apparatus**, a bipolar structure composed of organized arrays of microtubules. During nuclear division, the chromosomes will move toward the two spindle poles.

Remember that microtubules are components of the cell's internal framework, the cytoskeleton. When a nucleus is about to divide, nearly all of the cell's existing microtubules disassemble into their protein subunits.

Then the subunits reassemble into new microtubules, just outside the nucleus. These are the microtubules that become arranged as a spindle.

In many cells, a microtubule organizing center dictates *where* the new microtubules will arise. As mentioned on page 69, the center often includes a pair of centrioles. The center (including its centriole pair) is duplicated during interphase. Then, while the spindle is forming, the two centers separate and become positioned at opposite spindle poles (Figure 9.5). Their positioning and orientation seem to influence the organization of the cytoskeleton that will form in each daughter cell. And proper cell functioning depends on that organization.

Figure 9.6 Mitosis: the nuclear division mechanism that maintains the parental chromosome number in daughter cells. Shown here, a diploid animal cell (with pairs of homologous chromosomes, derived from two parents).

For the sake of clarity, only two pairs of homologous chromosomes are shown in the diagram and the spindle apparatus is simplified. With rare exceptions, the picture is more involved than this, as indicated by the micrographs of mitosis in a whitefish cell.

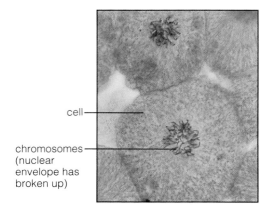

cell

chromosomes (nuclear envelope has broken up)

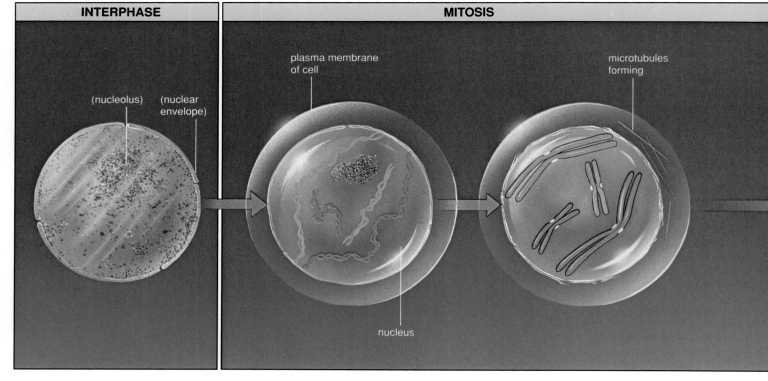

INTERPHASE	MITOSIS

(nucleolus) (nuclear envelope)

plasma membrane of cell

microtubules forming

nucleus

Nucleus at Interphase
The DNA is duplicated, then the cell prepares for division.

Early Prophase
The DNA and associated proteins start condensing into the threadlike chromosome form. (Chromosomes are already duplicated.) Two chromosomes derived from the male parent are in green; their homologues from the female are pink.

Late Prophase
Chromosomes continue to condense. Microtubules start to assemble outside the nucleus; they will form the spindle. Centrioles (if present) are moved by the microtubules toward opposite poles. The nuclear envelope starts breaking up during the transition to metaphase.

Prophase: Mitosis Begins

Each molecule of eukaryotic DNA has many proteins attached to it like beads on a string. **Prophase**, the first stage of mitosis, is evident when the "beaded strings" begin to fold and twist into condensed chromosome structures. At this stage the chromosomes are visible in the light microscope as threadlike forms. ("Mitosis" comes from the Greek *mitos*, meaning thread.)

Each chromosome was duplicated earlier, during interphase. In other words, it already consists of two sister chromatids joined at the centromere. By late prophase, the chromatids of each chromosome have condensed into thicker, rodlike forms (Figure 9.6).

Microtubules of the cytoskeleton disassemble toward the end of prophase. Then new microtubules begin to reassemble, and these will eventually form the spindle. At this time, however, the nuclear envelope prevents

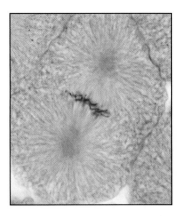
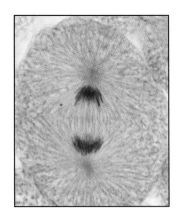
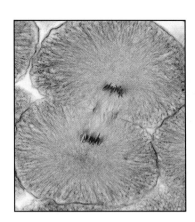

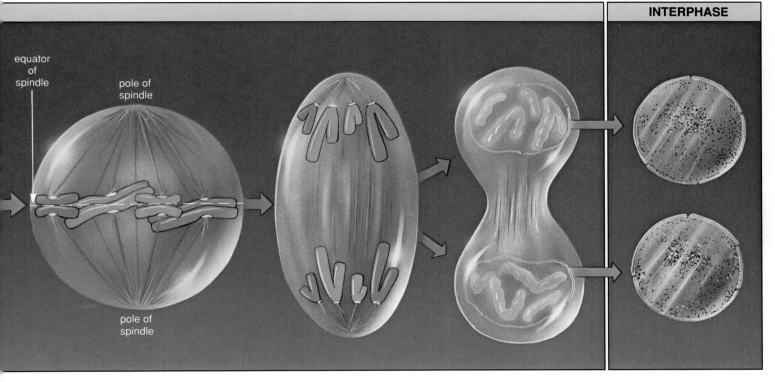

equator
of
spindle

pole of
spindle

pole of
spindle

Metaphase

Sister chromatids of each chromosome
are attached to the spindle. All
chromosomes are now lined up at
the spindle equator.

Anaphase

Sister chromatids of
each chromosome
will now be separated
from each other and
moved to opposite poles.

Telophase

Chromosomes decondense.
New nuclear membranes
start forming. Most often,
cytokinesis occurs before
the end of telophase.

Interphase

Two daughter nuclei
are formed, each with
a diploid number of
chromosomes (the
same as the parental
nucleus).

the microtubules from interacting with the chromosomes
inside the nucleus.

Metaphase

The nuclear envelope abruptly breaks up during the
transition from prophase to metaphase. Now the chromosomes are free to interact with microtubules that
had been assembled earlier, outside the nucleus. The

chromosome-microtubule interactions culminate in the
formation of a bipolar spindle. These events occur early
in **metaphase**, the second stage of mitosis. (Many
researchers now refer to early metaphase as "prometaphase.") While these interactions are proceeding,
fragments of the nuclear envelope become small membranous vesicles.

To gain insight into the interaction between the
chromosomes and microtubules, take a look at the meta-

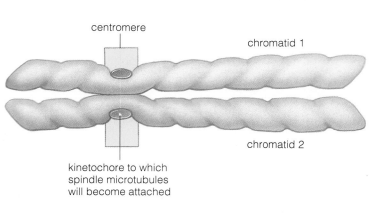

centromere

chromatid 1

chromatid 2

kinetochore to which
spindle microtubules
will become attached

a

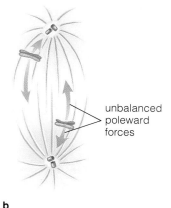

unbalanced
poleward
forces

b

balanced
poleward
forces

c

Figure 9.7 (**a**) Location of the kinetochore on each of the two sister chromatids of a duplicated chromosome. (**b**) Microtubules extending from the poles of the forming spindle become attached to the kinetochores during prometaphase. The sister chromatids become oriented toward and attached to opposite poles. (**c**) Each chromosome is aligned at the spindle equator when pulling forces acting at its two kinetochores are balanced and when pushing forces directed away from the two spindle poles are balanced.

phase chromosomes shown in Figures 9.6 and 9.7. Notice the centromere, the constricted region where the two sister chromatids of each chromosome are held together. At this region, each chromatid has a **kinetochore**, a specialized grouping of proteins and DNA to which several spindle microtubules become attached.

Early in metaphase, the chromosomes seem to go into a frenzy. This happens when kinetochores randomly start harnessing certain microtubules that extend toward them from the spindle poles. With the first random contacts, the chromosomes become attached to the forming spindle. Each chromosome is yanked back and forth until its sister chromatids become firmly oriented toward and attached to opposite poles. Now, forces work to pull the kinetochores of sister chromatids toward opposite poles. Other forces push the poles away from each other (Figure 9.7). The opposing forces are balanced by late metaphase. At that point, all the chromosomes lie halfway between the poles, at the spindle equator.

In sum, three events dominate metaphase. *First,* the spindle becomes fully formed. *Second,* as the spindle forms, the sister chromatids of each chromosome become oriented toward and attached to opposite spindle poles. *Third,* all chromosomes become aligned halfway between the poles. This alignment is crucial for the chromosomal movements that follow.

Anaphase

During **anaphase**, sister chromatids of each chromosome are separated from each other, and those former partners are moved toward opposite poles. Once they do separate, the partners are no longer referred to as chromatids. Each is now an independent chromosome:

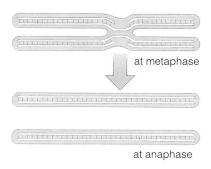

at metaphase

at anaphase

A chemical signal prods the sister chromatids into separating. Two mechanisms apparently account for the subsequent anaphase movements. First, microtubules attached to the kinetochores shorten as the chromosomes approach the spindle poles. Second, the spindle elongates when forces push the two spindle poles far-

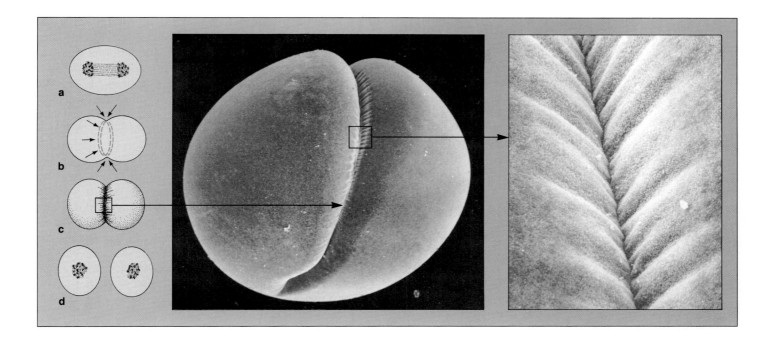

Figure 9.8 Cytokinesis in an animal cell. The scanning electron micrographs show the furrowing of the plasma membrane caused by the contraction of a microfilament ring just beneath it. (**a**) Nuclear division is complete; the spindle is disassembling. (**b**) Microfilament rings at the former spindle equator contract, like a purse string closing. (**c**) Contractions cause furrowing at the cell surface. (**d**) The cytoplasm is pinched in two.

ther apart. Although metaphase may last a long time, anaphase begins abruptly and commonly ends within a few minutes. All the duplicated chromosomes split apart at about the same time and move to opposite poles at the same rate.

Telophase

Telophase begins once the separated chromosomes arrive at opposite spindle poles. Now the kinetochores no longer have microtubules harnessed to them, and the chromosomes decondense into the threadlike form. The vesicles of the old nuclear envelope fuse together to form patches of membrane around the chromosomes. Patch joins with patch, and eventually a new, continuous nuclear envelope separates the hereditary material from the cytoplasm. Once the nucleus is completed, telophase is completed—and so is mitosis.

CYTOKINESIS

Cytoplasmic division, or cytokinesis, usually coincides with the period from late anaphase through telophase. For most animal cells, deposits accumulate and form a layer around microtubules at the cell midsection. A shallow, ringlike depression appears above the layer, at the cell surface (Figure 9.8). At this depression, the **cleavage furrow**, contractile microfilaments pull the plasma membrane inward and cut the cell in two. The contractile force is so strong, it can bend a fine glass needle that is inserted into a dividing cell.

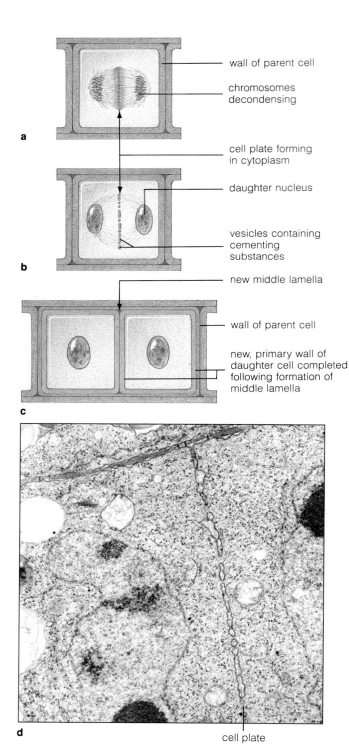

- wall of parent cell
- chromosomes decondensing

a

- cell plate forming in cytoplasm
- daughter nucleus
- vesicles containing cementing substances

b

- new middle lamella
- wall of parent cell
- new, primary wall of daughter cell completed following formation of middle lamella

c

d

cell plate

Figure 9.9 Cytokinesis following mitosis in a plant cell. (**a-c**) Vesicles form at the spindle equator and gradually fuse to form a cell plate. The cell plate grows outward until it reaches the parent cell wall. The vesicles contain substances that will form the middle lamella, which will cement together the primary walls of the daughter cells. (**d**) The membrane of the vesicles is used in forming the plasma membrane on both sides of the cell plate.

A different form of cytokinesis, **cell plate formation**, occurs in most land plants. (Plant cells typically have fairly rigid walls that preclude the formation of cleavage furrows.) Vesicles filled with wall-building material fuse with remnants from the spindle, forming a disklike structure (the "cell plate"). Here, cellulose deposits form a crosswall between the two daughter cells (Figure 9.9).

This concludes our introduction to mitosis. What is the take-home lesson? Simply this: You see the results of mitotic cell divisions whenever you look at *any* eukaryotic organism. Just look at your hands—and think of all the cells that are arranged in precise ways to form your palms, your thumbs, your fingers. Look at a tissue sample through a microscope and you will see individual cells, just as a special microscope might have revealed their forerunners when you were developing early on, inside your mother (Figure 9.10). And be grateful for the absolutely astonishing precision that led to their formation—because the alternatives can be terrible indeed (see the *Commentary*).

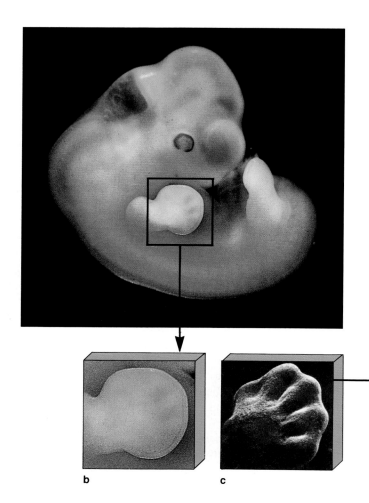

b   **c**

Commentary

Henrietta's Immortal Cells

Each human starts out as a single cell, a zygote, carrying 23 chromosomes from its mother and 23 from its father. At birth a human body has about a trillion cells, each carrying 46 copies of the chromosomes that were present in the zygote. Between conception and birth, then, a phenomenal number of cell divisions must have occurred.

Even in an adult, trillions of cells are still dividing. In the lining of the stomach, for example, cells divide every day. In the liver, cells usually do not divide—but if part of the liver is lost to injury or disease, cells start dividing and continue to do so until the damaged part is replaced. Only then does division stop.

In 1951, George and Margaret Gey of Johns Hopkins University were trying to develop a way to keep human cells dividing outside the body. Researchers could use those cells to gain insight into basic life processes. They also could use them to study diseases, including cancer, without having to experiment directly on humans.

The Geys obtained both normal and diseased human cells from local physicians, who had taken the cells from patients during normal medical procedures. But the Geys just couldn't stop the cell lines from dying out within a few weeks. Mary Kubicek, one of their assistants, was about to give up after dozens of failed attempts. When she received yet another sample of cancer cells on February 9, 1951, she expected to fail again.

Still, she took the sample and prepared it for culture. The sample was code-named HeLa, for the first two letters of the patient's first and last names.

The cells in this new sample began to divide. And divide. And divide again. By the fourth day there were so many cells that they had to be subdivided into more tubes. As months passed, the culture continued to thrive.

Unfortunately for the patient, the tumor cells inside her body were just as vigorous. Six months after she was first diagnosed as having cancer, tumor cells had spread through her body. Two months later, *H*enrietta *L*acks, a young woman from Baltimore, was dead.

Although Henrietta was gone, some of her cells continued to live in the Geys' laboratory. As the first successful human cell culture, HeLa cells were soon being shipped to researchers. Recipients passed cells onto others, and soon HeLa cells were growing in laboratories all over the world. Some even traveled into space aboard the *Discoverer XVII* satellite. Every year, researchers publish hundreds of scientific papers based on work with HeLa cells.

Henrietta was only thirty-one years old when runaway cell divisions killed her in 1951. Now, more than forty years later, her legacy is still benefiting humans everywhere, in cells that are still alive and dividing, day after day after day.

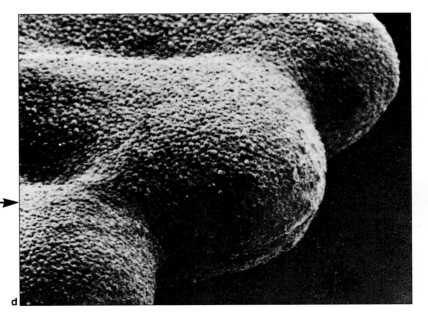

d

e

Figure 9.10 Development of the human hand by way of cell divisions and other processes. Individual cells resulting from the mitotic cell divisions are clearly visible in (**d**). The hand is turned palm upward in (**e**).

SUMMARY

1. Each cell of a new generation must receive the required hereditary instructions (in the form of parental DNA) and enough cytoplasmic machinery to start up its own operation.

2. Cells of multicelled eukaryotes commonly have a diploid number of chromosomes. That is, they have two of each type of chromosome characteristic of the species. The pairs of "homologous" chromosomes generally are alike in length, centromere location, and which heritable traits they deal with.

3. Eukaryotic chromosomes are duplicated during interphase (between cell divisions). Whereas each was one DNA molecule (and associated proteins), now it consists of two, which temporarily stay attached as sister chromatids.

4. Eukaryotes employ different division mechanisms that serve different functions:

   a. Mitosis is a type of nuclear division that maintains the parental number of chromosomes in each of two daughter nuclei. Thus if the parental cell's nucleus is diploid, the nucleus in each daughter cell also will be diploid.

   b. Mitosis is the basis of bodily growth for multicelled eukaryotes. It also is the basis of asexual reproduction for some eukaryotes.

   c. Meiosis is a type of nuclear division that reduces the parental chromosome number by half (to the haploid number) in each of four daughter nuclei.

   d. Meiosis is the basis of sexual reproduction. It is a necessary precursor to the formation of gametes.

   e. For most organisms, actual cytoplasmic division, or cytokinesis, occurs toward the end of nuclear division or at some point afterward.

5. A cell destined to divide by mitosis spends about 95 percent of the cell cycle in interphase, a period between mitotic divisions. During interphase, the cell increases in mass, doubles its number of cytoplasmic components, and duplicates its chromosomes.

6. Mitosis proceeds through five continuous stages:

   a. Prophase. Duplicated, threadlike chromosomes start to condense. New microtubules start to assemble in organized arrays near the nucleus; they will form a spindle apparatus.

   b. Metaphase. The nuclear envelope breaks up during the transition to metaphase. The kinetochores of chromosomes are free to capture microtubules of the forming spindle. Sister chromatids of each chromosome become oriented toward and attached to opposite spindle poles. All chromosomes are moved to and become aligned at the equator of the fully formed spindle.

   c. Anaphase. The two sister chromatids of each chromosome separate. Both are now independent chromosomes and they move to opposite poles.

   d. Telophase. Chromosomes decondense to the threadlike form. A new nuclear envelope forms around the two parcels of chromosomes. Mitosis is completed.

Review Questions

1. Define the two types of nuclear division mechanisms that occur in eukaryotes. What is cytokinesis? *138–139*

2. Define somatic cell and germ cell. Which type of cell can undergo meiosis? *138*

3. What is a chromosome? What is a chromosome called in its unduplicated state? In its duplicated state (that is, with two sister chromatids)? *138*

4. Define homologous chromosomes. Do homologous chromosomes pair during mitosis, meiosis, or both? *139*

5. Describe the spindle apparatus and its general function in nuclear division processes. *141*

6. Name the four main stages of mitosis, and describe the main features of each stage. *140–145*

Self-Quiz *(Answers in Appendix IV)*

1. Eukaryotic DNA is distributed to daughter cells by _____ or _____, both of which are nuclear division mechanisms.

2. Each kind of organism contains a characteristic number of _____ in each cell; each of those structures is composed of a _____ molecule with its associated proteins.

3. A pair of chromosomes that are similar in length, shape, and the traits they govern are called _____.
   a. diploid chromosomes   c. homologous chromosomes
   b. mitotic chromosomes   d. germ chromosomes

4. Somatic cells of multicelled eukaryotic organisms usually have a _____ number of chromosomes, whereas gametes have a _____ number.
   a. haploid; haploid   c. diploid; diploid
   b. haploid; diploid   d. diploid; haploid

5. Interphase is the stage when _____.
   a. nothing occurs
   b. a germ cell forms its spindle apparatus
   c. a cell grows and duplicates its DNA
   d. cytokinesis occurs

6. Following mitosis, a daughter cell will end up with genetic instructions that are _____ and with a chromosome number that is _____ the parent cell.
   a. identical to the parent cell's; the same as
   b. identical to the parent cell's; one-half
   c. rearranged; the same as
   d. rearranged; one-half

7. Cytokinesis is a term that describes _____.
   a. doubling the chromosome number
   b. nuclear division
   c. cytoplasmic division
   d. reducing the chromosome number

8. During interphase, a cell _____.
   a. grows
   b. doubles the number of cytoplasmic components
   c. duplicates its chromosomes
   d. all of the above

9. All of the following are stages of mitosis *except* _____.
   a. prophase
   b. interphase
   c. metaphase
   d. anaphase

10. A duplicated chromosome has _____.
    a. one chromatid
    b. two chromatids
    c. three chromatids
    d. four chromatids

11. Match each stage of mitosis with the following key events.
    _____ metaphase
    _____ prophase
    _____ telophase
    _____ anaphase

    a. sister chromatids of each chromosome separate and move to opposite poles
    b. threadlike chromosomes start to condense
    c. chromosomes decondense, daughter nuclei re-form
    d. spindle is fully formed and all chromosomes are aligned at its equator

Selected Key Terms

anaphase *144*
cell cycle *140*
cell plate formation *146*
centromere *138*
chromosome *138*
cleavage furrow *145*
cytokinesis *138*
diploid *139*
haploid *139*
homologous chromosome *139*
interphase *140*

kinetochore *144*
life cycle *137*
meiosis *138*
metaphase *143*
mitosis *138*
prophase *142*
reproduction *137*
sister chromatid *138*
spindle apparatus *141*
telophase *145*

Readings

Alberts, B., et al. 1989. *Molecular Biology of the Cell*. Second edition. New York: Garland Publishing.

John, B., and K. Lewis. 1980. *Somatic Cell Division*. Burlington, North Carolina: Carolina Biological Supply.

Prescott, D. 1988. *Cells: Principles of Molecular Structure and Function*. Boston: Jones and Bartlett. Chapter 7.

Smith-Klein, C., and V. Kish. 1988. *Principles of Cell Biology*. New York: Harper & Row.

10 A CLOSER LOOK AT MEIOSIS

Octopus Sex and Other Stories

The couple clearly are interested in each other. He caresses her first with one tentacle, then another—then another, another, and another. She reciprocates. This goes on for hours; a tentacled hug here, an enveloping squeeze there. Finally the male reaches deftly under his mantle and removes a packet of sperm, which he inserts into the cavity under the female's mantle. For every one of his sperm that successfully performs its function, a fertilized egg can develop into a new octopus.

For the octopus, sex is an occasional event, preceded by a courtship ritual involving intermingled tentacles. Sex for the slipper limpet is a lifelong group activity. Slipper limpets are marine animals, relatives of land snails. Before a slipper limpet becomes a sexually mature adult, it passes through a free-living larval stage. When the time comes for the larva to become transformed into the adult form, it settles onto a substrate. If it settles down all by itself, the slipper limpet will develop into a female. If another larva settles down on the female, it will develop into a male. If another larva settles down on that male it, too, will become a male. Adult slipper limpets almost always live in such piles, with the one on the bottom invariably being female (Figure 10.1a). All of the males continually contribute sperm to the task of sexual reproduction. When the one female finally dies, the male at the bottom of the pile becomes transformed into a female, and so it goes.

Strawberry plants, too, engage in sexual reproduction, but a strawberry plant also can do something you could not even begin to do except in your wildest imagination. It can reproduce all by itself. Through mitosis, aboveground stems called runners grow outward from the plant—and brand new plants sprout up along the runners. Similarly, the entire body of a flatworm can split into two roughly equivalent parts—then, through mitosis, each part can grow into a whole flatworm.

Or consider the aphid, an insect common to gardens. All summer long, nearly every aphid is female, and nearly every one is reproducing asexually. Inside her reproductive organs, unfertilized egg cells are chemically stimulated to develop into female embryos, which develop into young aphids—which are born alive (Figure 10.1b). All of this occurs so rapidly that young aphids are born pregnant—they carry embryos inside! As autumn approaches, males are produced and they do their part in sexual reproduction. Still, females that survive the winter can do without males, and come summer they begin another round of producing offspring all by themselves.

Figure 10.1 (**a**) Limpets busily perpetuating the species by way of sexual reproduction. (**b**) Live birth of an aphid, a type of insect that shows variations on the sexual theme.

a

Regardless of how—or how often—it takes place among plants and animals, reproduction involves predictable events. For asexually reproducing organisms, chromosome duplications and mitotic cell divisions provide each new cell with all required genetic instructions. For sexually reproducing organisms also, chromosome duplications precede cell divisions by which gametes (sex cells) are produced. In this case, however, *two* gametes fuse at fertilization, bearing genetic instructions from *two* parental cells. Mitosis won't work here. Only through *meiosis* will each gamete end up with half the parental number of chromosomes. Only then will a fertilized egg end up with the proper chromosome number, no more, no less. *Meiosis and fertilization, then, are the unifying theme of sexual reproduction, regardless of the species.* Intermingled tentacles and communal sex and pregnant newborns are simply variations on that theme.

b

KEY CONCEPTS

1. Sexual reproduction of multicelled plants and animals depends on these events: meiosis, gamete formation, and fertilization. In plants, other reproductive events, including the formation of spores, may occur between meiosis and gamete formation.

2. The somatic cells of most animals and many plants have a diploid number of chromosomes (*two* of each type characteristic of the species), half of which are from one parent and half from another parent organism. In germ cells, which are a subpopulation of cells set aside for sexual reproduction, every two chromosomes that are alike (homologues) will pair with each other during meiosis.

3. Meiosis, a nuclear division process in germ cells only, reduces the diploid number of chromosomes by half for the forthcoming gametes. Each gamete produced is haploid; it has only *one* of each type of chromosome that was present in the germ cell. The union of two gametes at fertilization restores the diploid number in the new individual.

4. During meiosis, homologues exchange segments (through crossing over and recombination), then shufflings occur that will give rise to different mixes of chromosomes from two parent organisms in gametes.

ON ASEXUAL AND SEXUAL REPRODUCTION

Strawberry plants, aphids, and transversely dividing flatworms give us a sense of how different organisms might rely, one way or another, on **asexual reproduction**. By this process, *one* parent always passes on a duplicate of all of its genes to offspring. "Genes" are specific portions of a DNA molecule, and they contain the inherited instructions for producing or influencing a trait in offspring. This means that, rare mutations aside, asexually produced offspring can only be genetically identical copies, or clones, of the parent.

Inheritance is much more interesting with sexual reproduction. Commonly, **sexual reproduction** involves two parents, each with two genes for nearly every trait. Both parents pass on one of each gene to offspring by way of meiosis, gamete formation, and fertilization. Thus the first cell of a new individual inherits two genes for every trait—one from each parent.

If the instructions in every pair of genes were identical down to the last detail, then sexual reproduction would produce clones, also. Just imagine—you, everyone you know, every member of the entire human population would be clones and might all end up looking exactly alike.

But it happens that the molecular structure of a gene can change; this is what we mean by "mutation." As a result of past mutations, different individuals of a species might be carrying different molecular forms of a gene that "say" slightly different things about how a trait will be expressed in offspring. Whenever there are different molecular forms of the same gene, each form is called an **allele**. Admittedly, this is not a word that is easy to warm up to. It may help to know that it is short for an even worse word—allelomorph, after the Greek *allos* (meaning other) and *morphē* (meaning form).

As Figure 10.2 indicates, a gene that affects how the human chin develops can vary this way. One molecular form of that gene says "put a dimple in it" and another says "no dimple." Different genes govern thousands of different traits. And this brings us to a key reason why individuals of any sexually reproducing species don't all look alike. *Sexual reproduction puts together new combinations of alleles in offspring.*

This chapter provides us with a closer look at meiosis, the foundation for sexual reproduction. More importantly, it starts us thinking about some far-reaching consequences of the gene shufflings possible with sexual reproduction. New gene combinations among offspring lead to variations in their physical and behavioral traits. *Such variation is acted upon by agents of natural selection—and so it is a basis of evolutionary change.*

Figure 10.2 (**a**) The chin fissure, a heritable trait arising from a rather uncommon form of a gene. Actor Kirk Douglas received a gene that influences this trait from each of his parents. One gene called for a chin fissure and the other didn't, but one is all it takes in this case. (**b**) This photograph shows what Mr. Douglas' chin might have looked like if he had inherited two ordinary forms of the gene instead.

Through meiosis and fertilization, old gene combinations are broken up and new ones are put together. The immediate consequence is variation in the physical and behavioral traits of offspring. The long-term consequence can be evolutionary change.

a   b

Figure 10.3 Examples of the location of germ cells that give rise to sperm and eggs.

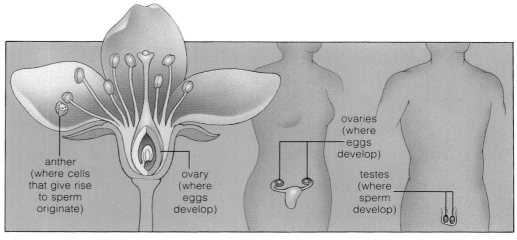

anther (where cells that give rise to sperm originate)

ovary (where eggs develop)

ovaries (where eggs develop)

testes (where sperm develop)

flowering plant   human female   human male

OVERVIEW OF MEIOSIS

Think "Homologues"

The preceding chapter made a few points about meiosis, and now we can put those points together in the following picture. Meiosis is a nuclear division mechanism. It sorts out the DNA present in a cell nucleus into parcels that can be distributed to forthcoming daughter cells. Meiosis happens only in *germ cells,* the cell lineage destined to give rise to the gametes (sperm or eggs) used in sexual reproduction. Germ cells develop in a variety of reproductive structures and organs (Figure 10.3). Like somatic cells, germ cells commonly have a diploid number of *chromosomes,* these being structures composed of DNA and proteins.

"Diploid" means there are two chromosomes of each type in a cell. The two are **homologous chromosomes**. Generally, homologues have the same length, the same centromere location, and the same genes—and they line up with each other during meiosis. Only the sex chromosomes, designated X and Y in humans and many other organisms, differ in form and in which genes they carry—but they still function as homologues during meiosis.

Meiosis reduces the diploid number ($2n$) by half, to the "haploid" number (n). And not just any half: *Each gamete ends up with one member of each pair of homologous chromosomes.* To give an example, the diploid number for humans is 46—that is, 23 + 23 homologues. A human gamete ends up with 23 chromosomes, one of each type.

Overview of the Two Divisions

Meiosis resembles mitosis in some respects, even though the outcome is different. While a germ cell is still in interphase, each chromosome is duplicated by a process called DNA replication. The chromosome hangs onto its duplicate at a small, constricted region (the centromere), and as long as the two remain attached, they are called **sister chromatids**:

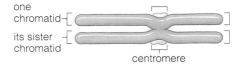

one chromatid — its sister chromatid — centromere

As in mitosis, microtubules of a spindle apparatus harness each chromosome and take part in its movement during nuclear division. Unlike mitosis, however, there are *two divisions,* which ultimately lead to the formation of four nuclei. The two different divisions are called meiosis I and II:

DNA duplication during interphase

MEIOSIS I
Prophase I
Metaphase I
Anaphase I
Telophase I

No DNA duplication between divisions

MEIOSIS II
Prophase II
Metaphase II
Anaphase II
Telophase II

During meiosis I, each duplicated chromosome lines up with its partner, *homologue to homologue,* then the partners are separated from each other. Here we show just one pair of homologous chromosomes, but the same thing happens to all pairs in the nucleus:

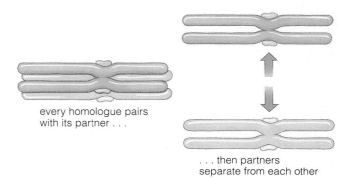

every homologue pairs with its partner . . .

. . . then partners separate from each other

Cytokinesis typically follows. At this point each daughter nucleus contains only one of each type of chromosome. But each chromosome is still in the duplicated state, consisting of two sister chromatids.

During meiosis II, *the sister chromatids of each chromosome are separated from each other:*

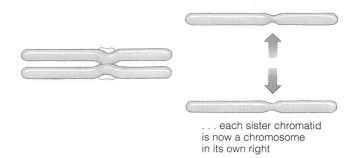

. . . each sister chromatid is now a chromosome in its own right

Cytokinesis typically follows this separation, also. Putting the preceding example in perspective, every chromosome in a diploid germ cell was duplicated before the onset of meiosis. Two nuclear divisions and two cytoplasmic divisions later, the outcome was four cells— that is, each with a haploid number of unduplicated chromosomes.

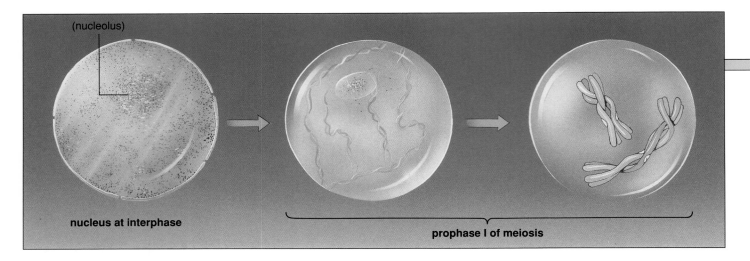

(nucleolus)

nucleus at interphase

prophase I of meiosis

Early in Prophase I, chromosomes are like thin threads, with each "thread" being two sister chromatids in close alignment. Shown here, two chromosomes from the male parent (green) and two from the female parent (pink).

Before prophase I ends, each chromosome pairs with its homologue; and *nonsister* chromatids undergo crossing over.

Figure 10.4 Prophase I of meiosis, when crossing over occurs between homologous chromosomes.

STAGES OF MEIOSIS

Prophase I Activities

The first stage of meiosis, **prophase I**, is a time of major gene shufflings between homologous chromosomes. Homologues begin to pair at the onset of prophase I. At this stage a duplicated chromosome looks like a very long, thin thread, so closely are its two sister chromatids aligned with each other.

Each threadlike chromosome and its homologue are drawn together during a process called *synapsis*. It is as if they become stitched point by point along their entire length, with little space between them. (The X and Y chromosomes pair at one end only.)

The intimate parallel arrangement between homologues favors **crossing over**. By this mechanism, *nonsister* chromatids break at one or more sites along their length and exchange corresponding segments—that is, genes—at the breakage points.

Figure 10.4 shows only a single crossover. On the average, between two and three crossovers are thought to occur between each pair of homologues in human germ cells undergoing meiosis.

Gene-swapping would be rather pointless if each type of gene never varied from one chromosome to the next. But remember a gene can come in alternative forms—alleles. You can safely bet that all the genes running down the length of one chromosome will not be an identical match with those on the homologue. With each crossover, then, there is a chance that homologues may be swapping *different* instructions for some traits.

We will look at the mechanism of crossing over in later chapters. For now it is enough to know that crossing over leads to **genetic recombination**, which in turn leads to variation in the traits of offspring.

Crossing over is an event by which old combinations of alleles in a chromosome are broken up and new ones put together during meiosis.

After segments have been exchanged, all four chromatids thicken and can spread apart somewhat—but they remain joined at a few places where nonsister chromatids extend across each other (Figure 10.5). Each crosslike, temporary attachment between two nonsister chromatids is a *chiasma* (plural, chiasmata). Such attachments will play a role in aligning the homologues during metaphase I.

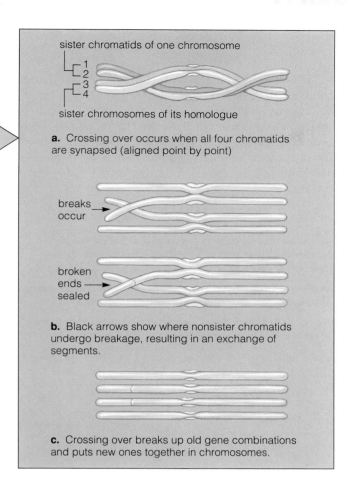

sister chromatids of one chromosome

1
2
3
4

sister chromosomes of its homologue

a. Crossing over occurs when all four chromatids are synapsed (aligned point by point)

breaks occur

broken ends sealed

b. Black arrows show where nonsister chromatids undergo breakage, resulting in an exchange of segments.

c. Crossing over breaks up old gene combinations and puts new ones together in chromosomes.

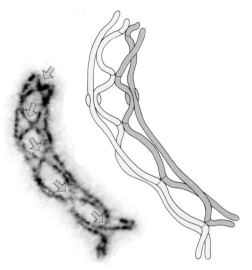

Figure 10.5 Two homologous chromosomes held together by chiasmata (gold arrows). A chiasma is not the same thing as a crossover. Although it indicates that crossing over occurred earlier, it is not always located at the site of breakage and exchange. The reason is that chiasmata are not stationary attachments; they are forced toward the chromatid tips as prophase I draws to a close.

Separating the Homologues

Metaphase I, the second stage of meiosis I, is a time of major shufflings of whole chromosomes, before their distribution into daughter nuclei. Suppose the shufflings are proceeding right now in one of your germ cells. We can call that cell's homologous chromosomes "maternal" and "paternal." (This is a short way of saying that one of each type was inherited from your mother and their homologues were inherited from your father.) Figure 10.6 shows how many maternal and paternal chromosomes are present in that cell, so you can get an idea of how complicated it would be to track their movements. To keep things simple, let's imagine that we are tracking only *two* pairs of homologous chromosomes.

During prophase I, microtubules had started to assemble outside the nucleus of the germ cell. As in mitosis, they ultimately will form a spindle apparatus (page 141). During the transition to metaphase, the nuclear envelope broke apart, so now the microtubules are free to interact with the chromosomes. The chromosomes become attached to microtubules extending from both spindle poles, then they are moved into position at the spindle equator. At **metaphase I**, the spindle is fully formed and all chromosomes are aligned at its equator. Then, during **anaphase I**, each homologue moves away

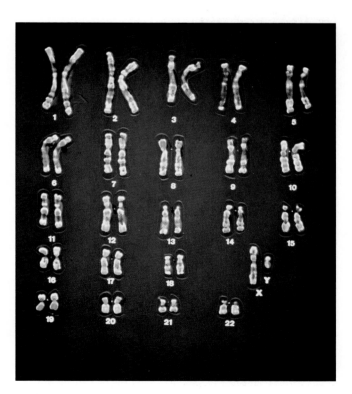

Figure 10.6 The 23 pairs of homologous chromosomes from a human male.

from its partner and the two head toward opposite spindle poles, as shown in Figure 10.7.

Are all the maternal chromosomes destined to move to one pole and the paternal chromosomes to the other? Maybe, but probably not. The positioning of each pair of homologues at the spindle equator and their subsequent direction of movement are random events. *It doesn't matter which partner moves to which pole.*

Consider Figure 10.8, which shows how just three pairs of homologues can be shuffled into any one of four possible positions at metaphase I. In this case, 2^3 or 8 combinations of maternal and paternal chromosomes are possible for the forthcoming gametes. A human germ cell has 23 pairs of homologous chromosomes, not just three. So 2^{23} or *8,388,608 combinations* of maternal and paternal chromosomes are possible every time a germ cell gives rise to sperm or eggs! (Are you beginning to get an idea of why such splendid mixes of inherited traits show up even in the same family?)

Typically, anaphase I proceeds to telophase I and **interkinesis**, which often are fleeting stages before the final nuclear division. There is no DNA duplication between the two meiotic divisions. But remember, each chromosome was duplicated earlier (during interphase). And it is still in the duplicated form when meiosis II gets under way.

One or the other member of each pair of homologous chromosomes may end up at a given spindle pole during meiosis I. It doesn't matter which one arrives at which pole.

This means each pair of homologues is assorted into gametes *independently* of the other pairs present in the cell.

As a result of independent assortment, gametes end up with different mixes of maternal and paternal chromosomes.

Figure 10.7 Meiosis: the nuclear division mechanism by which the parental number of chromosomes is reduced by half (to the haploid number) for forthcoming gametes. Only two pairs of homologous chromosomes are shown. The green ones are derived from one parent, and the pink ones are their homologues from the other parent.

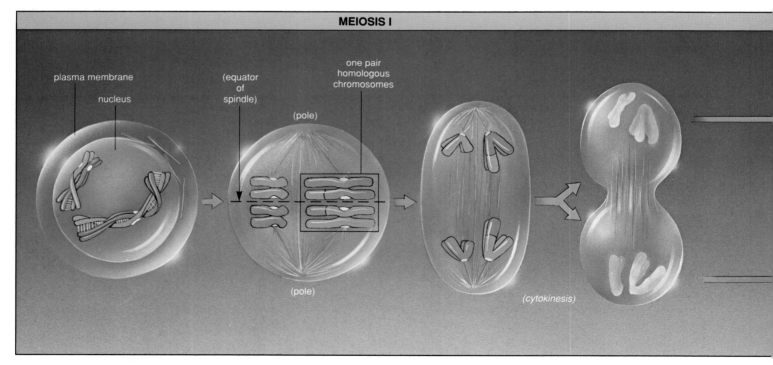

MEIOSIS I

Prophase I	**Metaphase I**	**Anaphase I**	**Telophase I**
Each chromosome condenses, then pairs with its homologue. Crossing over and recombination occur.	Spindle apparatus forms, nuclear envelope (not shown) breaks down during transition to metaphase I. Homologous pairs align randomly at spindle equator.	Each homologue is separated from its partner, and the two are moved to opposite poles.	A haploid number of chromosomes (still duplicated) ends up at each pole.

Separating the Sister Chromatids

Meiosis II has one overriding function: separation of the two sister chromatids of each chromosome. Here again, microtubules of a spindle apparatus have roles in the separation.

At metaphase II, the chromosomes attach to the microtubules of the forming spindle and are moved to its equator. Each duplicated chromosome becomes aligned at the equator. At anaphase II, each is split, and its (formerly) sister chromatids are now unduplicated chromosomes in their own right.

Each spindle pole is the destination of half the parental number of chromosomes. But that haploid number includes one of each type of chromosome characteristic of the species. During telophase II, new nuclear membranes form around the chromosomes after they have become clustered at the two poles. Two nuclei are thus formed. Meiosis is completed.

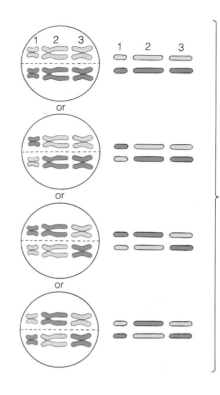

eight different haploid combinations of the maternal and paternal chromosomes are possible in the forthcoming gametes

Figure 10.8 (*Right*) Possible outcomes of the random alignment of three pairs of homologous chromosomes at metaphase I of meiosis. The three types of chromosomes are labeled 1, 2, and 3. Maternal chromosomes are pink; paternal ones are green.

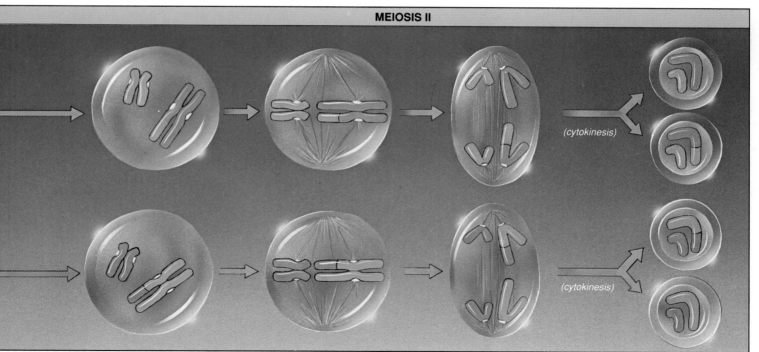

MEIOSIS II

Prophase II
There is no DNA replication between divisions. Sister chromatids of each chromosome are still attached at the centromere.

Metaphase II
Each chromosome is aligned at the spindle equator.

Anaphase II
Each chromosome splits; what were once sister chromatids are now chromosomes in their own right and are moved to opposite poles.

Telophase II
Four daughter nuclei form. Following cytokinesis, each gamete has a haploid number of chromosomes, all in the unduplicated state.

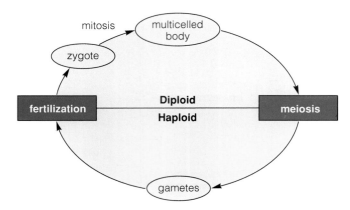

Figure 10.9 Generalized life cycle for animals.

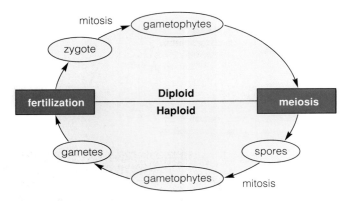

Figure 10.10 Generalized life cycle for most plants. During the diploid stage of the life cycle, a *sporophyte* ("spore-producing plant body") forms. During the haploid stage, a *gametophyte* ("gamete-producing plant body") forms. A pine tree is an example of a sporophyte. Gametophytes develop in its pine cones (page 398).

MEIOSIS AND THE LIFE CYCLES

Meiosis precedes the formation of mature gametes during the life cycle of all sexually reproducing organisms, as Figures 10.9 and 10.10 suggest. In later chapters, we will consider the details of some of those cycles, including the one for humans. Here we simply will make a few points that can help you keep the details in perspective.

Gamete Formation

Although all sexually reproducing organisms produce gametes, keep in mind that the gametes do not all look alike. For example, human sperm have one tail, opossum sperm have two, and roundworm sperm have none. Crayfish sperm look like pinwheels. Most eggs are microscopic, yet ostrich eggs (tucked inside a protective shell) are as large as a softball. In outward appearance, the gametes of plants are not recognizably like the gametes of mammals.

Gamete Formation in Animals. Animal life cycles typically proceed from meiosis to gamete formation, fertilization, then growth by way of mitosis. Formation of the male gametes, called spermatogenesis, is shown in Figure 10.9. Inside the reproductive tract of *male* animals, a diploid germ cell increases in size. The resulting large, immature cell (a primary spermatocyte) undergoes meiosis. Following cytokinesis, the four resulting cells eventually develop into four haploid spermatids (Figure 10.11). The spermatids change in form, develop a tail, and become **sperm**, the mature male gametes.

Inside the reproductive tract of *female* animals, meiosis and gamete formation is called oogenesis, and it differs from what goes on in males in two important features. Compared to a primary spermatocyte, many more cytoplasmic components accumulate in the female germ cell, the primary oocyte. Also, the cells formed after meiosis differ in size and function (Figure 10.12).

Following meiosis I, one cell (the secondary oocyte) receives nearly all the cytoplasm. The other, much smaller cell is the first "polar body." Both cells may undergo meiosis II, and the outcome is one large cell and three extremely small polar bodies.

The large cell develops into the mature female gamete, or **ovum**. This also is known rather loosely as "the egg." The polar bodies do not function as gametes. In effect, they serve as dumping grounds for three sets of parental chromosomes, so that the egg ends up with the necessary haploid number. Besides this, polar bodies do not receive much cytoplasm, so they do not get much in the way of nutrients and metabolic machinery. Thus they are destined to degenerate.

Gamete Formation in Plants. For pine trees, roses, and other familiar plants, the life cycle proceeds in the same general way. As Figure 10.10 indicates, however, some additional events occur between meiosis and gamete formation. Among other things, plants form **spores**, which are haploid cells that are resistant to dry periods or other adverse environmental conditions. When favorable conditions return, spores germinate and develop into some kind of haploid body or structure that will go on to produce the gametes. In short, the life cycles of flowering plants include both spore-producing bodies and gamete-producing bodies.

More Gene Shufflings at Fertilization

The diploid number of chromosomes is restored at **fertilization**, the fusion of nuclei of two gametes in the zygote (the first cell of a new multicelled individual). Here we see why meiosis is so necessary. If gametes ended up diploid, fertilization would double the number of chromosomes in cells of every new generation.

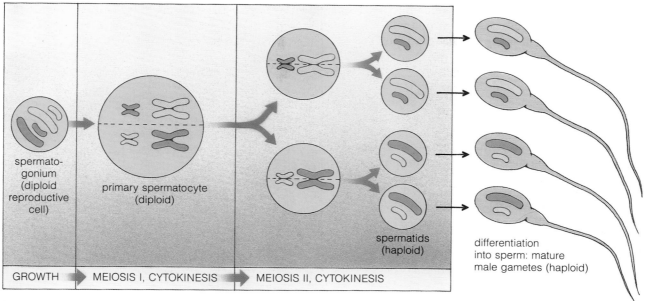

Figure 10.11 Generalized picture of spermatogenesis in male animals. (For the sake of clarity, the nuclear envelopes are not shown in Figures 10.11 through 10.14.)

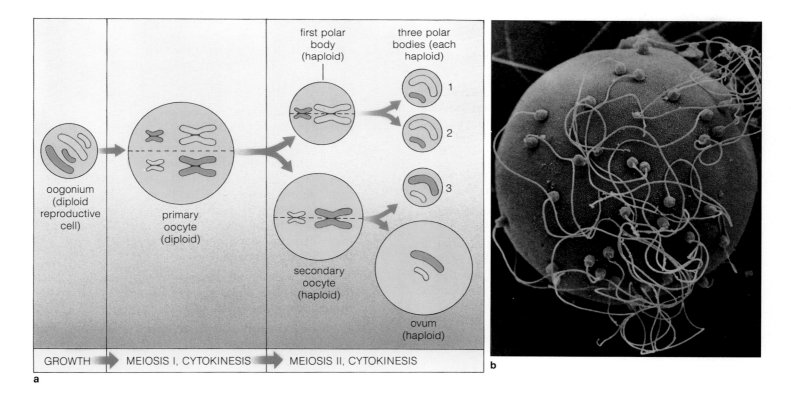

Like meiosis, fertilization also contributes to variation in the traits of offspring. Think about the possibilities for humans. First, genes are shuffled during prophase I, when each chromosome takes part in two or three crossovers, on the average. Second, random alignments at metaphase I lead to one of 8,388,608 possible combinations of maternal and paternal chromosomes in

Figure 10.12 Generalized picture of oogenesis in female animals. This sketch is not drawn to the same scale as Figure 10.11. A primary oocyte is *much* larger than a primary spermatocyte, as suggested by the scanning electron micrograph of the egg and sperm of a clam to the right. Also, the polar bodies are extremely small compared to an ovum, as shown in Figure 43.11.

each gamete. Third, of all the genetically diverse male and female gametes that are produced, *which* two will get together is a matter of chance. As you can see, the sheer number of new combinations brought together at fertilization is staggering!

MEIOSIS COMPARED WITH MITOSIS

In this unit, our main focus has been on mitosis and meiosis—two nuclear division mechanisms used in the reproduction of eukaryotic cells. Mitosis underlies asexual reproduction of single-celled eukaryotes as well as growth of multicelled eukaryotes. Meiosis occurs only in germ cells, which give rise to the haploid gametes used in sexual reproduction.

The major difference between them is this: Mitotic cell division produces clones—genetically identical copies of the parent cell. Meiosis and fertilization give rise to novel combinations of alleles in offspring which, as a consequence, vary from the parents and one another in the details of their traits. As we have seen, three events are responsible for the variation:

1. Crossing over and genetic recombination occur during prophase I of meiosis.

2. During anaphase I of meiosis, the two members of each pair of homologous chromosomes assort independently of the other pairs. Thus the forthcoming gametes can end up with different mixes of maternal and paternal chromosomes.

3. Fertilization is a chance mix of different combinations of alleles from two different gametes (Figure 10.13).

In later chapters, we will see how the variation in traits made possible by meiosis and fertilization is a testing ground for agents of natural selection. As such, both contribute to the evolution of sexually reproducing populations.

SUMMARY

1. The life cycle of multicelled plants and animals generally includes meiosis and gamete formation, fertilization, and growth by way of cell divisions.

   a. In animals, meiosis produces haploid gametes (sperm in males, eggs in females). Fusion of a sperm nucleus and an egg nucleus at fertilization produces a diploid cell (zygote), which develops into the multicelled individual by way of mitosis and cytokinesis.

   b. In plants, meiosis gives rise to spores, which

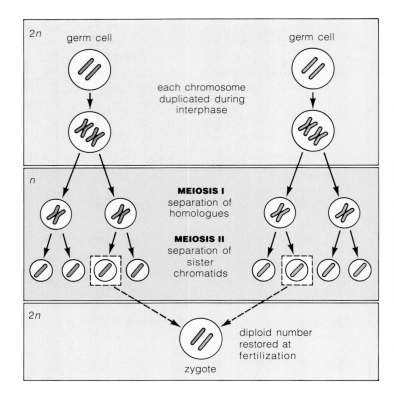

Figure 10.13 How fertilization restores a diploid (2*n*) number of chromosomes that has been reduced by half during meiosis.

eventually undergo mitotic cell divisions that give rise to gamete-producing bodies.

2. Sexually reproducing organisms commonly have a diploid number of chromosomes (2*n*), or two of each type characteristic of the species. The two are "homologous" chromosomes, with the same length, centromere location, and gene sequence. Homologues pair with each other during meiosis. The X and Y chromosomes vary in length, shape, and which genes they carry, but they still pair as homologues.

3. Before meiosis, all chromosomes in a germ cell are duplicated. The duplicates remain attached (as sister chromatids) at the centromere.

4. Meiosis consists of two consecutive divisions that sort out the chromosomes in a germ cell. In meiosis I, each chromosome pairs with and then separates from its homologue. In meiosis II, the sister chromatids of each chromosome separate from each other. In both cases, microtubules of a spindle apparatus help move the chromosomes.

5. The following key events occur during meiosis I:

   a. At prophase I, each chromosome comes into point-by-point alignment with its homologue. Crossing over occurs between nonsister chromatids. Crossing over breaks up old combinations of alleles and puts together new ones in the chromosomes. This genetic recombination leads to variation in traits among offspring.

b. Late in metaphase I, all pairs of homologous chromosomes are aligned at the spindle equator.

c. At anaphase I, each maternal chromosome is separated from its paternal homologue and moved to the opposite spindle pole.

6. The following key events occur during meiosis II:

a. At metaphase II, all chromosomes are moved to the spindle equator.

b. At anaphase II, the sister chromatids of each chromosome are separated for movement to opposite poles. Once separated, they are chromosomes in their own right. And each is in the unduplicated state.

7. Following cytokinesis (cytoplasmic division), there are four haploid cells, one or all of which may function as gametes (or give rise to gametes, in the case of spore-producing plants).

Figure 10.14 summarizes the similarities and differences between mitosis and meiosis.

Review Questions

1. Define sexual reproduction. How does it differ from asexual reproduction? *151*

2. Refer to Table 9.2, which gives the diploid number of chromosomes in the body cells of a few organisms. What would be the *haploid* number for the gametes of humans? For the garden pea? *153*

3. Suppose the diploid cells of an organism have four pairs of homologous chromosomes, designated AA, BB, CC, and DD. How would its haploid set of chromosomes be designated? *153*

4. When, and in which type of cells, does meiosis occur? *153*

5. Define meiosis and characterize its main stages. In what respects is meiosis like mitosis? In what respects is it unique? *153–157, 160*

6. Does crossing over occur during mitosis, meiosis, or both? At what stage of nuclear division does it occur, and what is its significance? *154*

7. Outline the steps involved in spermatogenesis and oogenesis. *158*

Self-Quiz *(Answers in Appendix IV)*

1. The somatic (body) cells of sexually reproducing organisms commonly have a _____ number of chromosomes, or _____ of each type characteristic of that species.

2. Two homologous chromosomes generally contain the same _____.
   a. genes in reverse order   c. alleles in the same order
   b. alleles in reverse order   d. none of the above

3. Prior to meiosis, all the chromosomes in a diploid germ cell are _____.
   a. paired   c. duplicated
   b. randomly mixed   d. separated

4. Crossing over _____.
   a. alters the chromosome alignments at metaphase
   b. occurs between sperm DNA and egg DNA at fertilization
   c. leads to genetic recombination
   d. occurs only rarely

5. Because of the _____ alignment of homologous chromosomes at metaphase, gametes can end up with _____ mixes of maternal and paternal chromosomes.
   a. unvarying; different   c. random; duplicate
   b. unvarying; duplicate   d. random; different

6. Variation in the traits of offspring is increased by the mix of _____ allele combinations from two _____ gametes at fertilization.
   a. similar; similar   c. different; different
   b. different; similar   d. similar; different

7. Prior to the meiotic divisions, duplicated chromosomes remain attached as sister _____ at the area of the chromosome called the _____.
   a. chromosomes; centromere   c. chromosomes; centriole
   b. chromatids; centriole   d. chromatids; centromere

8. Following meiosis and cytokinesis, there are _____ haploid cells, one or all of which may function as _____.
   a. two; body cells   c. four; gametes
   b. two; gametes   d. four; body cells

9. The net result of meiosis is that the _____ chromosome number is _____.
   a. diploid; doubled   c. haploid; doubled
   b. diploid; halved   d. haploid; halved

Selected Key Terms

allele *152*
anaphase I *155*
anaphase II *157*
asexual
   reproduction *151*
crossing over *154*
fertilization *158*
gene *151*
genetic
   recombination *154*
germ cell *153*
homologous
   chromosome *153*
interkinesis *156*
metaphase I *155*
oogenesis *158*
ovum *158*
prophase I *154*
sexual
   reproduction *151*
sister chromatid *153*
sperm *158*
spermatogenesis *158*
telophase I *156*
telophase II *157*

Readings

Cummings, M. 1988. *Human Heredity: Principles and Issues.* New York: West.

Strickberger, M. 1985. *Genetics.* Third edition. New York: Macmillan. Contains excellent introduction to chromosomes and meiosis.

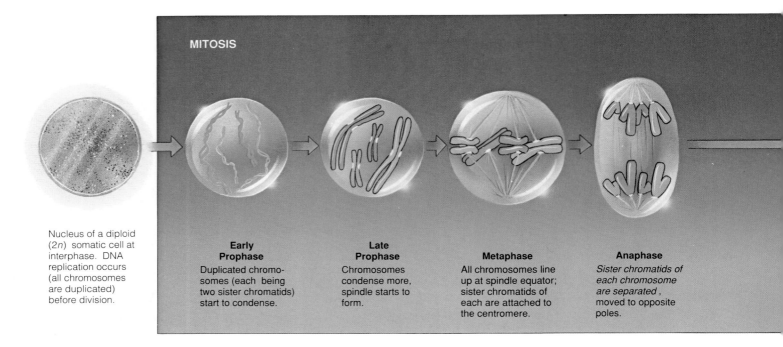

MITOSIS

Early Prophase
Duplicated chromosomes (each being two sister chromatids) start to condense.

Late Prophase
Chromosomes condense more, spindle starts to form.

Metaphase
All chromosomes line up at spindle equator; sister chromatids of each are attached to the centromere.

Anaphase
Sister chromatids of each chromosome are separated, moved to opposite poles.

Nucleus of a diploid (2n) somatic cell at interphase. DNA replication occurs (all chromosomes are duplicated) before division.

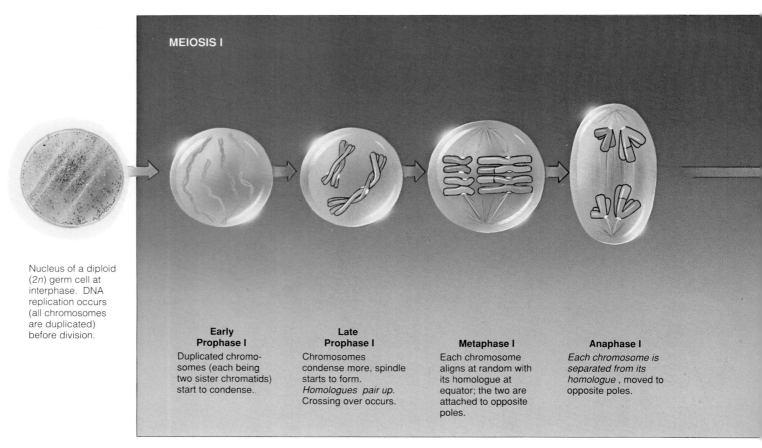

MEIOSIS I

Early Prophase I
Duplicated chromosomes (each being two sister chromatids) start to condense.

Late Prophase I
Chromosomes condense more, spindle starts to form. *Homologues pair up.* Crossing over occurs.

Metaphase I
Each chromosome aligns at random with its homologue at equator; the two are attached to opposite poles.

Anaphase I
Each chromosome is separated from its homologue, moved to opposite poles.

Nucleus of a diploid (2n) germ cell at interphase. DNA replication occurs (all chromosomes are duplicated) before division.

Figure 10.14 Summary of mitosis and meiosis, using a diploid (2n) animal cell as the example. The diagram is arranged to help you compare the similarities and differences between the two division mechanisms. (Chromosomes derived from the male parent are green; their homologues from the female parent are pink.)

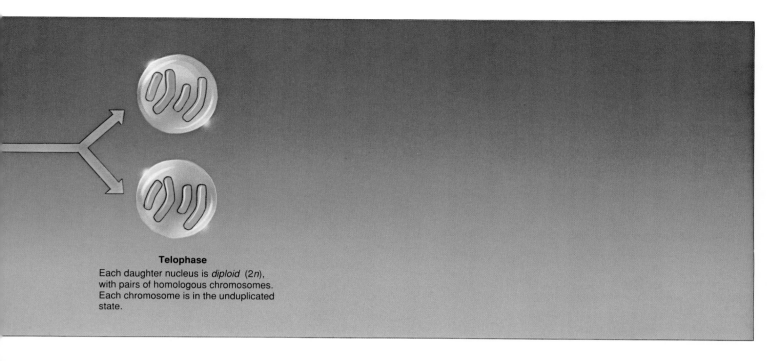

Telophase
Each daughter nucleus is *diploid* (2*n*), with pairs of homologous chromosomes. Each chromosome is in the unduplicated state.

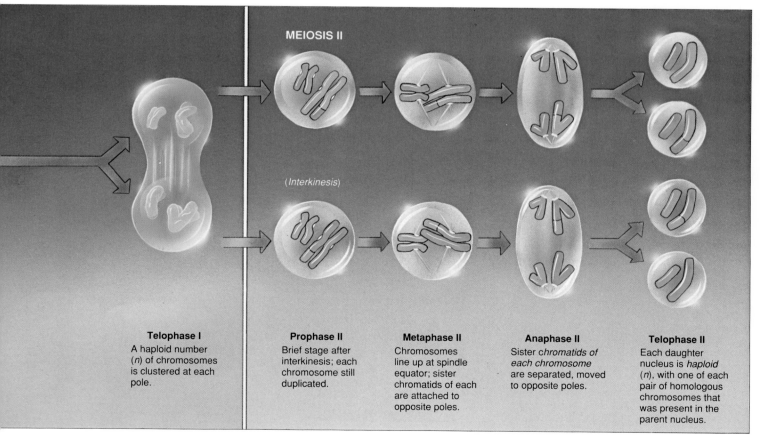

MEIOSIS II

(*Interkinesis*)

Telophase I
A haploid number (*n*) of chromosomes is clustered at each pole.

Prophase II
Brief stage after interkinesis; each chromosome still duplicated.

Metaphase II
Chromosomes line up at spindle equator; sister chromatids of each are attached to opposite poles.

Anaphase II
Sister *chromatids of each chromosome* are separated, moved to opposite poles.

Telophase II
Each daughter nucleus is *haploid* (*n*), with one of each pair of homologous chromosomes that was present in the parent nucleus.

Sickled Cells and Garden Peas

Parties and champagne were the last things on Ernest Irons' mind. It was New Year's Eve, 1904, and Irons, a medical intern, was doodling sketches of peculiarly elongated red blood cells. He and his supervisor, James Herricks, had never seen anything like them. The cells were reminiscent of sickles, a type of short-handled farm tool having a crescent-shaped blade. They were present in a blood sample from a new patient.

The patient had complained of weakness, dizziness, and painful skin sores. He had been gasping for air during bouts of wheezing. Already his father and two sisters had died from mysterious ailments that had damaged their lungs or kidneys. Did those deceased family members also have sickled cells in their blood? Was there a connection between the abnormal cells and the ailments? How did the cells become sickled in the first place?

The medical problems that baffled Irons and Herricks killed their patient when he was only thirty-two. The symptoms themselves are characteristic of a genetic disorder now called *sickle-cell anemia*. The disorder arises when a person receives two copies of a mutated gene (one from each parent) that codes for the protein hemoglobin. Hemoglobin is an oxygen-transporting protein in red blood cells—and the oxygen is vital for aerobic respiration in the body's living cells. The skewed genetic instructions result in skewed hemoglobin molecules.

Consider what happens when blood pumped from the heart reaches capillaries, the blood vessels having the smallest diameter and the thinnest wall. There, red blood cells move along single file. And there they give up oxygen, which diffuses across the capillary wall, through the surrounding tissue fluid, and into individual cells. The blood concentration of oxygen is lower in capillaries than in the rest of the body's blood vessels. Low oxygen levels cause the abnormal hemoglobin molecules to stick together and form long, rodlike structures. Whereas normal red blood cells are shaped like a doughnut without the hole, those containing clumped-together hemoglobin molecules become distorted into a rigid, crescent shape (Figures 11.1a and 11.1b).

Rigid, sickled cells cannot move through the blood capillaries. They clog the tiny passages, even rupture them. The blood cells themselves are easily ruptured. Thus the tissues served by the capillaries become starved for oxygen and saturated with waste products of metabolism. The resulting symptoms range from shortness of breath and fatigue to irreparably damaged organs (Figure 11.16).

You will be reading more about sickle-cell anemia in this chapter and others. It is a disorder that has been studied in great detail at both the molecular and the ecological level. You may find it curious, however, that

Figure 11.1 (**a**) Normal red blood cells. (**b**) Sickled cells, trademark of a well-known human genetic disorder. (**c**) A mere fifty years before such distorted cells were observed, Gregor Mendel identified rules that turned out to be the starting point for modern genetic analysis. Through such analysis, we have gained insight into many aspects of inheritance—including the chromosomal and molecular events that give rise to sickle-cell anemia.

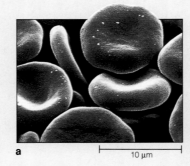
a   10 μm

b   10 μm

our understanding of sickle-cell anemia—and so many other special cases of heritable traits—actually began with studies of thousands of pea plants in a monastery garden.

Fifty years before Ernest Irons doodled red blood cells, a scholarly monk named Gregor Mendel started using peas to study *patterns* of inheritance among sexually reproducing organisms. To test his hypotheses about inheritance, which were novel at the time, Mendel bred generation after generation of pea plants. And he garnered indirect but *observable* evidence of how parents transmit discrete units of information about traits—genes—to offspring.

Mendel's findings seem simple enough today. The cells of pea plants carry two genes for each trait. Following meiosis, the two genes end up in different gametes. When two gametes combine at fertilization, each new plant again has two genes for each trait. At the time, however, no one—not even Mendel himself—knew that he had discovered some near-universal rules governing inheritance. As you will see, his insights still have the power to explain many of the puzzling and sometimes devastating aspects of inheritance that occupy our attention today.

1. Genes, the units of instructions for producing heritable traits in offspring, have specific locations on chromosomes. The molecular form of the gene at a given location may be slightly different from one individual to the next. All of the different molecular forms of a gene are called alleles.

2. Diploid cells, which have pairs of homologous chromosomes, have pairs of genes. Gregor Mendel's monohybrid crosses of pea plants provided indirect evidence that the two genes of each pair segregate from each other during meiosis and end up in different gametes.

3. Mendel's dihybrid crosses of pea plants provided indirect evidence that a gene pair tends to assort into gametes independently of gene pairs that are located on other (nonhomologous) chromosomes.

4. Traits may be influenced by dominance relations. Here, the two genes of a pair are not identical, and one exerts more pronounced effects compared to its partner. Traits also can be influenced by interactions among different gene pairs, by single genes that affect more than one structure or function in the body, and by environmental conditions.

MENDEL'S INSIGHTS INTO THE PATTERNS OF INHERITANCE

When Charles Darwin first proposed his theory of evolution by natural selection, he offered the world a new way of looking at life's diversity. In Darwin's view, members of a population vary in heritable traits. Variations that improve chances of surviving and reproducing show up more often in each generation. Those that don't become less frequent. In time the population changes—it evolves.

Not everyone accepted the theory, partly because it did not fit with a prevailing view of inheritance. It was common knowledge that instructions for heritable traits reside in sperm and eggs—but how were the

c

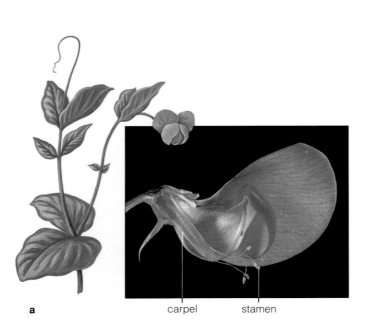

b Pollen from a plant that breeds true for purple flowers is brushed onto a floral bud of a plant that breeds true for white flowers and that had its own stamens snipped off.

c The cross-fertilized plant produces seeds, each of which is allowed to grow into a new plant.

d Flower color of new plants can be used as evidence of patterns in how hereditary material is transmitted from each parent.

a

carpel   stamen

Figure 11.2 The garden pea plant (*Pisum sativum*), the focus of Mendel's experiments. The photograph shows a flower that has been sectioned to reveal the location of the stamens (where pollen grains develop) and the carpel. (Eggs develop, fertilization takes place, and seeds mature inside the carpel.)

instructions combined at fertilization? Many thought they blended together, like cream into coffee.

Yet if blending were true, why aren't distinctive traits diluted out of a population? Why do children with freckles keep turning up among nonfreckled generations? If it were true, why aren't all the descendants of a herd of white stallions and black mares uniformly gray? Blending scarcely explained what people could see with their own eyes, but it was considered a rule anyway. According to the blending theory, populations "had to be" uniform—and without variation for selective agents to act upon, evolution simply could not occur.

Even before Darwin presented his theory, however, someone was gathering evidence that eventually would support its premise about variation in heritable traits. In a monastery garden in Brünn, now in Czechoslovakia, a scholarly monk named Gregor Mendel was beginning to identify the rules governing inheritance.

The monastery of St. Thomas was somewhat removed from the European capitals, which were then the centers of scientific inquiry. Yet Mendel was not a man of narrow interests who simply stumbled by chance onto principles of great import. Having been raised on a farm, he

was well aware of agricultural principles and their application. He kept abreast of breeding experiments and developments described in the available journals. Mendel was a founder of the regional agricultural society. He won several awards for developing improved varieties of fruits and vegetables. After entering the monastery, he spent two years studying mathematics at the University of Vienna.

Shortly after his university training, Mendel began experiments on the nature of plant diversity. Through his combined talents in plant breeding and mathematics, he perceived patterns in the emergence of traits from one generation to the next.

Mendel's Experimental Approach

Mendel experimented with the garden pea plant, *Pisum sativum* (Figure 11.2). This plant can fertilize itself. Its flowers produce male *and* female gametes, and fertilization can occur in the same flower. (To keep things simple, we will call those gametes "sperm" and "eggs," even though they bear little obvious resemblance to the sperm and eggs of animals.) Some pea plants are **true-**

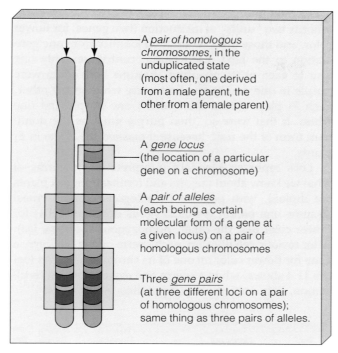

A *pair of homologous chromosomes,* in the unduplicated state (most often, one derived from a male parent, the other from a female parent)

A *gene locus* (the location of a particular gene on a chromosome)

A *pair of alleles* (each being a certain molecular form of a gene at a given locus) on a pair of homologous chromosomes

Three *gene pairs* (at three different loci on a pair of homologous chromosomes); same thing as three pairs of alleles.

Figure 11.3 A few genetic terms illustrated. In Mendel's time, no one knew about meiosis or chromosomes, but it was clear that offspring received hereditary material from parents by way of sperm and eggs. As we now know, the hereditary material (genes) is packaged in homologous chromosomes (one from the male, one from the female parent). Thus at each gene locus along the chromosomes, one allele has come from the male parent and its partner has come from the female parent.

breeding. Successive generations are exactly like the parents in one or more traits. For example, all offspring may "breed true" for white flowers. Of course, when left to their own devices, true-breeding pea plants show a rather monotonous and uninformative pattern of inheritance.

However, pea plants also lend themselves to artificial **cross-fertilization,** in which sperm from one plant are used to fertilize eggs from another. In some experiments, Mendel stopped plants from self-fertilizing by opening their flower buds and removing the stamens. (Stamens bear pollen grains, in which sperm develop.) Then he promoted cross-fertilization by brushing pollen from another plant on the "castrated" floral bud.

Why did Mendel tinker with plants this way? He wanted cross-fertilization to occur between two true-breeding plants that exhibited different forms of the same trait. For example, he crossed a white-flowered with a purple-flowered plant. If their offspring bore white *or* purple flowers, he could identify one plant or the other as the source of the hereditary material for that trait. If there *were* patterns in the way hereditary material is transmitted from parents to offspring, the use of variations in traits might be a way to identify them.

It will be useful to retrace a few of Mendel's experiments. The conclusions he drew from them have turned out to apply, with some modification, to all sexually reproducing organisms.

Some Terms Used in Genetics

Having read the chapter on meiosis, you already have insight into the mechanisms of sexual reproduction—which is more than Mendel had. He did not know about chromosomes and so could not have known that the parental chromosome number is reduced by half in gametes, then restored at fertilization. Yet Mendel had some hunches about what was going on. As we follow his thinking, let's simplify things by substituting a few modern terms used in studies of inheritance (see also Figure 11.3):

1. **Genes** are units of instructions for producing or influencing a specific trait in offspring. Each gene has its own particular location (*locus*) on a chromosome.

2. Diploid cells have inherited a pair of genes for each trait, one on each of two homologous chromosomes.

3. Although both genes of a pair deal with the same trait, they may vary in their information about it. This happens when they have slight molecular differences, as when one gene for flower color specifies "red" and another specifies "white." All of the different molecular forms of a gene that exist are called **alleles.**

4. Gene shufflings during meiosis and fertilization can put together different mixes of alleles in offspring. If it turns out that the two alleles of a pair are the same, this is a *homozygous* condition. If different, this is a *heterozygous* condition.

5. Often one allele of a pair is "dominant," meaning its effect on a trait masks the effect of its "recessive" partner. We use capital letters for dominant alleles and lowercase letters for recessive ones (for example, alleles *A* and *a*).

6. Putting this together, we say a **homozygous dominant** individual has two dominant alleles (*AA*) for the trait being studied. A **homozygous recessive** individual has two recessive alleles (*aa*). A **heterozygous** individual has two different alleles (*Aa*).

7. To keep the distinction clear between genes and the traits they specify, we can use **genotype** when referring to the genes present in an individual, and **phenotype** when referring to an individual's observable traits.

The Concept of Segregation

Mendel's first crosses were *monohybrid*. This means two parents that bred true for contrasting forms of a single trait were crossed in order to produce heterozygous offspring. Mendel tracked the trait through two generations of offspring, which can be designated as follows:

P parental generation

F_1 first-generation offspring

F_2 second-generation offspring

In one case, Mendel crossed a purple-flowered plant with a white-flowered one. *All* the F_1 offspring from that cross had purple flowers. Then he allowed F_1 plants to self-fertilize and produce seeds—and some of the F_2 offspring had white flowers!

Mendel interpreted the results this way. Each plant inherits two "units" of instruction (two genes) for flower color, and those units retain their identity from one generation to the next. Each parent contributes only one unit to each of its offspring. Assume both units were purple in one parent and both were white in the other. Each F_1 plant would then inherit one purple and one white. If that were so, then purple must be the dominant form of the trait, because it masked the white in F_1 plants.

Let's rephrase Mendel's interpretation in terms of what we know about meiosis and fertilization. Pea plants are diploid, with pairs of homologous chromosomes. Assume one parent is homozygous dominant (AA) for flower color and the other, homozygous recessive (aa). After meiotic cell division, each sperm or egg will carry a gene for flower color on one of its chromosomes. As Figure 11.4 shows, when a sperm and egg combine at fertilization, only one outcome is possible: $A + a = Aa$.

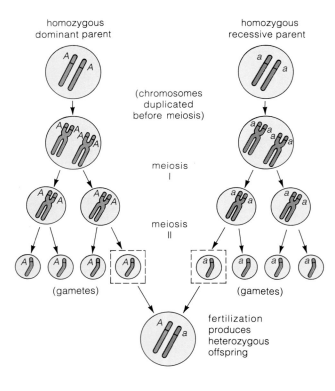

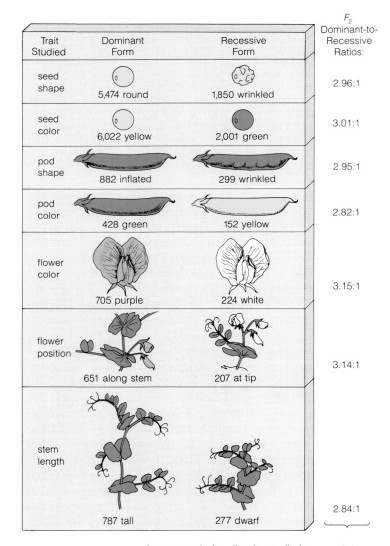

Figure 11.4 Segregation of alleles in a monohybrid cross. Two parents that are true-breeding for contrasting forms of a single trait can give rise only to heterozygous offspring.

Figure 11.5 (*Right*) Results from Mendel's monohybrid cross experiments with the garden pea. The numbers are his counts of F_2 plants showing the dominant or recessive form of the trait. On the average, the dominant-to-recessive ratio was 3:1.

Trait Studied	Dominant Form		Recessive Form		F_2 Dominant-to-Recessive Ratios:
seed shape		5,474 round		1,850 wrinkled	2.96:1
seed color		6,022 yellow		2,001 green	3.01:1
pod shape		882 inflated		299 wrinkled	2.95:1
pod color		428 green		152 yellow	2.82:1
flower color		705 purple		224 white	3.15:1
flower position		651 along stem		207 at tip	3.14:1
stem length		787 tall		277 dwarf	2.84:1

Average ratio for all traits studied: 3:1

But what about the F_2 results? To understand this aspect of Mendel's monohybrid crosses, you have to know he crossed hundreds of plants and kept track of thousands of first- and second-generation offspring. He also *counted* and *recorded* the number of dominant and recessive offspring for each cross. An intriguing ratio emerged from his records. On the average, of every four F_2 plants, three showed the dominant form of the trait and one showed the recessive (Figure 11.5).

Mendel used his knowledge of mathematics to explain the 3:1 phenotypic ratio. He began by assuming each particular sperm is not precommitted to combining with one particular egg; fertilization has to be a chance event. This meant the monohybrid crosses could be interpreted according to rules of **probability**, which apply to chance events. "Probability" simply means the number of times a particular outcome will occur, divided by the total number of all possible outcomes.

A simple way to predict the probable outcome of a cross between two F_1 plants is the **Punnett-square method**, shown in Figure 11.6. Assume each F_1 plant produced two kinds of sperm or eggs in equal proportions: half were *A*, and half were *a*. If any sperm is equally likely to fertilize any egg, there are four possibilities for each encounter:

Possible Event:	Probable Outcome:
sperm *A* meets egg *A*	1/4 *AA* offspring
sperm *A* meets egg *a*	1/4 *Aa*
sperm *a* meets egg *A*	1/4 *Aa* } or 1/2 *Aa*
sperm *a* meets egg *a*	1/4 *aa*

Thus, as Figure 11.7 shows, a randomly selected F_2 plant had three chances in four of carrying at least one dominant allele and developing purple flowers. It had only one chance in four of carrying two recessive alleles and developing white flowers. As a result, the probable phenotypic ratio was 3 purple to 1 white, or 3:1.

Results from his monohybrid crosses led Mendel to formulate a principle. Stated in modern terms,

Mendelian principle of segregation. **Diploid organisms inherit a pair of genes for each trait (on a pair of homologous chromosomes). The two genes segregate from each other at meiosis, so each gamete formed after meiosis has an equal chance of receiving one or the other gene, but not both.**

Keep in mind that Mendel's observed ratios weren't *exactly* 3:1 (see, for example, the numerical results in Figure 11.5). To understand why, flip a coin a few times.

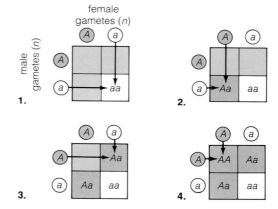

Figure 11.6 Punnett-square method of predicting the probable ratio of traits that will show up in offspring of self-fertilizing individuals known to be heterozygous (*Aa*) for a trait. The circles represent gametes. The letters inside gametes represent the dominant or recessive form of the gene being tracked. Each square depicts the genotype of one kind of offspring.

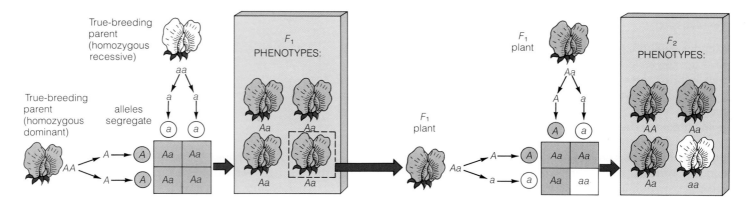

Figure 11.7 Results from one of Mendel's monohybrid crosses to produce heterozygous offspring. Notice that the dominant-to-recessive ratio is 3:1 for the second-generation (F_2) plants.

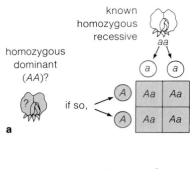

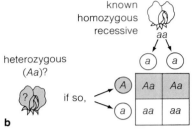

Figure 11.8 Punnett-square method of predicting the outcomes of a testcross between an individual known to be homozygous recessive for a trait (here, white flower color) and an individual showing the dominant form of the trait. (**a**) If the individual of unknown genotype is homozygous dominant, all offspring will show the dominant form of the trait. (**b**) If the individual is heterozygous, about half the offspring will show the recessive form.

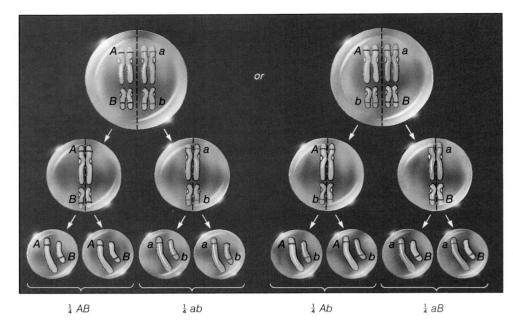

$\frac{1}{4}\ AB$ $\qquad$ $\frac{1}{4}\ ab$ $\qquad$ $\frac{1}{4}\ Ab$ $\qquad$ $\frac{1}{4}\ aB$

Figure 11.9 Example of independent assortment, showing just two pairs of homologous chromosomes. The different combinations of alleles possible in gametes arise in part through the random alignment of homologues during metaphase I of meiosis (page 156).

We all know that a coin is just as likely to end up heads as tails. But often it ends up heads, or tails, several times in a row. When you flip the coin only a few times, the observed ratio may differ greatly from the predicted ratio of 1:1. Only when you flip the coin many times will you come close to the predicted ratio. Almost certainly, Mendel's reliance on a large number of crosses and his understanding of probability kept him from being confused by minor deviations from the predicted results.

Testcrosses

Mendel gained support for his concept of segregation through the **testcross**. In this type of cross, first-generation hybrids are crossed to an individual known to be true-breeding for the same recessive trait carried by the recessive parent. (In other words, that individual is homozygous recessive.)

For example, Mendel crossed purple-flowered F_1 plants with true-breeding, white-flowered plants. If his concept were correct, there would be about as many recessive as dominant plants in the offspring from the testcross. That is exactly what happened (Figure 11.8). As predicted, about half the testcross offspring were purple-flowered (*Aa*) and half were white-flowered (*aa*).

The Concept of Independent Assortment

In another series of experiments, Mendel crossed true-breeding pea plants having contrasting forms of two traits. In such *dihybrid crosses*, the F_1 offspring inherit two gene pairs, neither of which consists of identical alleles. In the following example of a dihybrid cross, *A* and *B* stand for dominance in flower color and height, respectively; *a* and *b* stand for their recessive counterparts:

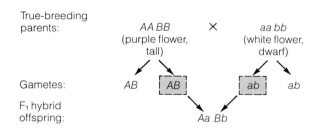

Mendel anticipated (correctly) that all the F_1 offspring of the cross would be purple-flowering and tall. But what would happen when those *Aa Bb* offspring formed sperm and eggs of their own? Would a gene for flower color and a gene for height *travel together or independently of each other* into the gametes?

Figure 11.10 Results from Mendel's dihybrid cross between true-breeding parent plants differing in two traits (flower color and height).

Here, *A* and *a* represent the dominant and recessive alleles for flower color. *B* and *b* represent the dominant and recessive alleles for height.

On the average, the phenotypic combinations in the F_2 generation occur in a 9:3:3:1 ratio. Keep in mind that working a Punnett square is really a way to show *probabilities* of certain combinations occurring as a result of allele shufflings during meiosis and fertilization.

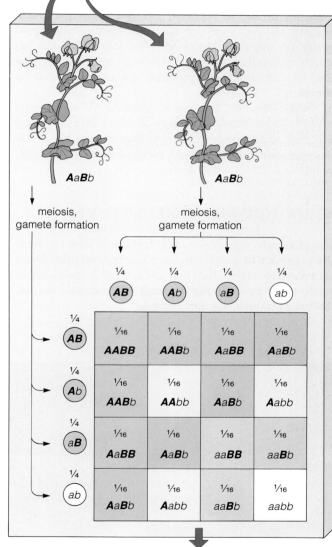

As we now know, a particular gene pair tends to segregate into gametes independently of other gene pairs—when the others are located on *non*homologous chromosomes. In this case, four combinations of alleles are possible:

$$1/4\ AB \qquad 1/4\ Ab \qquad 1/4\ aB \qquad 1/4\ ab$$

These possibilities arise through the random alignment of maternal and paternal chromosomes during metaphase I of meiosis (page 156 and Figure 11.9).

Now think about the way alleles can be mixed at fertilization. Simple multiplication (four kinds of sperm times four kinds of eggs) tells us that *sixteen* gametic combinations are possible in dihybrid F_2 plants. Figure 11.10 lays out the possibilities, using the Punnett-square method. The sixteen gametic combinations result in nine different genotypes and four different phenotypes. What accounts for this numerical difference? As you can see from Figure 11.10, more than one combination of gametes can give rise to the same genotype, and some genotypes give rise to the same phenotype.

When we total the possible outcomes by phenotype, we get 9/16 tall purple-flowered, 3/16 dwarf purple-flowered, 3/16 tall white-flowered, and 1/16 dwarf white-flowered plants. That is a probable phenotypic ratio of 9:3:3:1.

Results from all of Mendel's dihybrid crosses were close to a 9:3:3:1 ratio, and they led to the formulation of another principle. In modern terms, here was evidence that the gene pairs located on two homologous chromosomes tend to travel into gametes independently of the gene pairs that are located on other homologous chromosomes.

Mendelian principle of independent assortment. **Each gene pair tends to assort into gametes independently of other gene pairs that are located on nonhomologous chromosomes.**

The variety resulting from independent assortment and hybrid crossing is staggering. In a monohybrid cross (involving a single gene pair), only three genotypes are possible (AA, Aa, and aa). We can represent this as 3^n, where n is the number of gene pairs. When more gene pairs are involved, the number of possible combinations increases dramatically. Even if parents differ in only ten gene pairs, almost 60,000 different genotypes are possible among their offspring. If they differ in twenty gene pairs, the number is close to 3.5 billion!

On the basis of his experimental results, Mendel was convinced that the hereditary material comes in units that retain their physical identity from one generation to the next (despite being segregated and assorted during meiosis). In 1865 he reported this idea before the Brünn Society for the Study of Natural Science. His report made no impact whatsoever. The following year his paper was published. Apparently it was read by few and its near-universal applicability was understood by no one.

In 1871 Mendel became an abbot of the monastery, and his experiments gave way to administrative tasks. He died in 1884, never to know that his work was the starting point for the development of modern genetics.

VARIATIONS ON MENDEL'S THEMES

It was Mendel's genius or good fortune to limit his studies to genes that were expressed in completely dominant or recessive ways. But the phenotypic expression of many other genes is not as straightforward, as the following examples will demonstrate.

Dominance Relations

Different degrees of dominance may exist between a pair of alleles. One or both may be fully dominant, or one may be incompletely dominant over the other.

In **incomplete dominance**, the phenotype of a heterozygote is intermediate between that of the homozygous dominant or recessive types. Suppose you cross true-breeding red-flowered and white-flowered snapdragons. All of the F_1 offspring will have *pink* flowers (Figure 11.11). This might appear to be an outcome of "blending"—until you perform a cross between two F_1 plants. You will discover that the "red" allele was not blended away, because the F_2 offspring of those first-generation plants will have red, pink, or white flowers in a predictable 1:2:1 ratio. The dominant allele calls for a red pigment, and it takes two alleles to produce enough pigment to give the flowers a red color. With their single dominant allele, heterozygotes can only produce enough pigment to give the flowers a pinkish cast.

In **codominance**, two alleles of a pair are not identical, yet the expression of *both* can be discerned in heterozygotes. Each gives rise to a different phenotype. For example, you probably have heard of **ABO blood typing**. It refers to a method of characterizing blood according to which forms of a certain protein occur at the surface of a person's red blood cells. (The protein is like a molecular fingerprint; it identifies the cell as being of a certain type.) Three alleles influence the protein's structure. Two, I^A and I^B, are codominant when paired with each other. A third allele, i, is recessive. When paired with either I^A or I^B, its expression is masked.

Both I^A and I^B code for enzymes that attach a sugar group to the protein after it has been synthesized, but

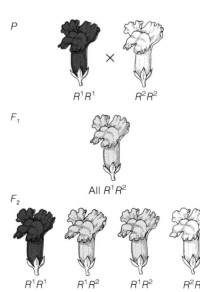

Figure 11.11 Incomplete dominance at one gene locus. Red-flowering and white-flowering homozygous snapdragons produce pink-flowering plants in the first generation. The red allele (R^1) is only partially dominant over the white allele (R^2) in the heterozygous state.

the enzymes make the attachment in different ways. Two forms of the completed protein, called A and B, result from expression of alleles I^A and I^B.

Whenever more than two forms of alleles exist for a given gene locus, we call them a **multiple allele system**. In this case, four blood types are possible, depending on which two of the codominant and recessive alleles of the system are present in a person's cells. As Figure 11.12 indicates, the four types are A, B, AB, and O. Both $I^A I^A$ and $I^A i$ individuals are type A. Both $I^B I^B$ and $I^B i$ individuals are type B. However, $I^A I^B$ individuals are type AB. Red blood cells of homozygous recessive people are neither A nor B; that is what the "O" means.

Interactions Between Different Gene Pairs

On thinking about the traits described so far, you might conclude that each trait arises from the expression of only one pair of genes. However, one gene pair often influences the expression of other gene pairs.

Comb Shape in Poultry. In some interactions, two gene pairs *cooperate* to produce a phenotype that neither can produce alone. W. Bateson and R. Punnett identified two gene pairs that cooperate to produce comb shape in chickens. The allelic combination *rr* at one locus together with *pp* at the other gives rise to the most common phenotype, the single comb. Other phenotypes occur when the two dominant alleles, *R* and *P*, are present. Depending on the allelic combinations, the two gene pairs can produce rose, pea, or walnut combs as well as the single comb (Figure 11.13).

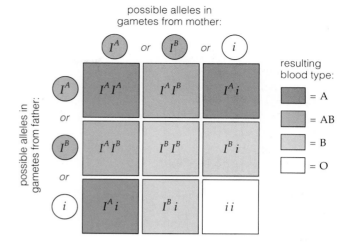

Figure 11.12 Possible combinations of alleles associated with ABO blood typing.

Figure 11.13 (*Below*) Interaction between two genes affecting the same trait in domestic breeds of chickens. The initial cross is between a Wyandotte (with a rose comb on the crest of its head) and a brahma (with a pea comb). With complete dominance at the gene locus for pea comb and at the gene locus for rose comb, the products of these two nonallelic genes interact and give a walnut comb (**a**). With complete recessiveness at both loci, the products interact and give rise to a single comb (**d**).

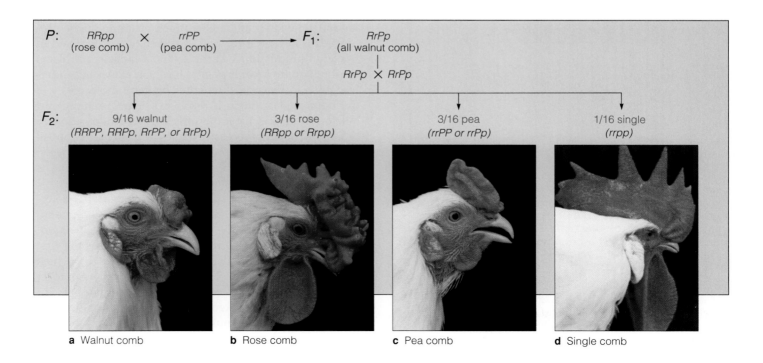

P: RRpp × rrPP ⟶ F₁: RrPp
    (rose comb) (pea comb)   (all walnut comb)

                    RrPp × RrPp

F₂: 9/16 walnut   3/16 rose   3/16 pea   1/16 single
     (RRPP, RRPp, RrPP, or RrPp) (RRpp or Rrpp)   (rrPP or rrPp)   (rrpp)

a Walnut comb   b Rose comb   c Pea comb   d Single comb

a Black Labrador   b Yellow Labrador   c Chocolate Labrador

P homozygotes: *BB EE* x *bb ee*

All *F₁*:   *Bb Ee*

F₂ combinations possible:

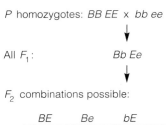

	BE	*Be*	*bE*	*be*
BE	*BB EE*	*BB Ee*	*Bb EE*	*Bb Ee*
Be	*BB Ee*	*BB ee*	*Bb Ee*	*Bb ee*
bE	*Bb EE*	*Bb Ee*	*bb EE*	*bb Ee*
be	*Bb Ee*	*Bb ee*	*bb Ee*	*bb ee*

RESULTING
PHENOTYPES:

▨ 9/16 or 9 black

▨ 3/16 or 3 brown

☐ 4/16 or 4 yellow

Figure 11.14 Interaction among alleles of two gene pairs affecting the coat color of Labrador retrievers. At one gene locus concerned with melanin production, allele B (black) is dominant to allele b (brown). At a different gene locus, allele E allows melanin pigment to be deposited in individual hairs. But two recessive alleles at that locus (ee) prevent deposition and a yellow coat results (this is a case of recessive epistasis).

The F_1 offspring of a dihybrid cross would produce F_2 offspring in a 9:3:4 ratio, as shown above. The yellow Labrador in (**b**) probably has genotype *BB ee*. It has the capacity to produce melanin but not to deposit the pigment in hairs. (Looking at that photograph, can you say why?)

Hair Color in Mammals. In other interactions, two alleles of a gene *mask* the expression of alleles of another gene, and some expected phenotypes do not appear at all. Such interactions, called **epistasis**, are common among the gene pairs that affect the color of fur or skin in mammals.

Figure 11.14 shows a case of epistasis among Labrador retrievers, which may have black, brown, or yellow hair. The different colors result from variations in the amount and distribution of a brownish-black pigment called melanin. Many gene pairs and their alleles function in the metabolic processes leading to melanin production and its deposition in certain body regions. Alleles at one gene locus affect how much melanin will be produced. (They code for one of the enzymes involved in melanin production.) At this locus, a *B* allele (for black) is dominant to *b* (for brown). However, alleles at a different gene locus control whether pigment molecules will be deposited at all in a retriever's hairs. A dominant allele (*E*) at this locus permits pigment deposition in hairs. But a pair of recessive alleles (*ee*) will block deposition, resulting in a yellow coat color. Interaction between the *B* gene and two recessive alleles of the *E* gene is a case of recessive epistasis.

Epistasis involving still another gene locus (*C*) determines whether an individual will be an *albino*, a phenotype resulting from the complete absence of melanin. Alleles at this gene locus code for tyrosinase, the first enzyme required in the series of reactions by which melanin is produced. The *C* allele is dominant to *c*. Melanin can be produced in a *CC* or *Cc* individual. But its production is blocked in a *cc* individual, which will end up with the phenotype characteristic of albinism. Figure 11.15 shows an example.

Figure 11.15 A rare albino rattlesnake, showing the pink eyes and white coloration of animals that are unable to produce the pigment melanin. (Eyes are pink because the absence of melanin allows red light to be reflected from blood vessels in the snake's eyes.)

Surface coloration in birds and mammals is due almost entirely to the color of feathers or fur. In fishes, amphibians, and reptiles, it is due to color-bearing cells in the skin. Some of the cells contain melanin (a brownish-black pigment) or red to yellow pigments. Others contain crystals that reflect light and alter the effect of other pigments present.

The mutation affecting melanin production in the snake shown here had no effect on the production of yellow-to-red pigments and light-reflecting crystals. So the snake's skin appears iridescent yellow as well as white.

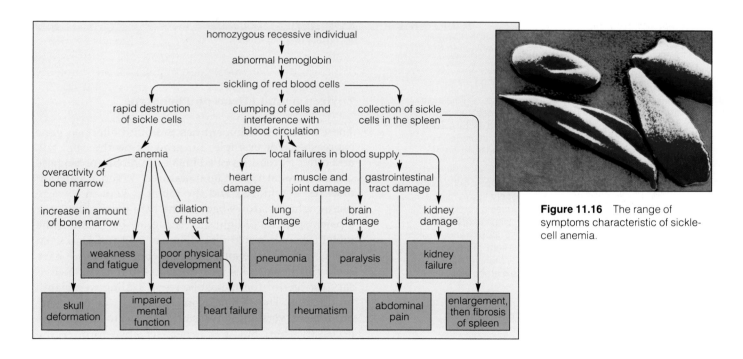

Figure 11.16 The range of symptoms characteristic of sickle-cell anemia.

Multiple Effects of Single Genes

A single gene can exert effects on seemingly unrelated aspects of an individual's phenotype. This form of gene expression is called **pleiotropy**.

For example, sickle-cell anemia, described at the start of this chapter, arises through a mutation at a single gene locus. The mutated allele gives rise to a defective form of hemoglobin, the oxygen-transporting pigment in red blood cells. When the defective pigment molecules give up oxygen, they stack together—and the cells become rigid and distorted. Blood flow through capillaries becomes impaired, the delivery of oxygen to cells is hampered, and so is the removal of carbon dioxide and other metabolic products from them.

As Figure 11.16 shows, the metabolic disruption leads to severe damage of many tissues and organs. The *symptoms* of this underlying genetic disorder are what we call sickle-cell anemia. Heterozygotes still have one functional allele and show few symptoms of the disorder. But homozygous recessives show severe phenotypic consequences.

Figure 11.17 An environmental effect on phenotype. A Himalayan rabbit normally has black hair only on its long ears, nose, tail, and lower limbs of its legs. In one experiment, a patch of a rabbit's white fur was plucked clean, then an icepack was secured over the hairless patch. For as long as the artificially cold conditions were maintained, the hairs that grew back were black.

Himalayan rabbits are homozygous for the c^h allele of a gene that codes for tyrosinase, one of the enzymes necessary to synthesize the pigment melanin. This particular allele produces a version of tyrosinase that is heat sensitive. It functions only when temperatures are below about 33°C. When the rabbit hair grows under warmer conditions, no melanin is produced, so the hair is light. Light fur normally covers the body regions that are warmer than the ears and other slender extremities (which tend to lose metabolically generated heat more rapidly).

Siamese cats that are homozygous for the same heat-sensitive allele also display this environmental effect on phenotype (Figure 6.12).

Environmental Effects on Phenotype

The external environment has profound effect on gene expression. For example, fur growing on the ears, tail, and other extremities of a Himalayan rabbit (or Siamese cat) darkens at cooler temperatures. When a patch of dark fur is removed and the rabbit is kept in a warm place, light fur grows back in the bare patch. Similarly, a patch stripped of light hair will grow back dark if the rabbit is kept in a cold place (Figures 6.12 and 11.17). An allele of the C gene affects this phenotype. As we have seen, this gene codes for tyrosinase, an enzyme necessary for one of the metabolic steps leading to melanin production. The "Himalayan" allele (c^h) of this gene produces a version of tyrosinase that can catalyze the necessary step. But it is a heat-sensitive form of the enzyme—it is less active at warm temperatures. Thus, for homozygotes ($c^h c^h$), fur color depends on what the temperature is at the time hair is growing.

The water buttercup (*Ranunculus aquatilis*) provides another example of environmental effects on phenotype. This plant grows in shallow ponds, and some of its leaves develop underwater. The submerged leaves are finely divided, compared with the leaves growing in air. When a leaf-bearing stem is half in and half out of the water, its leaves display both phenotypes (Figure 11.18). The genes responsible for leaf shape produce very different phenotypes under different environmental conditions.

The internal environment (that is, conditions within the tissues surrounding the body's cells) also influences gene expression. At puberty, for example, the male body steps up its production of testosterone, a sex hormone. Among other things, increased levels of testosterone affect genes that govern the development of cartilage in the larynx. The voice deepens as a result of those changes. Without the hormonal increase, the voice will remain higher pitched later in life.

Variable Gene Expression in a Population

Penetrance and Expressivity. Reflect, for a moment, on some of the concepts described so far. Many genes, such as the one responsible for purple flowers in pea plants, are expressed in a regular, consistent pattern. Others are expressed to varying degrees among the individuals of a population. Two terms, penetrance and expressivity, are used in describing variable gene expression in a population.

Penetrance is the frequency with which a dominant or homozygous recessive gene is actually expressed in the phenotype of the carriers of that gene. It's an all-or-none measure; either the gene is expressed in an individual or it is not. The dominant gene responsible for purple flowers is completely penetrant—if plants have the gene, the phenotype is fully expressed. By contrast, a dominant gene responsible for *campodactyly* is incompletely penetrant. This human genetic disorder is characterized by immobile, bent fingers. It results when muscles are improperly attached to bone when the little fingers are forming in an embryo. Because the gene is dominant, you might expect it to be expressed in all homozygotes and heterozygotes, but this is not the case. In one study, only eight of nine people who were known to carry the dominant gene exhibited the trait. Thus penetrance was 8/9, or 88 percent.

Expressivity is the *degree* to which a characteristic phenotype is exhibited among individuals of a population. In the case of campodactyly, the trait didn't show up at all in one individual. It showed up on both hands of some individuals but on only one hand of others.

Variable gene expression is not necessarily a property of the gene itself. Rather, it arises through gene interactions and through environmental influences. Keep in mind that *most* observable traits are the end products of a series of small, distinct metabolic steps. Some of those steps begin when certain genes first become active in a developing embryo. Other steps are taken later in life. As we have seen, a gene that is required for a step leading to a particular phenotype may be influenced by interactions with other genes. Some number of those genes occur in different allelic combinations in different indi-

Figure 11.18 Variable expressivity resulting from variation in the external environment. Leaves of the water buttercup (*Ranunculus aquatilis*) show dramatic phenotypic variation, depending on whether they grow underwater or above it. This variation occurs even in the same leaf if it develops half in and half out of water. Compare Figure 20.5.

Figure 11.19 A sampling from the range of continuous variation in human eye color. Many different genes are required to produce and deposit melanin in the iris of the eye, and the effects of those genes are cumulative. Slight differences in those genes and their gene products lead to small differences in the eye color trait, so that the frequency distribution for eye color appears to be continuous over the range from black to light blue.

Figure 11.20 (*Right*) Continuous variation in traits.

In a given population, most traits show continuous variation. For example, humans are not all tall or short. They grow to various heights, ranging from very short to very tall, with average heights being more common than either extreme.

The following process is used to describe a population showing continuous variation in some trait. First, the full range of phenotypic variation is divided into measurable categories. Then individuals falling within a given category are counted. This process shows how the relative frequency of different categories is distributed across the possible range of measurable values for the phenotype.

Suppose you wish to determine the frequency distribution for height in a specific group of men, such as those shown in (**a**). First you decide on the degree to which you should divide up the range of possible heights, then you measure each individual and assign him to the appropriate category. Once this is done, you can divide the number in each category by the total number of all individuals in all categories.

(**b**) Often a bar graph is used to illustrate continuous variation. Here, the proportion of individuals falling into each category is plotted against the range of measured phenotypes.

viduals, and their products (proteins) may affect a given gene in different ways, at different times. Besides this, nutrition and other environmental factors may affect the activity of a given gene, and those factors may not be the same from one individual to the next.

Continuous Variation. Think about an observable trait such as eye color and hair color. These traits result from the cumulative expression of many different genes that are involved in the stepwise production and distribution of melanin. For the reasons given above, those genes are expressed to different degrees in different individuals in a population. Black eyes have abundant melanin deposits in the iris. Dark-brown eyes have less melanin, and light-brown or hazel eyes have still less (Figure 11.19). Green, gray, and blue eyes don't have green, gray, or blue pigments. They have so little mela-

nin that we readily see blue wavelengths of light being reflected from the iris.

For eye color and many other traits, a population of humans and other organisms shows **continuous variation**. The individuals within the population exhibit a range of smaller, less pronounced differences in some trait. Continuous variation is especially evident in traits that are easily measurable, such as height (Figure 11.20). As the number of gene pairs affecting a given phenotype increases, the expected phenotypic distribution appears to be more and more continuous.

Continuous variation is consistent with Mendel's views about inheritance. "Continuous" does not mean "blending," a term that suggests loss of the original identity of the genes governing heritable traits. The genes are still intact, regardless of the phenotype produced by their combined positive and negative effects.

(number of individuals)

1 0 0 1 5 7 7 22 25 26 27 17 11 17 4 4 1

(height, inches)

58 59 60 61 62 63 64 65 66 67 68 69 70 71 72 73 74

a Height distribution in a group of 175 U.S. Army recruits about the turn of the century.

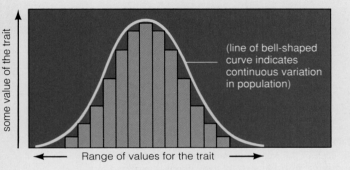

(line of bell-shaped curve indicates continuous variation in population)

Range of values for the trait

b Generalized bell-shaped curve typical of populations showing continuous variation in some trait.

The phenotypic expression of a given gene may vary, by different degrees, from one individual to the next in a population.

The degree to which the same gene is expressed in different individuals depends on gene interactions. It depends also on the physical and chemical environment in which that gene or its products must function.

SUMMARY

1. Mendel's hybridization studies with garden pea plants demonstrated that diploid organisms have two units of hereditary material (genes) for each trait, and that genes retain their identity when passed on to offspring.

2. Mendel conducted monohybrid crosses, which are crosses between two true-breeding individuals having contrasting forms of a single trait. Interpreting the results in modern terms, we can say that the crosses provided indirect evidence that a gene at a given location on a chromosome can exist in slightly different molecular forms (alleles), some of which are dominant over other, recessive forms.

3. Homozygous dominant individuals have two dominant alleles (AA) for the trait being studied. Homozygous recessives have two recessive alleles (aa). Heterozygotes have two different alleles (Aa).

4. In Mendel's monohybrid crosses ($AA \times aa$), all F_1 offspring were Aa. Crosses between plants produced these combinations of alleles in F_2 offspring:

	A	a
A	AA	Aa
a	Aa	aa

AA (dominant)
Aa (dominant)
Aa (dominant)
aa (recessive)

This produced an expected phenotypic ratio of $3:1$.

5. Results from such monohybrid crosses supported the Mendelian principle of segregation: Diploid organisms have a pair of genes for each trait, on a pair of homologous chromosomes. The two genes segregate from each other during meiosis, such that each gamete formed will end up with one or the other, but not both.

6. Mendel also performed dihybrid crosses, which are crosses between two true-breeding individuals having contrasting forms of two different traits. Results from many experiments were close to a $9:3:3:1$ phenotypic ratio:

9 dominant for both traits
3 dominant for A, recessive for b
3 dominant for B, recessive for a
1 recessive for both traits

7. On the basis of such dihybrid crosses, the Mendelian principle of independent assortment was formulated: Each gene pair (located on a pair of homologous chromosomes) tends to assort into gametes independently of other gene pairs that are located on other pairs of homologous chromosomes.

8. Since Mendel's time, we have learned that (1) degrees of dominance may exist between some gene pairs, (2) gene pairs can interact to produce some positive or negative effect on phenotype, (3) a single gene can have effects on many seemingly unrelated traits, and (4) the internal and external environments influence gene expression.

Review Questions

1. State the Mendelian principle of segregation. Does segregation occur during mitosis or meiosis? *169*

2. Distinguish between the following terms: (a) gene and allele, (b) dominant trait and recessive trait, (c) homozygote and heterozygote, (d) genotype and phenotype. *167*

3. Give an example of a self-fertilizing organism. What is cross-fertilization? *166–167*

4. Distinguish between monohybrid and dihybrid crosses. What is a testcross, and why is it valuable in genetic analysis? *168, 170*

5. State the Mendelian principle of independent assortment. Does independent assortment occur during mitosis or meiosis? *171*

Self-Quiz *(Answers in Appendix IV)*

1. Alleles are _____.
   a. alternative molecular forms of a gene
   b. alternative molecular forms of a chromosome
   c. self-fertilizing, true-breeding homozygotes
   d. self-fertilizing, true-breeding heterozygotes

2. A heterozygote is _____.
   a. one of at least two forms of a gene
   b. a condition in which both alleles of a pair are the same
   c. a condition in which both alleles of a pair are different
   d. a haploid condition in genetic terms

3. The observable traits of an organism are called its _____.
   a. phenotype   c. genotype
   b. sociobiology   d. pedigree

4. In the monohybrid cross $AA \times aa$, the F_1 offspring are _____.
   a. all AA d. 1/2 AA and 1/2 aa
   b. all aa e. none of the above
   c. all Aa

5. The second generation of offspring from a genetic cross is called the _____.
   a. F_1 generation   c. hybrid generation
   b. F_2 generation   d. none of the above

6. In the genetic cross $Aa \times Aa$ where A is completely dominant over a, the next generation will show a phenotypic ratio of _____.
   a. 1:2:1   d. 9:3:3:1
   b. 1:1:1:1   e. none of the above
   c. 3:1

7. Which of the following statements most accurately explains Mendel's principle of segregation?
   a. particular units of heredity are transmitted to offspring
   b. two genes on a pair of homologous chromosomes segregate from each other in meiosis
   c. members of a population become segregated

8. Assuming simple dominance and independent assortment, dihybrid crosses between two true-breeding organisms with con-trasting forms of two traits (as in $Aa\ Bb \times Aa\ Bb$) produce offspring phenotypic ratios close to _____.
   a. 1:2:1
   b. 1:1:1:1
   c. 3:1
   d. 9:3:3:1

9. "Each pair of genes tends to assort into gametes independently of other gene pairs located on nonhomologous chromosomes" is a statement of Mendel's _____.
   a. principle of dominance
   b. principle of segregation
   c. principle of independent assortment
   d. none of the above

10. Match each genetic term appropriately.
   _____ dihybrid cross   a. $AA \times aa$
   _____ monohybrid cross   b. Aa
   _____ homozygous condition   c. $Aa\ Bb \times Aa\ Bb$
   _____ heterozygous condition   d. $Aa \times Aa$
   _____ true-breeding parents   e. aa

Genetics Problems *(Answers in Appendix III)*

1. One gene has alleles A and a; another gene has alleles B and b. For each of the following genotypes, what type(s) of gametes will be produced? (Independent assortment is expected.)
   a. $AA\ BB$ c. $Aa\ bb$
   b. $Aa\ BB$ d. $Aa\ Bb$

2. Still referring to the preceding problem, what genotypes will be present in the offspring from the following matings? (Indicate the frequencies of each genotype among the offspring.)
   a. $AA\ BB \times aa\ BB$ c. $Aa\ Bb \times aa\ bb$
   b. $Aa\ BB \times AA\ Bb$ d. $Aa\ Bb \times Aa\ Bb$

3. In one experiment, Mendel crossed a true-breeding pea plant having green pods with a true-breeding pea plant having yellow pods. All of the F_1 plants had green pods. Which trait (green or yellow pods) is recessive? Can you explain how you arrived at your conclusion?

4. In addition to the two genes mentioned in Problem 1, assume you now study a third independently assorting gene having alleles C and c. For each of the following genotypes, indicate what type (or types) of gametes will be produced:
   a. $AA\ BB\ CC$ c. $Aa\ BB\ Cc$
   b. $Aa\ BB\ cc$ d. $Aa\ Bb\ Cc$

5. A man is homozygous dominant for ten different genes, which assort independently. How many genotypically different types of sperm could he produce? A woman is homozygous recessive for eight of these ten genes, and she is heterozygous for the other two. How many genotypically different types of eggs could she produce? What can you conclude regarding the relationship between the number of different gametes possible and the number of heterozygous and homozygous genes that are present?

6. Recall that Mendel crossed a true-breeding tall, purple-flowered pea plant with a true-breeding dwarf, white-flowered plant. All the F_1 plants were tall and purple-flowered. If an F_1 plant is now self-pollinated, what is the probability of obtaining an F_2 plant heterozygous for the genes controlling height and flower color?

7. Being able to curl up the sides of your tongue into a U-shape is under the control of a dominant allele at one gene locus. (When there is a recessive allele at this locus, the tongue cannot be rolled.) Having free earlobes is a trait controlled by a dominant allele at a different gene locus. (When there is a recessive allele at this locus, earlobes are attached at the jawline.) The two genes controlling tongue-rolling and free earlobes assort independently. Suppose a woman who has free earlobes and who can roll her tongue marries someone who has attached earlobes and who cannot roll his tongue. Their first child has attached earlobes and cannot roll the tongue.

a. What are the genotypes of the mother, the father, and the child?

b. If this same couple has a second child, what is the probability that it will have free earlobes and be unable to roll the tongue?

8. Assume that a gene was recently identified in canaries. One allele at this gene locus produces a yellow feather color. A second allele produces brown feathers. Suppose you are asked to determine the dominance relationship between these two alleles. (Is it one of simple dominance, incomplete dominance, or codominance?) What types of crosses would you make to find the answer? On what types of observations would you base your conclusions?

9. The ABO blood system has often been employed to settle cases of disputed paternity. Suppose, as an expert in genetics, you are called to testify in a case where the mother has type A blood, the child has type O blood, and the alleged father has type B blood. How would you respond to the following statements of the attorneys:

a. "Since the mother has type A blood, the type O blood of the child must have come from the father, and since my client has type B blood, he obviously could not have fathered this child." (*Made by the attorney of the alleged father*)

b. "Further tests revealed that this man is heterozygous and therefore he must be the father." (*Made by the mother's attorney*)

10. In mice, at one gene locus, the dominant allele (B) produces a dark-brown pigment; and the recessive allele (b) produces a light-brown, or tan, pigment. An independently assorting gene locus has a dominant allele (C) that permits the production of all pigments. Its recessive allele (c) makes it impossible to produce any pigment at all. The pigmentless condition is called "albino."

a. A homozygous *bb cc* albino mouse mates with a homozygous *BB CC* brown mouse. Assuming independent assortment, in what ratios would the phenotypes and genotypes be expected in the F_1 and F_2 generations?

b. If an F_1 mouse from part (a) above were backcrossed to its albino parent, what phenotypic and genotypic ratios would be expected?

11. Certain dominant alleles are so important for normal development that the mutant recessive alleles, when homozygous, lead to the death of the organism. However, such recessive alleles can be perpetuated as heterozygotes (*Ll*), which in many cases are not phenotypically different from homozygous (*LL*) normals. (In some cases, individuals carrying a recessive lethal allele do have a mutant phenotype.) Consider the mating of two such heterozygotes, *Ll* × *Ll*. Among their *surviving* progeny, what is the probability that any individual will be heterozygous?

12. In corn, a series of three independent pairs of gene loci (*A, C,* and *R*) affect the production of pigment that leads to kernel color. If any one of the three pairs is in the homozygous recessive state, then no pigment will form in the kernels. However, if at least one dominant allele of each locus is present, then pigment can form in the kernel. Two corn plants with the following genotypes were crossed:

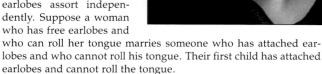

$$Aa\ cc\ Rr \times aa\ Cc\ Rr$$

What fraction of the progeny kernels will be pigmented? (Note: each kernel represents a separate (potential) individual; it will exhibit the pigment phenotype of the plant that can be grown from it.)

Selected Key Terms

ABO blood typing *172*	homozygous recessiveness *167*
allele *167*	incomplete dominance *172*
codominance *172*	independent assortment *171*
continuous variation *178*	monohybrid cross *168*
dihybrid cross *170*	multiple allele system, *173*
epistasis *174*	penetrance, *177*
expressivity *177*	phenotype *167*
gene locus *167*	pleiotropy *175*
gene *167*	probability, *169*
genotype *167*	Punnett-square method *169*
heterozygous condition *167*	segregation *169*
homozygous condition *167*	testcross *170*
homozygous dominance *167*	true-breeding organism *166*

Readings

Dunn, L. 1965. *A Short History of Genetics.* New York: McGraw-Hill.

Mendel, G. 1959. "Experiments in Plant Hybridization." Translation in J. Peters (editor), *Classic Papers in Genetics.* Englewood Cliffs, New Jersey: Prentice-Hall.

Suzuki, D., et al. 1989. *An Introduction to Genetic Analysis.* Fourth edition. New York: Freeman. Chapters 2 and 4 are clear introductions to Mendelian analysis.

12 CHROMOSOME VARIATIONS AND HUMAN GENETICS

Too Young To Be Old

Imagine being ten years old, trapped in a body that each day becomes a bit more shriveled, a bit more frail, *old*. You are just tall enough to peer over the top of the kitchen counter, and you weigh less than thirty-five pounds. Already you are bald, and your nose has become crinkled and beaklike.

Most likely, you have only a few more years to live. Yet in spite of this cruel twist of nature, you still have not lost your courage or your childlike curiosity about life. Like Mickey Hayes and Fransie Geringer (Figure 12.1), you play, laugh, celebrate birthdays, and hug your friends.

Of every 8 million humans born, one is destined to grow old far too soon, compared to the normal timetable for our species. Something has gone wrong with one gene on just one of the forty-six chromosomes brought together by chance at conception. From that moment on, the mistake is perpetuated each time cells of the embryo—then of the child—duplicate their chromosomes and divide. The outcome of that rare mistake will be an acceleration of the aging process and a greatly reduced life expectancy. This is the *Hutchinson-Gilford progeria syndrome*, once called the "leprechaun's disease." There is no cure.

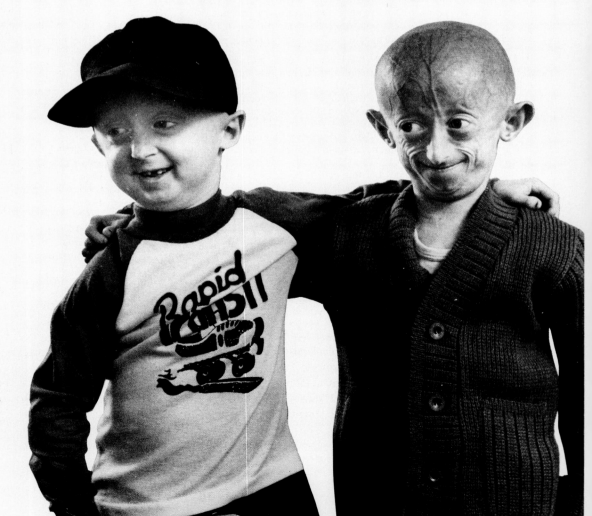

Figure 12.1 Two boys, both less than ten years old, who met at Disneyland, California, during a gathering of progeriacs. Progeria is a genetic disorder characterized by accelerated aging and extremely reduced life expectancy.

Usually, symptoms begin to appear before affected individuals are two years old. In some unknown way, the interaction of the altered gene with other genes has absolutely devastating effects on normal cell division, growth, and development. The rate of growth quickly declines to abnormally low levels. Skin becomes thinner and muscles become flaccid. Limb bones that otherwise should lengthen and become stronger start to soften. Most of the time, all hair is lost and the individuals become bald. Progeriacs never reach puberty, the onset of sexual maturation. Most die in their early teens from a stroke or heart attack brought on by an earlier hardening of the arteries, a condition typical of advanced age.

Outwardly, none of the chromosomes of affected individuals appears to be defective when viewed with a microscope. Simple analysis tells us that the condition probably arises through a dominant gene mutation that strikes arbitrarily. The mutated gene is always expressed. It does not occur on a sex chromosome, because the resulting disorder can occur in either boys or girls. There are no documented cases of progeria running in families. As is the case for some other dominant gene mutations, there is a "paternal age effect," in that the father is four or five years older than the mother at the time of conception, on the average.

We began this unit of the book by looking at the mechanisms of cell division, the starting point of inheritance. Then we started thinking about the genes on chromosomes, and we began to analyze some of the phenotypic consequences of gene shufflings at meiosis and at fertilization. With this chapter we delve more deeply into the inheritance of chromosomes and the genes they carry, with emphasis on the patterns of inheritance in humans. At times the methods of analysis might seem abstract. But keep in mind that we are talking about normal and abnormal gene expression in yourself and in other human beings. At this writing, Mickey Hayes is close to eighteen years old and is the oldest living progeriac. Fransie was seventeen when he died.

1. Genes are arranged linearly along chromosomes. Although genes on the same chromosome tend to stay together during meiosis, crossing over disrupts such linkages. The farther apart two genes are along a chromosome, the greater will be the frequency of crossing over and recombination between them.

2. Crossing over during meiosis leads to genetic variation—hence to variation in traits among offspring. So does independent assortment of homologous chromosomes during meiosis, the outcome of which is the random mix of maternal and paternal chromosomes in gametes.

3. Abnormal changes in the structure or number of chromosomes occur on rare occasions. Such events often cause genetic disorders. Together with crossing over and independent assortment, they also may influence the course of evolution. They lead to different combinations of traits in offspring, and the differences can be acted upon by agents of selection.

CHROMOSOMAL THEORY OF INHERITANCE

Return of the Pea Plant

The year was 1884. Mendel's paper on hybridization of pea plants had been gathering dust in a hundred libraries for nearly two decades, and Mendel himself had just passed away. Ironically, the experiments described in that forgotten paper were about to be devised all over again, as a way to test ideas emerging from another line of research. Cytology, the study of cell structure and function, was about to converge with genetic analysis.

Improvements in microscopy had rekindled efforts to locate the cell's hereditary material, and researchers were zeroing in on the nucleus. By 1882, Walther Flemming had observed threadlike bodies—chromosomes—in the nuclei of dividing cells. By 1884, a question was taking shape: Could those threadlike chromosomes be the hereditary material?

Then researchers realized each gamete has half as many chromosomes as a fertilized egg. In 1887, August Weismann proposed that a special division process must reduce the chromosome number by half before gametes form. Sure enough, in that same year meiosis was discovered. Weismann now began to promote his theory of heredity. In essence, he argued that the chromosome number is halved during meiosis, then restored when sperm and egg combine at fertilization. Thus half the hereditary material in offspring is paternal in origin, and half is maternal. His views were hotly debated, and the debates drove researchers into testing the theory. Throughout Europe there was a flurry of experimental crosses—just like the ones Mendel had carried out.

Finally, in 1900, researchers came across Mendel's paper while checking for literature related to their own hybridization studies. To their chagrin, their results merely confirmed what Mendel already had said. *Diploid cells have two units of instruction (genes) for each heritable trait, and the units segregate prior to gamete formation.*

This chapter covers some observations and experiments that unfolded in the decades after the rediscovery of Mendel's work. They lend impressive support to what is now called the chromosomal theory of inheritance. From preceding chapters, we are already familiar with many of the points of this theory:

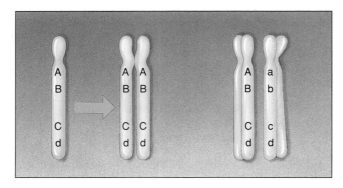

a Genes occur in linear sequence on a chromosome. Before a cell divides, each chromosome is copied; the copy has the same gene sequence. The capital and lowercase letters denote alleles— alternative forms of a gene at a particular location (locus) on the chromosome.

b Each pair of homologous chromosomes carries identical or different alleles at corresponding gene loci. Green represents paternal chromosomes; pink represents maternal ones.

1. Genes, the units of instruction for heritable traits, are arranged one after the other along chromosomes.

2. From meiosis through fertilization, sexual reproduction assures that each new individual will have the same chromosome number as its parents.

3. Diploid ($2n$) cells have pairs of homologous chromosomes. Except for sex chromosomes, the homologues have the same length, shape, and gene sequence. All of them, including sex chromosomes, line up with their partner at meiosis.

4. Genes located on different chromosomes are inherited independently of each other, in accordance with the Mendelian principle of independent assortment (page 171).

5. Genes on the same chromosome tend to stay together when chromosomes move about during meiosis. But crossing over (breakage and exchange of corresponding segments of homologues) can disrupt such linkages and lead to genetic recombination (Figure 12.2).

6. Chromosomal abnormalities sometimes occur. A chromosome segment may be deleted, duplicated, inverted, or moved to a new location. Chromosomes also may not separate properly at meiosis, and so gametes end up with an abnormal chromosome number.

7. Independent assortment, crossing over, and chromosomal abnormalities play roles in evolution. By changing the genotype (genetic makeup), they lead to variations in phenotype (observable traits) upon which selective agents can act.

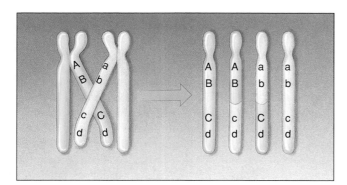

c Crossing over and genetic recombination typically occur during meiosis I. Each chromosome that is affected by a crossover event will carry both maternal and paternal alleles at different loci along its length.

Figure 12.2 Crossing over and other key events described by the chromosomal theory of inheritance.

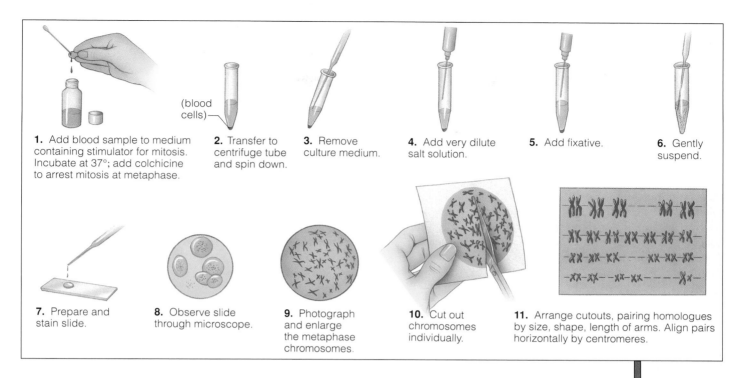

1. Add blood sample to medium containing stimulator for mitosis. Incubate at 37°; add colchicine to arrest mitosis at metaphase.

2. Transfer to centrifuge tube and spin down.

3. Remove culture medium.

4. Add very dilute salt solution.

5. Add fixative.

6. Gently suspend.

7. Prepare and stain slide.

8. Observe slide through microscope.

9. Photograph and enlarge the metaphase chromosomes.

10. Cut out chromosomes individually.

11. Arrange cutouts, pairing homologues by size, shape, length of arms. Align pairs horizontally by centromeres.

Figure 12.3 Simplified picture of karyotype preparation, in this case a karyotype of a human male. Human cells have a diploid chromosome number of 46. The nucleus contains 22 pairs of autosomes and 1 pair of sex chromosomes (X and Y). Each chromosome of a given type has already undergone DNA replication.

Autosomes and Sex Chromosomes

In the early 1900s, microscopists discovered a chromosomal difference between the sexes. Most of the chromosomes *are* the same number and the same type in both sexes; these were named **autosomes**. Depending on the species, however, one or two chromosomes *are not* the same in males and females; these were named sex chromosomes. **Sex chromosomes** carry hereditary instructions about *gender*—that is, whether a new individual will be male or female. For example, human females have two sex chromosomes designated "X." Males have one X chromosome and another, physically different chromosome designated "Y." (That is the most familiar pattern, but notable exceptions include birds, bees, moths, butterflies, turtles, and crocodiles.)

Today, human autosomes and sex chromosomes can be precisely characterized at metaphase, when they are in their most condensed form. Then, each type has a certain length, banding pattern, and so on. (The bands appear because some regions condense more than others and absorb stain differently. Figure 10.6 is an example.) These features are used to create a **karyotype:** a visual representation in which the chromosomes of a cell at metaphase are arranged in order, from largest to smallest (Figure 12.3).

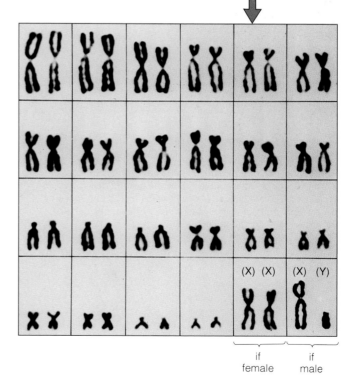

(X) (X)

(X) (Y)

if female

if male

Through microscopic studies, we know that each normal gamete produced by a female (XX) carries an X chromosome. Half the gametes produced by a male (XY) carry an X and half carry a Y chromosome. When a sperm and an egg both carry an X chromosome and combine at fertilization, the new individual will develop into a female. Conversely, when the sperm carries a Y chromosome, the new individual will develop into a male (Figure 12.4).

The Y chromosome of humans carries very few genes, but among them is a "male-determining gene." The presence of its product causes a new individual to develop testes. In the absence of that gene product, ovaries will develop automatically, as shown in the *Commentary*. The hormones secreted from testes and ovaries trigger the development of male *or* female characteristics.

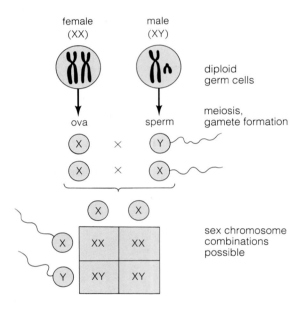

Figure 12.4 Sex determination in humans. This same pattern occurs in many animal species. Only the sex chromosomes, not the autosomes, are shown. Males transmit their Y chromosome to their sons, but not to their daughters. Males receive their X chromosome only from their mother.

Commentary

Girls, Boys, and the Y Chromosome

The formation of male or female parts in a human embryo depends on gene interactions and on physical and chemical conditions in the mother's uterus. From the time of conception, the embryo that develops from the zygote is normally XX or XY. For the first month or so of development, the embryo is neither male nor female. However, ducts and other structures start forming that can go either way, so to speak (Figure *a*).

As the next four to six weeks of development unfold, testes—the primary male reproductive organs—begin to

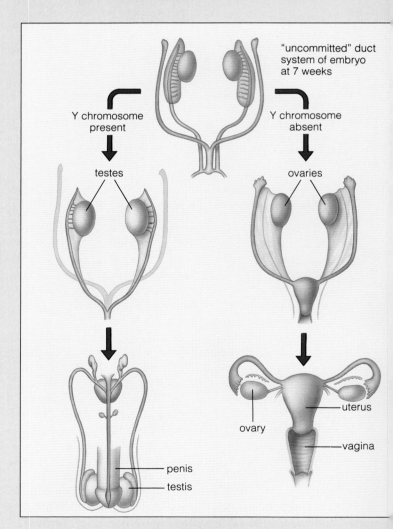

a Duct system in the early embryo that develops into a male *or* female reproductive system. Compare Figures 43.2 and 43.7.

develop in XY embryos. Apparently, a gene region on the Y chromosome governs the fork in the developmental road that can lead either to maleness or femaleness. If the embryo is XX, it has no such region, and it develops ovaries—the primary female reproductive organs. It does so automatically, in the absence of the Y chromosome.

Once their development is under way, the testes start producing sex hormones, including testosterone. These hormones influence the development of assorted parts that make up the male reproductive system, as described in Chapter 43. The newly forming ovaries also produce sex hormones, and these influence the development of the female reproductive system.

A newly identified region of the Y chromosome appears to be the master gene for sex determination. It has been called SRY (for sex-determining region of the Y chromosome). So far, the same gene has been identified in the DNA of human males and in male chimpanzees, rabbits, pigs, horses, cattle, and tigers. In all females tested, the gene was absent. Other tests with mice indicate that the gene region becomes active about the time that testes start developing.

In molecular structure, the SRY gene resembles the gene regions that are known to produce regulatory proteins. Such proteins can bind to certain parts of DNA and so turn genes on and off (page 234). Do the protein products of the SRY gene turn off the genes required for female development? Researchers are attempting to find out. For example, they are inserting the SRY gene into female mouse embryos. If those embryos go on to develop male characteristics, this will be evidence of a master regulatory role for the SRY gene.

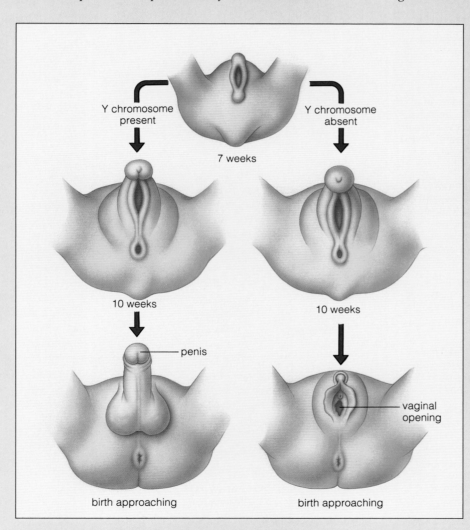

Y chromosome present

Y chromosome absent

7 weeks

10 weeks

10 weeks

penis

vaginal opening

birth approaching

birth approaching

b External appearance of developing reproductive organs.

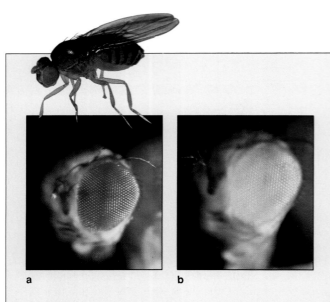

a   b

Figure 12.5 X-linked genes: clues to patterns of inheritance.

In the early 1900s, the embryologist Thomas Hunt Morgan began work to explain the apparent connection between gender and certain nonsexual traits. For example, blood clotting in humans occurs in both males and females. Yet for centuries, it was known that hemophilia (a blood-clotting disorder) shows up most often in the males, not females, of a family lineage. This phenotypic outcome was not like anything Mendel saw in his hybrid crosses between pea plants. In those crosses, either one parent plant or the other could carry a recessive allele. It made no difference *which* parent carried it; the phenotypic outcome was the same.

Morgan studied eye color and other nonsexual traits in the fruit fly, *Drosophila melanogaster*. These small flies can be grown in bottles on bits of cornmeal, molasses, and agar. A female lays hundreds of eggs in a few days, and her offspring can reproduce in less than two weeks. Morgan could track hereditary traits through nearly thirty generations of thousands of flies in a year's time.

At first, all the flies were wild-type for eye color; they had brick-red eyes, as in (a). ("Wild-type" simply means the normal or most common form of a trait in a population.) Then, through an apparent mutation in a gene controlling eye color, a *white-eyed* male appeared (b).

Morgan established true-breeding strains of white-eyed males and females. Then he did a series of *reciprocal crosses*. (These are pairs of crosses. In the first, one parent displays the trait in question; in the second, the other parent displays the trait.)

White-eyed males were mated with true-breeding (homozygous) red-eyed females. All the F_1 offspring of the cross had red eyes—but of the F_2 offspring, only some of the *males* had white eyes. Then white-eyed females were mated with true-breeding red-eyed males. Of the F_1 offspring of that second cross, half were red-eyed females and half were white-eyed males. Of the F_2 offspring, 1/4 were red-eyed females, 1/4 were white-eyed females, 1/4 red-eyed males, and 1/4 white-eyed males!

The seemingly odd results indicated the gene for eye color was related to gender. Probably it was located on one of the sex chromosomes. But which one? Since females (XX) could be white-eyed, the recessive allele would have to be on one of their X chromosomes. Suppose white-eyed males (XY) also carry the recessive allele on their X chromosome—and

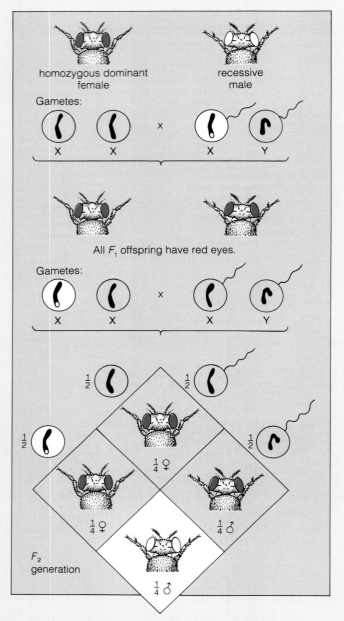

c Correlation between sex and eye color in *Drosophila*. Given the genetic makeup of the second generation, the recessive allele (depicted by the white dot) must be carried on the X chromosome only.

suppose there is no corresponding eye-color allele on their Y chromosome. Those males would have white eyes because the recessive allele would be the only eye-color gene they had!

In (c) are the results we can expect when the idea of an X-linked gene is combined with Mendel's concept of segregation. By proposing that a specific gene occurs on the X but not the Y chromosome, Morgan was able to explain the outcome of his reciprocal crosses. The results of the experiments matched the predicted outcomes.

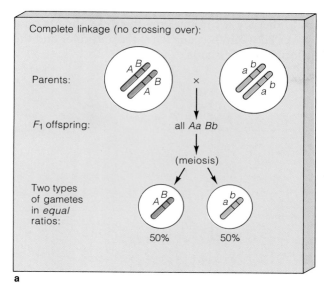

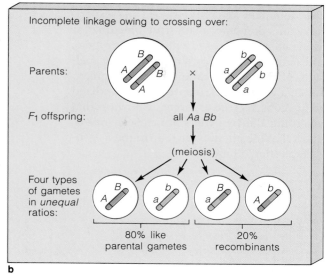

a   **b**

Figure 12.6 How crossing over can affect gene linkage, using two gene loci as the example.

The human X chromosome carries at least 100 and possibly more than 200 genes. Like all the other chromosomes, it carries some genes associated with sexual traits (such as the distribution of body fat). But most of the genes on the X chromosome are concerned with *nonsexual* traits, such as eye color.

Any gene on the X or Y chromosome may be called a "sex-linked gene." However, researchers now use the more precise designations, **X-linked** or **Y-linked genes**.

Some time ago, Thomas Hunt Morgan and his coworkers performed a series of hybridization experiments with the fruit fly, *Drosophila melanogaster*. As described in Figure 12.5, their work led to the discovery of X-linked genes, and it reinforced a major concept: *Each gene is located on a specific chromosome.*

Linkage and Crossing Over

Through their studies of the fruit fly, researchers came to realize that many traits were being inherited *as a group* from one parent or the other. They identified four groups that apparently corresponded to the haploid number of chromosomes (four) in the fruit fly's gametes. They suspected that the genes on any one of those chromosomes were probably staying together during meiosis and gamete formation.

The term **linkage** is now used to describe the tendency of genes located on the same chromosome to end up together in the same gamete. Linkage is not inevitable, however. It can be disrupted by **crossing over**— the breakage and exchange of segments between homologous chromosomes (Figure 12.2).

Think about the location of any two genes on the same chromosome. The probability of a crossover occurring between those two genes is proportional to the distance separating them along the chromosome. Suppose two genes *A* and *B* are twice as far apart as two other genes, *C* and *D*:

We would expect crossing over to disrupt the linkages between *A* and *B* much more frequently than those between *C* and *D*.

Two genes located physically close together on a chromosome nearly always end up in the same gamete; they are very closely linked (Figure 12.6a). Two genes relatively far apart are more vulnerable to crossing over and recombination, in comparison with closely linked genes (Figure 12.6b). Two genes very far apart on the same chromosome are affected by crossing over so often that they may appear to assort independently.

The farther apart two genes are on a chromosome, the greater will be the frequency of crossing over and recombination between them.

As an example of this, think about a watermelon with a green rind and one with a striped rind. The gene responsible for this trait happens to be closely linked to the gene for melon shape (round or oblong). Green and round are the dominant phenotypes. In one experiment, a homozygous dominant plant (true-breeding for green, round melons) was crossed with a homozygous recessive plant (true-breeding for striped, oblong melons). All plants of the F_1 generation produced green, round melons. In a testcross, F_1 plants were hybridized with the homozygous recessive parent. Knowing that the genes governing the two traits are closely linked, you might expect that the phenotypic ratio of the F_2 generation was 1:1, as shown here:

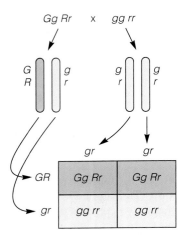

However, these were the observed F_2 phenotypes resulting from the cross:

46 green, round	4 green, oblong
47 striped, oblong	3 striped, round

We can explain the results by recognizing that 7 percent of the progeny (the 4 green, oblong and the 3 striped, round) inherited a chromosome that had undergone a crossover between the gene locus for rind color and the gene locus for rind shape:

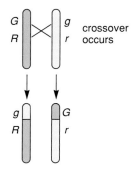

The patterns in which these and other genes are distributed into gametes tell us something about their organization on chromosomes. The patterns are so regular that they can be used to determine the *positions* of genes relative to one another. Plotting the positions of genes on a given chromosome is called linkage mapping. Figure 12.7 gives an example of a linkage map.

Of the several thousand known genes in the four chromosomes of *Drosophila* gametes, the positions of about a thousand have been mapped. There are many more genes in human chromosomes. As you will read in a later chapter, the locations of all human genes may be identified through an ambitious, long-term research project that is now under way. We have a long way to go in mapping the physical basis of heredity. But it is clear that genes are carried linearly, one after another, in human chromosomes—just as they are in fruit flies and all other organisms.

Figure 12.7 Genetic mapping of genes on a segment of chromosome 2 in *D. melanogaster*. Such maps don't show actual physical distances between genes. Rather they show relative distance between gene locations that undergo crossing over and other chromosomal rearrangements. Only if the probability of crossing over were equal along the chromosome's length (which it is not) would it be possible to calculate physical distance exactly.

Here, distances between genes are measured in map units, based on the frequency of recombination between the genes. (One genetic map unit = 1 percent recombination.) Thus, if the frequency turns out to be 10 percent, the genes are said to be separated by 10 map units. The amount of recombination to be expected between "vestigial wings" and "curved wings," for instance, would be 8.5 percent (75.5−67).

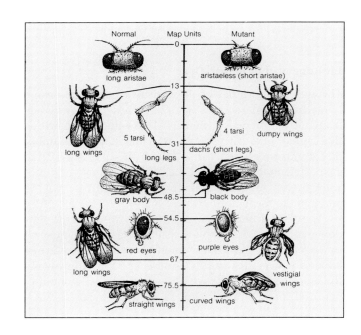

CHROMOSOME VARIATIONS IN HUMANS

Pea plants and fruit flies lend themselves to genetic analysis. They grow and reproduce rapidly in small spaces, under controlled conditions. Because flies or pea plants have a much shorter life span than the geneticist who studies them, their traits can be tracked through many generations in relatively little time.

Humans are another story. We humans live under variable conditions in diverse environments. Typically we find a mate by chance and reproduce if and when we want to. Human subjects live just as long as the geneticists who study them, so tracking traits through generations is rather tedious. And human families are generally so small, there aren't enough numbers for easy statistical inferences about inheritance patterns.

Even so, human genetics is a rapidly growing field. Researchers use standardized methods for constructing **pedigrees**, or charts of genetic relationships of individuals (Figure 12.8). With pedigrees, they can identify inheritance patterns and track genetic abnormalities through several generations. Studying the same trait in many families increases the numerical base for analysis.

Keep in mind that "abnormality" and "disorder" are not necessarily the same thing. *Abnormal* means deviation from the average. An abnormality is a rare or less common occurrence, as when a person is born with six toes on each foot instead of five. Whether such a trait is viewed as disfiguring or merely interesting is subjective—there is nothing inherently life-threatening or even ugly about it. Other abnormalities cause mild to severe medical problems, and genetic *disorder* is the more appropriate word here.

Figure 12.8 (**a**) Some symbols used in constructing pedigree diagrams. (**b**) Example of a pedigree for *polydactyly*, a condition in which an individual has extra fingers, extra toes, or both. The number of fingers on each hand is shown in black numerals, and the number of toes on each foot is shown in blue. The phenotype of female 1 is uncertain.

Polydactyly is an example of how gene expression can vary. As a human embryo develops, a dominant allele *D* controls how many sets of bones will form within the body regions destined to become hands and feet. The *Dd* genotype varies in how it is expressed. The pedigree shown here indicates the kind of variation possible.

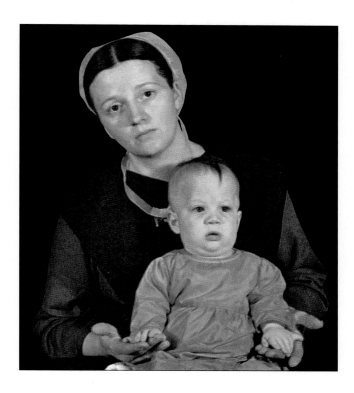

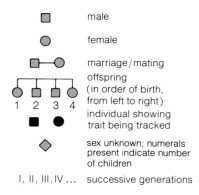

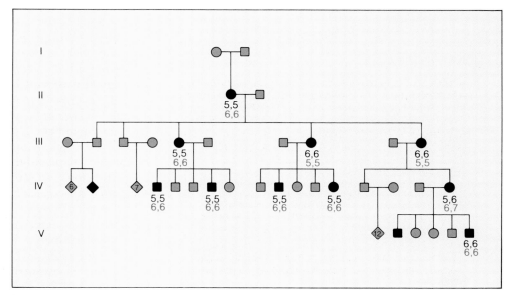

a

b

The next section of this chapter gives examples of the genetic disorders we deal with as individuals and as members of society. Table 12.1 lists some of them. The examples serve as a framework for considering some practical and ethical aspects of screening, counseling, and treatment programs.

Autosomal Recessive Inheritance

Sometimes a mutation produces a recessive allele on an autosome and gives rise to **autosomal recessive inheritance**. This condition has the following characteristics:

1. Males or females can carry the recessive allele on an autosome (not on a sex chromosome).

2. Heterozygotes generally are symptom-free. Homozygotes are affected.

Table 12.1	Examples of Human Genetic Disorders
Disorder (or Abnormality)*	Main Consequences
Autosomal Recessive Inheritance:	
Albinism (174)	Absence of pigmentation (melanin)
Sickle-cell anemia (164, 175, 284)	Severe tissue, organ damage
Galactosemia (192)	Brain, liver, eye damage
Phenylketonuria (199)	Mental retardation
Autosomal Dominant Inheritance:	
Achondroplasia (193)	A type of dwarfism
Compodactyly (177)	Rigid, bent little fingers
Huntington's disorder (193)	Progressive, irreversible degeneration of nervous system
Polydactyly (191)	Extra digits
Progeria (192)	Premature aging
X-Linked Inheritance:	
Hemophilia A (194)	Deficient blood-clotting
Testicular feminizing syndrome (595)	Absence of male organs, sterility
Changes in Chromosome Structure:	
Cri-du-chat (195)	Mental retardation, skewed larynx
Changes in Chromosome Number:	
Down syndrome (198)	Mental retardation, heart defects
Turner syndrome (201)	Sterility, abnormal development of ovaries and sexual traits
Klinefelter syndrome (201)	Sterility, mental retardation
XYY condition (201)	Mild mental retardation in some cases; no symptoms in others

*Number in parentheses indicates the page(s) on which the disorder is described.

3. When both parents are heterozygous, there is a 50 percent chance each child born to them will be heterozygous and a 25 percent chance it will be homozygous recessive (Figure 12.9). When both parents are homozygous, all of their children will be affected.

Galactosemia, an autosomal recessive condition, affects about 1 in 100,000 newborns. The condition arises when a breakdown product of lactose, or milk sugar, cannot be metabolized. Normally, lactose is first broken down to glucose and galactose, then ultimately to glucose-1-phosphate, which can be degraded by glycolysis. But galactosemics cannot produce molecules of an enzyme required for one of the conversion steps. They have two recessive alleles, coding for a defective form of the enzyme:

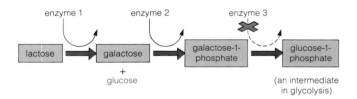

Galactose builds up in the blood, and in large concentrations it can damage the eyes, liver, and brain. Early symptoms include malnutrition, diarrhea, and vomiting. Without treatment, galactosemics usually die in childhood. However, unusually high concentrations can be detected in urine samples of homozygous recessive infants. If the disorder is detected early enough, infants can be put on a diet that includes milk substitutes and can grow up symptom-free.

Autosomal Dominant Inheritance

Recessive alleles that cause genetic disorders can persist at low but constant frequencies, because heterozygotes may still survive and reproduce. (Their one normal allele may yield enough of the required gene product.) But what if a *dominant* allele causes the disorder? In **autosomal dominant inheritance**, a dominant allele is usually expressed to some extent. If its expression reduces the chance of surviving and reproducing, its frequency among individuals in the population will decrease.

Even so, a few dominant alleles that cause pronounced disorders do remain in populations. How? Mutations can replenish the supply of defective alleles. Also, *some dominant alleles do not affect reproduction or they are only expressed after reproductive age*. Figure 12.10 shows an inheritance pattern for an autosomal dominant condition.

One autosomal dominant allele causes *achondroplasia*, a type of dwarfism, in about 1 in 10,000 individuals. When limb bones develop in affected children, cartilage forms in ways that lead to disproportionately short arms

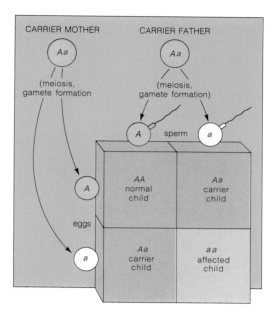

Figure 12.9 Possible phenotypic outcomes for autosomal recessive inheritance when both parents are heterozygous carriers of the recessive allele (shaded red here).

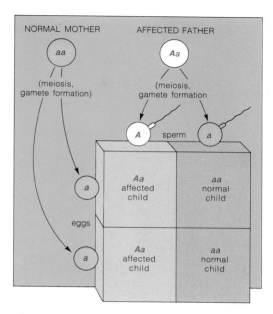

Figure 12.10 Possible phenotypic outcomes for autosomal dominant inheritance, assuming the dominant allele is fully expressed in the carriers. (The dominant allele is shaded red.)

and legs. Affected persons are less than 4 feet, 4 inches tall. The dominant allele often has no other phenotypic effects in heterozygotes, which normally are fertile and so may reproduce. Homozygous dominant fetuses usually are stillborn.

Huntington's disorder, a rare form of autosomal dominant inheritance, causes progressive degeneration of the nervous system. In about half the cases, symptoms emerge from age forty onward—after most people have already had children. In time, movements become convulsive, brain function deteriorates rapidly, and death follows.

Progeria, or early aging is an example of autosomal dominant inheritance. Children affected by this rare disorder start to show signs of advanced aging when they are only five or six years old (figure 12.1). Their skin wrinkles, their hair thins, they start to suffer arthritis, and their blood vessels show arteriosclerosis. Frenquently, affected indivduals die of heart disease before they are ten years old.

X-Linked Recessive Inheritance

Some genetic disorders fall in the category of **X-linked recessive inheritance**, which has these characteristics:

1. The mutated gene occurs on the X (not the Y) chromosome (Figure 12.11).

2. Heterozygous females are phenotypically normal; the nonmutated allele on their other X chromosome covers

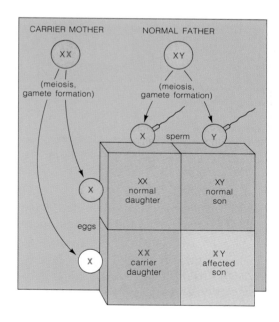

Figure 12.11 Possible phenotypic outcomes for X-linked inheritance when the mother carries a recessive allele on one of her X chromosomes (shaded red here).

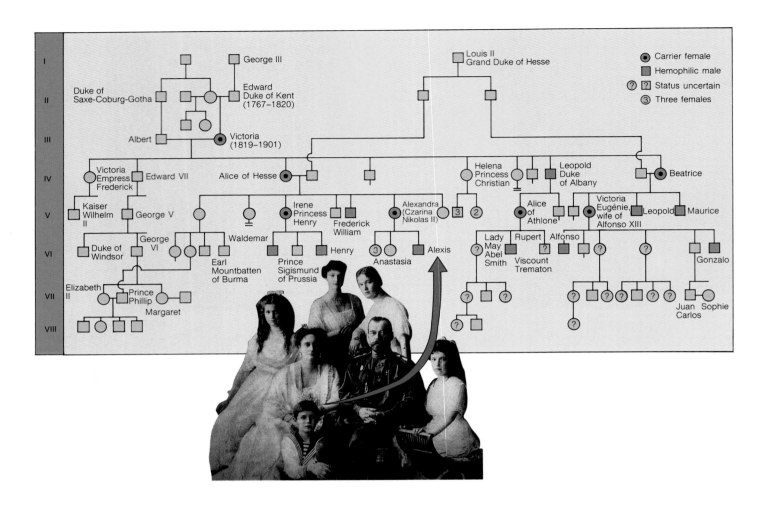

Figure 12.12 Descendants of Queen Victoria, showing carriers and affected males that possessed the X-linked gene conferring the disorder hemophilia A. The photograph shows the Russian royal family members. The mother was a carrier of the mutated gene; Crown Prince Alexis was afflicted with the disorder. Many individuals of later generations are not included in the family pedigree.

the required function. Males typically are affected. They have only one allele for the trait on the X chromosome, and it is recessive.

3. When the male is normal but the female is heterozygous, there is a 50 percent chance each daughter born to them will be a carrier, and a 50 percent chance each son will be affected (Figure 12.11). When the female is homozygous recessive and the male is normal, all daughters will be carriers and all sons will be affected.

Hemophilia A is an example of X-linked recessive inheritance. Normally, a blood-clotting mechanism quickly stops bleeding from minor injuries. Some people bleed for an unusually long time because the mechanism is defective. The reactions leading to clot formation depend on the products of several genes. If any of the genes is mutated, its defective product can cause one of several bleeding disorders.

In hemophilia A, the gene for a protein called clotting factor VIII is mutated. Males with a recessive allele on their X chromosome are always affected. They run the risk of dying from untreated bruises, cuts, or internal bleeding. Blood-clotting time is more or less normal in heterozygous females. The nonmutated gene on their other X chromosome produces enough factor VIII to cover the required function.

Hemophilia A affects only about 1 in 7,000 human males. The frequency of the recessive allele was unusually high among the royal families of nineteenth-century Europe, whose members often intermarried. Queen Victoria of England was a carrier, as were two of her daughters (Figure 12.12). At one time, eighteen of her sixty-nine descendants were affected males or female carriers.

Crown Prince Alexis of Russia was one of Victoria's hemophilic descendants. His affliction drew together an explosive cast of characters—Czar Nicholas II, Czarina Alexandra (a granddaughter of Victoria and a carrier), and the power-hungry monk Rasputin, who manipulated the aggrieved family to his political advantage. Indirectly, this hemophilic child helped catalyze events that brought an end to dynastic rule in the Western world.

Changes in Chromosome Structure

On rare occasions, chromosome structure becomes abnormally rearranged. Deletions, duplications, inversions, and translocations are examples of such rearrangements.

A **deletion** is a loss of a chromosome segment. The loss occurs when an end segment of a chromosome breaks off or when viral attack, irradiation, or chemical action causes breaks in a chromosome region. The loss almost always means problems, for genes influencing one or more traits may be missing. For example, one deletion from human chromosome 5 leads to mental retardation and a malformed larynx. When affected infants cry, the sounds produced are more like meowing—hence the name of the disorder, *cri-du-chat* (meaning cat-cry). Figure 12.13 shows an infant affected by the disorder.

A **duplication** is a gene sequence in excess of its normal amount in a chromosome. This happens, for example, when a deletion from one chromosome is inserted into its homologue (Figure 12.14a). Duplications probably have been important in evolution. Cells require specific gene products, so mutations that alter the function of most genes probably would be selected against. But *duplicates* of a gene could change through mutation (the normal gene would still provide the required product). In time, they could yield products with related or even new functions. This apparently happened in chromosome regions coding for the polypeptide chains of the hemoglobin molecule. In humans and other primates, those regions have multiple copies of strikingly similar gene sequences and they produce whole families of slightly different chains.

An **inversion** is a chromosome segment that separated from the chromosome and then was inserted at the same place—but in reverse. The reversal alters the position and relative order of the genes on the chromosome (Figure 12.14b).

Most often, a **translocation** is the transfer of part of one chromosome to a *non*homologous chromosome. In some types of human cancer, for example, a segment of chromosome 8 has been transferred to chromosome 14. Genes on that segment had been precisely regulated at their normal chromosomal location, but controls over their expression apparently are lost at the new location.

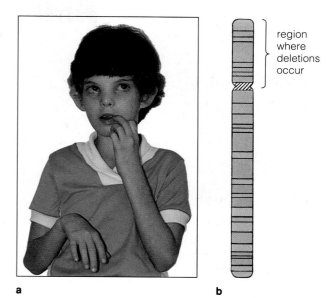

a   b

Figure 12.13 Cri-du-chat syndrome. Affected infants have an improperly developed larynx, and their cries sound like a cat in distress. They show severe mental retardation. Outward symptoms include a rounded face, small cranium, and misshapen ears (**a**). The deletions occur in the short arm of chromosome 5, which must have a fragile region because breaks occur more frequently than at any other part of the chromosome (**b**).

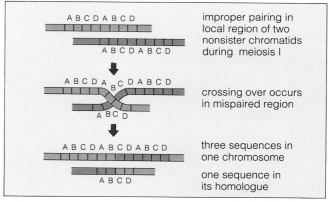

a Duplication

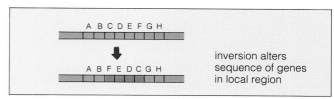

b Inversion

Figure 12.14 Two kinds of changes in chromosome structure: a duplication (**a**) and an inversion (**b**).

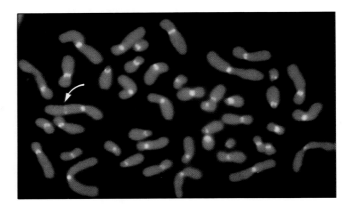

Figure 12.15 Translocation. These metaphase chromosomes were stained with substances that react preferentially with kinetochore proteins, which show up as fluorescent yellow in micrographs. The arrow points to the inactivated kinetochores of chromosome 9—which fused with chromosome 11. (The normal kinetochores of chromosome 11 are visible to the right of the fusion point.)

Once in a while, translocation occurs in such a way that almost an entire chromosome becomes fused to a nonhomologous chromosome—and the two continue to function. This structural rearrangement changes the chromosome number. Figure 12.15 shows the result of such a fusion in a cell from a human female. Her diploid chromosome number is 45 instead of the normal 46.

Chromosomes have undergone structural changes during the evolution of humans and their closest primate relatives. Eighteen of the twenty-three pairs of human chromosomes are nearly identical to their counterparts in chimpanzees and gorillas. However, karyotype analysis of banding patterns shows that inversions and translocations occurred in the others.

Changes in Chromosome Number

Sometimes new individuals end up with the wrong chromosome number. The phenotypic effects range from minor physical changes to devastating or lethal disruption of organ systems. More often, the affected individuals are miscarried—that is, spontaneously aborted before birth. There are two general categories of change in chromosome number. One is aneuploidy, the other is polyploidy.

Categories of Change.   Gametes or cells of a new individual may end up with one extra or one less than the

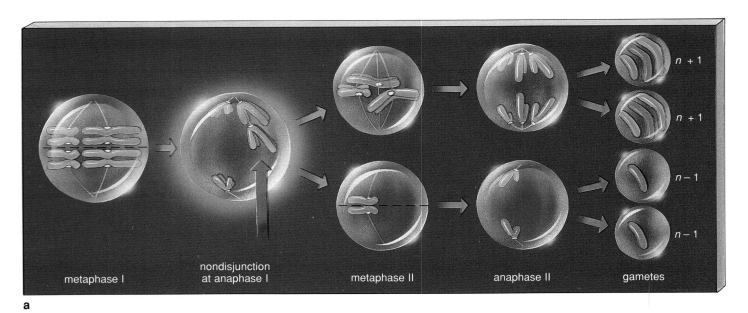

metaphase I   nondisjunction at anaphase I   metaphase II   anaphase II   gametes

$n + 1$

$n + 1$

$n - 1$

$n - 1$

a

Figure 12.16 Two examples of nondisjunction, an event that can change the chromosome number in gametes (hence in offspring).

parental number of chromosomes. This abnormal condition is called **aneuploidy**. It is a major cause of reproductive failure in humans. For reasons not fully understood, its rate of occurrence among humans is higher (by as much as ten times) than it is in other mammals. On the average, aneuploidy probably affects one of every two newly fertilized human eggs. Of the human embryos that have been miscarried and autopsied each year, about 70 percent were aneuploids.

Gametes or cells of a new individual also may end up with three or more of each type of chromosome characteristic of the parental stock. This condition is called **polyploidy**. As described in Chapter 18, polyploidy is common in the plant kingdom—in fact, it has given rise to about half of all flowering plant species. Polyploidy is lethal for humans. All but 1 percent of human polyploids die before birth, and the rare few who are born die within a month. Most likely, the resulting imbalance between the genetic instructions of autosomes and sex chromosomes disrupts key steps in the long, complex pathways of development and reproduction.

Mechanisms of Change. An abnormal chromosome number may arise during mitotic or meiotic cell divisions or during the fertilization process.

For example, even before meiosis occurs, mitotic cell divisions produce the germ cells that give rise to sperm and eggs. Suppose DNA replication and mitosis itself proceed as usual—but something prevents cytoplasmic division. The result will be a "tetraploid" germ cell with *four* of each type of chromosome instead of two. If meiotic cell division follows, the resulting gametes will be diploid instead of haploid—and so on down the line for an unfortunate embryo.

As another example, one or more pairs of chromosomes may fail to separate during mitosis or meiosis, an event called **nondisjunction**. Perhaps a chromosome does not separate from its homologue at anaphase I of meiosis. Or perhaps sister chromatids of a chromosome do not separate at anaphase II. As Figure 12.16 shows, some or all of the gametes can end up either with one extra chromosome or with one less than the parental number.

Suppose a human gamete has one *extra* chromosome ($n + 1$). If it combines with a normal gamete at fertilization, the diploid cells of the new individual will have three of one type of chromosome ($2n + 1$). Such a condition is called *trisomy*. If the gamete is *missing* a chromosome, the new individual will have a chromosome number of $2n - 1$. One chromosome in its diploid cells will not have a homologue—a condition called *monosomy*.

After asking whether their new baby is a girl or a boy, most parents apprehensively ask, "Is our baby normal?" They naturally want their baby to be free of genetic disorders, and most of the time it is. Chapter 43 describes the story of human reproduction and development when all goes well. Here, let's briefly consider some of the rare cases when it does not.

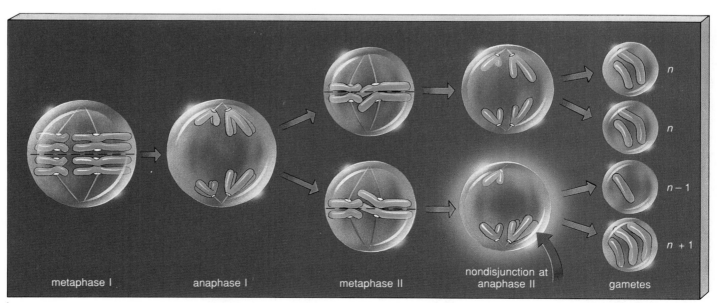

| metaphase I | anaphase I | metaphase II | nondisjunction at anaphase II | gametes |

n

n

$n - 1$

$n + 1$

b

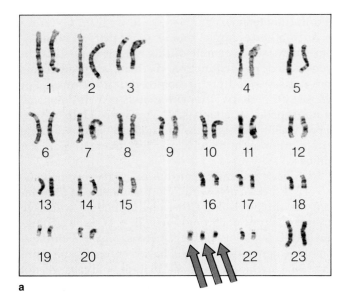

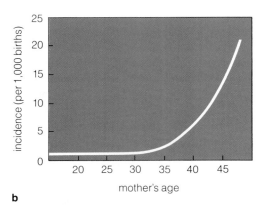

a

b

Figure 12.17 (**a**) Karyotype of a girl with Down syndrome; red arrows identify the trisomy of chromosome 21. (**b**) Relationship between the frequency of Down syndrome and the mother's age. Results are from a study of 1,119 children with the disorder who were born in Victoria, Australia, between 1942 and 1957.

Down Syndrome. Sometimes the cells of an individual have three copies of chromosome 21, which is one of the smallest chromosomes in human cells. That condition, called trisomy 21, leads to Down syndrome. ("Syndrome" simply means a set of symptoms characterizing a particular disorder; typically the symptoms occur together.) Figure 12.17 shows a karyotype of a girl with Down syndrome.

Trisomic 21 embryos are often miscarried, but about 1 of every 1,000 liveborns in North America alone will develop the disorder. Most affected children show moderate to severe mental retardation, and about 40 percent have heart defects. Skeletal development is slower than normal and muscles are rather slack. Older children are shorter than normal and have distinguishing facial features, including a small skin fold over the inner corner of the eyelid. With special training, affected individuals often participate in normal activities, and they enjoy life to the fullest extent allowed by their condition (Figure 12.18). Down syndrome is one of the genetic disorders that can be detected by prenatal diagnosis (see the *Commentary*).

Figure 12.18 Children with Down syndrome.

Commentary

Prospects and Problems in Human Genetics

Chances are, you know of someone who has a genetic disorder. Of all newborns, possibly 1 percent will have pronounced problems arising from a chromosomal aberration. Between 1 and 3 percent more will have problems because of mutant genes that produce defective proteins or none at all. Of all patients in children's hospitals, 10 to 25 percent are treated for problems arising from genetic disorders.

Human geneticists work to diagnose and treat heritable disorders. However, we apparently cannot approach the disorders in the same way we approach infectious diseases (such as influenza, measles, and polio). Infectious agents are enemies from the environment, so to speak. We have had no qualms about mounting counterattacks with immunizations and antibiotics that can eliminate the agents or bring them under control. With genetic disorders, the problem is inherent in the chromosomes of individual human beings.

How do we attack an "enemy" within? Do we institute regional, national, or global programs to identify affected persons? Do we tell them they are "defective" and run a risk of bestowing their disorder on their children? Who decides which alleles are "harmful"? Should society bear the cost of treating disorders such as Down syndrome? If so, should society also have some say in whether affected fetuses will be born at all, or aborted? These questions are only the tip of an ethical iceberg, and answers have not been worked out in universally acceptable ways.

Phenotypic Treatments

Genetic disorders cannot be permanently cured, but sometimes we can get around their phenotypic consequences. Treatments include diet modifications, environmental adjustments, surgery, and chemical modification of gene products.

Controlling the diet can suppress or minimize the outward symptoms of several disorders. Galactosemia is controlled this way (page 192). So is *phenylketonuria*, or PKU. Normally, a gene product (an enzyme) converts one amino acid to another (phenylalanine to tyrosine). However, the first amino acid builds up in people who are homozygous recessive for a mutated form of the gene. If the excess is diverted into other metabolic pathways, phenylpyruvic acid and other compounds may be produced. At high levels, phenylpyruvic acid can lead to mental retardation. A diet that provides the minimum required amount of phenylalanine will alleviate the symptoms of PKU. When the body is not called upon to dispose of excess amounts, affected persons can lead normal lives.

Diet soft drinks and many other products often are artificially sweetened with aspartame, which contains phenylalanine. Such products carry warning labels directed at phenylketonurics.

Environmental adjustments can alleviate the outward symptoms of other genetic disorders. True albinos, for example, can avoid direct sunlight. Individuals affected by sickle-cell anemia can avoid strenuous activity when oxygen levels are low, as at high altitudes.

Surgical reconstructions can correct or minimize many phenotypic defects. One type of *cleft lip* is a genetic abnormality of the upper lip. A vertical fissure cuts through the lip midsection and often extends into the roof of the mouth. Surgery can usually correct the lip's appearance and function.

Phenotypic treatments also include chemical modification of gene products. *Wilson's disorder* arises from an inability to use copper. The body requires trace amounts of copper, which serves as a cofactor for several enzymes. Excess copper can damage the brain and liver, leading to convulsions and death. One drug binds with the copper, and the excess is eliminated by the urinary system.

Genetic Screening

In some cases, genetic disorders can be detected early enough to start preventive measures *before* symptoms can develop. In other cases, carriers who show no outward symptoms can be identified before giving birth to affected children. "Genetic screening" refers to large-scale programs to detect affected persons or carriers in a population. Most hospitals in the United States routinely screen all newborns for PKU, for example, so it is becoming less common to see people with symptoms of the disorder.

Genetic Counseling

Sometimes prospective parents suspect they are very likely to produce a severely afflicted child. Either their first child or a close relative shows an abnormality and they now wonder if future children will be affected the same way. In such cases, clinical psychologists, geneticists, social workers, and other consultants may be brought in to give emotional support to parents at risk.

Counseling begins with accurate diagnosis of parental genotypes; this may reveal the potential for a specific disorder. Biochemical tests can be used to detect many metabolic disorders. Detailed family pedigrees can be constructed to aid the diagnosis.

For disorders showing simple Mendelian inheritance patterns, it is possible to predict the chances of having an affected child—but not all disorders follow Mendelian patterns. Even ones that do can be influenced by other factors, some identifiable, others not. Even when the extent of risk has been determined with some confidence, prospective parents must know the risk is the same for *each* pregnancy. For example, if a pregnancy has one chance in four of producing a child with a genetic disorder, the same odds apply to every subsequent pregnancy, also.

Removal of about 20 ml of amniotic fluid containing suspended cells sloughed off from the fetus

Centrifugation

Amniotic fluid: a few biochemical analyses possible

Fetal cells

Quick determination of fetal sex and analysis of purified DNA

Growth for weeks in culture medium

Biochemical analysis for the presence of about 40 metabolic disorders

Karyotype analysis for chromosomal aberration

Prenatal Diagnosis

What happens when a woman is already pregnant? Suppose a woman forty-five years old wants to know if the child she is bearing will develop Down syndrome. Through prenatal diagnosis, this and more than a hundred other genetic disorders can be detected before birth.

One detection procedure is based on *amniocentesis:* sampling the contents of the fluid-filled sac (amnion) that contains the fetus in the mother's uterus (Figure *a*). During the fourteenth to sixteenth week of pregnancy, the thin needle of a syringe is inserted through the mother's abdominal wall and into the amnion. Epidermal cells shed from the fetus float about in the amniotic fluid. The syringe withdraws some fluid—along with its sample of fetal cells. The cells are cultured and allowed to undergo mitosis. Abnormalities can be diagnosed by karyotype analysis and other tests that can be completed within weeks. Cells obtained by amniocentesis also can be tested for many biochemical defects, such as the one causing sickle-cell anemia.

Amniocentesis carries a risk: Care must be taken not to puncture the fetus or cause infection.

Chorionic villi sampling (CVS), a newer procedure, uses cells drawn from the chorion (a membranous sac surrounding the amnion). This procedure can be used earlier in pregnancy (by the eighth week), and results often are available in one or two days. However, a greater risk is associated with CVS, compared with amniocentesis.

What happens if an embryo is diagnosed as having a severe disorder? Unfortunately, there are no known cures for changes in chromosome number or structure. Prospective parents might decide on an *induced abortion* (an induced expulsion of the embryo from the uterus). Such decisions are bound by ethical considerations. We can expect the medical community to provide prospective parents with the information they need to make their own choice. That choice must be consistent with their own values, within the broader constraints imposed by society.

Turner Syndrome.   About 1 in every 5,000 newborns is destined to have Turner syndrome. Through a nondisjunction, affected individuals have a chromosome number of 45 instead of 46 (Figure 12.19). They are missing a sex chromosome (XO), this being a type of sex chromosome abnormality.

Turner syndrome occurs less often than other sex chromosome abnormalities, probably because most XO embryos are miscarried early in pregnancy. Affected persons have a distorted female phenotype. Their ovaries are nonfunctional, they are sterile, and secondary sexual traits fail to develop at puberty. Often they age prematurely and have shortened life expectancies.

Klinefelter Syndrome.   Nondisjunction can give rise to XXY males who show Klinefelter syndrome (Figure 12.19). This sex chromosome abnormality occurs at a frequency of about 1 in 1,000 liveborn males. Symptoms do not develop until after the onset of puberty. XXY males show low fertility and some degree of mental retardation. Their testes are much smaller than normal, body hair is sparse, and there may be some breast enlargement. Injections of the hormone testosterone can reverse the feminized phenotype but not the sterility or mental retardation.

A number of Klinefelter males are "mosaics," with both XY and XXY cell lineages. This condition commonly is related to nondisjunction during gamete formation in the individual's mother, especially in cases where pregnancy occurred late in life. Nondisjunction also has produced XXYY, XXXY, and XXXXY forms of Klinefelter syndrome. Symptoms are severe in these individuals, and mental retardation is pronounced.

XYY Condition.   About 1 in every 1,000 males has one X and two Y chromosomes, this being called an XYY condition. It is probably inappropriate to apply the term "syndrome" to this sex chromosome abnormality. XYY males tend to be taller than average and some may show mild mental retardation, but most are phenotypically normal.

At one time, XYY males were thought to be genetically predisposed to become criminals. But a comprehensive study in Denmark showed that the number who do end up in prison is no more notable than the percentage of other tall men. Compared to normal (XY) males, their rate of conviction was indeed greater: 41.7 percent compared to 9.3 percent. But this is not necessarily proof of a predisposition to crime. With their moderately impaired mental ability, XYY males simply may be easier to catch.

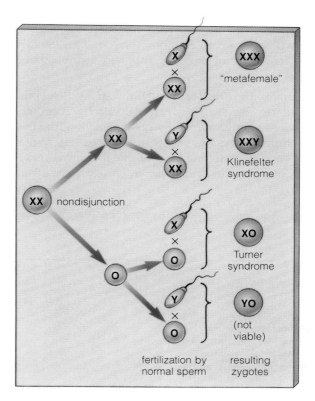

Figure 12.19   Genetic disorders that result from nondisjunction of X chromosomes followed by fertilization involving normal sperm.

SUMMARY

1. Genes, the units of instruction for heritable traits, are arranged linearly on a chromosome.

2. Sexual reproduction begins with meiosis and gamete formation and ends at fertilization. By this process, offspring receive the same number and type of chromosomes found in the body cells of their parents.

3. Human diploid cells have two chromosomes of each type (one from the mother, one from the father). The two are homologues; they pair at meiosis. The type called a sex chromosome is physically different in females (XX) and males (XY). All other pairs of chromosomes are autosomes; they are the same in males and females.

4. Homologues assort independently during meiosis.

5. Genes on one chromosome segregate from their partners on the homologous chromosome and end up in separate gametes.

6. Genes on the *same* chromosome tend to stay together during meiosis and end up in the same gamete. But

crossing over can disrupt such linkages. The farther apart two genes are on a chromosome, the greater will be the frequency of crossing over and recombination between them.

7. Chromosome *structure* can be altered by deletions, duplications, inversions, or translocations. Chromosome *number* can be altered by nondisjunction (the failure of chromosomes to separate during meiosis), so that gametes end up with one extra or one missing chromosome. Gene mutations, changes in chromosome structure, and changes in chromosome number cause many genetic disorders.

8. Variation in a population arises not only through mutation, but also through independent assortment, crossing over, and changes in the structure or number of chromosomes. These events can play roles in evolution. They change the genotypes (genetic makeup of individuals) and so lead to differences in phenotype (physical and behavioral traits) upon which selective agents can act.

Self-Quiz *(Answers in Appendix IV)*

1. _____ segregate during _____.
   a. Homologues; mitosis
   b. Genes on one chromosome; meiosis
   c. Homologues; meiosis
   d. Genes on one chromosome; mitosis

2. Two genes of a pair on homologous chromosomes end up in separate _____.
   a. body cells
   b. gametes
   c. nonhomologous chromosomes
   d. offspring
   e. both b and d are possible

3. Genes on the same chromosome tend to remain together during _____ and end up in the same _____.
   a. mitosis; body cell
   b. mitosis; gamete
   c. meiosis; body cell
   d. meiosis; gamete
   e. both a and d

4. The probability of a crossover occurring between two genes on the same chromosome is _____.
   a. unrelated to the distance between them
   b. increased if they are closer together on the chromosome
   c. increased if they are farther apart on the chromosome
   d. impossible

5. Chromosome structure can be altered by _____.
   a. deletions
   b. duplications
   c. inversions
   d. translocations
   e. all of the above

6. Nondisjunction can be caused by _____.
   a. crossing over in meiosis
   b. segregation in meiosis
   c. failure of chromosomes to separate during meiosis
   d. multiple independent assortments

7. A gamete affected by nondisjunction would have _____.
   a. a change from the normal chromosome number
   b. one extra or one missing chromosome
   c. the potential for a genetic disorder
   d. all of the above

8. Genetic disorders can be caused by _____.
   a. gene mutations
   b. changes in chromosome structure
   c. changes in chromosome number
   d. all of the above

9. Which of the following contributes to variation in a population?
   a. independent assortment
   b. crossing over
   c. changes in chromosome structure and number
   d. all of the above

10. Match the chromosome terms appropriately.
    _____ crossing over
    _____ deletion
    _____ nondisjunction
    _____ translocation
    _____ gene mutation
    a. a change in DNA which may affect genotype and phenotype
    b. movement of a chromosome segment to a nonhomologous chromosome
    c. disrupts gene linkages during meiosis
    d. causes gametes to have abnormal chromosome numbers
    e. loss of a chromosome segment

Genetics Problems *(Answers in Appendix III)*

1. Recall that human sex chromosomes are XX for females and XY for males.
   a. Does a male child inherit his X chromosome from his mother or father?
   b. With respect to an X-linked gene, how many different types of gametes can a male produce?
   c. If a female is homozygous for an X-linked gene, how many different types of gametes can she produce with respect to this gene?
   d. If a female is heterozygous for an X-linked gene, how many different types of gametes can she produce with respect to this gene?

2. One human gene, which may be Y-linked, controls the length of hair on men's ears. One allele at this gene locus produces nonhairy ears; another allele produces rather long hairs (hairy pinnae).

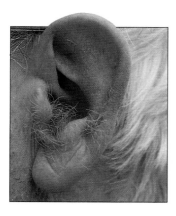

a. Why would you *not* expect females to have hairy pinnae?

b. If a man with hairy pinnae has sons, all of them will have hairy pinnae; if he has daughters, none of them will. Explain this statement.

3. Suppose that you have two linked genes with alleles *A,a* and *B,b* respectively. An individual is heterozygous for both genes, as in the following:

```
        A                B
    ———————————————————————
        A                B

        a                b
    ———————————————————————
        a                b
```

If the crossover frequency between these two genes is 0 percent, what genotypes would be expected among gametes from this individual, and with what frequencies?

4. In *D. melanogaster*, a gene influencing eye color has red (dominant) and purple (recessive) alleles. Linked to this gene is another that determines wing length. A dominant allele at this second gene locus produces long wings; a recessive allele produces vestigial (short) wings. Suppose a completely homozygous dominant female having red eyes and long wings mates with a male having purple eyes and vestigial wings. First-generation females are then crossed with purple-eyed, vestigial-winged males. From this second cross, offspring with the following characteristics are obtained:

```
    252 red eyes, long wings
    276 purple eyes, vestigial wings
     42 red eyes, vestigial wings
     30 purple eyes, long wings
────────────────────────────────────
    600 offspring total
```

Based on these data, how many map units separate the two genes?

5. Suppose you cross a homozygous dominant long-winged fruit fly with a homozygous recessive vestigial-winged fly. Shortly after mating, the fertilized eggs are exposed to a level of x-rays known to cause mutation and chromosomal deletions. When these fertilized eggs subsequently develop into adults, most of the flies are long-winged and heterozygous. However, a few are vestigial-winged.

Provide possible explanations for the unexpected appearance of these vestigial-winged adults.

6. Individuals affected by Down syndrome typically have an extra chromosome 21, so their cells have a total of 47 chromosomes. However, in a few cases of Down syndrome, 46 chromosomes are present. Included in this total are two normal-appearing chromosomes 21 and a longer-than-normal chromosome 14. Interpret this observation and indicate how these few individuals can have a normal chromosome number.

7. The mugwump, a type of tree-dwelling mammal, has a reversed sex-chromosome condition. The male is XX and the female is XY. However, perfectly good sex-linked genes are found to have the same effect as in humans. For example, a recessive, X-linked allele *c* produces red-green color blindness. If a normal female mugwump mates with a phenotypically normal male mugwump whose mother was color blind, what is the probability that a son from that mating will be color blind? A daughter?

8. One type of childhood *muscular dystrophy* is a recessive, X-linked trait in humans. A slowly progressing loss of muscle function leads to death, usually by age twenty or so. Unlike color blindness, this disorder is restricted to males, not ever having been found in a female. Suggest why.

Selected Key Terms

aneuploidy *197*
autosomal dominant inheritance *192*
autosomal recessive inheritance *192*
autosome *185*
chromosomal deletion *195*
chromosomal duplication *195*
chromosomal inversion *195*
chromosomal translocation *195*
crossing over *189*
family pedigree *191*

karyotype *185*
linkage *189*
monosomy *197*
nondisjunction *197*
polyploidy *197*
sex chromosome *185*
trisomy *197*
X-linked gene *189*
X-linked recessive
   inheritance *193*

Readings

Cummings, M. 1991. *Human Heredity: Principles and Issues.* St. Paul, Minnesota: West.

Edlin, G. 1988. *Genetic Principles: Human and Social Consequences.* Second edition. Portola Valley, California: Jones & Bartlett.

Fuhrmann, W., and F. Vogel. 1986. *Genetic Counseling.* Third edition. New York: Springer-Verlag.

Holden, C. 1987. "The Genetics of Personality." *Science* 237: 598–601. For students interested in human behavioral genetics.

Patterson, D. August 1987. "The Causes of Down Syndrome." *Scientific American* 257(2):52–60.

Weiss, R. November 1989. "Genetic Testing Possible Before Conception." *Science News* 136(21):326.

13 DNA STRUCTURE AND FUNCTION

Cardboard Atoms and Bent-Wire Bonds

Linus Pauling in 1951 did something no one had done before. He figured out the three-dimensional shape of one of the main biological molecules. Through solid training in biochemistry, a talent for model building, and a few lucky hunches, he discovered the structure of the protein collagen. At the time, proteins were known to be composed of amino acids, strung together in polypeptide chains. Pauling perceived that each polypeptide chain of collagen twists helically, like a spiral staircase, and that hydrogen bonds hold the chain in its helical shape.

His discovery electrified the scientific community. If the structural secrets of proteins could be unraveled, why not other biological molecules? Further, wouldn't structural details about those molecules provide clues to how they function? And which would turn out to be the biggest prize of all—*the molecule that serves as a book of genetic information in every living cell*? Scientists all around the world started scrambling to be the first to find out.

Proteins seemed like good candidates. After all, heritable traits are spectacularly diverse. The molecules containing the information for those traits surely were structurally diverse also. With their potentially limitless combinations of amino acid subunits, proteins almost certainly could function as the sentences (genes) in each cell's book of inheritance.

Yet there was something about another substance—DNA—that tugged at more than a few good minds.

Certainly Pauling thought the possibility was there. Just when researchers in England were hot on the DNA trail, he wrote to Maurice Wilkins, one of the trailblazers. He requested copies of intriguing x-ray pictures that had been developed in Wilkins' lab. Wilkins' response was lukewarm, perhaps a delaying tactic. Why should he help a formidable competitor win the race to glory?

As it turned out, neither won the race. James Watson, a young postdoctoral student from Indiana University, had teamed up with Francis Crick, an energetic Cambridge University researcher. Watson and Crick spent long hours arguing over everything they had read about the size, shape, and bonding requirements of the known components of DNA. They fiddled with cardboard cutouts of those components. They badgered chemists to identify bonds they might have overlooked. They assembled models of bits of metal, held together with wire "bonds" bent at chemically correct angles.

In 1953, they finally put together a model that fit all the pertinent biochemical rules and all the facts about DNA that had been gleaned from other sources (Figure 13.1). Watson and Crick had discovered the structure of DNA. More than this, the breathtaking simplicity of that structure enabled them to solve a long-standing riddle about life—how it can show unity at the molecular level and yet give rise to so much diversity at the level of whole organisms.

With this chapter, we turn to some investigations that led to our current understanding of DNA structure and function, for they are revealing of how ideas are generated in science. On the one hand, having a shot at enduring fame and fortune quickens the pulse of competitive men and women in any profession, and scientists are no exception. On the other hand, science proceeds as a community effort, with individuals sharing not only what they can explain but also what they do not understand. Thus, even if an experiment "fails," it may turn up information that others can use or lead to questions that others can answer. Unexpected results, too, might be clues to something important about the natural world.

Figure 13.1 James Watson and Francis Crick posing in 1953 by their newly unveiled model of DNA structure. Behind this photograph is a recent computer-generated model. It is more sophisticated in appearance, yet basically the same as the prototype that was built nearly four decades before.

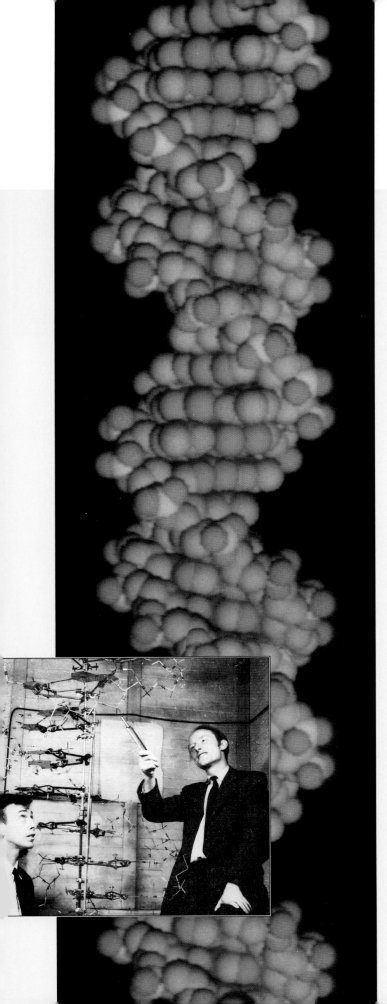

1. Hereditary instructions in living cells are encoded in the linear sequence of nucleotides that make up DNA molecules. The four kinds of nucleotides in DNA differ in which nitrogen-containing base they contain. The bases are adenine, guanine, thymine, or cytosine.

2. In each DNA molecule, two strands of nucleotides are twisted together like a spiral staircase; they form a double helix. Hydrogen bonds occur between the bases of the two strands. As a rule, adenine pairs (hydrogen-bonds) only with thymine, and guanine pairs only with cytosine.

3. Before a cell divides, its DNA is replicated with the help of enzymes and other proteins. Each double-stranded DNA molecule starts unwinding, and a new, complementary strand is assembled on the exposed bases of each parent strand according to base-pairing rules.

4. There is only one DNA molecule (one double helix) in a chromosome. Except for bacterial cells, the chromosome also consists of many histones and other proteins that have roles in the structural organization of DNA.

DISCOVERY OF DNA FUNCTION

One might have wondered, in the spring of 1868, why Johann Friedrich Miescher was collecting cells from the pus of open wounds and, later, from the sperm of a fish. Miescher wanted to identify the chemical composition of the nucleus, and he was interested in those cells because they are composed mostly of nuclear material, with very little cytoplasm. He succeeded in isolating an acidic substance, one with a notable amount of phosphorus. Miescher called it "nuclein." He had discovered what came to be known as deoxyribonucleic acid, or **DNA**.

The discovery caused scarcely a ripple through the scientific community. At the time, no one knew much about the physical basis of inheritance—that is, *which* chemical substance in cells actually encodes the instructions for reproducing parental traits in offspring. Only a

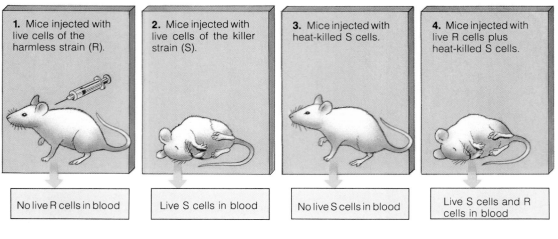

1. Mice injected with live cells of the harmless strain (R).

No live R cells in blood

2. Mice injected with live cells of the killer strain (S).

Live S cells in blood

3. Mice injected with heat-killed S cells.

No live S cells in blood

4. Mice injected with live R cells plus heat-killed S cells.

Live S cells and R cells in blood

DNA
protein coat
sheath
baseplate
tail fiber

a

Figure 13.2 Summary of the results from Griffith's experiments with harmless (R) strains and disease-causing (S) strains of *Streptococcus pneumoniae*, as described in the text. You may be wondering why the S form is deadly and the R form harmless. The disease-causing strain produces a thick external capsule that protects the bacterial cells from attack by the host's immune system. Cells of the R strain form no such capsule. The host's defense system has the chance to destroy those cells before they can cause disease.

few researchers suspected that the nucleus might hold the answer. In fact, seventy-five years passed before DNA was recognized as having profound biological importance.

A Puzzling Transformation

In 1928 an army medical officer, Fred Griffith, attempted to develop a vaccine against the bacterium *Streptococcus pneumoniae*, which causes the lung disease pneumonia. (Many vaccines are preparations of killed or weakened bacterial cells which, when introduced into the body, can mobilize the body's defenses against a later attack.) He never did create a vaccine, but his experiments unexpectedly opened a door to the molecular world of heredity.

Griffith isolated two strains of the bacterium, which he designated *S* and *R*. (When grown in culture, bacterial colonies of one strain have a *S*mooth surface appearance and colonies of the other have a *R*ough surface.) He used the strains in four experiments and came up with the results shown in Figure 13.2 and listed below:

1. Laboratory mice were injected with live R cells. They did not develop pneumonia; the R strain was harmless.

2. Mice were injected with live S cells. The mice died, and blood samples from them teemed with live S cells. The S strain was pathogenic, or disease-causing.

3. S cells were killed by exposure to high temperature. Mice injected with the heat-killed cells did not die.

4. Live R cells were mixed with heat-killed S cells and injected into mice. Oddly, the mice died and blood samples from them teemed with live S cells!

What was going on in the fourth experiment? Maybe the heat-killed pathogens in the mixture were not really dead. But if that were true, the group of mice injected with heat-killed pathogenic cells alone would have contracted the disease, too. Maybe the harmless R cells in the mixture had mutated into the killer S form. But if that were true, the group of mice injected with R cells alone would have died.

The simplest explanation was this: Although heat did kill the pathogenic cells, it did not damage the chemical substance containing their hereditary information—including the part that specified "how to cause infection." Somehow the substance had been liberated from those dead cells, and it entered living cells of the harmless strain—where its instructions were expressed.

Further experiments made it clear that the harmless cells had indeed picked up instructions for causing infection and had been permanently transformed into pathogens because of it. Hundreds of generations of bacteria descended from the transformed cells also caused infections!

A few years later, researchers found that extracts of the killed pathogenic bacteria also could cause hereditary transformation. The microbiologist Oswald Avery and his colleagues began work to purify and experiment with the chemical substances of those extracts. This was the time when most biochemists believed that hereditary instructions were encoded in proteins, not DNA. That prevailing belief was challenged in 1944, when Avery's group reported that DNA probably was the substance of heredity.

Avery's key experiments could not be explained away. He could *block* transformation of harmless bacteria by adding an enzyme, pancreatic deoxyribonuclease, to

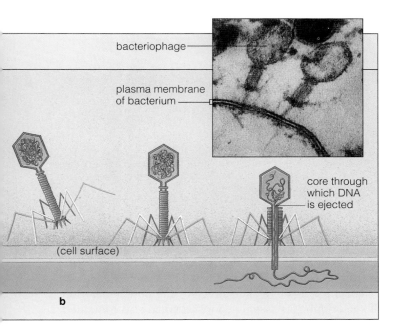

Figure 13.3 (a) Components of a T4 bacteriophage. (b) When a T4 bacteriophage makes contact with the cell surface of *Escherichia coli*, proteins in its tail fibers chemically recognize molecules at the bacterial cell surface. The sheath contracts, and the DNA contained within the protein coat is injected into the cell.

extracts of the pathogenic strain. The enzyme degrades DNA molecules but has no effect on proteins. In contrast, protein-degrading enzymes had no effect at all on the transforming activity.

Yet how were Avery's impressive findings received? Many (if not most) biochemists refused to give up on the proteins. His experimental results, they said, probably applied only to bacteria.

Bacteriophage Studies

While work was going on in Avery's laboratory, Max Delbrück, Alfred Hershey, and Salvador Luria were studying a class of viruses called **bacteriophages**, which infect bacterial cells. The infectious cycle starts when bacteriophages latch onto a target host cell. Within sixty seconds, an infected cell starts making the nucleic acids and proteins, including enzymes, necessary to build new bacteriophages. Then *lysis* occurs; the cell comes under chemical attack and dies. In this case, viral enzymes degrade the bacterial cell wall and the cell membrane becomes grossly leaky, thereby liberating a new infectious generation.

By 1952, researchers knew that some bacteriophages contain only DNA and protein. Electron micrographs had revealed that the main part of a virus particle remains at the surface of an infected host cell (Figure 13.3).

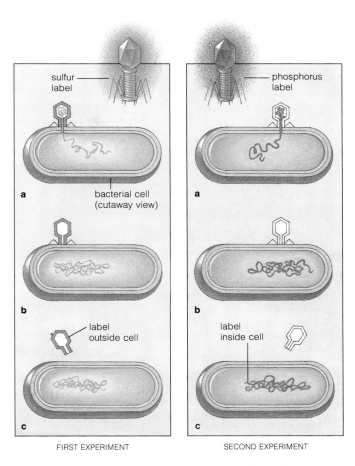

FIRST EXPERIMENT   SECOND EXPERIMENT

Figure 13.4 Hershey-Chase bacteriophage studies pointing to DNA as the substance of heredity.

Bacteriophages, a class of viruses that infect bacteria, consist only of protein and DNA (Figure 13.3). When bacteriophages attach to a host cell, they inject their DNA into it. Soon the cell starts making viral nucleic acids and proteins (including enzymes) necessary to build new bacteriophages. Then viral enzymes degrade the bacterial cell wall. The cell bursts, releasing the new infectious generation.

(a) Bacteriophage proteins contain sulfur (S) but no phosphorus (P)—and the DNA contains phosphorus but no sulfur. In one experiment, bacteriophages were labeled with a radioisotope (^{35}S) to tag their proteins. In a second experiment, they were labeled with a radioisotope (^{32}P) to tag their DNA.

(b) Labeled bacteriophages were allowed to infect unlabeled cells suspended in fluid. Hershey and Chase whirred the fluid in a kitchen blender to remove the bacteriophage bodies from the cells. (c) Labeled protein remained in the fluid; it was associated with the bacteriophage bodies. Labeled DNA remained with the bacterial cells—it had to contain the hereditary instructions for producing new bacteriophages.

Clearly, genetic information was being injected *into* the cell body. Now Hershey and his colleague Martha Chase asked the question: Is the injected material DNA, protein, or both? They devised a way to track both substances through the infectious cycle (Figure 13.4).

Bacteriophage proteins contain sulfur but no phosphorus, and DNA contains phosphorus but no sulfur. Both elements have radioisotopes (^{35}S and ^{32}P) that can

All chromosomes contain DNA. What does DNA contain? Only four kinds of nucleotides. Each nucleotide has a five-carbon sugar (shaded red). That sugar has a phosphate group attached to the fifth carbon atom of its ring structure. It also has one of four kinds of nitrogen-containing bases (shaded blue) attached to its carbon atom. The nucleotides differ <u>only</u> in which base is attached to that atom:

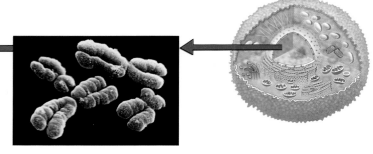

Figure 13.5 The nucleotide subunits of DNA. The small numerals on the structural formulas identify the carbon atoms to which other parts of the molecule are attached.

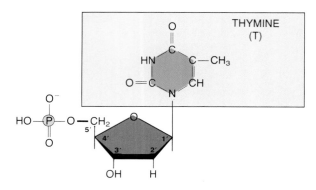

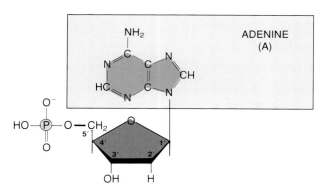

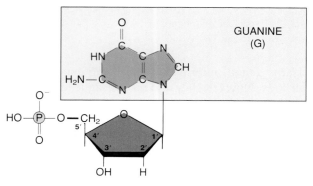

be used as tracers (page 22). Bacterial cells were grown on a culture medium containing ^{35}S. When they synthesized proteins, the isotope became incorporated into the proteins and so made them radioactively labeled. When the cells were later infected, the new generation of bacteriophages assembled inside them also contained labeled protein. Similarly, cells grown on a culture medium containing ^{32}P ended up with labeled DNA.

As Figure 13.4 shows, labeled bacteriophages were now allowed to infect unlabeled cells. After the cycle of infection was under way, Hershey and Chase determined that the new bacteriophage generation incorporated labeled DNA—not the labeled protein.

Many bacteriophage studies confirmed Avery's conclusion that DNA is the hereditary substance. In fact, many different experiments have since been performed on cells of a variety of species from all five kingdoms. All confirm the following statement:

Information for producing the heritable traits of single-celled and multicelled organisms are encoded in DNA.

DNA STRUCTURE

Components of DNA

Long before the studies just described were proceeding, biochemists had shown that DNA contains only four kinds of nucleotides, the building blocks of nucleic acids. A **nucleotide** consists of a five-carbon sugar called deoxyribose, a phosphate group, and one of the following nitrogen-containing **bases**:

adenine	guanine	thymine	cytosine
(A)	**(G)**	**(T)**	**(C)**

Each type of nucleotide in DNA has its component parts joined together in much the same way as the others. But notice in Figure 13.5 that T and C are smaller, single-ring structures (called *pyrimidines*). A and G are larger, double-ring structures (called *purines*).

By 1949, the biochemist Erwin Chargaff had added these crucial insights about DNA structure. *First*, the four kinds of nucleotide bases making up a DNA molecule differ in relative amounts from species to species. *Second*, the amount of adenine always equals the amount of thymine (A = T), and the amount of guanine equals the amount of cytosine (G = C).

Now, here was something to think about! Could it be that the arrangement of the four kinds of bases in a DNA molecule represented the hereditary instructions? Maurice Wilkins, Rosalind Franklin, and others thought they might be able to identify that arrangement through *x-ray diffraction methods*. The atoms of any crystallized substance will disperse an x-ray beam. If the atoms in a crystal occur in a regular order, they will disperse the beam in a regular pattern. Such patterns show up as dots and streaks on a piece of film placed behind the crystal. By itself, the pattern on the exposed film does *not* reveal molecular structure. But it can be used to calculate the position of groups of atoms relative to one another in the crystal.

The researchers knew that DNA molecules are too large to be crystallized. But a suspension of DNA could be spun rapidly, spooled onto a rod, and gently pulled into gossamer fibers, like cotton candy. DNA molecules would be oriented in a regular pattern in such fibers, and those could be subjected to x-ray diffraction analysis.

Franklin obtained the best x-ray diffraction images, which provided convincing evidence that DNA had the following features. First, DNA had to be long and thin, with a uniform 2-nanometer diameter. Second, its structure had to be highly repetitive: Some part of the molecule was repeated every 0.34 nanometer, and a different part was repeated every 3.4 nanometers. Third, DNA might be helical, with a shape like a circular stairway.

Patterns of Base Pairing

While work was proceeding in Wilkins' laboratory, Watson and Crick joined the search for the structure of DNA. According to Chargaff's data, the amount of adenine in DNA always equals the amount of thymine, and the amount of guanine always equals that of cytosine. According to Franklin's data, DNA has a uniform diameter. The sugar components of the different nucleotides were probably covalently bonded one after another along the length of the molecule. Thus DNA probably had some sort of sugar-phosphate backbone.

Watson and Crick reasoned that the *double*-ringed A and G bases of the nucleotides were probably paired with the *single*-ringed T and C bases along the entire length of DNA. Otherwise, DNA would bulge where two double rings were linked and narrow down where two

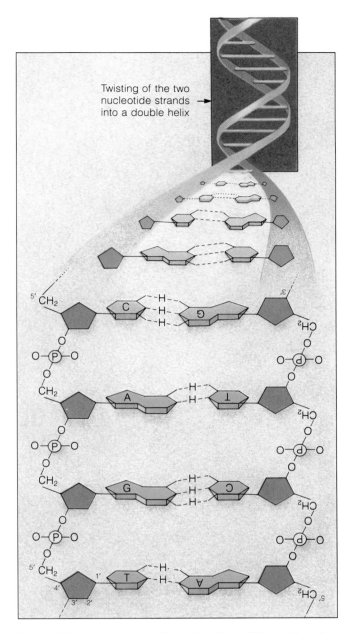

Figure 13.6 Arrangement of bases (blue) in a DNA double helix.

single rings were linked. Watson and Crick shuffled and reshuffled paper cutouts of the nucleotides. They realized that in certain orientations, A and T could become linked by two hydrogen bonds, and G and C could become linked by three. Suppose there were *two strands* of nucleotides, with their bases facing each other. The hydrogen bonds could easily bridge the gap between them, like rungs of a ladder.

Scale models were constructed of how the "ladder" might look. The only model that fit all available data had A-T and G-C pairs. And those pairs formed the proper hydrogen bonds only when two sugar-phosphate backbones of two DNA strands ran in *opposing directions* and were twisted together to form a *double helix* (Figure 13.6).

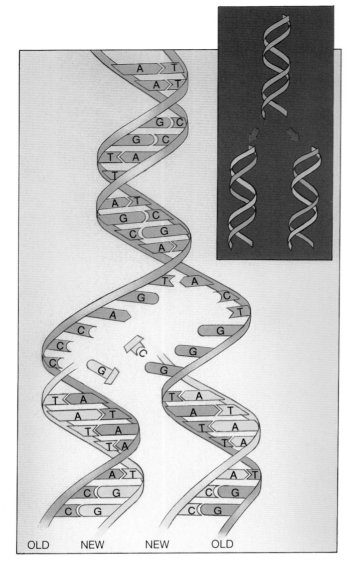

Figure 13.7 Semiconservative nature of DNA replication. The original two-stranded DNA molecule is shown in blue. A new strand (yellow) is assembled on each of the two original strands.

OLD   NEW   NEW   OLD

In the Watson-Crick model, then, hydrogen bonds join the bases of one strand with bases of the other. For the entire length of a DNA molecule, adenine always pairs with thymine, and cytosine always pairs with guanine. However, the *order* of bases in a nucleotide strand can vary greatly from one species to the next. In even a tiny stretch of DNA from a rose, gorilla, human, or any other organism, the base sequence might be:

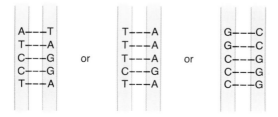

As you can see, DNA molecules show both constancy and variation in their structure. This is the molecular foundation for the unity and diversity of life.

Base pairing between the two nucleotide strands in DNA is *constant* for all species (adenine to thymine, guanine to cytosine).

The base sequence (that is, which base follows the next in a nucleotide strand) is *different* from species to species.

DNA REPLICATION

Assembly of Nucleotide Strands

The discovery of DNA structure was a turning point in studies of inheritance. Until then, no one had any idea of how the hereditary material is replicated (that is, duplicated) prior to cell division. The Watson-Crick model suggested at once how this might be done.

Hydrogen bonds hold together the two nucleotide strands making up the DNA double helix, and those weak bonds are readily broken. Enzymes acting on a given region of the DNA molecule can cause one strand to unwind from the other, leaving bases exposed in the unwound region. Cells have stockpiles of free nucleotides, and these pair with exposed bases. Thus, each parent strand remains intact and a companion strand is assembled on each one according to the base-pairing rule.

As replication proceeds, each parent strand is twisted into a double helix with its new, partner strand. Because the parent strand is conserved, each "new" DNA molecule is really half-old, half-new. That is why the process is called **semiconservative replication** (Figure 13.7).

Prior to cell division, the double-stranded DNA molecule unwinds and is replicated. Each parent strand remains intact—it is conserved—and a new, complementary strand is assembled on each one.

A Closer Look at Replication

Origin and Direction of Replication. Where does replication of the DNA molecule actually begin? The two strands of the double helix start to unwind at one or more distinct sites, each being a short, specific base sequence called the "origin." A viral or bacterial DNA molecule usually has one origin; a eukaryotic DNA molecule has many. Unwinding usually proceeds simul-

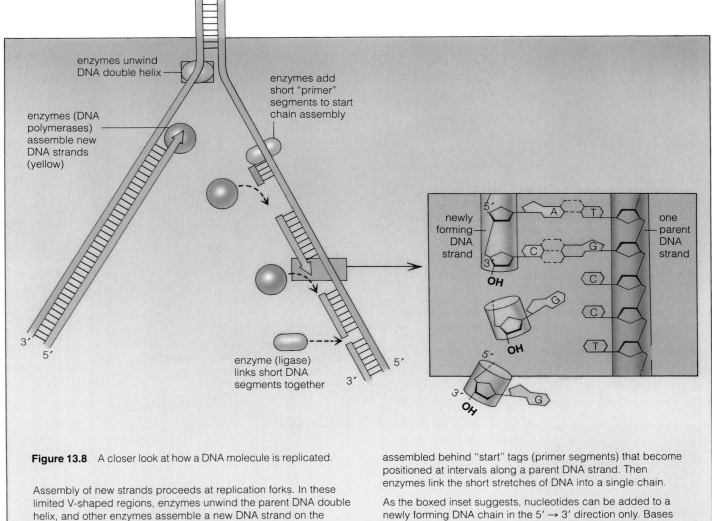

Figure 13.8 A closer look at how a DNA molecule is replicated.

Assembly of new strands proceeds at replication forks. In these limited V-shaped regions, enzymes unwind the parent DNA double helix, and other enzymes assemble a new DNA strand on the exposed regions of each parent strand, which serves as a template. As discovered by Reiji Okazaki, DNA assembly is usually *continuous* on one parent template but *discontinuous* on the other. In discontinuous synthesis, short stretches of nucleotides are assembled behind "start" tags (primer segments) that become positioned at intervals along a parent DNA strand. Then enzymes link the short stretches of DNA into a single chain.

As the boxed inset suggests, nucleotides can be added to a newly forming DNA chain in the 5′ → 3′ direction only. Bases projecting from a parent template dictate which kind of nucleotide can be added next. But an exposed —OH group must be present on the growing end of a DNA strand if enzymes are to catalyze the addition of more nucleotides to it.

taneously in both directions away from an origin. Strand assembly occurs behind each "fork" that continues to advance as the double helix is being unwound:

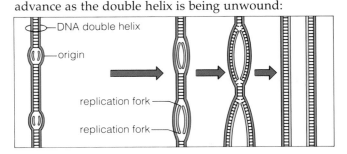

Energy and Enzymes for Replication.   A DNA double helix does not unwind all by itself during replication. It takes a battery of enzymes and other proteins to unwind the molecule, keep the two strands separate behind the replication forks, and assemble a new strand on each one. Even while one DNA region is being unwound, enzymes are winding up the replicated regions.

DNA polymerases are major replication enzymes. They govern nucleotide assembly on a parent strand (Figure 13.8). They also "proofread" the growing strands for mismatched base pairs, which are replaced with correct bases. The proofreading function is one reason DNA is replicated with such accuracy. On the average, for every 100 million nucleotides added to a growing strand, only *one* mistake slips through the proofreading net.

Where does the energy come from to drive replication? It happens that the free nucleotides brought up for strand assembly are not quite in the form shown in Figure 13.5. They are triphosphates, meaning they have

Figure 13.9 (*Below*) Structural organization of DNA in mammalian metaphase chromosomes.

(**a**) Chromosomes in two chicken cells at mitosis, made visible by fluorescence microscopy. Blue-stained regions indicate where a fluorescent dye bound with DNA. Pink indicates where the dye bound with a molecule that can attach specifically to a type of chromosomal protein. The protein, topoisomerase II, is an enzyme that cuts and reseals DNA during replication. Its action helps prevent tangles and counteracts torsional stress when the DNA molecule twists about.

Notice how DNA loops fan out from the protein scaffold. The cells were treated with a buffer that caused the metaphase chromosomes to loosen up. The same kind of looping can be seen in electron micrographs of metaphase chromosomes from a HeLa cell (page 147). The chromosome in (**b**, **c**) had its histones removed by treating the cell with a mild detergent. The shape of the metaphase scaffolding is still evident. Topoisomerase II is the major protein of this scaffold.

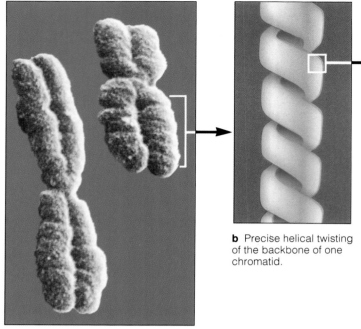

b Precise helical twisting of the backbone of one chromatid.

a Two metaphase chromosomes, each in the duplicated state.

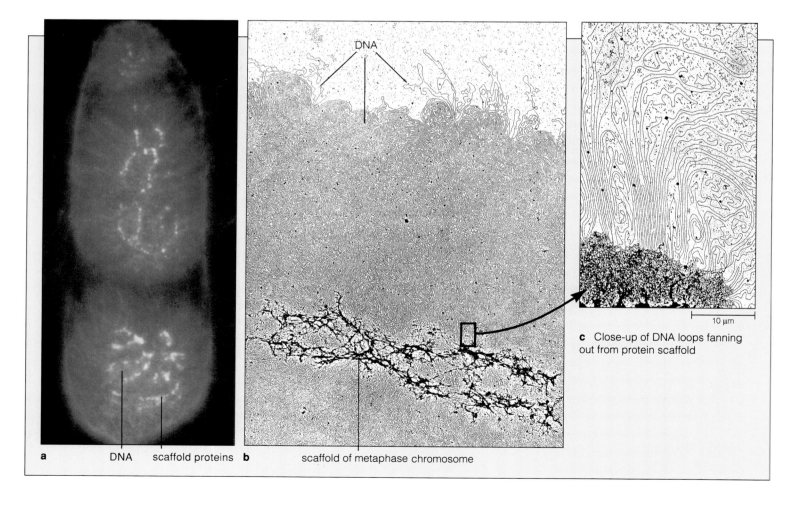

a DNA   scaffold proteins   **b** scaffold of metaphase chromosome

c Close-up of DNA loops fanning out from protein scaffold

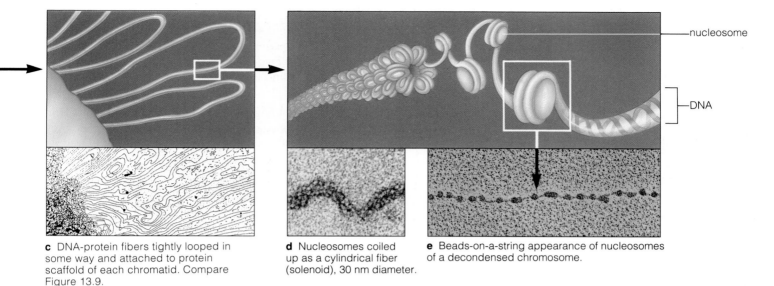

c DNA-protein fibers tightly looped in some way and attached to protein scaffold of each chromatid. Compare Figure 13.9.

d Nucleosomes coiled up as a cylindrical fiber (solenoid), 30 nm diameter.

e Beads-on-a-string appearance of nucleosomes of a decondensed chromosome.

Figure 13.10 Levels of organization of DNA in a eukaryotic chromosome.

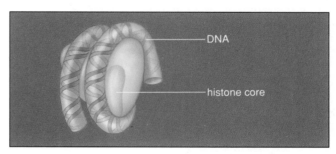

f Nucleosome (a double loop of DNA around a core of histone molecules).

three phosphate groups attached, not one. Triphosphates readily give up phosphate groups and so transfer energy to specific reactions. During replication, DNA polymerases use the energy released when two of the phosphate groups are split away. That energy drives the addition of nucleotides to a growing DNA strand. The unwinding process runs on energy provided by ATP.

ORGANIZATION OF DNA IN CHROMOSOMES

There is one DNA molecule in each chromosome. If you put the DNA of all forty-six chromosomes of a human cell into a single line, end to end, it would extend over a meter. All that DNA might be a tangled mess if it were not for its precise organization (Figure 13.9).

The DNA of humans and all other eukaryotes is tightly bound with many proteins, including **histones.**

Some histones are like spools for winding up small stretches of DNA. Each histone-DNA spool is a **nucleosome** (Figure 13.10). Another histone (H1) stabilizes the arrangement and plays a role in higher levels of organization. The chromosome becomes coiled repeatedly through interactions between the histones and DNA; this process greatly increases its diameter.

Further folding results in a series of loops. Proteins other than histones may serve as a structural scaffold for the loops. Experiments suggest that at interphase the DNA and scaffold proteins remain in association. They also suggest that the scaffold regions occur *between* genes, not in regions that contain the information for building proteins. If this is true, it may be that the scaffold proteins organize the chromosome in functional "domains" that facilitate protein synthesis as well as DNA replication. The looped regions are known to vary in size. Does each one contain one or more gene sequences? This is the kind of question being asked by the new generation of molecular detectives.

SUMMARY

1. Deoxyribonucleic acid, or DNA, is the master blueprint of hereditary instructions in cells. It is assembled from small organic molecules called nucleotides.

2. All nucleotides have a five-carbon sugar (deoxyribose) and a phosphate group. They also have one of four nitrogen-containing bases: adenine, thymine, guanine, or cytosine.

3. In a DNA molecule, two nucleotide strands are twisted together into a double helix. The bases of one strand pair with bases of the other strand (by hydrogen bonding).

4. There is constancy in base pairing in a DNA molecule. Adenine always pairs with thymine (A = T), and guanine always pairs with cytosine (G = C).

5. There is variation in the *sequence* of base pairs in the DNA molecules of different species.

6. DNA replication is semiconservative. The two strands of the DNA double helix unwind from each other, and a new strand is assembled on each one according to base-pairing rules. Two double-stranded molecules result, in which one strand is "old" (it is conserved) and the other is "new."

7. A variety of enzymes and other proteins take part in DNA replication. DNA polymerases are examples; they catalyze strand assembly and also perform proofreading functions.

8. A eukaryotic chromosome consists of one DNA molecule bound with many proteins. Interactions between the DNA and proteins give rise to a structural organization that is most pronounced at metaphase.

9. DNA replication usually is bidirectional. It proceeds simultaneously from each origin, which is a specific, short sequence of bases that functions as an initiation site for replication.

Review Questions

1. How did Griffith's use of control groups help him deduce that the transformation of harmless *Streptococcus* strains into deadly ones involved a change in the hereditary material of the harmless forms? *206*

2. What is a bacteriophage? In the Hershey-Chase experiments, how did bacteriophages become labeled with radioactive sulfur and radioactive phosphorus? Why were these particular elements used instead of, say, carbon or nitrogen? *207–208*

3. DNA is composed of only four different kinds of nucleotides. Name the three molecular parts of a nucleotide. Name the four different kinds of nitrogen-containing bases that may occur in the nucleotides of DNA. *208–209*

4. What kind of bond holds two DNA chains together in a double helix? Which nucleotide base-pairs with adenine? Which pairs with guanine? Do the two DNA chains run in the same or opposite directions? *209–210*

5. The four bases in DNA may differ greatly in relative amounts from one species to the next—yet the relative amounts are always the *same* among all members of a single species. How does the concept of base pairing explain these twin properties—the unity and diversity—of DNA molecules? *210*

6. When regions of a double helix are unwound during DNA replication, do the two unwound strands join back together again after a new DNA molecule has formed? *210–211*

Self-Quiz (Answers in Appendix IV)

1. _____ bonds hold the bases of one nucleotide strand to bases of the other nucleotide strand of a DNA double helix.

2. Which of the following is *not* a nitrogenous base of DNA?
   a. adenine   d. cytosine
   b. thymine   e. uracil
   c. guanine

3. Base pairing in the DNA molecule follows which configuration?
   a. A--G, T--C
   b. A--C, T--G
   c. A--U, C--G
   d. A--T, C--G

4. A single strand of DNA with the base sequence C–G–A–T–T–G would be complementary to the sequence _____.
   a. C–G–A–T–T–G
   b. G–C–T–A–A–G
   c. T–A–G–C–C–T
   d. G–C–T–A–A–C

5. The DNA of one species differs from others in its _____.
   a. sugars   c. base-pair sequence
   b. phosphate groups   d. all of the above

6. When DNA replication begins, _____.
   a. the two strands of the double helix unwind from each other
   b. the two strands condense tightly for base-pair transfers
   c. two DNA molecules bond
   d. old strands move to find new strands before bonding

7. DNA replication produces _____.
   a. two half-old, half-new double-stranded molecules
   b. two double-stranded molecules, one with the old strands and one with newly assembled strands
   c. three new double-stranded molecules, one with both strands completely new and two that are discarded
   d. none of the above

8. The process of DNA replication requires _____.
   a. a supply of new nucleotides
   b. forming of new hydrogen bonds
   c. many enzymes and other proteins
   d. all of the above

9. DNA polymerase has _____ functions.
   a. strand assembly
   b. phosphate attachment
   c. proofreading
   d. a and c are both correct

10. Match these DNA concepts appropriately.
    _____ base pair   a. two nucleotide strands twisted
    sequences   together
    _____ metaphase   b. A = T, G = C
    chromosome   c. one strand old (conserved), the
    _____ constancy in   other new
    base pairing   d. accounts for differences
    _____ replication   among species
    _____ double helix   e. structure results from
                                   interactions between DNA and
                                   proteins

Selected Key Terms

adenine (A) *208*
bacteriophage *207*
base *208*
cytosine (C) *208*
DNA *205*
DNA polymerase *211*
guanine (G) *208*
histone *213*

nucleosome *213*
nucleotide *208*
purine *209*
pyrimidine *209*
semiconservative replication *210*
thymine (T) *208*
x-ray diffraction *209*

Readings

Cairns, J., G. Stent, and J. Watson (editors). 1966. *Phage and the Origins of Molecular Biology*. Cold Spring Harbor, New York: Cold Spring Harbor Laboratories. Collection of essays by the founders of and converts to molecular genetics. Gives a sense of history in the making—the emergence of insights, the wit, the humility, the personalities of the individuals involved.

Darnell, J., et al. 1990. *Molecular Cell Biology*. Second edition. New York: Scientific American Books.

Felsenfeld, G. October 1985. "DNA." *Scientific American* 253(4): 58–67. Describes how the DNA double helix may change its shape during interactions with regulatory proteins.

Radman, M., and R. Wagner. August 1988. "The High Fidelity of DNA Duplication." *Scientific American* 259(2).

Taylor, J. (editor). 1965. *Selected Papers on Molecular Genetics*. New York: Academic Press.

Watson, J. 1978. *The Double Helix*. New York: Atheneum. Highly personal view of scientists and their methods, interwoven into an account of how DNA structure was discovered.

14 FROM DNA TO PROTEINS

Beyond Byssus

For the unattached mussel, creeping across a wave-scoured rock, time is of the essence. At any moment the pounding waves can whack it loose, hurl it repeatedly against the rock with shell-shattering force, and so offer up one more gooey lunch for gulls. That marine mussel —soft of body, nearly brainless—literally must hold on for dear life.

The mussel is fortunate. By chance, its muscular, probing foot comes across a suitable anchoring site—a small crevice in the rock. Now the mussel moves its foot, broomlike, and sweeps the site clean. Next it presses the foot down, like a rubber plunger, and expels water trapped beneath it. Then the flattened foot arches upward, creating a vacuum-sealed chamber (Figure 14.1).

A fluid flows into the chamber from ducts in the mussel's body. The fluid, which mussels stockpile in a special gland, contains enzymes, the protein keratin, and a resinous protein. Inside the chamber, the fluid bubbles into a sticky foam that gives the mussel a pre-liminary foothold on the rock. Now the foot flattens out, then curves into a deep groove. Like a spider spinning threads for a web, the mussel pumps the foam through the groove, converting it into a fine thread. Such threads, about as wide as a human whisker, become varnished with another protein. Together, they form an adhesive material called byssus.

Byssus is the world's premier underwater adhesive. Nothing that humans have manufactured comes close. (Sooner or later, water chemically degrades or deforms all synthetic adhesives.) Byssus fascinates biochemists, adhesive manufacturers, dentists, and surgeons looking for better ways to do tissue grafts and to rejoin severed nerves. Even now, genetic engineers are inserting a bit of mussel DNA into yeast cells, which reproduce in large numbers and serve as "factories" for translating the genes of mussels into useful amounts of proteins. This exciting work, like the mussel's own byssus-building, starts with one of life's universals. *Every protein is synthesized according to instructions that have been copied from DNA.*

You are about to trace the steps of protein synthesis, beginning with a linear sequence of code words in a strand of DNA. You will see how this becomes translated into a linear sequence of amino acids in a polypeptide chain. One or more of those newly crafted chains goes on to become a protein. Many enzymes and other proteins are players as well as products in this story, as are molecules of RNA. As you will see, it takes the same kinds of steps to produce all of the world's proteins, from mussel-inspired adhesives to the keratin of your own fingernails to the insect-digesting enzymes of a Venus flytrap.

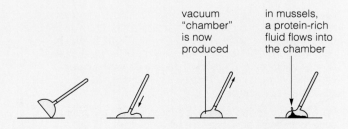

vacuum "chamber" is now produced

in mussels, a protein-rich fluid flows into the chamber

Figure 14.1 Mussels (*Mytilus edulis*) busily demonstrating the importance of proteins for survival. When mussels come across a suitable anchoring site on a rocky shoreline, they use their foot like a plumber's plunger to create a vacuum chamber. In this chamber they manufacture the world's best underwater adhesive from a wonderful mix of proteins synthesized by some of their specialized cells.

KEY CONCEPTS

1. Life cannot exist without enzymes and other proteins. Genes, which are specific regions of DNA, contain the information required to build proteins. The "code words" of genes are sequences of nucleotide bases, read three at a time.

2. The path from genes to proteins has two steps. In transcription, an RNA strand is assembled on exposed bases of an unwound gene region. In translation, the code-word sequence of the RNA that was transcribed from DNA is converted into the amino acid sequence of a polypeptide chain. A protein molecule consists of one or more such chains.

3. Thus, DNA is used to build RNA, then RNA is used to build proteins—some of which take part in building DNA and RNA. This flow of information is the "central dogma" of molecular biology.

4. Replication and repair enzymes work to preserve genes. On rare occasions, however, one to several bases in a gene sequence may be deleted, added, or replaced. Such gene mutations are the original source of genetic variation in populations.

PROTEIN SYNTHESIS

The Central Dogma

DNA is like a book of instructions in each cell. As we have seen, the alphabet used to create the book is simple enough: A, T, G, and C. But how is the alphabet arranged into the sentences (genes) that become expressed as proteins? How does a cell skip through the book, reading only those genes that will provide certain proteins at certain times? Answers to these questions begin with the structure of DNA.

Each DNA molecule consists of two long strands, twisted together into a double helix (Figure 13.6). The four kinds of nucleotide subunits making up the strands differ only in their nitrogen-containing base (adenine, thymine, guanine, or cytosine). Which base follows the

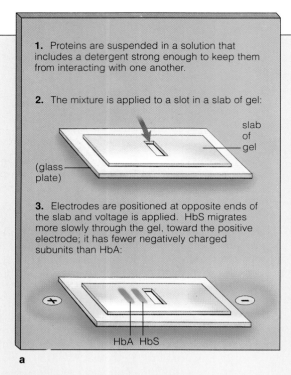

1. Proteins are suspended in a solution that includes a detergent strong enough to keep them from interacting with one another.

2. The mixture is applied to a slot in a slab of gel:

slab of gel

(glass plate)

3. Electrodes are positioned at opposite ends of the slab and voltage is applied. HbS migrates more slowly through the gel, toward the positive electrode; it has fewer negatively charged subunits than HbA:

+   −

HbA HbS

a

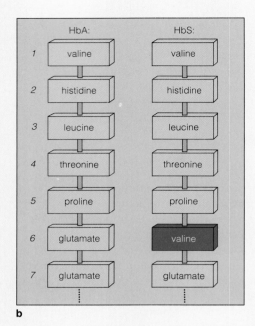

b

Figure 14.2 The connection between genes and proteins.

In the early 1900s a physician, Archibald Garrod, was tracking metabolic disorders that seemed to be heritable (they kept recurring in the same families). Blood or urine samples from affected persons contained abnormally high levels of a substance known to be produced at a certain step in a metabolic pathway. Most likely, the enzyme at the *next* step in the pathway was defective and could not use that substance. Because the pathway was blocked from that step onward, unused molecules of the substance accumulated in the body:

$$A \longrightarrow B \overset{\overset{C\,C\,C}{}}{\underset{C}{\longrightarrow}} C \;\;\times\;\; D$$

pathway is blocked

Only one thing distinguished affected persons from normal ones. They had inherited one metabolic defect. Thus, Garrod concluded, specific "units" of inheritance (genes) function through the synthesis of specific enzymes.

Thirty-three years later, George Beadle and Edward Tatum were using the bread mold *Neurospora crassa* to study gene function. *N. crassa* will grow on a medium containing only sucrose, mineral salts, and biotin, one of the B vitamins. It synthesizes all other nutrients it requires, including other vitamins, and the steps of those synthesis pathways were known.

Suppose an enzyme of a synthesis pathway is defective as a result of a gene mutation. Beadle and Tatum suspected this had happened in some *N. crassa* strains. One strain grew only when supplied with vitamin B_6, another with vitamin B_1, and so on. Chemical analysis of cell extracts revealed a different defective enzyme in each mutant strain. *Each*

inherited mutation corresponded to a defective enzyme. Here was evidence favoring Garrod's "one gene, one enzyme" hypothesis.

The hypothesis was refined through studies of sickle-cell anemia (page 175). This heritable disorder arises from the presence of abnormal hemoglobin in red blood cells. The abnormal molecule is designated HbS instead of HbA. In 1949 Linus Pauling and Harvey Itano subjected HbS and HbA molecules to *electrophoresis*. This is a way to measure how fast and in what direction an organic molecule will move in response to an electric field.

As shown in (**a**), suppose you place a mixture of different proteins in a slab of gel. Each type will move toward one end of the slab or the other when voltage is applied to it. The rate and direction of movement depend partly on a molecule's net surface charge.

Electrophoresis studies showed that HbS and HbA molecules move toward the positive pole of the field—but HbS does so more slowly. HbS, it seemed, has fewer negatively charged amino acids.

Later, Vernon Ingram pinpointed the difference. Hemoglobin, recall, consists of four polypeptide chains (page 45). Two are designated alpha and the other two, beta. As (**b**) shows, in each beta chain of HbS, one amino acid (valine) has replaced another (glutamate). Glutamate carries a negative charge; valine has no net charge. Thus HbS behaved differently in the electrophoresis studies.

More importantly, this discovery suggested that *two* genes code for hemoglobin—one for each kind of polypeptide chain—and that genes code for proteins in general, not just for enzymes.

And so a more precise hypothesis emerged: *One gene codes for the amino acid sequence of one polypeptide chain—the structural unit of proteins.*

next in a strand—that is, the **base sequence**—differs to some degree from one species of organism to the next.

Before a cell divides, its DNA is replicated, so that each of its daughter cells will end up with a full complement of the required genetic information. At this time, the two strands of a DNA molecule unwind from each other and the exposed base sequence of each serves as a structural pattern, or **template**, upon which a new strand is built. At other times in a cell's life, however, *DNA regions are unwound so that the cell gains access to specific genes.*

For now, think of a **gene** as a region of DNA that calls for the assembly of specific amino acids into a polypeptide chain. Such chains are the basic structural units of proteins. Evidence for the connection between genes and proteins comes from many studies. Some of the classic research leading to this understanding is described in Figure 14.2.

The path from genes to proteins has two steps, called transcription and translation. Here our main focus will be on how the two steps occur in eukaryotic cells. In **transcription**, single-stranded molecules of ribonucleic acid, or **RNA**, are assembled on DNA templates in the nucleus. In **translation**, the RNA molecules are shipped from the nucleus into the cytoplasm, where they are used as templates for assembling polypeptide chains. (In bacterial cells, which have no nucleus, RNA molecules start getting translated while they are still peeling off the DNA.) Following translation, one or more chains become folded into the three-dimensional shape of protein molecules.

A circular relationship exists between DNA, RNA, and proteins. DNA is used in the synthesis of RNA, then RNA directs the synthesis of proteins. Those proteins have structural and functional roles in cells. They also have roles in transmitting genetic information from one cell generation to the next. Among them are all the enzymes and other proteins that take part in DNA replication, RNA synthesis, and protein synthesis. This flow of information in cells is the **central dogma** of molecular biology:

With this simple picture in mind, we are ready to expand our picture of the gene—and our description of RNA.

Overview of the RNAs

Genes are transcribed into three different types of RNA molecules which are composed of nucleotide subunits of the sort shown in 14.3. Only *one* of the three types even-

tually becomes translated into a protein product. By contrast, the other two types of RNA molecules have specific roles during the process of translation.

ribosomal RNA (rRNA)	a type of molecule that combines with certain proteins to form the *ribosome,* the structural "workbench" on which a polypeptide chain is assembled
messenger RNA (mRNA)	the "blueprint" (a linear sequence of nucleotides) delivered to the ribosome for translation into a polypeptide chain
transfer RNA (tRNA)	an adaptor molecule; it can pick up a specific amino acid *and* pair with an mRNA code word for that amino acid

Therefore, three types of RNA are transcribed from DNA. All are shipped from the nucleus into the cytoplasm, and all take part in translation, the second stage of protein synthesis. *But only the mRNA molecules carry protein-building instructions out of the nucleus.*

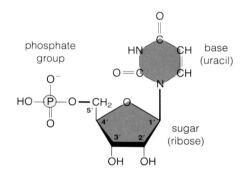

Figure 14.3 Structure of one of the four nucleotides of RNA. The other three differ only in their component base (adenine, guanine, or cytosine instead of the uracil shown here). Compare Figure 13.5, which shows the nucleotides of DNA.

TRANSCRIPTION OF DNA INTO RNA

How RNA Is Assembled

Let's now consider how RNA is transcribed from a gene that codes for a specific polypeptide chain. A strand of RNA is almost, but not quite, like a strand of DNA. Its nucleotides consist of a sugar (ribose), a phosphate group, and a nitrogen-containing base. The bases in RNA are adenine, cytosine, guanine, and **uracil** (Figure 14.3). Like the thymine in DNA, uracil can base-pair with adenine. Thus a new RNA strand can be assembled on a DNA template according to base-pairing rules, in a manner similar to DNA replication:

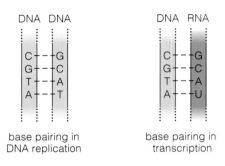

base pairing in DNA replication   base pairing in transcription

Transcription is similar to DNA replication in another respect. The nucleotides are added to a growing RNA

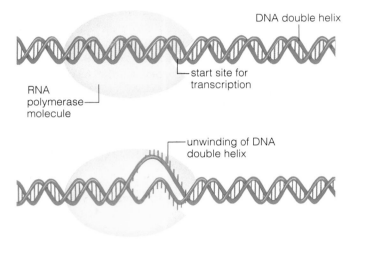

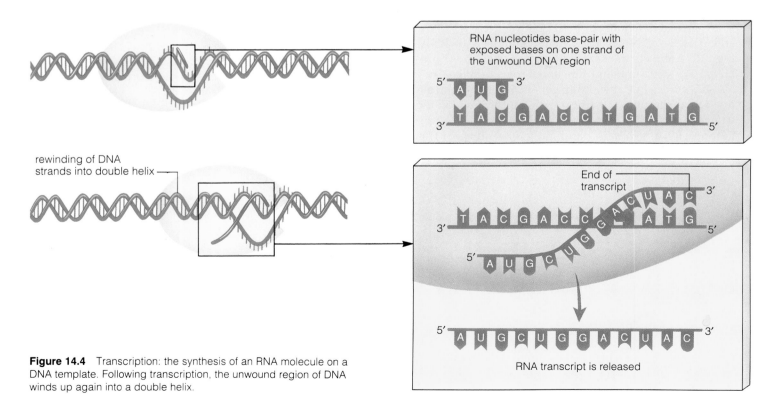

Figure 14.4 Transcription: the synthesis of an RNA molecule on a DNA template. Following transcription, the unwound region of DNA winds up again into a double helix.

strand one at a time, in the 5′ → 3′ direction. (Here you may wish to refer to Figure 13.8.)

Transcription *differs* from DNA replication in three key respects. First, only one region of one DNA strand—not the whole strand—serves as the template. Typically, the genes coding for polypeptide chains are confined to a stretch of DNA that is between 70 and 10,000 nucleotides long. Second, transcription requires different enzymes. Three types of **RNA polymerases** put together the strands of rRNA, mRNA, and tRNA. Third, unlike DNA replication, the result is a single-stranded molecule, not a double-stranded one.

Transcription starts at a **promoter**, a specific sequence of bases on one of the two DNA strands that signals the start of a gene. An RNA polymerase does not find the location of a promoter all by itself. Rather, it recognizes and binds with one or more small proteins that have become positioned on the DNA strand a few nucleotides "up the road" from a promoter. After this happens, the enzyme can bind with the promoter and open up a local region of the DNA double helix.

As Figure 14.4 shows, the enzyme moves stepwise along the exposed nucleotides of one DNA strand, unwinding a bit more of the double helix as it goes. The RNA strand continues to grow until the enzyme encounters a base sequence that serves as a termination signal. Then the RNA is released from the template as a free, single-stranded transcript.

Messenger-RNA Transcripts

Again, of the three classes of RNA, only mRNA carries protein-building instructions from the nucleus into the cytoplasm. However, a newly formed mRNA transcript cannot even depart from the nucleus without first undergoing modification. Just as a dressmaker might snip off some threads or add bows on a dress before it leaves the shop, so does a eukaryotic cell tailor its mRNA.

For one thing, the first end of the mRNA to be synthesized (the 5′ end) quickly gets capped. The cap is simply a nucleotide covalently bonded to a methyl group and phosphate groups. It seems to be recognized by other molecules (initiation factors) as the "start" signal for translation. Besides having a cap, most mature mRNA transcripts acquire a tail at the opposite end. This "poly-A tail" consists of about 100 to 200 adenine-containing molecules and seems to help prevent mRNA from being degraded in the cytoplasm.

For another thing, newly transcribed mRNA contains more than the code for a string of amino acids. The actual coding portions, which will become translated into proteins, are called **exons**. But new mRNA also contains **introns**. These are "noncoding" portions; they have no information about the amino acid sequence. Before the mRNA leaves the nucleus, its introns are snipped out and the exons spliced together, in the manner shown in Figure 14.5.

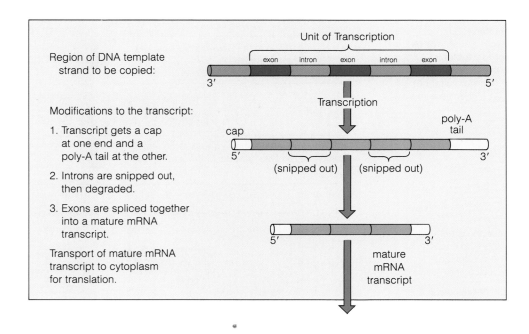

Figure 14.5 Transcription and modification of newly formed mRNA in the nucleus of eukaryotic cells.

First Letter	Second Letter				Third Letter
	U	C	A	G	
U	phenylalanine	serine	tyrosine	cysteine	U
U	phenylalanine	serine	tyrosine	cysteine	C
U	leucine	serine	stop	stop	A
U	leucine	serine	stop	tryptophan	G
C	leucine	proline	histidine	arginine	U
C	leucine	proline	histidine	arginine	C
C	leucine	proline	glutamine	arginine	A
C	leucine	proline	glutamine	arginine	G
A	isoleucine	threonine	asparagine	serine	U
A	isoleucine	threonine	asparagine	serine	C
A	isoleucine	threonine	lysine	arginine	A
A	(start) methionine	threonine	lysine	arginine	G
G	valine	alanine	aspartate	glycine	U
G	valine	alanine	aspartate	glycine	C
G	valine	alanine	glutamate	glycine	A
G	valine	alanine	glutamate	glycine	G

Figure 14.6 The genetic code. The codons in an mRNA molecule are nucleotide bases, read in blocks of three. Each of those base triplets will call for a specific amino acid during mRNA translation. In this diagram, the first nucleotide of any triplet is given in the left column. The second is given in the middle columns; the third, in the right column. Thus we find (for instance) that trytophan is coded for by the triplet U G G. Phenylalanine is coded for by both U U U and U U C.

Figure 14.7 (*Right*) Genetic code in action. Notice the green-shaded blocks extending down through all three parts of this diagram (**a-c**). They show the relation between the nucleotide sequence of DNA and the amino acid sequence of proteins.

During transcription, the region of the DNA double helix shown here was unwound, and the exposed bases on one strand served as a template for assembling the mRNA strand. The bases of every three nucleotides in an mRNA strand equal one codon. Here, each codon called for one of the amino acids in this polypeptide chain. Using Figure 14.6 as a guide, can you fill in the blank codon for threonine in the chain?

TRANSLATION

The Genetic Code

Like a DNA strand, an mRNA transcript is a linear sequence of nucleotides. So we are still left with a central question: What are the protein-building "words" encoded in that sequence?

Francis Crick, Sidney Brenner, and others came up with the answer. They deduced the nature of the **genetic code**—that is, how the nucleotide sequence of DNA and then mRNA corresponds to the amino acid sequence of a polypeptide chain. The bases of the nucleotides in mRNA are read three at a time, and each of these *base triplets* calls for an amino acid. GGU, for example, calls for glycine (Figures 14.6 and 14.7). A start signal built into the mRNA strand establishes the correct "reading frame" for selecting three bases at a time in the sequence. Each base triplet in mRNA is now called a **codon**.

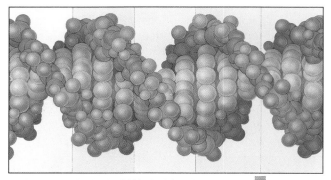

a A gene region in the DNA double helix.

b Part of an mRNA strand, transcribed from one of the two DNA strands of the double helix.

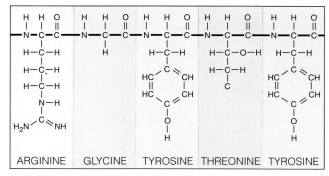

ARGININE GLYCINE TYROSINE THREONINE TYROSINE

c What the amino acid sequence will be when the mRNA is translated into a polypeptide chain.

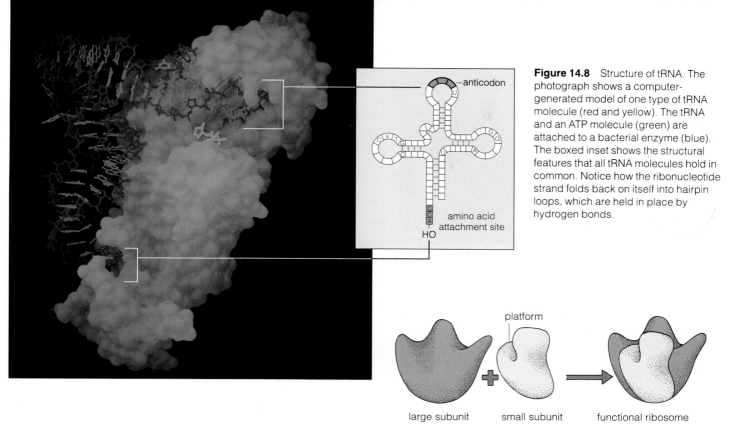

Figure 14.8 Structure of tRNA. The photograph shows a computer-generated model of one type of tRNA molecule (red and yellow). The tRNA and an ATP molecule (green) are attached to a bacterial enzyme (blue). The boxed inset shows the structural features that all tRNA molecules hold in common. Notice how the ribonucleotide strand folds back on itself into hairpin loops, which are held in place by hydrogen bonds.

anticodon

amino acid attachment site

platform

large subunit   small subunit   functional ribosome

Figure 14.9 A current model of the two-part structure of eukaryotic ribosomes.

H. Gobind Khorana, Marshall Nirenberg, Severo Ochoa, Robert Holley—these and so many others did the meticulous work to decipher the genetic code. Because there are 4 kinds of nucleotides in RNA and 3 bases in each codon, the researchers suspected that each mRNA strand must be assembled from a selection of 4^3 or 64 different codons. They discovered that 61 codons actually specify amino acids. The other three (UAA, UAG, UGA) are *stop codons* that act like stop signs in an mRNA strand. Their presence "tells" enzymes that the end of the gene region has been reached and no more amino acids are to be added to the growing polypeptide chain.

As Figure 14.6 shows, the codon AUG specifies methionine. An mRNA strand typically contains a number of AUG codons. However, starting at the capped end of mRNA, the first suitable AUG that occurs in the strand also serves as the *start codon* for translation.

One final point should be made here. As described on page 362, mitochondria and chloroplasts have their own DNA, and their genetic code is almost but not quite like the one just described. But these are organelles, not organisms. The genetic code shown in Figure 14.6 is the language of protein synthesis *for nearly all organisms*. Codons calling for certain amino acids in bacteria call for the same amino acids in protistans, fungi, plants, and animals.

Codon-Anticodon Interactions

Let's now turn to the fate of mRNA, with its string of codons, after it arrives in the cytoplasm. Sooner or later,

it will interact with its molecular relatives, the tRNAs and rRNAs.

Thirty-one kinds of tRNA molecules are pooled together in the cytoplasm of eukaryotic cells. Each tRNA has an **anticodon**, a sequence of three nucleotide bases that can base-pair with a specific mRNA codon. Each tRNA also has a molecular "hook," an attachment site for a particular amino acid (Figure 14.8).

If an anticodon is to interact with any codon, its first two bases must not violate the base-pairing rules. (Adenine must always pair with uracil, and cytosine must always pair with guanine.) The rules loosen up with respect to the third base. For example, notice in Figure 14.6 that CCU, CCC, CCA, and CCG all specify proline. Such freedom in codon-anticodon pairing at the third base is called the *wobble effect*. Through the wobble effect, *sixty-one* kinds of codons that may be present in an mRNA molecule can call up amino acids, using as few as *thirty-one* kinds of tRNAs.

Ribosome Structure

Polypeptide chains are assembled as a result of codon-anticodon interactions. And those interactions take place at specific binding sites on the surface of ribosomes. As Figure 14.9 shows, a **ribosome** has two subunits, each composed of rRNA and a number of proteins. The two subunits perform their function only during translation.

Figure 14.10 Simplified picture of protein synthesis.

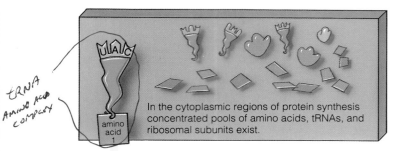

tRNA
Amino Acid
complex

In the cytoplasmic regions of protein synthesis concentrated pools of amino acids, tRNAs, and ribosomal subunits exist.

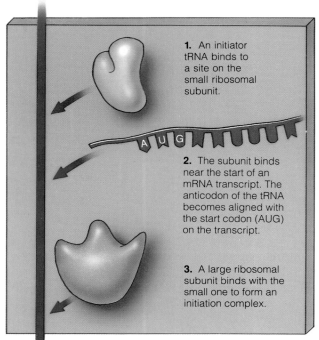

1. An initiator tRNA binds to a site on the small ribosomal subunit.

2. The subunit binds near the start of an mRNA transcript. The anticodon of the tRNA becomes aligned with the start codon (AUG) on the transcript.

3. A large ribosomal subunit binds with the small one to form an initiation complex.

a Initiation

Transfer RNA molecules deliver amino acids to two binding sites, called the P and A sites, which are very close together on the smaller of the two ribosomal subunits. That same subunit also has a binding site for mRNA.

Ten thousand ribosomes may be present in the cytoplasm of a single bacterial cell. A eukaryotic cell may contain many tens of thousands. The bacterial ribosome is only 25 nanometers wide. That's about a millionth of an inch. Although the components of a eukaryotic ribosome are larger and more numerous, both kinds of ribosomes have nearly the same shape and function.

Stages of Translation

Now that we have finally arrived at the ribosome, we are ready to consider how genes are actually translated into proteins. To keep things simple, we can portray the codon-anticodon interactions in this fashion:

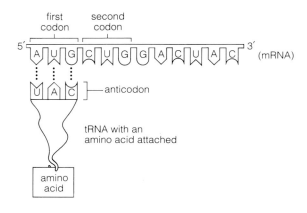

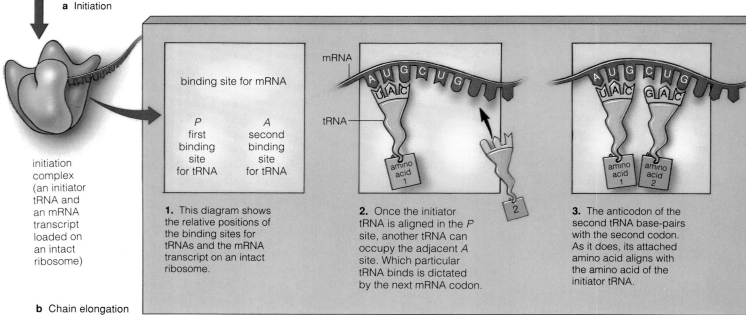

initiation complex (an initiator tRNA and an mRNA transcript loaded on an intact ribosome)

binding site for mRNA

P	A
first binding site for tRNA	second binding site for tRNA

1. This diagram shows the relative positions of the binding sites for tRNAs and the mRNA transcript on an intact ribosome.

2. Once the initiator tRNA is aligned in the P site, another tRNA can occupy the adjacent A site. Which particular tRNA binds is dictated by the next mRNA codon.

3. The anticodon of the second tRNA base-pairs with the second codon. As it does, its attached amino acid aligns with the amino acid of the initiator tRNA.

b Chain elongation

Translation proceeds through three stages, called initiation, chain elongation, and chain termination. Figure 14.10 is a step-by-step picture of the stages of translation in eukaryotic cells.

In *initiation,* both a tRNA that can start transcription and an mRNA transcript become loaded onto an intact ribosome. First, the "initiator" tRNA binds with the small ribosomal subunit. Next, the cap on the mRNA molecule binds with the small ribosomal subunit in such a way that the proper start codon (AUG) becomes positioned in front of the initiator tRNA. Finally, a large ribosomal subunit joins with the small one. Chain elongation can begin.

In *chain elongation,* amino acids are strung together in the sequence dictated by the codons of mRNA. The mRNA's start codon defines the reading frame for the sequence. The mRNA strand passes between the two ribosomal subunits, like a thread being moved through the eye of a needle.

Figure 14.10b shows the initiator tRNA in position at the ribosome's P site. Another tRNA binds to the adjacent A site. Its anticodon is able to base-pair with the next codon "down the road" from the start codon of the mRNA molecule. Now the two tRNAs are positioned in such a way that a peptide bond can readily form between their attached amino acids. (Here you may wish to refer to Figure 3.16.) The initiator tRNA gives up its amino acid entirely and is removed from the ribosome. This leaves two amino acids attached to the second tRNA—which is shifted over to the P site. The mRNA molecule

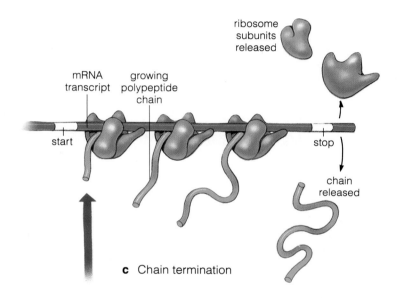

c Chain termination

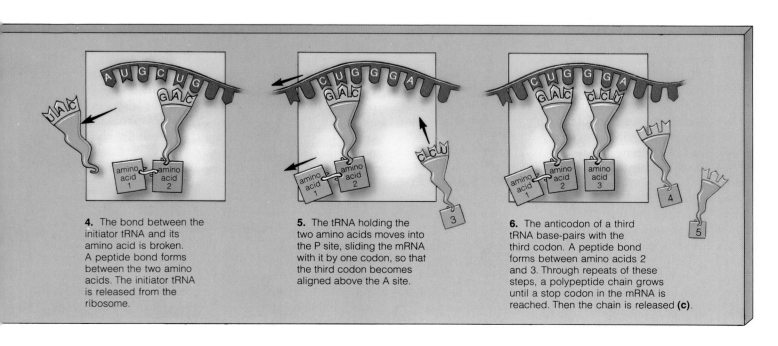

4. The bond between the initiator tRNA and its amino acid is broken. A peptide bond forms between the two amino acids. The initiator tRNA is released from the ribosome.

5. The tRNA holding the two amino acids moves into the P site, sliding the mRNA with it by one codon, so that the third codon becomes aligned above the A site.

6. The anticodon of a third tRNA base-pairs with the third codon. A peptide bond forms between amino acids 2 and 3. Through repeats of these steps, a polypeptide chain grows until a stop codon in the mRNA is reached. Then the chain is released **(c)**.

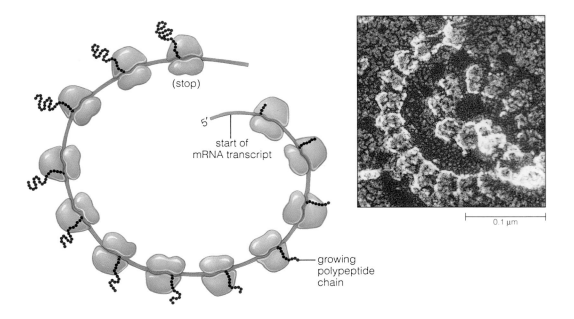

(stop)

5'
start of
mRNA transcript

growing
polypeptide
chain

0.1 µm

Figure 14.11 Micrograph and sketch of many ribosomes simultaneously translating the same mRNA molecule in a eukaryotic cell.

slides along with it. At this point *another* tRNA moves into the now-empty A site, and the steps are repeated. Enzymes built into the ribosome catalyze peptide bonds between every two amino acids delivered to the ribosome. In this way a polypeptide chain grows.

In *chain termination*, a stop codon in the mRNA signals that no more amino acids can be added to the polypeptide chain. Now, with the help of specific proteins called release factors, the ribosome and polypeptide chain are detached from the mRNA. The detached chain joins the pool of free proteins in the cytoplasm or enters the cytomembrane system for further processing. Page 63 outlines the final destinations of the completed proteins.

The steps just described can be repeated many times on the *same* mRNA transcript, with several copies of the polypeptide chain forming at the same time. What happens is this: A new ribosome hops onto the mRNA almost as soon as the preceding ribosome has translated enough of it to get out of the way. The word **polysome** refers to several ribosomes that are spaced closely together on the same mRNA. Figure 14.11 shows an example of a polysome. Such "assembly lines" for protein synthesis are a common feature of cells.

Figure 14.12 summarizes the flow of information along the path leading from genes to proteins.

MUTATION AND PROTEIN SYNTHESIS

In general, the base sequence in DNA must be preserved from one generation to the next, otherwise offspring might not be able to synthesize all the proteins required for their own survival and reproduction. Yet changes do occur in the DNA, and they may affect one or more traits of the individual. Said another way, they give rise to variations in phenotype.

For instance, as we saw in the preceding chapters, phenotypic variation arises through crossing over and recombination, which put new mixes of alleles in chromosomes. It also arises through changes in the structure and number of chromosomes, as brought about by nondisjunction.

Another kind of change is called the **gene mutation**. This is a deletion, addition, or substitution of one to several bases in the nucleotide sequence of a gene. Gene mutations are rare events. On the average, the mutation rate for a gene is only one in a million replications.

Some gene mutations are induced by **mutagens**, environmental agents that can attack a DNA molecule and modify its structure. Viruses, ultraviolet radiation, and certain chemicals are examples of mutagens. Other gene mutations are spontaneous; they are not induced

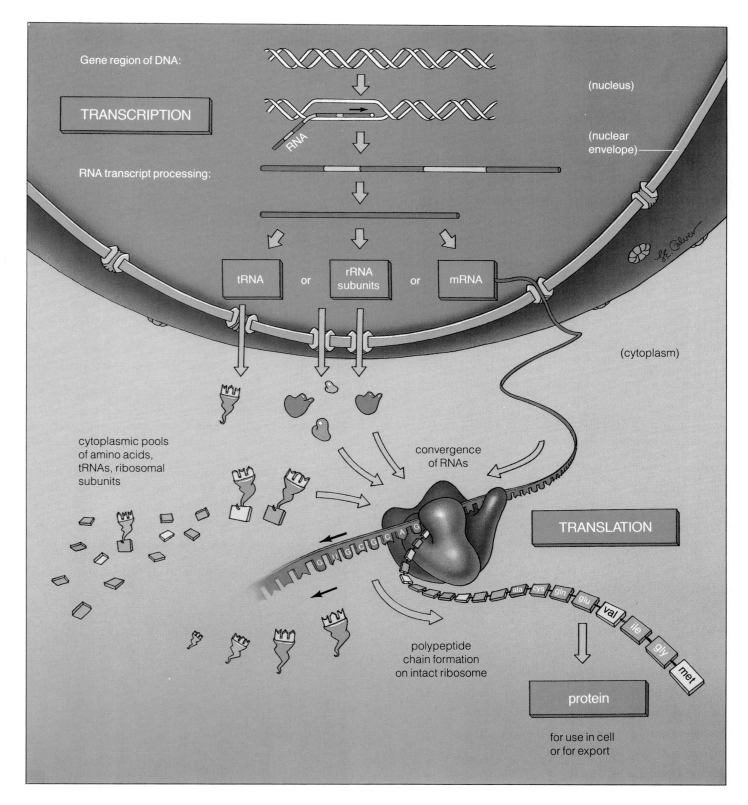

Figure 14.12 Summary of the flow of genetic information in protein synthesis in eukaryotic cells.

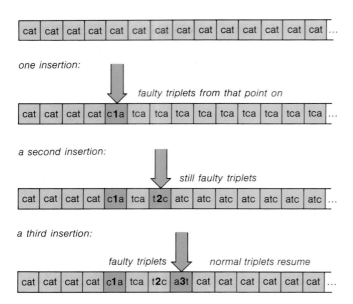

one insertion:

faulty triplets from that point on

a second insertion:

still faulty triplets

a third insertion:

faulty triplets / normal triplets resume

Figure 14.13 Simplified picture of frameshift mutation. Studies of such mutations in bacteriophage led to the discovery of the genetic code. During translation, the base sequence of mRNA is read in blocks of three (that is, as base triplets) according to the genetic code.

The word *cat* is used here to represent every base triplet. Suppose an extra nucleotide ("1") is inserted into a gene. When mRNA is transcribed off that gene, the insertion will put the reading frame out of phase and the wrong amino acids will be called up during translation. A second insertion ("2") would not improve matters. A third insertion ("3") will restore the reading frame, so only part of the resulting protein will be defective. Insertions of extra bases into a DNA molecule often give rise to mutant phenotypes.

Figure 14.14 Color variations in kernels of Indian corn. All the cells in a kernel have pigment-coding genes, so you might expect the whole kernel to be the same color. Some are indeed fully colored, but others are spotted or entirely colorless. Early in plant growth, transposable elements (movable DNA regions) were present in a gene necessary to produce the kernel pigment. The transposable elements caused a mutation that gave rise to colorless kernels. When the transposable element moved out of the gene, the gene's activity was restored. Thus, all cells that were descendants of the cells in which the transposition occurred were able to produce pigment. The outcome was pigmented spots on the kernel.

by agents outside the cell. During DNA replication, for example, an A might become paired with C instead of T. (Even enzymes make mistakes.) Proofreading enzymes might detect the mistake—but will it "fix" the A or the C? If the enzymes remove the wrong base from a mismatched pair, the result is a type of spontaneous mutation called a *base-pair substitution.*

You might be thinking that a change in a single base pair is insignificant. But as Figure 14.2 made clear, sickle-cell anemia has been traced to a single mutation in the DNA strand coding for the beta chain of hemoglobin. Only one amino acid is substituted for another in the resulting chain—yet the substitution can have severe consequences.

Another type of spontaneous change is the *frameshift mutation.* Here, the insertion or deletion of one to several base pairs in a DNA molecule puts the nucleotide sequence out of phase, so that the reading frame shifts during protein synthesis (Figure 14.13). Because genetic instructions are not read correctly, an abnormal protein is synthesized.

Barbara McClintock discovered still another type of spontaneous gene mutation. Through studies of Indian corn (maize), she realized that certain DNA regions frequently "jump" to new locations in the same DNA molecule or in a different one. These *transposable elements* often inactivate the genes into which they become inserted and give rise to observable changes in phenotype (Figure 14.14).

What is the point to remember about these mutations and others? The *Commentary* provides us with an answer.

Commentary

Gene Mutation and Evolution

Each gene is said to have a characteristic *mutation rate,* which simply is the probability of its mutating between or during DNA replications. On the average, the mutation rate for a gene is one in a million (10^6) replications.

Genes mutate independently of one another. To determine the probability of any two mutations occurring in a given cell, we would have to multiply the individual mutation rates for two of its genes. For example, if the rate for the first gene is one in a million per cell generation and the rate for the second is one in a billion (10^9), then there is only one chance in a million billion (10^{15}) that both genes will mutate in the same cell.

In the natural world, gene mutations are rare, chance events. It is impossible to predict exactly when, and in which organism, they will appear. They also are inevitable. They arise not only through mistakes in DNA replication. They arise also through the action of mutagens, including ultraviolet light, ionizing radiation, and various chemicals in our environment.

A mutation may turn out to be beneficial, neutral, or harmful. The outcome depends on how the protein specified by the mutated DNA region functions in the cells that require the protein, and on how the protein affects the coordinated workings of the entire individual (page 277). Because of this, most mutations do not bode well for the individual. No matter what the species, each organism generally inherits a combination of many genes that already are fine-tuned for a given range of operating conditions in the body. A mutant gene is likely to code for a protein that is less functional, not more so, under those conditions.

Over evolutionary time, agents of natural selection undoubtedly perpetuated those packages of genes having a history of survival value. They must have favored DNA polymerases that showed little tolerance of mismatched base pairs. They also must have favored enzymes that could chemically recognize, remove, and replace mismatched base pairs. What is clear is that replication enzymes have protected the overall stability of the vulnerable molecules of inheritance that have been replicated through billions of years.

Yet every so often, some mutations provided their bearers with advantages. We saw this with the example of the peppered moth, described briefly in Chapter 1. Selection processes worked to perpetuate mutations having adaptive value. Besides this, other mutations produced DNA regions with no known function—but they did their bearers no harm and they, too, have been perpetuated. After more than three billion years, molecular descendants of the first strands of DNA are replete with mutations.

What does all this mean? It means that every living thing on earth shares the same chemical heritage with all others. Your DNA has the same kinds of substances, and follows the same base-pairing rules, as the DNA of earthworms in Missouri and grasses on the Mongolian steppes. Your DNA is replicated in much the same way as theirs, and the same genetic code is followed in translating its messages into proteins. In the evolutionary view, the reason you don't look like an earthworm or a grass plant is largely a result of selection of different mutations that originated in different lineages of organisms. Thus the sequence of base pairs along the DNA molecule has come to be different in you, the plant, and the worm.

And so we have three concepts of profound importance. *First, DNA is the source of the unity of life. Second, mutations and other changes in the structure and number of DNA molecules are the source of life's diversity. Finally, the changing environment is the testing ground for the success or failure of the proteins specified by each novel DNA sequence and assortment that appears on the evolutionary scene.*

SUMMARY

1. Protein-building instructions are encoded in genes, each of which is a specific, linear sequence of nucleotides in one strand of a DNA double helix. The path leading from genes to proteins has two steps, called transcription and translation.

   a. In transcription, an exposed region of a DNA strand serves as the template for assembling an RNA strand.

   b. In translation, RNA molecules interact to convert the gene's message into a linear sequence of amino acids—that is, a polypeptide chain. Such chains are the structural units of proteins.

2. There are three classes of RNA:

   a. Molecules of rRNA are components of the ribosome on which polypeptide chains are assembled.

   b. An mRNA strand is the blueprint of genetic information for building a specific chain.

   c. Many different tRNA molecules deliver amino acids to the ribosome in a sequence dictated by their interaction with the mRNA.

3. Thus DNA is used to synthesize RNA, the RNA is used to synthesize proteins (some of which will take part in DNA and RNA synthesis). This flow of information is the central dogma of molecular biology:

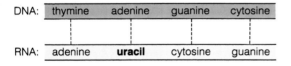

4. The genetic code is the relation between a linear sequence of nucleotides in DNA (then mRNA) and a linear sequence of amino acids in a polypeptide chain. The code words are a sequence of nucleotide bases that are read in blocks of three (base triplets). Each base triplet in mRNA is a codon; a complementary base triplet in tRNA is an anticodon.

5. Transcription follows essentially the same base-pairing rules that apply to DNA replication. However, uracil—not thymine—pairs with the adenine present in the DNA template strand:

6. In eukaryotes, a new mRNA transcript becomes modified in the nucleus before moving into the cytoplasm for translation. A cap and commonly a tail are added to it. Its introns (nucleotide sequences that do not code for parts of the polypeptide chain) are excised and its exons (coding sequences) are spliced together.

7. Translation proceeds through three stages:

   a. In initiation, a small ribosomal subunit binds with an initiator tRNA, then with an mRNA transcript. The small subunit then binds with a large ribosomal subunit to form the initiation complex.

   b. In chain elongation, tRNAs deliver amino acids to the ribosome. The tRNA anticodons pair appropriately with mRNA codons. Then peptide bonds form between their amino acids, forming the chain.

   c. In chain termination, a stop codon triggers events that cause the polypeptide chain to detach from the ribosome.

8. There is an underlying chemical unity among all organisms. Regardless of the species, DNA is composed of the same substances, follows the same base-pairing rules, and is replicated in much the same way. The genetic code by which its instructions are translated into proteins is nearly universal.

9. Overall, the protein-building instructions encoded in DNA are preserved through the generations. But crossing over and recombination, changes in chromosome structure or number, and gene mutations can change parts of those instructions. Such changes lead to phenotypic variation among individuals, and so provide grist for the evolutionary mill.

Review Questions

1. Are the proteins specified by eukaryotic DNA assembled *on* the DNA molecule? If so, state how. If not, tell where they are assembled, and on which molecules. *219*

2. Figure 14.12 shows the steps by which hereditary instructions are transcribed from DNA into RNA, which is then translated into proteins. Study this figure and then, on your own, write a description of this sequence, taking care to define the terms *transcription* and *translation*. *219, 220–226*

3. Define *genetic code*. Is the same basic genetic code used for protein synthesis in all living organisms? *222–223*

4. Define the three types of RNA. What is a codon? *219, 222, 223*

5. If sixty-one codons in mRNA actually specify amino acids, and if there are only twenty common amino acids, then more than one

codon combination must specify some of the amino acids. How do triplets that code for the same thing usually differ? *223*

6. Define intron and exon. What happens to introns before an mRNA transcript is shipped to the cytoplasm? *221*

7. If genetic information were transmitted precisely from generation to generation, organisms would never change. What are some mutations that give rise to phenotypic diversity? *226, 228*

10. Using the genetic code shown in Figure 14.6, translate the mRNA sequence UAUCGCACCUCAGGAUGAGAU. Which of the following polypeptide chains does this sequence specify?
    a. tyr-arg-thr-ser-gly-stop-asp...
    b. tyr-arg-thr-ser-gly...
    c. tyr-arg-tyr-ser-gly-stop-asp...
    d. none is correct

Self-Quiz *(Answers in Appendix IV)*

1. Nucleotide bases, read _____ at a time, serve as the "code words" of genes.

2. Genetic information in DNA is transferred to RNA strands during _____, the first step in protein synthesis.
    a. replication   c. multiplication
    b. duplication   d. transcription

3. The RNA molecule is _____.
    a. a double helix
    b. usually single-stranded
    c. always double-stranded
    d. usually double-stranded

4. During transcription, base-pairing is similar to that of DNA replication except _____.
    a. cytosine in DNA pairs with guanine in RNA
    b. adenine in DNA pairs with uracil in RNA
    c. thymine in DNA pairs with adenine in RNA
    d. guanine in DNA pairs with cytosine in RNA

5. _____ starts when two ribosomal subunits, an initiator tRNA, and an mRNA transcript come together.
    a. Transcription
    b. Replication
    c. Subduction
    d. Translation

6. The coded genetic instructions for forming polypeptide chains are carried to the ribosome by _____.
    a. DNA
    b. rRNA
    c. mRNA
    d. tRNA

7. The function of tRNA is to _____.
    a. deliver amino acids to the ribosome
    b. pick up genetic messages from rRNA
    c. synthesize mRNA
    d. all of the above

8. How many amino acids are coded for in this mRNA sequence: CGUUUACACCGUCAC?
    a. three
    b. five
    c. six
    d. seven
    e. more than seven

9. An anticodon pairs with the nitrogen-containing bases of _____.
    a. mRNA codon
    b. DNA codons
    c. tRNA anticodon
    d. amino acids

Selected Key Terms

anticodon *223*
base-pair substitution *228*
base sequence *219*
base triplet *222*
central dogma *219*
codon *222*
electrophoresis *218*
exon *221*
frameshift mutation *228*
gene *219*
gene mutation *226*
genetic code *222*
intron *221*
messenger RNA (mRNA) *219*
polysome *226*
promoter *221*
ribosomal RNA (rRNA) *219*
ribosome *223*
RNA *219*
RNA polymerase *221*
start codon *223*
template *219*
transcription *219*
transfer RNA (tRNA) *219*
translation *219*
transposable element *228*
uracil *220*

Readings

Alberts, B., et al. 1989. *Molecular Biology of the Cell.* Second edition. New York: Garland.

Amato, I. January 1991. "Stuck on Mussels." *Science News* 139: 8–15.

Darnell, J. October 1985. "RNA." *Scientific American* 253(4):68–78.

Nomura, M. January 1984. "The Control of Ribosome Synthesis." *Scientific American* 250(1):102–114.

Prescott, D. 1988. *Cells.* Boston: Jones and Bartlett. Chapter 8 of this textbook contains an excellent introduction to protein synthesis.

15 CONTROL OF GENE EXPRESSION

A Cascade of Proteins and Cancer

Every second of the day, millions of cells in your skin, gut lining, liver, and other body regions divide and replace their worn-out, dead, and dying predecessors. They do not divide willy-nilly. They cannot divide at all unless they first synthesize and stockpile molecules of cyclin, a protein. Inside the cell that makes it, cyclin binds with another protein, called cdc2, and sets the division machinery in motion.

The cdc2 is the first of a series of enzymes that catalyze the transfer of phosphate from ATP to the next enzyme in line. The enzyme molecules act more than once. Each molecule activates many others, which activate many others, and so on in a growing cascade of reactions that ripple through the cell.

The first enzymes go to work while the cell is still in interphase; they replicate its DNA. Later, other enzymes act directly on proteins that are part of the nuclear envelope and so trigger its breakdown. Others help assemble and operate a spindle of microtubules. That spindle harnesses and moves the cell's chromo-

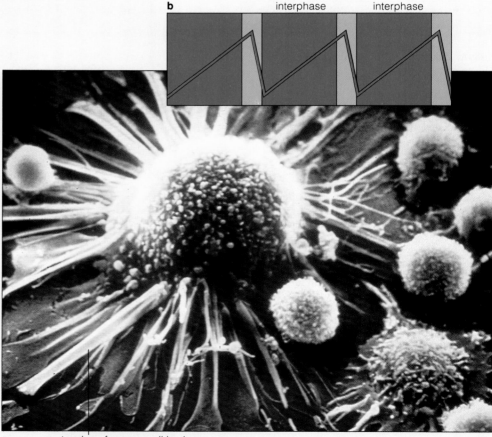

Figure 15.1 (**a**) Scanning electron micrograph of a cancer cell, surrounded by some of the body's white blood cells that may or may not be able to destroy it. Cancer cells are skewed in structure and function. Worse, they have lost the controls over the genes and gene products that can suppress cell division. Because cancer cells cannot stop dividing, they form abnormal masses of cells that can destroy surrounding tissues. They are graphic examples of why gene expression must be regulated with precision. (**b**) From one study, a chart of the controlled changes in the intracellular levels of cyclin (brown line) in normal cells. Cyclin is the protein that guides cells into mitosis (light-blue bands) during the cell cycle.

a   extension of cancer cell body

b   interphase   interphase

somes into two parcels before the cytoplasm divides in two. Still other enzymes orchestrate the split. Following division, cyclin-degrading enzymes are activated and destroy the cell's entire batch of cyclin. Without cyclin, the division machinery is put to rest. Now the daughter cells, starting life at interphase, start stockpiling cyclin. If they, too, go on to divide, cyclin will again be destroyed.

Of all the cell's proteins, only cyclin accumulates at a constant rate, disappears abruptly, then accumulates again during cell cycles (Figure 15.1). Through its interaction with cdc enzymes, cyclin guides cells into the division process.

How does a cell "know" when to start and stop building cyclin? Just as it takes a turn from an ignition key to start and stop the engine of your car, so does it take signals from regulator molecules, such as hormones, to control the cyclin-driven engine. *Such signals lift the controls that otherwise suppress cell division, then put on the brakes by reinstating controls when division is completed.*

Researchers are only now unraveling the mystery of how those signals work, and their sleuthing is of more than passing interest to us. Why? Sometimes the controls over cell division are lost. It is not that cells start dividing at a berserk pace. Rather, it is that the cell division cycle cannot stop. This is what has happened in the body of a family member, a friend, or an acquaintance who has been stricken with cancer.

Unless cancer cells are eradicated, their chronically dividing descendants will kill the individual. In fact, cancer is a leading cause of human death. In the United States, it is second only to heart disease. Cancer is not just a human affliction. It has been observed in most animal species that have been studied to date. Comparable abnormalities have even been observed in many plants. At the chapter's end, we will consider the nature of cancerous transformations. To gain insight into what is going wrong in affected individuals, however, we must start with how cells use and control their genes when things are going right.

1. All cells precisely control when, how, and to what extent their various genes are expressed.

2. Control is exerted through regulatory proteins and other molecules that interact with DNA, with RNA transcribed off the DNA, or with the resulting polypeptide chains. Especially among vertebrates, hormones have major roles in controlling gene expression.

3. In all cells, controls come into play during transcription, translation, and post-translation (when new polypeptide chains become modified, as by having simple sugars attached to them). In eukaryotes, controls also govern the processing of new RNA transcripts and their shipment out of the nucleus for translation in the cytoplasm.

4. In multicelled eukaryotes, all cells have the same genes, but they activate or suppress many of those genes in different ways. This selective gene expression leads to cell differentiation—to pronounced structural and functional variations among the cells that make up different tissues.

THE NATURE OF GENE CONTROL

All the different cells of your body carry the same genes, and they use most of them to synthesize proteins that are basic to any cell's structure and functions. That is why the protein subunits of microtubules are the same from one cell to the next, as are many of the enzymes used in metabolism.

Yet each type of cell also uses a small fraction of genes in highly specialized ways. Even though they all carry the genes for hemoglobin, only red blood cells activate those genes. Even though they all carry the genes for antibodies, which are protein "weapons" against specific agents of disease, only certain white blood cells activate them. *These and all other living cells control which genes are active and which gene products appear, at what times, and in what amounts.*

When Controls Come Into Play

In any organism, gene controls operate in response to chemical changes within the cell or its surroundings. In terms of your own cells, chemical conditions change when you vary your diet or level of activity. Conditions also have been changing in inevitable ways ever since you were a tiny mass of cells growing in your mother's body. Within each responding cell, gene activity has been changing appropriately, either to keep the cell itself alive or to contribute to your overall growth and development through time.

The elements of control operate in response to changing chemical conditions within a cell or its surroundings.

Control Agents and Where They Operate

Gene controls are exerted through the action of regulatory proteins and other molecules that interact with DNA, with mRNA transcribed from DNA, and with gene products resulting from mRNA translation. (Here you may wish to refer to page 219.) Transcriptional controls are the most common. They depend on two types of regulatory proteins, called repressors and activators, that can change the rates at which particular genes are transcribed:

repressor protein *prevents the enzymes of transcription (RNA polymerases) from binding to the DNA; affords negative control of gene activity*

activator protein *enhances the binding of RNA polymerases to the DNA; affords positive control of gene activity*

Many control agents bind to sites in the DNA, such as promoters. A **promoter**, recall, is a specific base sequence that signals the start of a gene in a DNA strand. Before enzymes can even assemble RNA on the DNA, they must first bind with a promoter. As you will see, some control agents bind to **operators**, which are short base sequences between a promoter and the start of a gene. Typically, a control agent does not stay permanently attached to a binding site. The binding may be reversed, for example, when the conditions that called for the synthesis of a particular protein have changed.

Among eukaryotes, **hormones** are major control agents. They are signaling molecules secreted from specific types of cells. They travel the bloodstream and affect gene expression in target cells somewhere in the body. (Any cell is a "target" if it has receptors to which a specific signaling molecule can bind.)

Gene expression is controlled through regulatory proteins, hormones, and other molecules that interact with DNA, RNA, and the protein products of genes.

GENE CONTROL IN PROKARYOTES

Studies of *Escherichia coli*, a type of bacterial cell, yielded the first insights into gene controls. As is the case for most prokaryotes, its gene activity depends largely on negative and positive controls over the rate of transcription.

Negative Control of Lactose Metabolism

E. coli makes its home in the intestines of mammals, and it survives on sugars and other nutrients of its host's diet. Some controls allow it to produce enzymes *only when needed* to degrade lactose, a sugar in milk.

After you drink a glass of milk, *E. coli* rapidly transcribes three adjacent genes that code for the lactose-degrading enzymes. A promoter and an operator precede those genes in *E. coli* DNA and have roles in the transcription of all three. This type of arrangement, in which the same promoter-operator sequence services more than a single gene, is called an **operon**:

A gene at a different location in the DNA codes for a type of repressor protein:

Depending on cellular conditions, this repressor can lock onto the operator *or* a lactose molecule. The repressor is part of a negative control mechanism, for it prevents the genes of the lactose operon from being transcribed. It binds with the operator when concentrations of lactose are low—as when you have not been drinking any milk. Being a rather large molecule, the repressor overlaps the promoter and so blocks RNA polymerase's access to the genes. Figure 15.2b illustrates this effect. Thus lactose-degrading enzymes are not produced when they are not required.

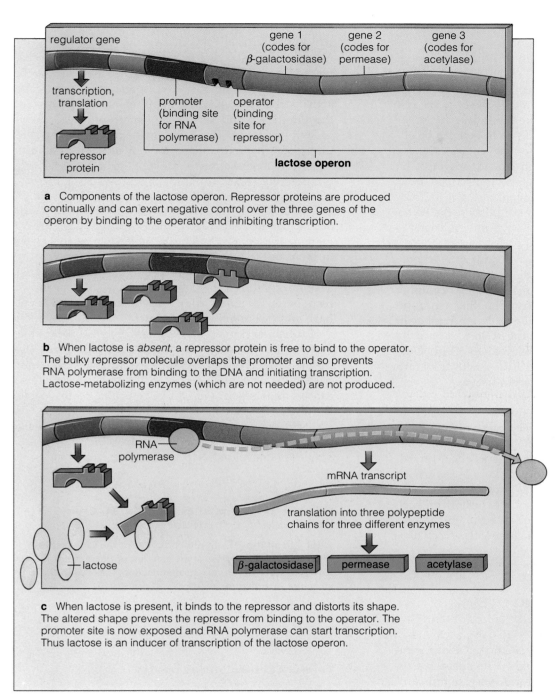

a Components of the lactose operon. Repressor proteins are produced continually and can exert negative control over the three genes of the operon by binding to the operator and inhibiting transcription.

b When lactose is *absent*, a repressor protein is free to bind to the operator. The bulky repressor molecule overlaps the promoter and so prevents RNA polymerase from binding to the DNA and initiating transcription. Lactose-metabolizing enzymes (which are not needed) are not produced.

c When lactose is present, it binds to the repressor and distorts its shape. The altered shape prevents the repressor from binding to the operator. The promoter site is now exposed and RNA polymerase can start transcription. Thus lactose is an inducer of transcription of the lactose operon.

Figure 15.2 Negative control of the lactose operon. The first gene of the operon codes for an enzyme that splits lactose into two subunits (glucose and galactose). The second one codes for an enzyme that transports lactose molecules across the plasma membrane and into the cytoplasm. The third plays a complex role in lactose metabolism.

When you have been drinking milk and lactose has entered your small intestines, the repressor does not block transcription. What happens is this: A lactose molecule binds with and alters the shape of the repressor. In its altered shape, the repressor cannot bind to the operator. As a result, RNA polymerase has access to the genes, which are transcribed and translated into proteins (Figure 15.2c). Thus lactose-degrading enzymes are produced when they are required.

Positive Control of Nitrogen Metabolism

The positive control mechanisms afforded by activator proteins resemble the negative control just described—but with opposite results. Here, the promoter sequence is such that RNA polymerase binds very inefficiently to it without assistance. The promoter becomes functional when an activator protein is bound to it. Then, RNA polymerase binds and initiates transcription with greater efficiency. The transcription rate slows when the activator protein is removed from the promoter.

Consider what happens when you have not eaten much protein. The food material moving through your gut is therefore low in nitrogen, a vital nutrient for *E. coli* as well as for yourself. At such times, the bacterial cell steps up its synthesis of glutamine synthetase and other enzymes of the pathways by which nitrogen can be obtained from the surroundings. The genes coding for these enzymes are called the **nitrogen-related operon**. The more mRNA molecules that can be transcribed from those genes, the more enzyme molecules will be produced. The more enzymes the cell produces, the better will be its chance of assimilating what little nitrogen *is* available at that time.

The drop in nitrogen triggers a cascade of reactions, similar to the one described at the start of this chapter. One type of enzyme activates many molecules of another type. Each of these activates many of another type, and so on. At the end of the cascade, a large number of activator molecules have been called into service by having a phosphate group transferred to them. Only when they are phosphorylated can these activators turn on the gene coding for glutamine synthetase.

These reactions permit *E. coli* to make a very big metabolic response to a dilute amount of nitrogen in its surroundings. When nitrogen becomes plentiful, the same enzyme that transferred phosphate to the activator protein removes it, so the response can now be reversed.

In this example, a regulatory protein is controlled by its conversions between active (phosphorylated) and inactive forms. Prokaryotes do not rely heavily on such interconversions as a means of gene control. Eukaryotes do. The cascade of reactions that culminates in cell division is but one example.

GENE CONTROL IN EUKARYOTES

 Selective Gene Expression

Compared to prokaryotes, less is known about gene controls in multicelled eukaryotes. The main reason is that patterns of gene expression vary within and between different body tissues. Consider that all the cells in your body are descendants of the same fertilized egg and so have the same genes. But cells of your brain, liver, and other tissues are **differentiated**: they have become specialized in composition, structure, and function.

Differentiation arises through *selective* gene expression in different cells. Depending on the cell type and the control agents acting on it, some genes might be turned on only at one particular stage of the life cycle. Others might be left on all the time or never activated at all. Still other genes might be switched on and off throughout an individual's life.

Think about hormones and other signaling molecules, which play crucial roles in selective gene expression. Hormones are secreted from glands or glandular cells, picked up by the bloodstream, and distributed throughout the body. Some have widespread effects on gene activity in many cell types. In vertebrates, for example, the pituitary gland secretes somatotropin (also called growth hormone). This hormone helps control the synthesis of proteins required for cell division and, ultimately, the body's growth. Most of the body's cells have receptors for somatotropin.

Other hormones affect gene expression only in certain cells at certain times. Prolactin is like this. Its target cells are in mammary glands. Beginning a few days after a mammalian female gives birth, prolactin activates genes in those cells alone—genes that have exclusive responsibility for milk production. Liver cells and heart cells also have those genes, but they have no means of responding to signals from prolactin and they never will have any role in milk production.

Explaining hormonal control of gene activity is like explaining a full symphony orchestra to someone who has never seen one or heard it perform. Many separate parts must be defined before their intricate interactions can be understood! We will be returning to this topic, starting with Chapter 34 on the endocrine system. As you will see in Chapters 42 and 43, some of the most elegant examples of hormonal controls are drawn from studies of animal reproduction and development.

Cell differentiation occurs in multicelled eukaryotes as a result of *selective gene expression*.

Although all the cells in the body inherit the same genes, they activate or suppress some fraction of those genes in different ways to produce pronounced differences in their structure or function.

Levels of Control in Eukaryotes

Let's now consider a few examples of the levels of control of gene expression in eukaryotes. As Figure 15.3 and the following list indicate, controls are exerted at different levels:

1. *Transcriptional controls* influence when and to what degree a particular gene will be transcribed (if at all).

2. *Transcript-processing controls* govern modification of the initial mRNA transcripts in the nucleus.

3. *Transport controls* dictate which mature mRNA transcripts will be shipped out of the nucleus and into the cytoplasm for translation.

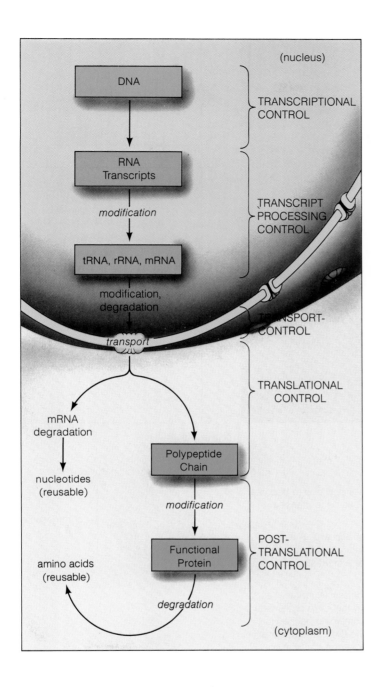

Figure 15.3 (*Left*) Control of eukaryotic gene expression: levels at which regulatory mechanisms can be brought into play. (Here, the steps are superimposed on a sketch of nuclear and cytoplasmic regions of a eukaryotic cell.)

4. *Translational controls* govern the rates at which mRNA transcripts that reach the cytoplasm will be translated into polypeptide chains at the ribosomes.

5. *Post-translational controls* govern how the polypeptide chains become modified into functional enzymes and other proteins. (For example, some chains have specific sugar or phosphate groups attached to them.) They also govern the enzymes themselves by enhancing or repressing their action.

We know the most about controls that operate during transcription and transcript processing. Among these are regulatory proteins, especially activators that are turned on and off (by the addition and removal of phosphate). Transcription also is controlled through the way eukaryotic DNA is packed up with proteins in chromosomes. Figure 15.4 shows the packing at the most basic level of chromosome structure—the nucleosome, which consists of DNA looped around a core of histone proteins. Besides histones, other chromosomal proteins may

Figure 15.4 (*Below*) One model of how DNA loosens up during transcription. As indicated earlier in Figure 13.10, nucleosomes are the basic packing unit of the eukaryotic chromosome. Each consists of a portion of the DNA double helix, looped twice around a core of histone proteins. The diagrams show a nucleosome from *Physarum polycephalum*. At one point in the life cycle of this slime mold, only the genes coding for rRNA are transcribed. The histones remain associated with the DNA in the gene regions being transcribed. But the tight packing becomes more relaxed.

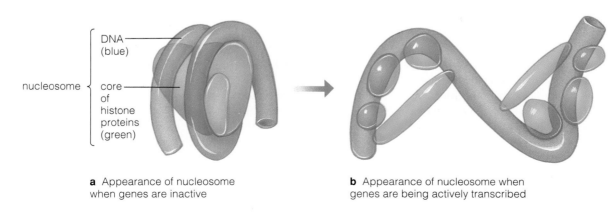

a Appearance of nucleosome when genes are inactive

b Appearance of nucleosome when genes are being actively transcribed

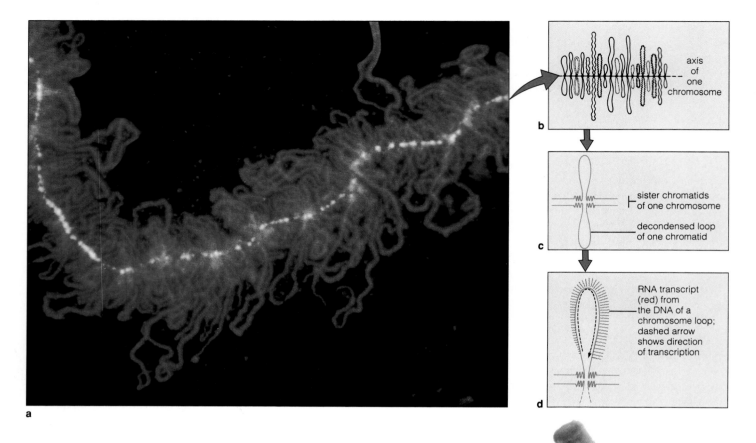

Figure 15.5 From microscopy, evidence of transcription in chromosomes.

During prophase I of meiosis, *lampbrush* chromosomes become visible in cells destined to become the eggs of amphibians and some other animals. At this time it is possible to see decondensed regions, where thousands of looped DNA regions have become uncoiled.

(a–d) This is part of a lampbrush chromosome from a newt, *Notophthalmus viridiscens*. A dye that binds with DNA caused the axis of the chromosome to fluoresce white. A dye that binds with a ribonucleoprotein caused the loops to fluoresce red. Ribonucleoproteins are a clear sign of transcriptional activity.

Intense transcription also can be observed in the *polytene* chromosomes of certain fly larvae. Years ago, staining techniques revealed their faint banding patterns. Today we know that DNA replication has occurred repeatedly in such chromosomes, with the duplicated strands remaining packed in parallel. A hormone (ecdysone) prods a regulatory protein into binding with the DNA, and binding promotes transcription. In regions being transcribed, the chromosome has loosened up, forming *chromosome puffs*. The degree of transcription correlates with how large and diffuse the puffs become. **(e)** Puffing in a polytene chromosome from a midge. The red-violet stain indicates transcriptionally active regions.

organize the chromosome into functional domains, which can be kept locked up or loosened up in a given cell at a given time. Certainly packing variations are known to occur during the life cycle of many organisms, and these affect the accessibility and activity of different genes. In an example to be described shortly, the genes on an entire chromosome are never allowed to be transcribed.

Evidence of Control Mechanisms

Chromosome Loops and Puffs. Evidence of controls over transcription can be observed in amphibian eggs and the larvae of certain insects, both of which grow very rapidly. It takes a great deal of mRNA and protein synthesis to sustain that rapid growth. Electron micrographs of amphibian eggs reveal visible changes in chromosome structure that can be correlated directly with transcription of the genes necessary for growth. During prophase I of meiosis, the chromosomes decondense into thousands of looped domains. The DNA has been selectively loosened, making specific gene regions accessible for transcription. The chromosomes look so bristly at this time, they are said to have a "lampbrush" configuration (Figure 15.5a).

As another example, many insect larvae are like feeding machines that munch incessantly on plant or animal tissues. It takes quite a bit of saliva to prepare chunks of food for digestion. Cells of their salivary glands contain rather unusual chromosomes in which the DNA molecule has been replicated repeatedly. Thus the cells have multiple copies of the same genes that can be transcribed at the same time to yield many copies of the protein components of saliva. These so-called "polytene" chromosomes puff out when the genes are being transcribed, as Figure 15.5e shows.

X Chromosome Inactivation. Each cell in a female mammal has two X chromosomes, one maternal and one paternal in origin. Only the genes of one are available for transcription. Most genes of the other chromosome are completely inaccessible. This condition is normal. Her cells function perfectly well with the gene products of only one X chromosome. It may even be that a double dose of gene products from two X chromosomes would prove lethal.

When each female is developing as an embryo in her mother's body, one X chromosome in each cell becomes condensed and transcription of most of its genes is permanently suppressed. Which of the two becomes condensed is a matter of chance. As Figure 15.6 shows, the condensed X chromosome is quite distinct in the interphase nucleus. It is called a "Barr body" after its discoverer, Murray Barr.

The embryo grows through cell divisions, and each daughter cell inherits the same pattern of X chromosome inactivation that occurred in its parent cell. *Either* the maternal *or* paternal X chromosome (never both) can be randomly inactivated in the cell lineage that gives rise to a given tissue region. Thus every adult female is a "mosaic" of X-linked traits. She has patches of tissue in which an allele on the maternal *or* paternal X chromosome is being expressed.

The mosaic tissue effect arising from random X chromosome inactivation is called *Lyonization* (after its discoverer, Mary Lyon). It is especially visible in female calico cats, which are heterozygous for black and yellow coat-color alleles on the X chromosome. Coat color in a given body region depends on which of the two X chromosomes is functioning and on which of the particular alleles is available for transcription (Figure 15.7).

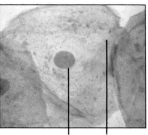

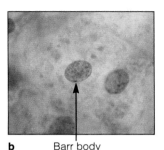

a nucleus cytoplasm

b Barr body

Figure 15.6 (**a**) Nucleus from a male cell, which has no Barr body.

(**b**) Nucleus from a female cell, showing a Barr body.

Figure 15.7 Why is this female calico cat "calico"? Each of her cells contains two X chromosomes. One chromosome carries an allele coding for black coat color and the other carries an allele coding for yellow. When she was an embryo developing in her mother's body, one of the two X chromosomes was inactivated at random in each of the cells that had formed by that time. In all the cellular descendants of each cell, the same chromosome remains inactivated, leaving only one functional allele for the coat-color trait.

And so we see different patches of color, depending on which allele was inactivated in cells making up the tissue in a given region. (The white patches result from interaction with another gene locus—the so-called spotting gene—that determines whether any color appears at all.)

The mosaic effect also is evident in human females affected by *anhidrotic ectodermal dysplasia*. A mutant allele on one X chromosome gives rise to this skin disorder, which is characterized by an absence of sweat glands. In patches of defective skin, the X chromosome bearing the normal allele has been inactivated and the X chromosome bearing the mutant allele is functional (Figure 15.8).

Transcript-Processing Controls. So far, we have considered evidence of transcriptional control. Keep in mind that other levels of gene control may be equally important for normal cell activities; we just don't know as much about them.

For example, we still have a long way to go in identifying the controls over transcript processing—that is, over how newly formed mRNA transcripts get modified. Even so, researchers are giving us some interesting things to think about. Experiments show that the primary mRNA transcript from a single gene is sometimes processed in alternative ways. For instance, transcripts from a gene that codes for a contractile protein (troponin-T) are processed differently in different cell types! The outcome is two or more different mRNAs, each coding for a distinct kind of protein. Although the proteins are very similar, each is unique in a certain region of its amino acid sequence (Figure 15.9). All of the resulting proteins still function in contraction. But they do so in different ways—which may account for subtle variations in the way different types of muscles in your body function.

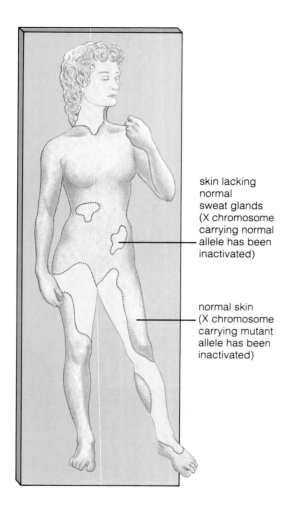

skin lacking normal sweat glands (X chromosome carrying normal allele has been inactivated)

normal skin (X chromosome carrying mutant allele has been inactivated)

Figure 15.8 (*Above*) Pattern of gene expression in a woman affected by anhidrotic ectodermal dysplasia, a disorder in which patches of skin do not have normal sweat glands. The pattern arises through random inactivation of the X chromosome during embryonic development. The mutant allele responsible for the disorder is on one X chromosome, and its corresponding allele on the other chromosome is normal. Depending on which of the two chromosomes is inactivated in an embryonic cell, all of the clonal descendants of that cell will display the same pattern of gene expression.

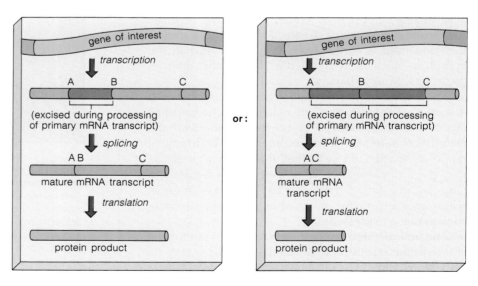

Figure 15.9 Alternative processing of a primary mRNA transcript from a single gene. The primary transcript can be processed in different ways to produce distinct mRNA molecules that code for similar but distinct proteins.

Commentary

Characteristics of Cancer

On rare occasions, certain cells in the body that should not be doing so divide again and again, until their offspring begin to crowd surrounding cells and interfere with tissue functions. The parent cell that started it all had become mutated, and it spawned a tumor. By definition, a **tumor** is any tissue mass composed of cells that are not responding to normal controls over cell growth and division.

The problem is not that tumor cells divide at a horrendous rate. Normal cells divide much faster when they replace a surgically removed portion of the liver. Rather, tumor cells have lost the controls telling them when to stop. They will not stop as long as conditions for growth remain favorable.

When a tumor is *benign,* it may continue to grow more rapidly than normal but it remains in the same place in the body. Surgical removal of the tissue mass removes its threat to health. When a tumor is *malignant,* its cells can migrate and then grow and divide in other organs.

Normally, recognition proteins at the plasma membrane allow cells to bind together in tissues and organs. When genes for those proteins are altered or suppressed, the cell can leave its proper place and travel (in blood or lymph) to other tissues, where it can form a new growth. This process of invasion is called **metastasis**.

There are many types of malignant tumors, but all are grouped into the general category of **cancer**. At the minimum, all cancer cells have these characteristics:

1. *Profound changes in the plasma membrane and cytoplasm.* Membrane permeability is amplified. Some membrane proteins are lost or altered, and new ones appear. The cytoskeleton shrinks, becomes disorganized, or both. Enzyme activity shifts (as in an amplified reliance on glycolysis).

2. *Abnormal growth and division.* Inhibitors of overcrowding in tissues are lost. Cell populations increase to unusually high densities. New proteins trigger an abnormal increase in small blood vessels that service the growing cell mass.

3. *Weakened capacity for adhesion.* Cells cannot become properly anchored in the parent tissue.

4. *Lethality.* Unless cancer cells are eradicated, they will kill the individual.

Any gene having the potential to induce cancerous transformations is called an **oncogene**. When introduced into a normal cell, an oncogene may transform it into a tumor cell. Such genes were first identified in retroviruses, a type of infectious agent (page 350). But nearly identical gene sequences *also* occur in the normal DNA of many species and rarely trigger cancer! These sequences are called **proto-oncogenes**.

Proto-oncogenes code for proteins necessary in normal cell function. They may become cancer-causing genes only on rare occasions, when specific mutations alter their structure or their expression. In other words, the *normal* expression of proto-oncogenes is vital, even though their abnormal expression may be lethal.

Insertion of viral DNA into the DNA of a host cell can skew transcription of a proto-oncogene. **Carcinogens** may do the same thing. Carcinogens bind to DNA and can cause a mutation. They include many natural and synthetic compounds (such as asbestos and components of cigarette smoke), x-rays, gamma rays, and ultraviolet radiation.

Yet cancer seems to be a multistep process, with mutations in more than one proto-oncogene required to bring it about. Look again at the listed characteristics of cancer cells. Now think about some of the products of proto-oncogenes. *They include growth factors (signals sent by one cell to trigger growth in other cells), regulatory proteins involved in cell adhesion, and the protein signals for cell division.*

When the Controls Break Down

This chapter has barely touched on the controls over gene expression in eukaryotes. How is it possible to leave you with a strong impression of their absolutely crucial importance? Perhaps with a close look at what happens when controls over some basic cell functions are disrupted.

As you saw at the start of this chapter, we are beginning to understand which genes govern cell growth and division. On rare occasions, controls over cell division are lost and cells become cancerous, a transformation that is described in the *Commentary.* Possibly more than any other example, this transformation brings home the critical extent to which you and all other organisms depend on controls over gene expression.

Chapter 15   Control of Gene Expression   **241**

SUMMARY

1. In both prokaryotes and eukaryotes, shifts in chemical conditions inside and outside cells can trigger changes in gene activity. In multicelled eukaryotes, for example, conditions vary as a result of changes in diet and levels of activity. They vary inevitably during growth and development.

2. Gene expression is controlled by many interacting elements, including control sites built into DNA molecules, regulatory proteins, enzymes, and hormones. Their interactions govern which gene products appear, at what times, and in what amounts.

3. The best understood gene controls are transcriptional. In prokaryotes, operon controls influence transcription rates. In eukaryotes, the timing and rate of transcription are influenced by chromosome organization as well as by control factors that bind to the DNA.

4. In prokaryotes especially, negative control of transcription is afforded by repressor proteins that can bind to control sites near specific genes. A bound repressor protein inhibits transcription by blocking the access of RNA polymerase to those genes.

5. Positive control of transcription is afforded by activator proteins that also bind to control sites near specific genes. Bound activator proteins promote transcription by helping RNA polymerase bind and start transcription. Positive controls may be the more common type in eukaryotes.

6. In complex eukaryotes, cells differentiate: They become different in appearance, composition, function, and often position. Differentiation arises through "selective gene expression." This term means that different types of cells activate and suppress some fraction of their genes in a variety of ways that lead to pronounced differences in cell structure and function.

Review Questions

1. Define these terms: promoter, operator, repressor protein, and activator protein. *234*

2. Cells depend on controls over which gene products are synthesized, at what times, at what rates, and in what amounts. Describe one type of control over transcription in *E. coli*, a type of prokaryote. Then list five general kinds of controls involved in eukaryotes. *234–236*

3. Define cell differentiation. How does it arise? *236*

4. A plant, fungus, or animal is composed of diverse cell types. How might this diversity arise, given that all of the body cells in each organism inherit the *same* set of genetic instructions? *236*

5. Somatic cells of human females have two X chromosomes. During what developmental stage are genes on *both* chromosomes active? Explain what happens to each of those chromosomes after that stage. *239*

6. What are the characteristics of cancer cells? Explain the difference between a benign tumor and one that is malignant. *241*

Self-Quiz *(Answers in Appendix IV)*

1. In all cells, gene activity may be altered in response to changes in _____ conditions or environmental conditions.

2. Selective gene expression results in cell _____, or changes in the cell's appearance, structure, and function.

3. The best-understood gene control mechanisms are the ones concerned with _____ .
   a. translation
   b. replication
   c. post-translation
   d. transcription
   e. mRNA transport

4. Gene expression is controlled by _____ .
   a. control sites built into DNA
   b. regulatory proteins
   c. enzymes
   d. hormones
   e. all of the above may have control functions

5. _____ is an aspect of protein synthesis that occurs *only* in eukaryotic cells.
   a. transcription in the cytoplasm
   b. translation in the nucleus
   c. post-translation in the nucleus
   d. processing mRNA transcripts

6. In bacteria (prokaryotes), _____ have the most important roles in the negative control of transcription.
   a. activator proteins
   b. repressor proteins
   c. RNA polymerase
   d. DNA polymerase
   e. hormones

7. Positive control of transcription appears to be more common in _____ ; this control involves the action of _____ .
   a. eukaryotes; repressor proteins
   b. prokaryotes; repressor proteins
   c. eukaryotes; activator proteins
   d. prokaryotes, activator proteins

8. Activator proteins promote RNA transcription by _____ .
   a. binding to repressor proteins
   b. binding near control sites to assist release of RNA polymerase

c. inhibiting the action of repressor proteins

d. assisting the binding of RNA polymerase to begin transcription

e. directly initiating transcription

9. Cell differentiation in complex eukaryotes arises through _____ .

   a. selective gene expression
   b. activating and suppressing the same genes in all cells
   c. mostly negative transcriptional controls
   d. operon controls

10. Match the gene control concepts.

   _____ bound repressor protein
   _____ cell differentiation
   _____ bound activator protein
   _____ transcript processing controls
   _____ post-translational controls

   a. governs modification of protein into functional form
   b. blocks access of RNA polymerase to the promoter
   c. assists binding of RNA polymerase to the promoter
   d. changes sequence of mRNA to specify different proteins
   e. occurs through selective gene expression

Selected Key Terms

activator protein *234*
Barr body *239*
cancer *241*
carcinogen *241*
chromosome puff *238*
differentiation *236*
hormone *234*
Lyonization *239*
metastasis *241*
oncogene *241*
operator *234*
operon *234*
post-translational control *237*
promoter *234*
proto-oncogene *241*
repressor protein *234*
selective gene expression *236*
transcriptional control *236*
transcript-processing control *236*
translational control *237*
transport control *236*
X chromosome inactivation *239*

Readings

Feldman, M., and L. Eisenbach. November 1988. "What Makes a Tumor Cell Metastatic?" *Scientific American* 259(5):60–85.

Kupchella, C. 1987. *Dimensions of Cancer.* Belmont, California: Wadsworth.

Murray, A., and M. Kirschner. March 1991. "What Controls the Cell Cycle." *Scientific American* 264(3):56–63.

Ptashne, M. January 1989. "How Gene Activators Work." *Scientific American* 260(1):41–47.

Weintraub, H. January 1990. "Antisense RNA and DNA." *Scientific American* 262(1):40–46.

16 RECOMBINANT DNA AND GENETIC ENGINEERING

Life and Death on the Threshold of a New Technology

For much of human history, we have been dealing with a world that is often harsh. In a bad year, an influenza virus might strike hundreds of thousands of us. In any year, wheat rusts, viruses, and other agents of disease might destroy millions of acres of crops and contribute to starvation on a global scale. Every year, some of our kind are born with crippling genetic disorders.

And yet, through research that began only a few decades ago, we soon may have more control over our individual and combined destinies. A new technology is giving us the means to alter, to our advantage, the DNA molecules of viruses, crop plants and any other kind of organism—including human beings.

In September 1990, for example, a four-year-old girl received what may be a historic genetic reprieve. Her problem is chillingly simple. Due to a single gene mutation, she was born without defenses against viruses, bacteria, and other agents of disease. She has

no immune system. Of the forty-six chromosomes she inherited from her parents, one copy of chromosome 20 bore a defective gene. That gene codes for an enzyme, adenosine deaminase (ADA).

Normally, the enzyme functions in a pathway by which excess adenosine monophosphate (AMP) is stripped of its phosphate group and degraded to uric acid, which the body excretes. In ADA-deficient individuals, a related nucleotide phosphate accumulates. For reasons not fully understood, the excess nucleotide phosphate is toxic to lymphoblasts, which are a type of stem cell in bone marrow. Lymphoblasts divide again and again. Their progeny later differentiate into new recruits and replacements for the immune system's army, the infection-fighting white blood cells.

The ADA-deficient girl is a victim of *severe combined immune deficiency* (SCID). This is a set of disorders

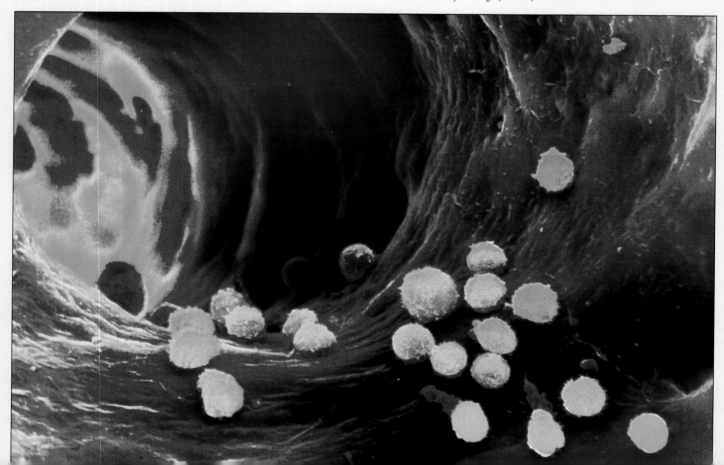

brought on by a drastic reduction or complete absence of two subpopulations of white blood cells—the B and T cells. In their absence, the girl risks death even from minor infections, even from bacteria that otherwise live harmlessly in the body.

Some individuals with SCID benefit from bone marrow transplants. If the transplants take hold, the donated stem cells may produce functional B and T cells. Similarly, ADA injections help some children. Both therapeutic approaches buy time, but neither solves the underlying problem of ADA deficiency. And neither apparently ends the threat of a killer infection.

Given the options, the parents of the four-year-old girl allowed her to participate in the first federally approved gene therapy test for humans. Using recombinant DNA methods of the sort described in this chapter, medical researchers introduced copies of the ADA gene into some of her T cells, which were then encouraged to divide repeatedly. Later, they placed about a billion copies of the genetically engineered cells in a saline solution and delivered them, through a plastic tube, into her bloodstream. Each month, the girl will receive another T cell infusion—and another chance at a longer life. In time, her doctors hope to extract lymphoblasts from her bone marrow, introduce functional ADA genes into them, and put them back in place. Only then might she be assured of a constant, lifelong supply of the crucial enzyme—and functional disease fighters.

As this example suggests, recombinant DNA technology has staggering potential for medicine. It has equally staggering potential for agriculture and industry. It does not come without risks. With this chapter, we consider some basic aspects of the new technology, and we address ecological, social, and ethical questions related to its application.

Figure 16.1 A few white blood cells that were on patrol inside a blood vessel. Individuals with a severely compromised immune system have drastically reduced numbers of these infection-fighting cells—or none at all. Such individuals are prime candidates for gene therapy, one of the beneficial applications of recombinant DNA research.

KEY CONCEPTS

1. Genetic experiments have been occurring in nature for billions of years as a result of gene mutations, crossing over and recombination, and other events. Humans are now engineering genetic changes by way of recombinant DNA technology.

2. With recombinant DNA technology, DNA molecules are cut into fragments, and the fragments of interest are inserted into cloning tools such as plasmids. Then they are amplified rapidly into quantities suitable for research and practical applications.

3. The new technology raises social, legal, ecological, and ethical questions regarding its benefits and risks.

For more than 3 billion years, nature has been conducting genetic experiments through mutation, chromosomal crossing over, and other events. Genetic messages have changed countless times; this is the source of life's diversity.

We humans have been changing the genetic character of species for thousands of years. Through artificial selection, we coaxed modern crop plants and new breeds of cattle, birds, dogs, and cats from wild ancestral stocks. We developed meatier turkeys, sweeter oranges, seedless watermelons, and flamboyant ornamental plants. We produced the tangelo (tangerine × grapefruit) and the mule (donkey × horse).

Today we are analyzing and even engineering genetic changes through **recombinant DNA technology**. With this technology, DNA from different species can be cut, spliced together, then inserted into bacteria or other types of rapidly dividing cells—which multiply in quantity the recombinant DNA molecules. Genes can be isolated, modified, and reinserted into the organism (or transplanted into a different one). In many cases, the engineered genes produce functional proteins. Before looking at how this work is done, let's start out by considering a few recombination mechanisms in nature that actually paved the way for the new technology.

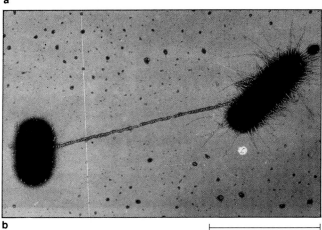

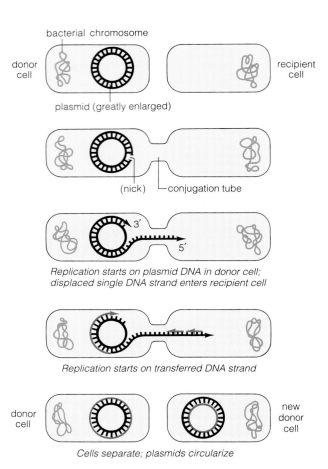

bacterial chromosome

donor cell

plasmid (greatly enlarged)

recipient cell

(nick) — conjugation tube

3'
5'

Replication starts on plasmid DNA in donor cell; displaced single DNA strand enters recipient cell

Replication starts on transferred DNA strand

donor cell

new donor cell

Cells separate; plasmids circularize

Figure 16.2 (**a**) A ruptured bacterial cell (*Escherichia coli*). Notice the larger bacterial chromosome and the several small, circular plasmids (blue arrows). (**b**) Early stage of conjugation between a recipient (F⁻) cell, at left, and a donor (F⁺) cell of *E. coli*. The long appendage joining the two bacteria is an F pilus; it will bring the two participants into close contact so that DNA can be transferred.

Figure 16.3 Transfer of a plasmid between two bacterial cells during conjugation. For clarity, the bacterial chromosomes are not drawn full size. (Each bacterial chromosome actually contains about forty times more genetic information than the largest plasmids do.)

RECOMBINATION IN NATURE: SOME EXAMPLES

Transfer of Plasmid Genes

Bacteria have played a central role in the development of recombinant DNA technology. In nature, as in the laboratory, they have characteristics that suit them to the task. All bacteria have a single chromosome, and many bacterial species also have plasmids. The bacterial chromosome, a circular DNA molecule, contains all the genes necessary for normal growth and development. A **plasmid**, a small, circular molecule of "extra" DNA or RNA, carries only a few genes and is self-replicating (Figure 16.2a). In other words, a plasmid can produce more copies of itself regardless of whether the bacterial chromosome is undergoing replication.

Plasmid genes are transmitted through successive generations of bacterial cells. They are even transferred to bacterial cells of different species. What might be the selective advantage of spreading them around? Although not essential for normal growth and development, the products of plasmid genes do serve useful functions under some circumstances, as we know from studies of the **F plasmids** (the F stands for *Fertility*).

Genes carried on an F plasmid permit *bacterial conjugation,* a process by which one bacterial cell transfers DNA to another. Only a cell having an F plasmid can be a donor; it is designated F⁺. Only a cell lacking an F plasmid (designated F⁻) can be a recipient. Sometimes the transfer is said to be a form of sexual reproduction between "male" and "female" bacterial cells, although comparing this to sex among the eukaryotes requires a rather breathtaking leap of the imagination.

Some F plasmid genes code for the proteins necessary to construct an F pilus, a long appendage that can latch onto a recipient cell and draw it right up against the donor (Figure 16.2b). The attachment apparently activates an enzyme, which cuts one strand of the plasmid DNA. The cut strand starts to unwind from the other strand and enters an F⁻ cell. As Figure 16.3 shows, DNA replication proceeds on the exposed bases of both strands in both cells. Once the transfer and replication are completed, the cells separate. Each is now an F⁺ cell.

Once in a great while, a donated plasmid becomes integrated into the recipient's chromosome through natural recombination mechanisms. Like all cells, bacteria have enzymes capable of cutting DNA strands during normal repair operations. One such enzyme recognizes a short nucleotide sequence that happens to occur on both the plasmid and the chromosome. It cuts both molecules at the sequence, then splices their cut ends together (Figure 16.4).

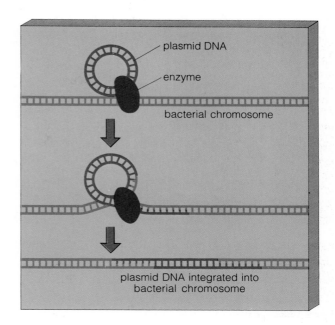

Figure 16.4 Integration of a plasmid into a bacterial chromosome. Only a small stretch of the circular bacterial chromosome is shown.

Transfer of Viral Genes

Among the viruses called bacteriophages are types that can integrate their DNA into the chromosome of bacterial host cells. Lambda bacteriophage is one of them. It can spread through a population of *Escherichia coli* through repeated cycles of infection. On rare occasions, a viral enzyme cuts the bacterial chromosome at a specific site, then the viral DNA is inserted between the cuts and sealed in place. The modified bacterial chromosome is replicated and the viral DNA is passed on in latent form to succeeding generations of *E. coli* (page 349). Later, the viral DNA may move out of the chromosome, and an infectious cycle begins again.

To date, transfer of genes by recombination has been discovered in many organisms, including a variety of bacteria, bacteriophages, yeasts, fruit flies, and mammals. It appears that gene transfer may be common to most, if not all, organisms. As we turn now to the kind of recombination techniques going on in the laboratory, keep in mind this basic point:

Gene transfer and recombination is a common occurrence in nature, made possible by specific enzymes that can make cuts in DNA molecules.

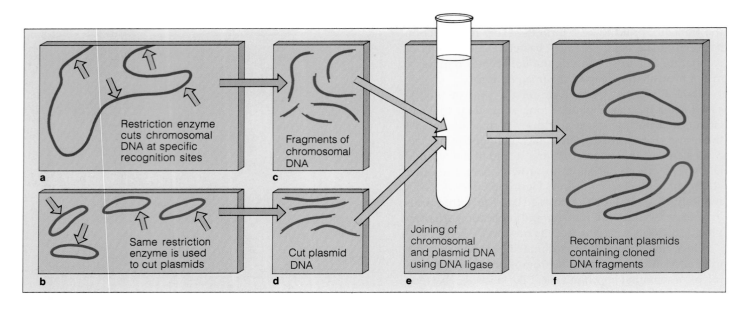

Figure 16.5 Formation of a DNA library.

RECOMBINANT DNA TECHNOLOGY

Recombinant DNA technology grew out of experiments with *E. coli* and the bacteriophages that infect it. In the late 1960s and early 1970s, researchers learned how to use different bacterial enzymes to cut DNA into fragments and "package" the fragments into plasmids for insertion into host cells. They developed ways to pinpoint DNA fragments of interest in dividing cells. They also started to identify nucleotide sequences of individual genes. And they used that information to map the positions of various genes in the DNA of different species.

Producing Restriction Fragments

Different bacterial species produce **restriction enzymes**, the sole function of which is to cut apart and destroy foreign DNA molecules that often are injected into the cell by viruses. Several hundred restriction enzymes have been identified. Each makes its cut only at sites having a short, specific nucleotide sequence. This feature enables researchers to select those enzymes that will produce a DNA fragment containing a particular gene.

Restriction enzymes are also useful in another respect. Many produce staggered cuts, so both ends of a DNA fragment end up with short, single-stranded tails:

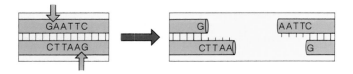

These "sticky ends" of the fragment have a terrific property. They can base-pair with any other DNA molecule cut by the same restriction enzyme.

Suppose we use the same restriction enzyme to cut plasmids and a chromosome. When the chromosomal fragments and cut plasmids are mixed together, they base-pair at the cut sites (Figure 16.5). The base-pairing can be made permanent by using another enzyme, **DNA ligase**. As shown earlier in Figure 13.8, DNA ligase is a replication enzyme that joins the short fragments of DNA formed by discontinuous synthesis.

We now have a **DNA library**—a collection of DNA fragments produced by restriction enzymes and incorporated into plasmids. The amount of DNA in a library is almost vanishingly small. To do anything useful with it, we must first amplify it—that is, allow the fragments to be copied again and again to produce huge numbers of them.

DNA Amplification

By Cloning. A DNA library of recombinant plasmids can be inserted into many host cells for propagation. (Bacteria and yeast cells are commonly used for such a purpose. Both can be grown easily in the laboratory, and their rapid reproduction rates are quite impressive.) When put to such use, a plasmid or any other self-replicating genetic element is called a "cloning vector." After repeated replications and divisions of the host cells, we end up with **cloned DNA**—multiple, identical copies of DNA fragments contained within their cloning vectors.

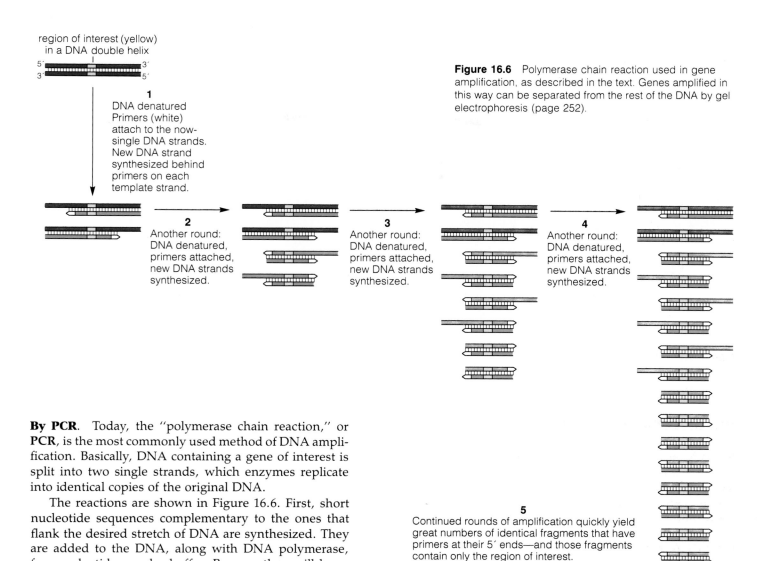

region of interest (yellow)
in a DNA double helix

5′ ▬▬▬▬▬▬▬▬▬▬ 3′
3′ ▬▬▬▬▬▬▬▬▬▬ 5′

1
DNA denatured
Primers (white)
attach to the now-
single DNA strands.
New DNA strand
synthesized behind
primers on each
template strand.

2
Another round:
DNA denatured,
primers attached,
new DNA strands
synthesized.

3
Another round:
DNA denatured,
primers attached,
new DNA strands
synthesized.

4
Another round:
DNA denatured,
primers attached,
new DNA strands
synthesized.

5
Continued rounds of amplification quickly yield
great numbers of identical fragments that have
primers at their 5′ ends—and those fragments
contain only the region of interest.

Figure 16.6 Polymerase chain reaction used in gene amplification, as described in the text. Genes amplified in this way can be separated from the rest of the DNA by gel electrophoresis (page 252).

By PCR. Today, the "polymerase chain reaction," or **PCR,** is the most commonly used method of DNA amplification. Basically, DNA containing a gene of interest is split into two single strands, which enzymes replicate into identical copies of the original DNA.

The reactions are shown in Figure 16.6. First, short nucleotide sequences complementary to the ones that flank the desired stretch of DNA are synthesized. They are added to the DNA, along with DNA polymerase, free nucleotides, and a buffer. Because they will base-pair with the flanking regions, those short nucleotide sequences can serve as "primers" for DNA polymerase. That enzyme, recall, cannot synthesize a new DNA strand unless a primer is already positioned on the existing one (Figure 13.8).

In this case, researchers use a DNA polymerase from *Thermus aquaticus,* a type of bacterium that thrives in hot springs, even in hot water heaters. The polymerase functions at high temperatures (for example, 72°C), so it remains stable at the elevated temperatures required to denature DNA.

The mixture of DNA, primers, and enzymes is subjected to multiple rounds of three temperatures. In each round, the DNA first is denatured by exposure to near-boiling temperatures (about 94°C). Then the temperature is lowered to between 37°C and 60°C, which is the range in which the primers base-pair most readily with the DNA. Then the temperature is raised to 72°C—and the DNA polymerases go to work. As the number of cycles increases, the newly synthesized DNA strands themselves become templates for primer base-pairing and copying.

After several rounds of PCR synthesis, nearly all of the DNA fragments will be the same length, corresponding to the distance between the ends of the two primers. This amplified DNA greatly outnumbers the original DNA template. It can be easily detected and analyzed by gel electrophoresis, as described in the *Doing Science* essay on page 252.

The advantage of PCR is that extremely small samples of DNA can be increased to high concentrations that can be studied easily, cloned, or both. A single DNA molecule can be quickly amplified to many *billions* of molecules. Thus it can be used to amplify samples with too little DNA, as might be found in a single hair left at the scene of a crime. Besides this, PCR by its very nature can be used to repair fragmented DNA. That is why it is being used in studies of samples that are too old to contain intact DNA, like those that might be found in an Egyptian mummy. In the near future, PCR may be used to identify genetic disorders in very early human embryos that consist only of eight cells!

Figure 16.7 Use of a DNA probe to identify the colony of transformed bacterial cells that have taken up the DNA of interest.

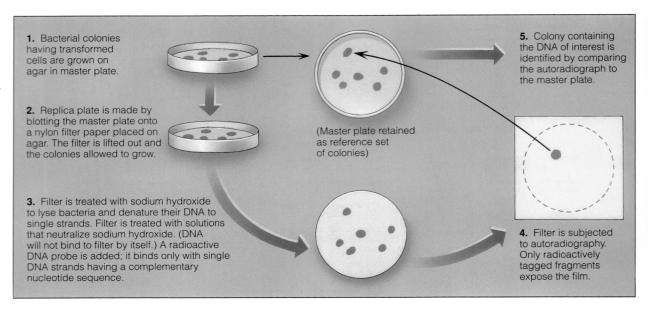

1. Bacterial colonies having transformed cells are grown on agar in master plate.

2. Replica plate is made by blotting the master plate onto a nylon filter paper placed on agar. The filter is lifted out and the colonies allowed to grow.

3. Filter is treated with sodium hydroxide to lyse bacteria and denature their DNA to single strands. Filter is treated with solutions that neutralize sodium hydroxide. (DNA will not bind to filter by itself.) A radioactive DNA probe is added; it binds only with single DNA strands having a complementary nucleotide sequence.

(Master plate retained as reference set of colonies)

4. Filter is subjected to autoradiography. Only radioactively tagged fragments expose the film.

5. Colony containing the DNA of interest is identified by comparing the autoradiograph to the master plate.

Identifying Modified Host Cells

Suppose we mix recombinant plasmids with potential host cells on a culture medium in a petri dish. Not all of the cells will take up a plasmid, but they all may grow and reproduce to form large colonies on the culture medium. The following procedure can be used to identify the colonies that do harbor the DNA of interest. This procedure is an example of **nucleic acid hybridization**.

First, a plasmid with antibiotic-resistance genes is selected as a cloning vector. If the culture medium has been supplemented with antibiotics, all the colonies *except* the ones with transformed cells will be destroyed or their growth inhibited. Only cells that take up and successfully express the antibiotic-resistance genes will survive.

Second, a specific probe is prepared so that the colonies can be analyzed by nucleic acid hybridization techniques, using DNA probes. A **DNA probe** is a short DNA sequence that is assembled from radioactively labeled nucleotides. Part of the sequence must be complementary to that of the desired gene.

Nucleic acid hybridization requires that a replica plate be made of the colonies growing in the petri dish (the master plate). The colonies are blotted onto a nylon filter and the transferred cells are allowed to grow into new colonies. Their locations correspond to the locations of the original colonies on the master plate (Figure 16.7). Then the filter is treated with solutions that cause the cells to rupture and that allow the released DNA to become permanently affixed to the nylon filter. Nucleic acid hybridization follows when the radioactive DNA probe is added to such filters. The probe will hybridize only with the DNA having the complementary base sequence and will thereby tag the location of the transformed bacterial colony or colonies.

Expressing the Gene of Interest

Even when a host cell has taken up a cloned gene, it may not be able to transcribe and translate it into functional protein. For example, human genes contain noncoding regions (introns) as well as coding regions (exons). The genes cannot be translated unless the introns are spliced out and the exons spliced together into a mature mRNA transcript (page 221). As it happens, bacterial host cells don't have enzymes that recognize the splice signals on eukaryotic genes. That is why cDNA typically is the choice of researchers who study the products of human genes.

The term **cDNA** refers to any DNA molecule that has been "copied" from a mature mRNA transcript by a process called *reverse transcription*. The single-stranded mRNA molecule serves as a template for assembling a DNA strand that is identical in sequence to the gene of interest. This seemingly "backwards" process (by which mRNA is transcribed into DNA) requires a special viral enzyme, *reverse transcriptase*. After the "hybrid" DNA/RNA molecule is assembled, the RNA strand is degraded. Other enzymes convert the remaining strand of DNA to double-stranded form (Figure 16.8). Once cDNA has been obtained, it typically is cloned into a vector.

Despite the obstacles, several human gene products already are being mass-produced and many more are being developed. Table 16.1 lists some of them. The availability of these gene products already has had major impact on genetic research.

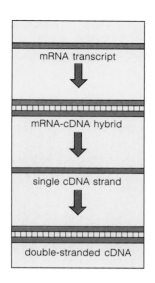

1

mRNA transcript of desired gene is used as template for the assembly of a DNA strand by enzyme action (reverse transcriptase). The result is an RNA-DNA hybrid molecule.

2

RNA polymerase nicks only the RNA in the hybrid molecule. DNA polymerase in the same sample uses the nicked RNA fragments as "primers" to start synthesis of DNA on the template strand. As it proceeds, the DNA polymerase displaces other sections of RNA from the template strand. DNA ligase in the same sample joins together the new lengths of DNA in the template strand, forming double-stranded cDNA.

Figure 16.8 Formation of cDNA from an mRNA transcript.

Table 16.1	Examples of Cloned Human Gene Products Approved for Use or Under Development
Protein	Used in Treating
Insulin	*Diabetes*
Somatotropin (growth hormone)	*Pituitary dwarfism*
Erythropoetin	*Anemia*
Factor VIII	*Hemophilia*
Factor IX	*Hemophilia*
Interleukin-2	*Cancer*
Tumor necrosis factor	*Cancer*
Interferons	*Some cancers, viral infections*
Atrial natriuretic factor	*High blood pressure*
Tissue plasminogen factor	*Heart attack, stroke*

RISKS AND PROSPECTS OF THE NEW TECHNOLOGY

Uses in Basic Biological Research

Centuries ago a new invention—the microscope—revealed the existence of a world of diverse organisms that live in, on, and around us. That invention eventually gave us great insight into the nature of life itself—first by allowing us to explore the secrets of cells, then of chromosomes, then finally of DNA. Now, through implementation of DNA technology, that long road of basic research is forking almost daily in new directions.

Think about Gregor Mendel growing pea plants, selecting and planting seeds, growing new plants, and forming hypotheses about what *might* be giving rise to their patterns of inheritance, season after season. Now think about how radioactive DNA probes can be used to study genes *directly* in cells, in chromosomes, in DNA. Think about those who are destined to develop a life-threatening disease such as cancer. Without the new technology, we never would have known about proto-oncogenes, oncogenes, and skewed controls over cell division.

Or think about Darwin and others who have puzzled over the meaning of life's diversity. Through DNA technology, we can study a selected nucleotide sequence in the DNA of a given species. Then we can compare the extent to which the equivalent sequence in the DNA of another species differs from it. Chapter 19 will describe how the degrees of divergence in DNA sequences are an important source of evidence of evolutionary relationships among organisms.

As one final example, consider the **human genome project**. In a truly ambitious undertaking, researchers are working to sequence the estimated 3 billion nucleotides present in human chromosomes. They also are sequencing the chromosomes of *E. coli*, yeasts, mice, and other organisms that are commonly used in genetic research.

The potential benefits from the project are enormous. Imagine being able to pinpoint the gene or genes responsible for any specific genetic disorder. Such information surely will enhance efforts to diagnose, prevent, or treat the disorder, as the parents of the ADA-deficient girl already know. Imagine how advantageous such information will be to researchers investigating the controls over gene expression or studying the evolution of life.

By current calculations, it will take researchers working in about 1,000 laboratories ten years to complete the project. Even when the sequence of every chromosome is deciphered, it may take *many* decades to decipher what the sequences mean. Even so, it is likely that the sequencing technology itself will advance rapidly. In the meantime, many laboratories are collaborating in the mapping attempt.

Applications of RFLPs

One recombinant DNA technique is being used to good advantage in the human genome project and other basic research efforts. It also is being put to use in some rather startling ways.

Earlier, we saw how restriction enzymes cut the DNA of an individual into fragments of specific lengths. The fragments separate from one another into distinct bands when subjected to gel electrophoresis (page 252). As it happens, certain DNA fragments show slight variations in the banding patterns from one person to the next.

Doing Science

Looking for Needles in Molecular Haystacks

Molecular biologists have ingenious tools available to them. Among the most useful in recombinant DNA work are gel electrophoresis and the Sanger method of sequencing DNA.

Gel Electrophoresis of DNA

Earlier, we saw how gel electrophoresis is used to separate proteins on the basis of their electric charge and their size (Figure 14.2). Similar methods can be used to separate restriction fragments of different lengths.

In this case, a mixture of DNA fragments is placed near one end of a thin slab of gel, which is mounted on a glass plate and bathed in an appropriate solution. An electric current is allowed to run through the gel. Because of their charged phosphate groups, the DNA fragments respond to the current by moving through the gel. The degree to which fragments of a given length move depends only on their size—larger ones cannot move as fast through the gel. This method actually can separate DNA fragments that differ in length by merely one nucleotide.

The photograph in Figure *a* shows columns, running from left to right, in which fragments have been separated. The gel was treated with a fluorescent dye that binds specifically to DNA; hence the luminous pink band for the fragments of a specific length.

DNA Sequencing

A method developed by Frederick Sanger can be used to determine the nucleotide sequence of DNA fragments of a given length.

First, the single-stranded fragments are incubated with primers, DNA polymerases, and the four kinds of nucleotide subunits of DNA (that is, A, T, C, and G). All of those components are required for DNA replication, as Figure 13.8 shows. Also, a trace of one radioactively labeled nucleotide (usually A) is added to the mixture.

Second, the fragments are divided into batches in four different test tubes. Each tube contains a modified form of one—and only one—of the four kinds of nucleotides. We can designate these A*, T*, C*, and G*.

Think about what happens in the tube that contains the A* subunits. As expected, the DNA polymerase recognizes a primer that has become attached to a fragment and starts assembling a new strand behind it. The assembly follows the base-pairing rules (A only to T, and C only to G).

Each time there is a T on the template strand, an A becomes base-paired with it. Sooner or later, the enzyme

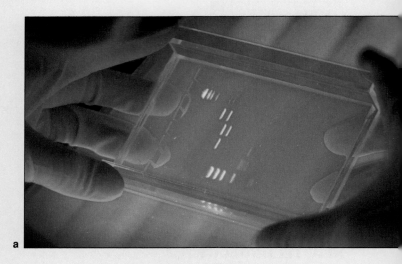

a

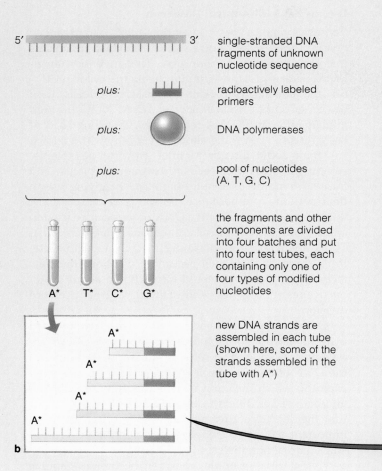

5′ |||||||||||||||||||||||| 3′ single-stranded DNA fragments of unknown nucleotide sequence

plus: radioactively labeled primers

plus: DNA polymerases

plus: pool of nucleotides (A, T, G, C)

A* T* C* G* the fragments and other components are divided into four batches and put into four test tubes, each containing only one of four types of modified nucleotides

A*
A*
A*
A*

b

new DNA strands are assembled in each tube (shown here, some of the strands assembled in the tube with A*) picks up an A* subunit. This modified nucleotide is actually a chemical roadblock—it prevents the enzyme from adding more nucleotides to the growing DNA strand.

In time, the tube contains radioactively labeled strands of different lengths, as dictated by the location of each A* in the sequence (Figure b). The same thing happens in the other three tubes, with strand lengths dictated by the location of T*, C*, and G*.

Now the DNA strands in all four batches are put in four parallel lanes in the same gel and subjected to electrophoresis. A nucleotide sequence can be read off the resulting bands in the gel.

Look at the numbers running down the side of the diagram in Figure c. Start with "1" and read across the four lanes (A, T, C, G). You can see that T is closest to the start of the nucleotide sequence; it has migrated farthest through the gel. At "2," you see that the next nucleotide is C, and so on. The entire sequence, read from the first nucleotide to the last, is

TCGTACGCAAGTTCACGT

Now, using the rules of base-pairing, you can deduce the sequence of the DNA fragment that served as the template.

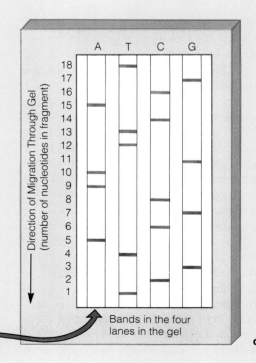

Bands in the four lanes in the gel

c

The variations arise because different people can carry nonidentical alleles at given gene locations in their chromosomes. In other words, some number of their genes vary in molecular form. The molecular differences cause slight shifts in the number and location of sites where restriction enzymes can make their cuts. Thus, each person may have a unique banding pattern if a large-enough number of DNA fragments are analyzed. (The only exceptions are identical twins, who have identical DNA.) Such differences in the banding patterns of DNA fragments are called "restriction fragment length polymorphisms." They also are called **RFLPs** (pronounced RIFF-lips) for short.

RFLPs have greatly increased the number of sites available for mapping the human genome. Also, mutant alleles responsible for genetic disorders sometimes have a unique restriction site that is *not* present in the DNA of normal persons. Sickle-cell anemia is an example. The mutant allele coding for the defective hemoglobin chain associated with this disorder can be detected through RFLP analysis.

RFLP analysis also is revolutionizing criminal investigations. Each person has a *genetic fingerprint*, a unique array of RFLPs inherited from each parent in a Mendelian pattern. Already, we can resolve paternity and maternity cases by comparing the genetic fingerprint of the child with that of the disputed parent.

Today, murderers are much easier to identify if they leave even a few drops of their blood at the scene of the crime or if a few drops of the victim's blood are found on their clothing. The bloodstain provides enough DNA for its genetic fingerprint to be compared with a suspect's DNA. Similarly, a genetic fingerprint from just a drop or two of semen recovered from a rape victim can identify the perpetrator of the crime. Courts have already ruled for conviction when a suspect's genetic fingerprint has matched that found in semen samples. Although genetic fingerprinting has been disputed in some cases, the accuracy of the laboratory work, not the approach itself, is being questioned.

Enterprising biologists also have applied RFLP analysis to DNA extracted from mummies and from the well-preserved remains of mammoths and plants that became extinct long ago. Information from such research may be extremely valuable in deciphering evolutionary relationships among living and extinct organisms.

Genetic Modification of Organisms: Some Examples

Paul Berg and his colleagues were the first to insert foreign DNA into a bacterial plasmid. Their pioneer work brought a question into sharp focus: Is it dangerous to transfer genes between different species? Let's look at the kind of work going on today with bacteria, plants, and animals.

Figure 16.9 Spraying an experimental strawberry patch in California with "ice-minus" bacteria. (Government regulations required that the sprayer use elaborate protective gear.)

Figure 16.10 Cultured cells from a carrot plant. Roots and shoots of embryonic plants are already forming.

Genetic Engineering of Bacteria. The bacterial strains used in many genetic engineering experiments are harmless to begin with, and they are also modified (by mutation) so they cannot survive outside the laboratory. Even so, there is concern about possible risks of introducing genetically engineered bacteria into humans or the environment.

Consider what happened when Steven Lindow genetically engineered a strain of *Pseudomonas syringae*. This common bacterium lives on leaves and stems, and it makes many crop plants susceptible to frost damage. Proteins at the bacterial cell surface encourage ice crystals to form even when the air temperature is several degrees above freezing. Lindow excised the "ice-forming" gene from some *P. syringae* cells, which thereafter could not synthesize the ice-forming protein. For one experiment, the so-called **ice-minus bacteria** were to be sprayed on strawberry plants in an isolated field just before a frost. The experiment would indicate whether the plant cells would resist freezing.

The proposed field trial involved an organism from which a harmful gene had been *deleted*, yet it triggered a bitter legal debate on the risks of releasing genetically engineered microbes in the environment. The courts finally ruled in favor of allowing the experiment to pro-

ceed, and a small patch of strawberries was sprayed (Figure 16.9). As predicted, no ecological disaster occurred—but a few environmental activists entered the patch at night and pulled up the plants.

The lessons of the ice-minus controversy are important. Rules governing the release of genetically engineered organisms have since been clarified. Environmental impact reports are filed first. And biotechnologists have learned they must communicate effectively with the public about their work.

Genetic Modification of Plants. Many years ago, Frederick Steward and his coworkers cultured some cells of carrot plants, much as bacterial cells are cultured. By altering the hormonal composition of the culture medium, they induced the cells to grow into small embryos, some of which grew into whole plants (Figure 16.10). Today, researchers routinely regenerate whole plants from cultured cells of many species, including some major crop plants.

Theoretically, all plants regenerated from cultured cells that were derived from a single plant should be genetically identical—that is, clones of one another. Yet the culturing apparently increases mutation rates—and so provides more genetic novelties than researchers might otherwise expect to get. A potentially useful mutation can be pinpointed among millions of cells. For example, the cells can be grown on a medium that contains a toxin associated with a certain plant disease. If a few cells carry a mutated gene that confers resistance to the disease, they will end up being the only live cells in the culture dish. Disease resistance might then be conferred on plants of other varieties by hybridizing them with plants regenerated from the cultured cells.

Today, researchers are successfully inserting genes into cultured plant cells. For example, they have inserted

Figure 16.11 Crown gall tumors on a willow tree, as caused by a tumor-inducing plasmid from *A. tumefaciens*.

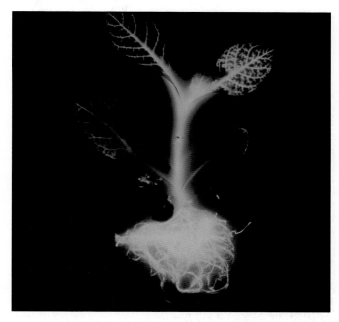

Figure 16.12 Genetically modified tobacco plant, made luminescent by the addition of a gene from fireflies, coding for the enzyme luciferase.

DNA fragments into the "Ti" plasmid from *Agrobacterium tumefaciens*, a bacterium that infects many flowering plants. Some of the plasmid genes become integrated into the DNA of infected plants, and they induce the formation of crown gall tumors (Figure 16.11). When the plasmid is used as a vector, however, the tumor-inducing genes are first removed and other, desired genes are substituted. Then the genetically modified bacterial cells are grown with cultured plant cells. Many of the plants that are regenerated from infected cultures contain the foreign genes within their DNA. In some cases, the foreign genes are expressed in the plant tissues, with observable effects on phenotype.

Vivid evidence of a successful gene transfer came from researchers who used *A. tumefaciens* to deliver a firefly gene into cultured tobacco plant cells. The gene codes for luciferase, an enzyme required for bioluminescence (page 101). Plants regenerated from the infected culture cells have the peculiar ability to glow in the dark (Figure 16.12).

Similar methods also have produced improved varieties of crop plants. For example, certain cotton plants have been genetically engineered for resistance to worm attacks (Figure 16.13). Gene insertions of this sort are ecologically safer than pesticide applications. They kill only the targeted pest. They do not interfere with beneficial insects, such as ladybird beetles that prey on aphids.

A. tumefaciens infects only dicots, which are broadleaf plants. Wheat, corn, rice, oats, and other major food

Figure 16.13 Examples of genetically engineered plants having commercial importance. (**a**) Part of a cotton plant modified for resistance to attack by worms, as shown on an unmodified plant in (**b**). A normal potato plant susceptible to viral attack is shown in (**c**); a genetically modified strain is shown in (**d**).

Figure 16.14 Ten-week-old mouse littermates, the one on the left weighing 29 grams, and the one on the right, 44 grams. The larger mouse grew from a fertilized egg into which the gene for human somatotropin (growth hormone) had been inserted.

crop plants are monocots. In some cases, genetic engineers use chemicals or electric shocks to deliver DNA directly into protoplasts (plant cells stripped of their walls). For some species, however, regenerating whole plants from protoplast cultures is not yet possible. Recently, someone came up with the idea to deliver genes into cultured plant cells by shooting them with a pistol. Instead of bullets, blanks are used to drive DNA-coated, microscopic tungsten particles into the cells. Although this might seem analogous to using a battleship cannon to light a match, the shooter is reporting some success.

Genetic Modification of Animals. In 1982, Ralph Brinster and Richard Palmiter introduced the rat gene for somatotropin (growth hormone) into fertilized mouse eggs. When the mice grew, it became clear that the rat gene had become integrated into the mouse DNA and was being expressed. The mice grew much larger than their normal littermates. Their cells had up to thirty-five copies of the gene, and blood concentrations of the hormone were several hundred times higher than normal values. More recently, the gene for human somatotropin was successfully introduced and expressed in mice, giving rise to the "super rodent" shown in Figure 16.14.

Similar experiments with large, domesticated animals have not been successful. For example, when the somatotropin gene was inserted into pigs, it was expressed—but the pigs developed arthritis-like symptoms and other disorders.

As these examples suggest, gene modification in animals is extremely unpredictable. Not only must new or modified genes be inserted in the body cells of an animal or in gametes, they also must end up in specific locations in chromosomes so that their expression will be properly regulated. And they must not disrupt the function of other genes.

For instance, genes can be inserted into a sperm nucleus just after it penetrates an egg. But the eggs are so sensitive to being poked that the procedure has a high rate of failure. Even when gene delivery is successful, researchers still cannot control *where* in the DNA the inserted gene will end up. In its new location, will the inserted gene activate an oncogene, with its potential to cause cancer? Or cause a mutation? Or alter expression of related genes?

Despite such obstacles, research is advancing on many fronts. The genes responsible for a few disorders have been cloned recently. DNA probes are being produced for use in prenatal diagnosis of heritable disorders. In combination with other technical advances, such as PCR, such probes are having dramatic impact on the incidence of many diseases, including sickle-cell anemia.

Human Gene Therapy

Inserting one or more normal genes into the body cells of an organism to correct a genetic defect is called **gene therapy**. The idea of doing this to offer relief from severe genetic disorders seems to be socially acceptable at present, even though the technology by which gene therapy might be accomplished has yet to be developed and refined.

In contrast, inserting genes into a normal human (or sperm or egg) in order to modify or enhance a particular trait is called many things, including **eugenic engineering**. Yet who decides which traits are "desirable"? What if prospective parents could pick the sex of their children by way of genetic engineering? (Three-fourths of one recently surveyed group said they would choose a boy. What would be the long-term social implications of a drastic shortage of girls?) If it is okay to engineer taller or blue-eyed individuals, would it be okay to engineer "superhuman" offspring with exceptional strength or intelligence? Actually, intelligence and most other traits arise through complex interactions among many genes, so this will put them outside the reach of genetic manipulation for some time.

We have only touched on some of the social and ethical issues raised by recombinant DNA technology and genetic engineering. Some individuals say that no matter what the species of organism, DNA should never be altered. But as an earlier discussion in the chapter made clear, nature itself alters DNA much of the time. The real argument, of course, is whether we have the wisdom to bring about beneficial changes without causing harm to ourselves or the environment.

When it comes to manipulating human genes, one is reminded of our very human tendency to leap before we look. When it comes to restricting genetic modifications of any sort, one also is reminded of an old saying: "If God had wanted us to fly, he would have given us wings." And yet, something about the human experience gave us the *capacity* to imagine wings of our own making—and that capacity carried us to the frontiers of space.

Where are we going from here with recombinant DNA technology, this new product of our imagination? To gain perspective on the question, spend some time reading the history of our species. It is a history of survival in the face of all manner of threats, expansions, bumblings, and sometimes disasters on a grand scale. It is also a story of increasingly intertwined interactions with the environment and with one another. The questions confronting you today are these: Should we be more cautious, believing that one day the risk takers may go too far? And what do we as a species stand to lose if the risks are *not* taken?

SUMMARY

1. Genetic "experiments" have been occurring in nature for billions of years. Gene mutation, crossing over and recombination at meiosis, and other natural events have all contributed to the current diversity among organisms.

2. Humans have been manipulating the genetic character of different species for thousands of years. The emergence of recombinant DNA technology in the past few decades has enormously expanded our capacity to modify organisms genetically.

3. Recombinant DNA technology is founded on procedures by which DNA molecules can be cut into fragments, inserted into plasmids or some other cloning tool, then propagated in a population of rapidly dividing cells.

4. Restriction enzymes and DNA ligase are used to insert DNA from a single species of interest into cloning vectors. A cloning vector is a plasmid, virus, or any other self-replicating genetic element that can be inserted into a host cell for propagation.

5. A collection of DNA fragments produced by restriction enzymes and incorporated into cloning vectors is called a DNA library. A DNA clone is any DNA sequence that has been amplified in dividing cells. DNA sequences also can be amplified in test tubes by the polymerase chain reaction.

6. Recombinant DNA technology and genetic engineering have enormous potential for research and applications in medicine, agriculture, and home and industry. As with any new technology, potential benefits must be weighed against potential risks, including ecological and social disruptions.

7. Although the new technology has not developed to the extent that human genes can be modified, it seems appropriate that the social, legal, ecological, and ethical questions should be explored in detail before such an application is possible.

1. What is a plasmid? What is a restriction enzyme? Do such enzymes occur naturally in organisms? *246, 248*

2. Recombinant DNA technology involves the following:
   a. Producing DNA restriction fragments. *248*
   b. Amplifying the DNA. *248–249*
   c. Identifying modified host cells. *250*

Briefly describe one of the methods used in each of these categories.

3. Having read about examples of genetic engineering in this chapter, can you think of some additional potential benefits of this technology? Can you envision other potential problems? *251–257*

Self-Quiz *(Answers in Appendix IV)*

1. Gene mutations, crossing over and recombination during meiosis, and other natural events are the basis of the _____ observed in present-day organisms.

2. Causing genetic change by deliberately manipulating DNA is known as _____ .

3. _____ are small circles of bacterial DNA that are separate from the bacterial chromosome.

4. Genetic researchers use plasmids as _____ .

5. Rejoining cut DNA fragments from any organism is best known as _____ .
   a. cloning genes   c. recombinant DNA technology
   b. mapping genes   d. conjugating DNA

6. Using the metabolic machinery of a bacterial cell to produce multiple copies of genes carried on hybrid plasmids is _____ .
   a. a way to create a DNA library   c. mapping a genome
   b. bacterial conjugation   d. DNA amplification

7. Any DNA sequence that has been amplified in dividing cells is a _____ .
   a. DNA clone   c. chunk of foreign DNA
   b. DNA library   d. gene map

8. The polymerase chain reaction _____ .
   a. is a natural reaction in bacterial DNA
   b. cuts DNA into fragments
   c. amplifies DNA sequences in test tubes
   d. inserts foreign DNA into bacterial DNA

9. Which may benefit from recombinant DNA technology?
   a. households   d. agriculture
   b. industry   e. all of the above
   c. medicine

10. Match the recombinant DNA information appropriately.
    _____ DNA clone   a. mutation, recombination
    _____ bacterial plasmid   b. raises social, legal, and
    _____ natural genetic   ethical questions
                "experiments"   c. a method of test tube gene
    _____ polymerase chain   amplification
                reaction   d. any cellular amplification
    _____ modification of   of DNA sequences
                human genes   e. a cloning tool

Selected Key Terms

bacterial conjugation *246*   nucleic acid hybridization *250*
cDNA *250*   plasmid *246*
cloned DNA *248*   polymerase chain reaction *249*
DNA library *248*   recombinant DNA technology *245*
DNA ligase *248*   restriction enzyme *248*
eugenic engineering *257*   reverse transcriptase *250*
gene therapy *257*   RFLP *253*
genetic fingerprint *253*

Readings

Alberts, B., et al. 1989. *Molecular Biology of the Cell.* Second edition. New York: Garland. Chapter 5 has excellent coverage of DNA recombination and genetic engineering.

Anderson, W. F. 1985. "Human Gene Therapy: Scientific and Ethical Considerations." *Journal of Medicine and Philosophy* 10:274–291.

Brill, W. 1985. "Safety Concerns and Genetic Engineering in Agriculture." *Science* 227:381–384.

Guyer, R., and D. Koshland, Jr. December 1989. "The Molecule of the Year." *Science* 246(4937):1543–1546.

Palmiter, R., et al. 1983. "Metallothionein-Human GH Fusion Genes Stimulate Growth of Mice." *Science* 222:809–814. Report on landmark experiments in mammalian gene transfers.

White, R., and J. Lalouel. February 1988. "Chromosome Mapping with DNA Markers." *Scientific American* 258(2):40–48.

FACING PAGE: *Millions of years ago, a bony fish died, and sediments gradually buried it. Today its fossilized remains are studied as one more piece of the evolutionary puzzle.*

Fire, Brimstone, and Human History

Lying in the Mediterranean waters between Greece and Crete is a crescent-shaped island, Thera, that wraps around a submerged crater 400 meters deep. It is the remnant of a volcano that blew apart around 3,500 years ago. The eruption was so violent, it generated seismic sea waves that were probably 100 meters high. Within twenty minutes, those giant waves would have reached the island of Crete to the south (Figure 17.1). Most likely, the catastrophic event brought about the abrupt collapse of the Minoan civilization on Crete, the earliest civilization in the history of Europe. Knossos, the Minoan capital, was virtually leveled at about the same time as the eruption. Prevailing winds carrying huge volumes of volcanic ash across the island would have darkened the skies for days, further terrifying those who survived the deluge.

a

Between the Red Sea to the south and the Dead Sea to the north is a long depression in the earth's crust, the Jordan Valley. Apparently the notorious cities of Sodom and Gomorrah flourished at the south end of the Dead Sea in this valley around 4,000 years ago. By biblical account, "brimstone and fire rained upon the cities . . . and the smoke of the land went up like the smoke of a furnace." To the everlasting terror of their residents, both cities were destroyed.

Today, geologists look at satellite images of the straight walls of the Jordan Valley. They see evidence of hot springs, past and present. They see evidence of ancient, great lava flows and violent earth tremors. Taken together, the straight valley walls, hot springs, lava flows, and earthquakes are signs of deep cracks (faults) in the earth's crust. Geologists postulate that severe earthquakes must have tilted part of the crust along the fault at the southern end of the Dead Sea. In what surely must have been one of the all-time nightmares, the earth heaved, incandescent lava poured out from the depths, hot springs spewed forth a sulfurous brew—and the violently displaced waters of the sea flooded Sodom and Gomorrah.

You can understand how catastrophic events of this sort could have impressed the people of the early Mediterranean civilizations. They still impress us. Even though our seismographs, satellite images, and numerous other technological advances give us a better picture of what causes the earth to tremble beneath our feet and skies to darken or giant waves to sweep across oceans when island volcanoes erupt, it doesn't make us any less uneasy.

Imagine yourself living 2,000 years ago, ignorant of the geologic forces responsible for earthquakes, giant waves, and other events. How would you have interpreted what was going on? Most likely then, as now, your interpretation would have been shaped by the prevailing beliefs of your society. Floods obviously occurred, and one sudden, devastating flood in particular was interpreted as punishment for bad human behavior. Bones and shells were observed in deep layers of the earth, and they were interpreted as

APPENDIX I
Units of Measure

Metric-English Conversions

Length

English		Metric
inch	=	2.54 centimeters
foot	=	0.30 meter
yard	=	0.91 meter
mile (5,280 feet)	=	1.61 kilometer

To convert	multiply by	to obtain
inches	2.54	centimeters
feet	30.00	centimeters
centimeters	0.39	inches
millimeters	0.039	inches

Weight

English		Metric
grain	=	64.80 milligrams
ounce	=	28.35 grams
pound	=	453.60 grams
ton (short) (2,000 pounds)	=	0.91 metric ton

To convert	multiply by	to obtain
ounces	28.3	grams
pounds	453.6	grams
pounds	0.45	kilograms
grams	0.035	ounces
kilograms	2.2	pounds

Volume

English		Metric
cubic inch	=	16.39 cubic centimeters
cubic foot	=	0.03 cubic meter
cubic yard	=	0.765 cubic meters
ounce	=	0.03 liter
pint	=	0.47 liter
quart	=	0.95 liter
gallon	=	3.79 liters

To convert	multiply by	to obtain
fluid ounces	30.00	milliliters
quart	0.95	liters
milliliters	0.03	fluid ounces
liters	1.06	quarts

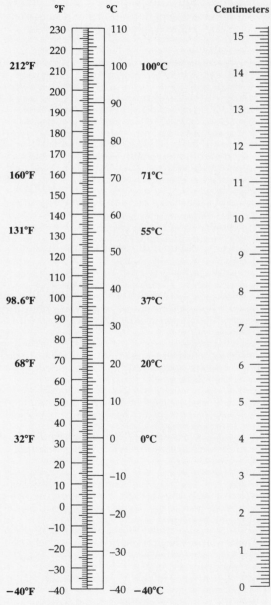

To convert temperature scales:

Fahrenheit to Celsius: $°C = 5/9 \, (°F - 32)$

Celsius to Fahrenheit: $°F = 9/5 \, (°C) + 32$

APPENDIX II
Brief Classification Scheme

This classification scheme is a composite of several used in microbiology, botany, and zoology. Although major groupings are more or less agreed upon, what to call them and (sometimes) where to place them in the overall hierarchy are not. As Chapter 20 indicated, there are several reasons for this. First, the fossil record varies in its quality and completeness, so certain evolutionary relationships are open to interpretation. Comparative studies at the molecular level are firming up the picture, but this work is still under way.

Second, since the time of Linnaeus, classification schemes have been based on perceived morphological similarities and differences among organisms. Although some original interpretations are now open to question, we are so used to thinking about organisms in certain ways that reclassification proceeds slowly. For example, birds and reptiles traditionally are considered separate classes (Reptilia and Aves)—even though there now are compelling arguments for grouping lizards and snakes as one class, and crocodilians, dinosaurs, and birds as another.

Finally, botanists as well as zoologists have inherited a wealth of literature based on schemes that are peculiar to their fields; and most see no good reason to give up established terminology and so disrupt access to the past. Thus botanists continue to use Division as a major taxon in the hierarchical schemes and zoologists use Phylum in theirs. Opinions are notably divergent with respect to an entire Kingdom (the Protista), certain members of which could just as easily be called single-celled forms of plants, fungi, or animals. Indeed, the term protozoan is a holdover from earlier schemes that ranked the amoebas and some other forms as simple animals.

Given the problems, why do we bother imposing hierarchical schemes on the natural history of life on earth? We do this for the same reason that a writer might decide to break up the history of civilization into several volumes, many chapters, and a multitude of paragraphs. Both efforts are attempts to impart structure to what might otherwise be an overwhelming body of information.

One more point to keep in mind: The classification scheme in this Appendix is primarily for reference purposes, and it is by no means complete (numerous phyla of existing and extinct organisms are not represented). Our strategy is to focus mainly on the organisms mentioned in the text, with numerals referring to some of the pages on which representatives are illustrated or described. A few examples of organisms are also listed under the entries.

SUPERKINGDOM PROKARYOTA. Prokaryotes (single-celled organisms with no nucleus or other membrane-bound organelles in the cytoplasm).

KINGDOM MONERA. Bacteria, either single cells or simple associations of cells; autotrophic and heterotrophic forms. *Bergey's Manual of Systematic Bacteriology*, the authoritative reference in the field, calls this "a time of taxonomic transition" and groups bacteria mainly on the basis of form, physiology, and behavior, not on phylogeny (Table 22.4 gives examples). The scheme presented here does reflect the growing evidence of evolutionary relationships for at least some bacterial groups.

SUBKINGDOM ARCHAEBACTERIA. Methanogens, halophiles, thermophiles. Strict anaerobes, distinct from other bacteria in their cell wall, membrane lipids, ribosomes, and RNA sequences. 356, 357

SUBKINGDOM EUBACTERIA. Gram-negative and Gram-positive forms. Peptidoglycan in cell walls. Photosynthetic autotrophs, chemosynthetic autotrophs, and heterotrophs. 358–359

DIVISION GRACILICUTES. Typical Gram-negative, thin wall. Autotrophs (photosynthetic and chemosynthetic) and heterotrophs. *Anabaena, Chlorobium, Escherichia, Shigella, Desulfovibrio, Agrobacterium, Pseudomonas, Neisseria.* 57, 354, 356, 781

DIVISION FIRMICUTES. Typical Gram-positive, thick wall. Heterotrophs. *Staphylococcus, Streptococcus, Clostridium, Bacillus, Actinomyces.* 346, 359, 557, 680

DIVISION TENERICUTES. Gram-negative, wall absent. Heterotrophs (saprobes, parasites, pathogens). *Mycoplasma.* 356

SUPERKINGDOM EUKARYOTA. Eukaryotes (single-celled and multicelled organisms; cells typically have a nucleus and other organelles).

KINGDOM PROTISTA. Mostly single-celled eukaryotes. Some colonial forms.

PHYLUM GYMNOMYCOTA. Heterotrophs.
   Class Acrasiomycota. Cellular slime molds. *Dictyostelium.* 362, 363
   Class Myxomycota. Plasmodial slime molds. *Physarum.* 362
PHYLUM EUGLENOPHYTA. Euglenoids. Mostly heterotrophic, some photosynthetic. Flagellated. *Euglena.* 364
PHYLUM CHRYSOPHYTA. Golden algae, yellow-green algae, diatoms. Photosynthetic. Some flagellated, others not. *Vaucheria.* 50, 364–368
PHYLUM PYRRHOPHYTA. Dinoflagellates. Mostly photosynthetic, some heterotrophs. *Gonyaulax.* 365
PHYLUM MASTIGOPHORA. Flagellated protozoans. Heterotrophs. *Trypanosoma, Trichomonas.* 366–367

PHYLUM SARCODINA.   Amoeboid protozoans. Heterotrophs. Amoebas, foraminiferans, heliozoans, radiolarians. 366–367

PHYLUM CILIOPHORA.   Ciliated protozoans. Heterotrophs. *Paramecium, Didinium.* 367–368

SPOROZOANS.   Parasitic protozoans, many intracellular. ("Sporozoans" is the common name for these diverse organisms; it has no formal taxonomic status.) *Plasmodium.* 368–369

KINGDOM FUNGI.   Mostly multicelled eukaryotes. Heterotrophs (mostly saprobes, some parasites). All rely on extracellular digestion and absorption of nutrients.

DIVISION MASTIGOMYCOTA.   All produce flagellated spores.
  Class Chytridiomycetes. Chytrids. 375
  Class Oomycetes. Water molds and related forms. *Plasmopora, Phytophthora, Saprolegnia.* 375, 376
DIVISION AMASTIGOMYCOTA.   All produce nonmotile spores.
  Class Zygomycetes. Bread molds and related forms. *Rhizopus, Pilobolus.* 376, 377
  Class Ascomycetes. Sac fungi. Most yeasts and molds; morels, truffles. *Saccharomyces, Morchella.* 376, 378
  Class Basidiomycetes. Club fungi. Mushrooms, shelf fungi, bird's nest fungi, stinkhorns. *Agaricus, Amanita.* 376, 378–380
FORM-DIVISION DEUTEROMYCOTA.   Imperfect fungi. All with undetermined affiliations because sexual stage unknown; if better known they would be grouped with sac fungi or club fungi. *Verticillium, Candida.* 376, 380–381

KINGDOM PLANTAE.   Nearly all multicelled eukaryotes. Photosynthetic autotrophs, except for a few saprobes and parasites. 385, 396

DIVISION RHODOPHYTA.   Red algae. *Porphyra.* 308, 388, 389, 869
DIVISION PHAEOPHYTA.   Brown algae. *Fucus, Laminaria.* 388–390
DIVISION CHLOROPHYTA.   Green algae. *Ulva, Spirogyra.* 16, 55, 308, 390–391, 817
DIVISION CHAROPHYTA.   Stoneworts.
DIVISION BRYOPHTYA.   Liverworts, hornworts, mosses. *Marchantia, Sphagnum.* 392–393
DIVISION RHYNIOPHYTA.   Earliest known vascular plants; extinct. *Cooksonia, Rhynia.* 310, 393
DIVISION PSILOPHYTA.   Whisk ferns. 310, 393–394
DIVISION LYCOPHYTA.   Lycopods, club mosses. *Lycopodium, Selaginella.* 394
DIVISION SPHENOPHYTA.   Horsetails. *Equisetum.* 280, 281, 394, 396
DIVISION PTEROPHYTA.   Ferns. 34, 395, 874
DIVISION PROGYMNOSPERMOPHYTA.   Progymnosperms. Ancestral to early seed-bearing plants; extinct. *Archaeopteris.* 294, 397
DIVISION PTERIDOSPERMOPHYTA.   Seed ferns (extinct fernlike gymnosperms).
DIVISION CYCADOPHYTA.   Cycads. *Zamia.* 311, 397
DIVISION GINKGOPHYTA.   Ginkgo. *Ginkgo.* 311, 398
DIVISION GNETOPHYTA.   Gnetophytes. *Ephedra, Welwitschia, Gnetum.* 398
DIVISION CONIFEROPHYTA.   Conifers. 309, 398, 852, 858, 860
  Family Pinaceae. Pines, firs, spruces, hemlock, larches, Douglas firs, true cedars. *Pinus.* 399, 483, 784, 878
  Family Cupressaceae. Junipers, cypresses, false cedars. 382
  Family Taxodiaceae. Bald cypress, redwood, Sierra bigtree, dawn redwood. *Sequoia.* 10, 820, 821
  Family Taxaceae. Yews.
DIVISION ANTHOPHYTA.   Flowering plants. 400, 469*ff.*
  Class Dicotyledonae. Dicotyledons (dicots). Some families of several different orders are listed: 402, 474
    Family Magnoliaceae. Magnolias, tulip trees. 478
    Family Ranunculaceae. Buttercups, delphinium. 177, 480, 814
    Family Nymphaeaceae. Water lilies. 400, 402
    Family Papaveraceae. Poppies, including opium poppy. 516
    Family Brassicaceae. Mustards, cabbages, radishes, turnips.
    Family Malvaceae. Mallows, cotton, okra, hibiscus. 461
    Family Solanaceae. Potatoes, eggplant, petunias. 114, 376, 818
    Family Salicaceae. Willows, poplars. 477
    Family Rosaceae. Roses, peaches, apples, almonds, strawberries. 503, 510, 511
    Family Fabaceae. Peas, beans, lupines, mesquite, locust. 166, 477, 489, 519, 854
    Family Cactaceae. Cacti. 512, 846, 854
    Family Euphorbiaceae. Spurges, poinsettia. 847
    Family Cucurbitaceae. Gourds, melons, cucumbers, squashes.
    Family Apiaceae. Parsleys, carrots, poison hemlock. 254, 479, 514
    Family Aceraceae. Maples. 109, 477, 511
    Family Asteraceae. Composites. Chrysanthemums, sunflowers, lettuces, dandelions. 10, 511, 384, 857
  Class Monocotyledonae. Monocotyledons (monocots). Some families of several different orders are listed: 402, 474
    Family Liliaceae. Lilies, hyacinths, tulips, onions, garlic. 401, 816
    Family Iridaceae. Irises, gladioli, crocuses. 474
    Family Orchidaceae. Orchids. 326, 400, 874
    Family Arecaceae. Date palms, coconut palms. 397, 477
    Family Cyperaceae. Sedges.
    Family Poaceae. Grasses, bamboos, corn, wheat, sugarcane. 115, 401, 479, 518, 856–857
    Family Bromeliaceae. Bromeliads, pineapple, Spanish moss. 511

KINGDOM ANIMALIA.   Multicelled eukaryotes. Heterotrophs (herbivores, carnivores, omnivores, parasites, decomposers, detritivores). 310, 407, 440

PHYLUM PLACOZOA.   Small, organless marine animal. *Trichoplax.* 369, 440
PHYLUM MESOZOA.   Ciliated, wormlike parasites, about the same level of complexity as *Trichoplax.*
PHYLUM PORIFERA.   Sponges. 408–409
PHYLUM CNIDARIA   410
  Class Hydrozoa. Hydrozoans. *Hydra, Obelia, Physalia.* 410, 412
  Class Scyphozoa. Jellyfishes. *Aurelia.* 410, 411
  Class Anthozoa. Sea anemones, corals. *Telesto.* 410, 565, 624, 867, 868
PHYLUM CTENOPHORA.   Comb jellies. *Pleurobrachia.* 413
PHYLUM PLATYHELMINTHES.   Flatworms. 414
  Class Turbellaria. Triclads (planarians), polyclads. *Dugesia.* 414, 642, 700
  Class Trematoda. Flukes. *Schistosoma.* 414–415, 416
  Class Cestoda. Tapeworms. *Taenia.* 414–415, 416, 417
PHYLUM NEMERTEA.   Ribbon worms. 418
PHYLUM NEMATODA.   Roundworms. *Ascaris, Trichinella.* 416–417, 418–419
PHYLUM ROTIFERA.   Rotifers. 419
PHYLUM MOLLUSCA.   Mollusks. 420
  Class Polyplacophora. Chitons. 421
  Class Gastropoda. Snails (periwinkles, whelks, limpets, abalones, cowries, conches, nudibranchs, tree snails, garden snail), sea slugs, land slugs. 150, 421, 422, 817
  Class Bivalvia. Clams, mussels, scallops, cockles, oysters, shipworms. 216, 423
  Class Cephalopoda. Squids, octopuses, cuttlefish, nautiluses. *Loligo.* 282, 310, 424–425, 553
PHYLUM BRYOZOA.   Bryozoans (moss animals).
PHYLUM BRACHIOPODA.   Lampshells.
PHYLUM ANNELIDA.   Segmented worms. 425
  Class Polychaeta. Mostly marine worms. 425, 426

APPENDIX III
Answers to Genetics Problems

Chapter Eleven

1. a. *AB*
   b. *AB* and *aB*
   c. *Ab* and *ab*
   d. *AB, aB, Ab,* and *ab*

2. a. *AaBB* will occur in all the offspring.
   b. 25% *AABB*; 25% *AaBB*; 25% *AABb*; 25% *AaBb*.
   c. 25% *AaBb*; 25% *Aabb*; 25% *aaBb*; 25% *aabb*.
   d. $\frac{1}{16}$ *AABB* (6.25%)

   $\frac{1}{8}$ *AaBB* (12.5%)

   $\frac{1}{16}$ *aaBB* (6.25%)

   $\frac{1}{8}$ *AABb* (12.5%)

   $\frac{1}{4}$ *AaBb* (25%)

   $\frac{1}{8}$ *aaBb* (12.5%)

   $\frac{1}{16}$ *AAbb* (6.25%)

   $\frac{1}{8}$ *Aabb* (12.5%)

   $\frac{1}{16}$ *aabb* (6.25%)

3. Yellow is recessive. Because the first-generation plants must be heterozygous and had a green phenotype, green must be dominant over the recessive yellow.

4. a. *ABC*
   b. *ABc* and *aBc*
   c. *ABC, aBC, ABc,* and *aBc*
   d. *ABC, aBC, AbC, abC, ABc, aBc, Abc,* and *abc*

5. Because the man can only produce one type of allele for each of his ten genes, he can only produce one type of sperm. The woman, on the other hand, can produce two types of alleles for each of her two heterozygous genes; she can produce 2 × 2 or 4 different kinds of eggs. As can be observed, as the number of heterozygous genes increases, more and more different types of gametes can be produced.

6. The first-generation plants must all be double heterozygotes. When these plants are self-pollinated, $\frac{1}{4}$ (25%) of the second-generation plants will be doubly heterozygous.

7. The most direct way to accomplish this would be to allow a true-breeding canary having yellow feathers to mate with a true-breeding canary having brown feathers. Such true-breeding strains could be obtained by repeated inbreeding (mating of related individuals; for example, a male and a female of the same nest) of yellow and brown strains. In this way, it should be possible to obtain homozygous yellow and homozygous brown canaries. When true-breeding yellow and true-breeding brown canaries are crossed, the progeny should all be heterozygous. If the progeny phenotype is either yellow or brown, then the dominance is simple or complete, and the phenotype reflects the dominant allele. If the phenotype is intermediate between yellow and brown, there is incomplete dominance. If the phenotype shows both yellow and brown, there is codominance.

8. a. Mother must be heterozygous for both genes; father is homozygous recessive for both genes. The first child is also homozygous recessive for both genes.
   b. The probability that the second child will not be able to roll the tongue and will have free earlobes is $\frac{1}{4}$ (25%).

9. a. The mother must be heterozygous ($I^{A}i$). The man having type B blood could have fathered the child if he were also heterozygous ($I^{B}i$).
   b. If the man is heterozygous, then he *could be* the father. However, because any other type B heterozygous male also could be the father, one cannot say that this particular man absolutely must be. Actually, any male who could contribute an O allele (*i*) could have fathered the child. This would include males with type O blood (*ii*) or type A blood who are heterozygous ($I^{A}i$).

10. a. F_1 genotypes and phenotypes: 100% *Bb Cc*, brown progeny. F_2 phenotypes: $\frac{9}{16}$ brown + $\frac{3}{16}$ tan + $\frac{4}{16}$ albino.

    F_2 genotypes: $\begin{cases} \frac{1}{16}\ BB\ CC + \frac{2}{16}\ BB\ Cc + \frac{2}{16}\ Bb\ CC + \\ \quad \frac{4}{16}\ Bb\ Cc;\ \left(\frac{9}{16}\ \text{brown}\right) \\ \frac{1}{16}\ bb\ CC + \frac{2}{16}\ bb\ Cc;\ \left(\frac{3}{16}\ \text{tan}\right) \\ \frac{1}{16}\ BB\ cc + \frac{2}{16}\ Bb\ cc + \frac{1}{16}\ bb\ cc;\ \left(\frac{4}{16}\ \text{albino}\right) \end{cases}$

    b. Backcross phenotypes: $\frac{1}{4}$ brown + $\frac{1}{4}$ tan + $\frac{2}{4}$ albino.

    Backcross genotypes: $\begin{cases} \frac{1}{4}\ Bb\ Cc;\ \left(\frac{1}{4}\ \text{brown}\right) \\ \frac{1}{4}\ bb\ Cc;\ \left(\frac{1}{4}\ \text{tan}\right) \\ \frac{1}{4}\ Bb\ cc + \frac{1}{4}\ bb\ cc;\ \left(\frac{1}{2}\ \text{albino}\right) \end{cases}$

11. The mating is *Ll* × *Ll*.
    Progeny genotypes: $\frac{1}{4}$ *LL* + $\frac{1}{2}$ *Ll* + $\frac{1}{4}$ *ll*

    Phenotypes: $\begin{cases} \frac{1}{4}\ \text{homozygous survivors (}LL\text{)} + \\ \frac{1}{2}\ \text{heterozygous survivors (}Ll\text{)} + \\ \frac{1}{4}\ \text{lethal (}ll\text{) nonsurvivors.} \end{cases}$

    Thus, among the survivors, there is a $\frac{2}{3}$ probability that any individual will be heterozygous.

12. To work this problem, consider the effect of each pair of genes separately.
    a. *Aa* × *aa*: The resulting progeny genotypes are $\frac{1}{2}$ *Aa* and $\frac{1}{2}$ *aa*. Thus, half of the progeny will receive an *A* allele, which permits kernel pigmentation.

b. $cc \times Cc$: Here, too, half of the progeny will receive a dominant C allele, which permits kernel pigmentation.

c. $Rr \times Rr$: In this case, $\frac{3}{4}$ of the progeny will receive at least one R allele, which permits kernel pigmentation. Remember that in order to have pigmented kernels, at least one dominant allele of each and every one of these three gene loci must be simultaneously present. Since the A, C, and R loci assort independently, to find the overall fraction of kernels that are pigmented, you must multiply together the fraction of pigmented kernels produced separately by the A, C, and R loci, and that is: $\frac{1}{2} \times \frac{1}{2} \times \frac{3}{4} = \frac{3}{16}$.

Chapter Twelve

1. a. Males inherit their X chromosome from their mothers.
   b. A male can produce two types of gametes with respect to an X-linked gene. One type will lack this gene and possess a Y chromosome. The other will have an X chromosome and the linked gene.
   c. A female homozygous for an X-linked gene will produce just one type of gamete containing an X chromosome with the gene.
   d. A female heterozygous for an X-linked gene will produce two types of gametes. One will contain an X chromosome with the dominant allele, and the other type will contain an X chromosome with the recessive allele.

2. a. Because this gene is only carried on Y chromosomes, females would not be expected to have hairy pinnae because they normally do not have Y chromosomes.
   b. Because sons always inherit a Y chromosome from their fathers and because daughters never do, a man having hairy pinnae will always transmit this trait to his sons and never to his daughters.

3. A 0% crossover frequency means that 50% of the gametes will be AB and 50% will be ab.

4. The first-generation females must be heterozygous for both genes. The 42 red-eyed, vestigial-winged and the 30 purple-eyed, long-winged progeny represent recombinant gametes from these females. Because the first-generation females must have produced 600 gametes to give these 600 progeny, and because $42 + 30$ of these were recombinant, the percentage of recombinant gametes is 72/600, or 12%, which implies that 12 map units separate the two genes.

5. The rare vestigial-winged flies could be explained by a deletion of the dominant allele from one of the chromosomes, due to the action of the x-rays. Alternatively, the radiation may have induced a mutation in the dominant allele.

6. If the longer-than-normal chromosome 14 represented the translocation of most of chromosome 21 to the end of a normal chromosome 14, then this individual would be afflicted with Down syndrome due to the presence of this attached chromosome 21 as well as two normal chromosomes 21. The total chromosome number, however, would be 46.

7. Using c as the symbol for color blindness and C for normal vision, then the cross can be diagrammed as follows:

$$C(Y) \text{ female} \times Cc \text{ male}$$

In mugwumps, a son receives one sex-linked allele from each of his parents, but a daughter inherits her unpaired sex-linked allele solely from her father. In this cross, half of the sons will be CC and half Cc, but none will be color blind. Of the daughters, half will be $C(Y)$ and half $c(Y)$. There is a 50% chance that a daughter will be color blind. Note: this answer is backward from the way it would be in humans; but it is correct not only for mugwumps but also for all birds and Lepidoptera (moths and butterflies), as well as a few other forms.

8. In order to produce a female suffering from childhood muscular dystrophy, not only must her mother be a carrier of the disease, but her father must have it. Few, if any, such males who survive to adulthood are capable of having children.

APPENDIX IV
Answers to Self-Quizzes

CHAPTER 1
1. DNA
2. energy
3. Metabolism
4. Homeostasis
5. adaptations
6. mutations
7. reproductive capacity
8. c
9. c
10. c

CHAPTER 2
1. electrons
2. d
3. electrons
4. Isotopes
5. c
6. b
7. d
8. b
9. b
10. d
11. b, e, a, c, d

CHAPTER 3
1. carbon
2. a
3. polysaccharides, lipids, proteins, nucleic acids
4. c
5. c
6. d
7. b
8. b
9. c
10. c, e, b, d, a

CHAPTER 4
1. Cells
2. e
3. c
4. c
5. d
6. d
7. b
8. d
9. cytoskeleton
10. c, h, g, e, d, b, a, f

CHAPTER 5
1. plasma membrane
2. c
3. a
4. a
5. b
6. transport, receptor, and recognition proteins
7. b
8. d
9. c
10. g, f, d, a, e, b, c

CHAPTER 6
1. metabolism
2. thermodynamics
3. c
4. d
5. e
6. d
7. d
8. d
9. c
10. c, e, d, a, b

CHAPTER 7
1. carbon
2. carbon dioxide; sunlight
3. d
4. c
5. b
6. e
7. c
8. c
9. c, d, e, b, a

CHAPTER 8
1. glucose, other organic compounds
2. fermentation; anaerobic electron transport
3. pyruvate; carbon dioxide; water
4. d
5. c
6. d
7. c
8. b
9. b
10. b, c, a, d

CHAPTER 9
1. mitosis; meiosis
2. chromosomes; DNA
3. c
4. d
5. c
6. a
7. c
8. d
9. b
10. b
11. d, b, c, a

CHAPTER 10
1. diploid; two
2. c
3. c
4. c
5. d
6. c
7. d
8. c
9. b

CHAPTER 11
1. a
2. c
3. a
4. c
5. b
6. c
7. b
8. d
9. c
10. c, d, e, b, a

CHAPTER 12
1. c
2. e
3. e
4. c
5. e
6. c
7. d
8. d
9. d
10. c, e, d, b, a

CHAPTER 13
1. Hydrogen
2. e
3. d
4. d
5. c
6. a
7. a
8. d
9. d
10. d, e, b, c, a

CHAPTER 14
1. three
2. d
3. b
4. b
5. d
6. c
7. a
8. b
9. a
10. a

CHAPTER 15
1. chemical
2. differentiation
3. d
4. e
5. d
6. b
7. c
8. d
9. a
10. b, e, c, d, a

CHAPTER 16
1. diversity
2. genetic engineering
3. Plasmids
4. cloning vectors
5. c
6. d
7. b
8. c
9. e
10. d, e, a, c, b

CHAPTER 17
1. body
2. evolve
3. fossils
4. population
5. b
6. c
7. e
8. e
9. c, d, e, b, a

CHAPTER 18
1. population
2. differences
3. e
4. a
5. e
6. d
7. c
8. c
9. d
10. d, c, e, a, b

Chapter 19

1. macroevolution
2. mass extinctions; adaptive radiations
3. gradual; bursts
4. e
5. e
6. c
7. b
8. e
9. b, d, e, a, c

Chapter 20

1. d
2. e
3. c
4. c
5. b
6. a
7. d
8. b
9. e, c, g, a, d, f, b

Chapter 21

1. e
2. b
3. c
4. c
5. a
6. b
7. d
8. c
9. d, a, e, c, b

Chapter 22

1. a living host cell
2. e
3. d
4. d
5. c
6. c
7. e
8. b
9. c
10. c, e, d, a, b

Chapter 23

1. decomposers
2. mycelium; hyphae
3. a
4. c
5. b
6. c
7. b
8. f
9. d
10. c, e, d, b, a

Chapter 24

1. autotrophs
2. green algae
3. b
4. d
5. a
6. b
7. c
8. d
9. c
10. d, e, c, f, b, a

Chapter 25

1. body symmetry, cephalization, type of gut, type of body cavity, segmentation
2. a
3. b
4. d
5. a
6. a
7. b
8. c

Chapter 26

1. bilateral
2. c
3. b
4. brain
5. c
6. c
7. b
8. c
9. e

Chapter 27

1. dermal, ground, vascular
2. apical; lateral
3. b
4. a
5. a
6. b
7. c
8. a
9. d
10. c, d, f, a, e, b

Chapter 28

1. hydrogen bonds
2. stomata
3. d
4. e
5. c
6. b
7. d
8. d
9. d
10. e, d, c, a, b

Chapter 29

1. a
2. pollinators
3. d
4. c
5. b
6. c
7. b
8. a
9. a, c, d, e, b

Chapter 30

1. c
2. c
3. e
4. d
5. d
6. a
7. c
8. c
9. a, e, c, d, b

Chapter 31

1. epithelial, connective, muscle, nervous
2. e
3. d
4. c
5. b
6. a
7. d
8. c
9. a
10. Receptors, integrator, effectors
11. d, e, c, b, a

Chapter 32

1. neurons
2. sensory neurons, interneurons, motor neurons
3. a
4. c
5. a
6. d
7. b
8. d
9. d
10. c, d, a, b, e

Chapter 33

1. nerve nets
2. cephalization; bilateral
3. a
4. b
5. c
6. a
7. d
8. c
9. a
10. b
11. e, d, b, c, a

Chapter 34

1. Hormones; transmitter substances; local signaling molecules; pheromones
2. hypothalamus; pituitary
3. negative feedback
4. receptors
5. a
6. d
7. e
8. b
9. d
10. d, b, e, c, a

Chapter 35

1. stimulus
2. sensation
3. perception
4. d
5. b
6. d
7. c
8. b
9. a
10. f, d, c, e, a, b

Chapter 36

1. integumentary
2. Skeletal; muscular
3. smooth; cardiac; skeletal
4. d
5. d
6. c
7. b
8. c
9. d, e, b, a, c

Chapter 37

1. digestive; circulatory; respiratory and urinary
2. digesting; absorbing; eliminating
3. caloric; energy
4. carbohydrates
5. essential amino acids; essential fatty acids
6. b
7. c
8. d
9. c
10. e, d, a, c, b

Chapter 38

1. circulatory; lymphatic
2. Arteries; veins; capillaries; venules; arterioles
3. d
4. d
5. c
6. b
7. c
8. b
9. e, b, a, d, c
10. d, e, a, f, c, b

Chapter 39

1. Antigens
2. antigen that is circulating or attached to pathogen; infected, cancerous, or mutated cells
3. e
4. d
5. b
6. b
7. b
8. a
9. c
10. b, a, e, d, c

Chapter 40

1. Oxygen; carbon dioxide
2. gas; partial pressure gradient
3. epithelium; diffuse
4. adaptations; air flow
5. d
6. e
7. b
8. d
9. d
10. e, a, d, b, c

Chapter 41

1. filtration; absorption; secretion
2. c
3. c
4. d
5. d
6. a
7. a
8. c
9. d
10. b, d, e, c, a

Chapter 42

1. d
2. a
3. a
4. a
5. c
6. morphogenesis
7. b
8. c
9. e
10. d, a, f, e, c, b

Chapter 43

1. hypothalamus
2. b
3. d
4. c
5. b
6. a
7. c
8. b
9. c
10. c, e, b, d, a

Chapter 44

1. Ecology
2. population
3. e
4. a
5. b
6. e
7. d
8. c
9. d
10. d, e, c, a, b

Chapter 45

1. Communities
2. niche
3. habitat
4. e
5. d
6. d
7. b
8. d
9. c
10. e, c, a, d, b

Chapter 46

1. ecosystem; energy; cycling
2. photosynthesis; one
3. energy; nutrients; energy; nutrients
4. biogeochemical cycles
5. d
6. d
7. e
8. c
9. e
10. d, c, b, a
11. c, d, e, b, a

Chapter 47

1. sunlight
2. oceans; weather systems
3. rainfall
4. d
5. c
6. e
7. a
8. a
9. e
10. f, a, c, d, e, b

Chapter 48

1. exponential
2. pollutants
3. e
4. c
5. b
6. c
7. d
8. d
9. e, a, c, b, d

Chapter 49

1. d
2. b
3. a
4. d
5. b
6. d
7. c
8. a
9. b
10. c

Chapter 50

1. c
2. c
3. d
4. c
5. d
6. a
7. c
8. c
9. d
10. d

APPENDIX V
A Closer Look at Some Major Metabolic Pathways

ENERGY-
REQUIRING
STEPS OF
GLYCOLYSIS

(two ATP
invested)

ENERGY-
RELEASING
STEPS OF
GLYCOLYSIS

(four ATP
produced)

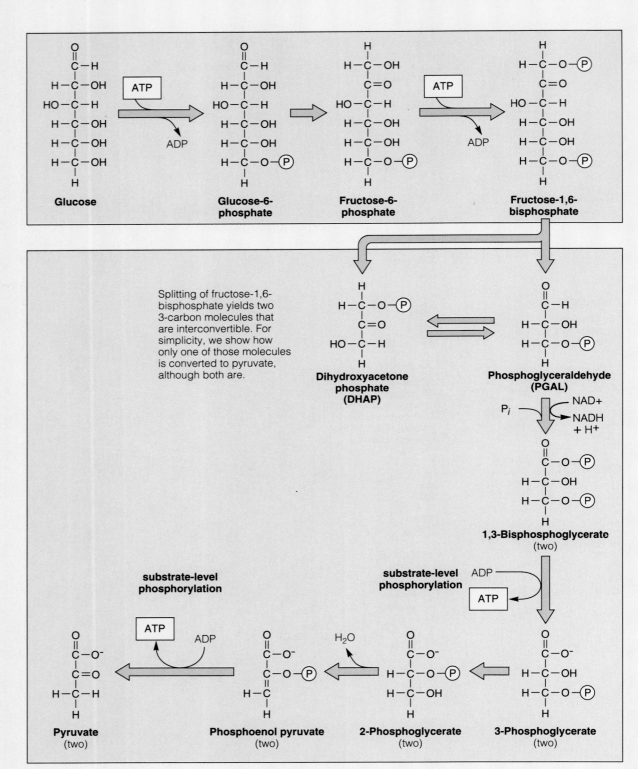

Figure a Glycolysis, ending with two 3-carbon pyruvate molecules for each 6-carbon glucose entering the reactions. The *net* energy yield is two ATP molecules (two invested, four produced).

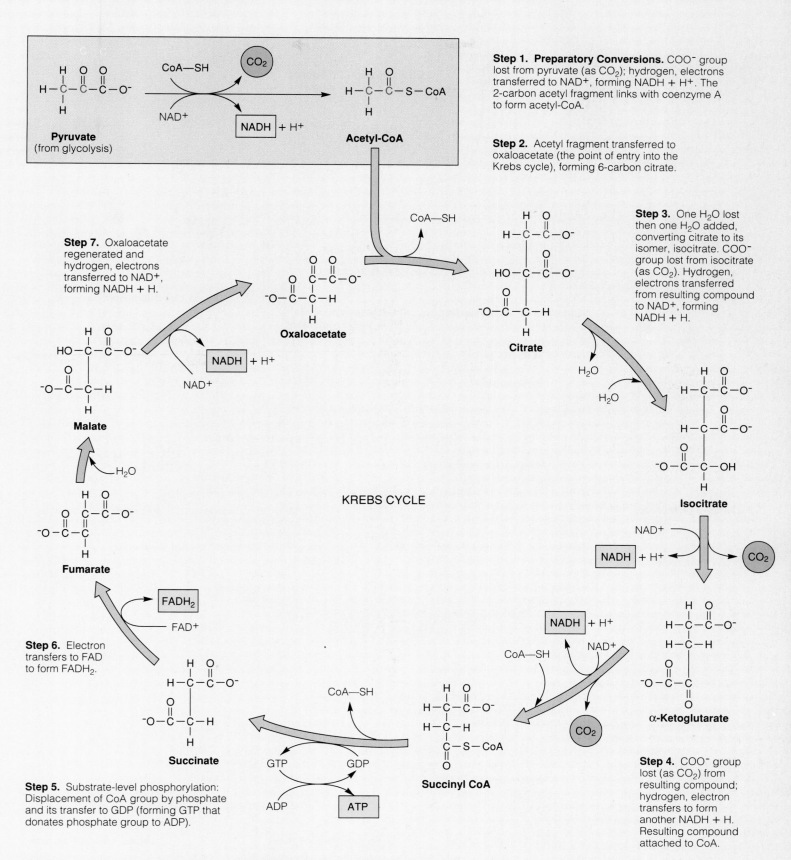

Step 1. Preparatory Conversions. COO^- group lost from pyruvate (as CO_2); hydrogen, electrons transferred to NAD^+, forming $NADH + H^+$. The 2-carbon acetyl fragment links with coenzyme A to form acetyl-CoA.

Step 2. Acetyl fragment transferred to oxaloacetate (the point of entry into the Krebs cycle), forming 6-carbon citrate.

Step 3. One H_2O lost then one H_2O added, converting citrate to its isomer, isocitrate. COO^- group lost from isocitrate (as CO_2). Hydrogen, electrons transferred from resulting compound to NAD^+, forming $NADH + H$.

Step 7. Oxaloacetate regenerated and hydrogen, electrons transferred to NAD^+, forming $NADH + H$.

Step 6. Electron transfers to FAD to form $FADH_2$.

Step 5. Substrate-level phosphorylation: Displacement of CoA group by phosphate and its transfer to GDP (forming GTP that donates phosphate group to ADP).

Step 4. COO^- group lost (as CO_2) from resulting compound; hydrogen, electron transfers to form another $NADH + H$. Resulting compound attached to CoA.

Pyruvate (from glycolysis)

Acetyl-CoA

KREBS CYCLE

Oxaloacetate

Citrate

Isocitrate

Malate

Fumarate

Succinate

Succinyl CoA

α-Ketoglutarate

Figure b Krebs cycle (citric acid cycle). Red identifies carbon entering the cycle by way of acetyl-CoA. Blue identifies carbon destined to leave the substrates (as carbon dioxide molecules).

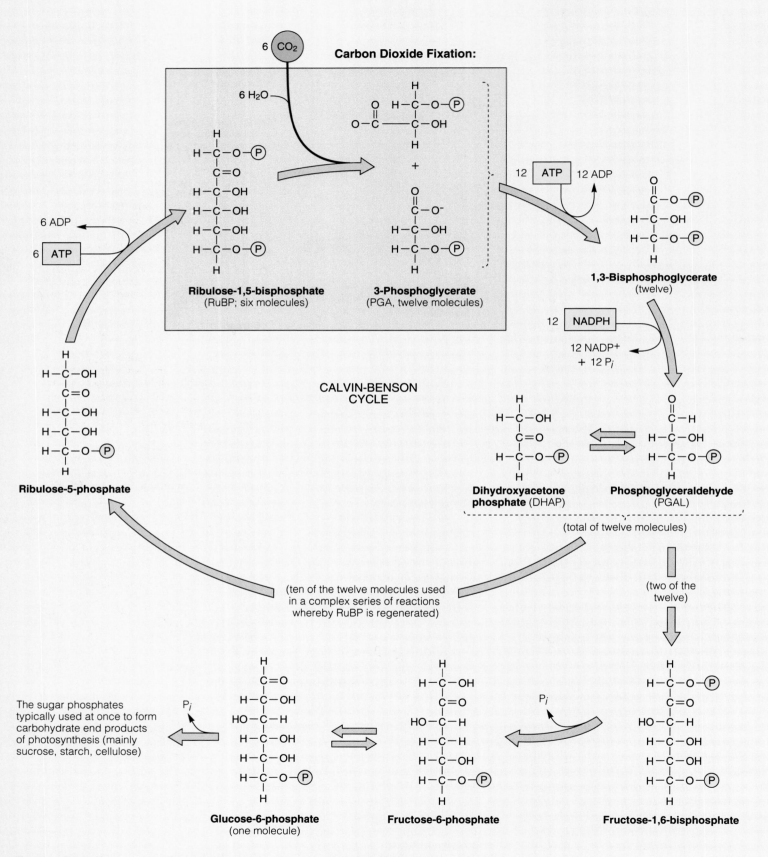

Figure c Calvin-Benson cycle of the light-independent reactions of photosynthesis.

GLOSSARY OF BIOLOGICAL TERMS

ABO blood typing Method of characterizing blood according to particular proteins at the surface of red blood cells.

abortion Spontaneous or induced expulsion of the embryo or fetus from the uterus.

abscisic acid (ab-SISS-ik) Plant hormone that promotes stomatal closure, bud dormancy, and seed dormancy.

abscission (ab-SIH-zhun) [L. *abscissus*, to cut off] The dropping of leaves, flowers, fruits, or other plant parts due to hormonal action.

absorption For complex animals, the movement of nutrients, fluid, and ions across the gut lining and into the internal environment.

acid [L. *acidus*, sour] A substance that releases hydrogen ions (H⁺) in solution.

acid deposition, dry The falling to earth of airborne particles of sulfur and nitrogen oxides.

acid deposition, wet The falling to earth of snow or rain that contains sulfur and nitrogen oxides.

acoelomate (ay-SEE-la-mate) Type of animal that has no fluid-filled cavity between the gut and body wall.

actin (AK-tin) A globular contractile protein. Within sarcomeres, the basic units of muscle contraction, actin molecules form two beaded strands, twisted together into filaments, that are pulled by myosin molecules during contraction.

action potential An abrupt but brief reversal in the steady voltage difference across the plasma membrane (that is, the resting membrane potential) of a neuron and some other cells.

activation energy The minimum amount of collision energy required to bring reactant molecules to an activated condition (the transition state) at which a reaction will proceed spontaneously. Enzymes enhance reaction rates by lowering the activation energy.

active site A crevice on the surface of an enzyme molecule where a specific reaction is catalyzed.

active transport The pumping of specific solutes through transport proteins that span the lipid bilayer of a plasma membrane, most often against their concentration gradient. The proteins act when they receive an energy boost, as from ATP.

adaptation [L. *adaptare*, to fit] In evolutionary biology, the process of becoming adapted (or more adapted) to a given set of environmental conditions. Of sensory neurons, a decrease in the frequency of action potentials (or their cessation) even when a stimulus is being maintained at constant strength.

adaptive radiation A burst of speciation events, with lineages branching away from one another as they partition the existing environment or invade new ones.

adaptive trait Any aspect of form, function, or behavior that helps an organism survive and reproduce under a given set of environmental conditions.

adaptive zone A way of life, such as "catching insects in the air at night." A lineage must have physical, ecological, and evolutionary access to an adaptive zone to become a successful occupant of it.

adenine (AH-de-neen) A purine; a nitrogen-containing base found in nucleotides.

adenosine diphosphate (ah-DEN-uh-seen die-FOSS-fate) ADP, a molecule involved in cellular energy transfers; typically formed by hydrolysis of ATP.

adenosine phosphate Any of several relatively small molecules, some of which function as chemical messengers within and between cells, and others as energy carriers.

adenosine triphosphate *See* ATP.

ADH Antidiuretic hormone, produced by the hypothalamus and released from the posterior pituitary; stimulates reabsorption in the kidneys and so reduces urine volume.

adrenal cortex (ah-DREE-nul) Outer portion of either of two adrenal glands; its hormones have roles in metabolism, inflammation, maintaining extracellular fluid volume, and other functions.

adrenal medulla Inner region of the adrenal gland; its hormones help control blood circulation and carbohydrate metabolism.

aerobic respiration (air-OH-bik) [Gk. *aer*, air, + *bios*, life] Degradative, oxygen-requiring pathway of ATP formation. Oxygen serves as the final acceptor of electrons stripped from glucose or some other organic compound. The pathway proceeds from glycolysis, then through the Krebs cycle and electron transport phosphorylation. Of all degradative pathways, aerobic respiration has the greatest energy yield, with 36 ATP typically formed for each glucose molecule.

agglutination (ah-glue-tin-AY-shun) Clumping of foreign cells, induced by the cross-linking of antigen-antibody complexes at their surface.

aging A range of processes, including the breakdown of cell structure and function, by which the body gradually deteriorates. Characteristic of all organisms showing extensive cell differentiation.

AIDS Acquired immune deficiency syndrome, a set of chronic disorders following infection by the human immunodeficiency virus (HIV), which destroys key cells of the immune system.

alcoholic fermentation Anaerobic pathway of ATP formation in which pyruvate from glycolysis is broken down to acetaldehyde; the acetaldehyde then accepts electrons from NADH to become ethanol.

aldosterone (al-DOSS-tuh-rohn) Hormone secreted by the adrenal cortex that helps regulate sodium reabsorption.

allantois (ah-LAN-twahz) [Gk. *allas*, sausage] A vascularized extraembryonic membrane; in reptiles and birds, it functions in excretion and respiration; in placental mammals, it functions in oxygen transport by way of the umbilical cord.

allele (uh-LEEL) One of two or more alternative forms of a gene at a given locus on a chromosome.

allele frequency The relative abundance of each kind of allele that can occur at a given gene locus for all individuals in a population.

allergy An abnormal, secondary immune response to a normally harmless substance.

allopatric speciation [Gk. *allos*, different, + *patria*, native land] Formation of new species as a result of geographic isolation.

allosteric control (AL-oh-STARE-ik) Control of enzyme functioning through the binding of a specific substance at a control site on the enzyme molecule.

altruistic behavior (al-true-ISS-tik) Self-sacrificing behavior; the individual behaves in a way that helps others but, in so doing, decreases its own chance to produce offspring.

alveolar sac (al-VEE-uh-lar) Any of the pouch-like clusters of alveoli in lungs; the major sites of gas exchange.

alveolus (ahl-VEE-uh-lus), plural **alveoli** [L. *alveus*, small cavity] Any of the many cup-shaped, thin-walled outpouchings of the respiratory bronchioles; a site where oxygen from air in the lungs diffuses into the bloodstream and where carbon dioxide from the bloodstream diffuses into the lungs.

amino acid (uh-MEE-no) A small organic molecule having a hydrogen atom, an amino group, an acid group, and an R group covalently bonded to a central carbon atom; amino acids strung together as polypeptide chains represent the primary structure of proteins.

ammonification (uh-moan-ih-fih-KAY-shun) A decomposition process by which certain bacteria and fungi break down nitrogen-containing wastes and remains of other organisms.

amnion (AM-nee-on) In land vertebrates, an extraembryonic membrane in the form of a fluid-filled sac around the embryo; it absorbs shocks and keeps the embryo from drying out.

anaerobic pathway (an-uh-ROW-bok) [Gk. *an*, without, + *aer*, air] Degradative metabolic pathway in which a substance other than oxygen serves as the final electron acceptor for the reactions.

analogous structures Occurrence of similar body parts being used for similar functions in evolutionarily remote lineages; evolutionary outcome of morphological convergence.

anaphase (AN-uh-faze) The stage of mitosis when sister chromatids of each chromosome separate and move to opposite poles of the spindle.

anaphase I and II Stages of meiosis when each chromosome separates from its homologous partner (anaphase I) and, later, when sister chromatids of each chromosome separate and move to opposite poles of the spindle (anaphase II).

angiosperm (AN-gee-oh-spurm) [Gk. *angeion*, vessel, + *spermia*, seed] A flowering plant.

animal A heterotroph that eats or absorbs nutrients from other organisms; is multicelled, usually with tissues arranged in organs and organ systems; is usually motile during at least part of the life cycle; and goes through a period of embryonic development.

Animalia The kingdom of animals.

annual plant Vascular plant that completes its life cycle in one growing season.

anther [Gk. *anthos*, flower] In flowering plants, the pollen-bearing part of the male reproductive structure (stamen).

antibody [Gk. *anti*, against] Any of a variety of Y-shaped receptor molecules with binding sites for a specific antigen (molecule that triggers an immune response); produced by B cells of the immune system.

anticodon In a tRNA molecule, a sequence of three nucleotide bases that can pair with an mRNA codon.

antigen (AN-tih-jen) [Gk. *anti*, against, + *genos*, race, kind] Any large molecule (usually a protein or polysaccharide) with a distinct configuration that triggers an immune response.

aorta (ay-OR-tah) [Gk. *airein*, to lift, heave] Main artery of systemic circulation; carries oxygenated blood away from the heart.

apical dominance The influence exerted by a terminal bud in inhibiting the growth of lateral buds.

apical meristem (AY-pih-kul MARE-ih-stem) [L. *apex*, top, + Gk. *meristos*, divisible] In most plants, a mass of self-perpetuating cells at a root or shoot tip that is responsible for primary growth, or elongation, of plant parts.

appendicular skeleton (ap-en-DIK-you-lahr) In vertebrates, bones of the limbs, pelvic girdle (at the hips), and pectoral girdle (at the shoulders).

arteriole (ar-TEER-ee-ole) Any of the blood vessels between arteries and capillaries; arterioles serve as control points where the volume of blood delivered to different body regions can be adjusted.

artery Any of the large-diameter blood vessels that conduct oxygen-poor blood to the lungs and oxygen-enriched blood to all body tissues; with their thick, muscular wall, arter-

ies are pressure reservoirs that smooth out pulsations in blood pressure caused by heart contractions.

asexual reproduction Mode of reproduction in which offspring arise from a single parent, and inherit the genes of that parent only.

atmosphere A region of gases, airborne particles, and water vapor enveloping the earth; 80 percent of its mass is distributed within seventeen miles of the earth's surface.

atmospheric cycle A biogeochemical cycle in which a large portion of the element being cycled between the physical environment and ecosystems occurs in a gaseous phase in the atmosphere; examples are the carbon cycle and nitrogen cycle.

atom The smallest unit of matter that is unique to a particular kind of element.

atomic number The number of protons in the nucleus of each atom of an element; differs for each element.

ATP Adenosine triphosphate, an energy carrier composed of adenine, ribose, and three phosphate groups; directly or indirectly transfers energy to or from nearly all metabolic pathways; produced during photosynthesis, aerobic respiration, fermentation, and other pathways.

australopith (OHSS-trah-low-pith) [L. *australis*, southern, + Gk. *pithekos*, ape] Any of the earliest known species of hominids; that is, the first species on the evolutionary branch leading to humans.

autoimmune response Abnormal immune response in which lymphocytes mount an attack against the body's own cells.

autonomic nervous system (auto-NOM-ik) Those nerves leading from the central nervous system to the smooth muscle, cardiac muscle, and glands of internal organs and structures; that is, to the visceral portion of the body.

autosomal dominant inheritance Condition in which a dominant allele on an autosome (not a sex chromosome) is always expressed to some extent.

autosomal recessive inheritance Condition in which a mutation produces a recessive allele on an autosome (not a sex chromosome); only recessive homozygotes show the resulting phenotype.

autosome Any of those chromosomes that are of the same number and kind in both males and females of the species.

autotroph (AH-toe-trofe) [Gk. *autos*, self, + *trophos*, feeder] An organism able to build all the organic molecules it requires using carbon dioxide (present in air and in water) and energy from the physical environment. Photosynthetic autotrophs use sunlight energy; chemosynthetic autotrophs extract energy from chemical reactions involving inorganic substances. Compare *heterotroph*.

auxin (AWK-sin) Any of a class of plant growth-regulating hormones; auxins promote stem elongation as one effect.

axial skeleton In vertebrates, the skull, backbone, ribs, and breastbone (sternum).

axon A long, cylindrical extension of a neuron with finely branched endings. Action potentials move rapidly, without alteration,

along an axon; their arrival at the axon endings can trigger the release of transmitter substances that may affect the activity of an adjacent cell.

bacterial flagellum Whiplike motile structure of many bacterial cells; unlike other flagella, it does not contain a core of microtubules.

bacteriophage (bak-TEER-ee-oh-fahj) [Gk. *baktērion*, small staff, rod, + *phagein*, to eat] Category of viruses that infect bacterial cells.

balanced polymorphism Of a population, the maintenance of two or more forms of a trait in fairly stable proportions over the generations.

Barr body In the cells of female mammals, a condensed X chromosome that was inactivated during embryonic development.

basal body A centriole which, after having given rise to the microtubules of a flagellum or cilium, remains attached to the base of either motile structure.

base A substance that, in solution, releases ions that can combine with hydrogen ions.

base pair A pair of hydrogen-bonded nucleotide bases; in the two strands of a DNA double helix, either A–T (adenine with thymine), or G–C (guanine with cytosine). When an mRNA strand forms on a DNA strand during transcription, uracil (U) base-pairs with the DNA's adenine.

behavior, animal A response to external and internal stimuli, following integration of sensory, neural, endocrine, and effector components. Behavior has a genetic basis, hence is subject to natural selection, and it commonly can be modified through experience.

benthic province All of the sediments and rocky formations of the ocean bottom; begins with the continental shelf and extends down through deep-sea trenches.

bilateral symmetry Of animals, a body plan with left and right halves that are basically mirror-images of each other.

biennial Flowering plant that lives through two growing seasons.

binary fission Of bacteria, a mode of asexual reproduction in which the parent cell replicates its single chromosome, then divides into two genetically identical daughter cells.

biogeochemical cycle The movement of carbon, oxygen, hydrogen, or some mineral element necessary for life from the environment to organisms, then back to the environment.

biogeographic realm [Gk. *bios*, life, + *geographein*, to describe the surface of the earth] In one scheme, one of six major land regions, each having distinguishing types of plants and animals.

biological clocks In many organisms, internal time-measuring mechanisms that have roles in adjusting daily and often seasonal activities in response to environmental cues.

biological magnification The increasing concentration of a nondegradable or slowly degradable substance in body tissues as it is passed along food chains.

biological systematics Branch of biology that assesses patterns of diversity based on information from taxonomy, phylogenetic reconstruction, and classification.

biomass The combined weight of all the organisms at a particular trophic (feeding) level in an ecosystem.

biome A broad, vegetational subdivision of some biogeographic realm, shaped by climate, topography, and composition of regional soils.

biosphere [Gk. *bios*, life, + *sphaira*, globe] the entire realm in which organisms exist; the lower regions of the atmosphere, the earth's waters, and the surface rocks, soils, and sediments of the earth's crust; the most inclusive level of biological organization.

biosynthetic pathway A metabolic pathway in which small molecules are assembled into lipids, proteins, and other large organic molecules.

biotic potential Of a population, the maximum rate of increase, per individual, under ideal conditions.

bipedalism A habitual standing and walking on two feet. Humans and ostriches are examples of bipedal animals.

blastocyst (BLASS-tuh-sist) [Gk. *blastos*, sprout, + *kystis*, pouch] In mammalian development, a modified blastula stage consisting of a hollow ball of surface cells and an inner cell mass.

blastula (BLASS-chew-lah) Among animals, an embryonic stage consisting of a ball of cells produced by cleavage.

blood A fluid connective tissue composed of water, solutes, and formed elements (blood cells and platelets).

blood-brain barrier Set of mechanisms that help control which blood-borne substances reach neurons in the brain.

blood pressure Fluid pressure, generated by heart contractions, that keeps blood circulating.

brainstem The vertebrate midbrain, pons, and medulla oblongata, the core of which contains the reticular formation that helps govern activity of the nervous system as a whole.

bronchus, plural **bronchi** (BRONG-CUSS, BRONG-kee) [Gk. *bronchos*, windpipe] Tubelike branchings of the trachea that lead to the lungs.

bud An undeveloped shoot of mostly meristematic tissue; often protected by a covering of modified leaves.

buffer A substance that can combine with hydrogen ions, release them, or both, in response to changes in pH.

bulk flow In response to a pressure gradient, a movement of more than one kind of molecule in the same direction in the same medium (as in blood, sap, or air).

C4 pathway Alternative light-independent reactions of photosynthesis in which carbon dioxide is fixed twice, in two different cell types. Carbon dioxide accumulates in the leaf and helps counter photorespiration (a "wasteful" process that reduces synthesis of sugar phosphates). The first compound formed is the 4-carbon oxaloacetate.

calorie (KAL-uh-ree) [L. *calor*, heat] The amount of heat needed to raise the temperature of 1 gram of water by 1°C. Nutritionists sometimes use "calorie" to mean kilocalorie

(1,000 calories), which is a source of confusion.

Calvin-Benson cycle Cyclic, light-independent reactions, the "synthesis" part of photosynthesis. For every six carbon dioxide molecules that become affixed to six RuBP molecules, twelve 3-carbon PGA molecules form; ten are used to regenerate RuBP and the other two to form a sugar phosphate. The cycle runs on ATP and NADPH formed in the light-dependent reactions.

CAM plant A plant that conserves water by opening stomata only at night, when carbon dioxide is fixed by way of the C4 pathway.

cambium, plural **cambia** (KAM-bee-um) In vascular plants, one of two types of meristems that are responsible for secondary growth (increase in stem or root diameter). Vascular cambium gives rise to secondary xylem and phloem; cork cambium gives rise to periderm.

camouflage An outcome of form, patterning, color, or behavior that helps an organism blend with its surroundings and escape detection.

cancer A type of malignant tumor, the cells of which show profound abnormalities in the plasma membrane and cytoplasm, abnormal growth and division, and weakened capacity for adhesion within the parent tissue (leading to metastasis), and, unless eradicated, lethality.

capillary [L. *capillus*, hair] A thin-walled blood vessel; component of capillary beds, the diffusion zones for exchanges of gases and materials between blood and interstitial fluid.

carbohydrate [L. *carbo*, charcoal, + *hydro*, water] A simple sugar or a polymer composed of sugar units, and used universally by cells for energy and as structural materials. A monosaccharide, oligosaccharide, or polysaccharide.

carbon cycle Biogeochemical cycle in which carbon moves from reservoirs in the land, atmosphere, and oceans, through organisms, then back to the reservoirs.

carbon dioxide fixation Initial step of the light-independent reactions of photosynthesis. Carbon dioxide becomes affixed to a specific carbon compound (such as RuBP) that undergoes rearrangements leading to regeneration of that carbon compound *and* to a sugar phosphate.

carcinogen (kar-SIN-uh-jen) Ultraviolet radiation and many other agents that can trigger cancer.

cardiac cycle [Gk. *kardia*, heart, + *kyklos*, circle] The sequence of muscle contractions and relaxation constituting one heartbeat.

cardiovascular system Of animals, an organ system composed of blood, one or more hearts, and blood vessels.

carnivore [L. *caro*, *carnis*, flesh, + *vovare*, to devour] An animal that eats other animals; a type of heterotroph.

carotenoid (kare-OTT-en-oyd) A light-sensitive pigment that absorbs violet and blue wavelengths but reflects yellow, orange, and red.

carpel (KAR-pul) One or more closed vessels that serve as the female reproductive parts of

a flower. The chamber within a carpel is the ovary where eggs develop and are fertilized, and seeds mature.

carrier protein Type of transport protein that binds specific substances and changes shape in ways that shunt the substances across a plasma membrane. Some carrier proteins function passively, others require an energy input.

carrying capacity The number of individuals of a given species that can be sustained indefinitely by the available resources in a given environment; births are balanced by deaths in a population at its carrying capacity.

Casparian strip In plant roots, a waxy band that acts as an impermeable barrier between the walls of abutting cells of the endodermis or exodermis.

cDNA Any DNA molecule copied from a mature mRNA transcript by way of reverse transcription.

cell [L. *cella*, small room] The basic *living* unit. A cell has the capacity to maintain itself as an independent unit and to reproduce, given appropriate conditions and resources.

cell cycle A sequence of events by which a cell increases in mass, roughly doubles its number of cytoplasmic components, duplicates its DNA, then undergoes nuclear and cytoplasmic division. The cycle extends from the time the cell forms until its own daughter cells form.

cell junction Of multicelled organisms, a point of contact that physically links two cells or that provides functional links between their cytoplasm.

cell plate Of plant cells undergoing mitotic cell division, a partition that forms at the spindle equator, between the two newly forming daughter cells.

cell theory A theory in biology, the key points of which are that (1) all organisms are composed of one or more cells, (2) the cell is the smallest unit that still retains a capacity for independent life, and (3) all cells arise from preexisting cells.

cell wall A rigid or semirigid supportive wall outside the plasma membrane; a cellular feature of plants, fungi, protistans, and most bacteria.

central nervous system Of vertebrates, the brain and spinal cord.

central vacuole Of living plant cells, a fluid-filled organelle that stores nutrients, ions, and wastes; its enlargement during cell growth has the effect of improving the cell's surface-to-volume ratio.

centriole (SEN-tree-ohl) A small cylinder of triplet microtubules near the nucleus in most animal cells. Centrioles occur in pairs; some give rise to the microtubular core of flagella and cilia, and some may govern the plane of cell division.

centromere (SEN-troh-meer) [Gk. *kentron*, center, + *meros*, a part] A small, constricted region of a chromosome having attachment sites for microtubules that help move the chromosome during nuclear division.

cephalization (sef-ah-lah-ZAY-shun) [Gk. *kephalikos*, head] Of an animal body, having sensory structures and nerve cells concentrated in the head.

cerebellum (ser-ah-BELL-um) [L. diminutive of *cerebrum*, brain] Hindbrain region that coordinates motor activity for refined limb movements, appropriate posture, and spatial orientation.

cerebral cortex Thin surface layer of the cerebral hemispheres. Some regions of the cortex receive sensory input, others integrate information and coordinate appropriate motor responses.

cerebrospinal fluid Clear extracellular fluid that surrounds and cushions the brain and spinal cord. The bloodstream exchanges substances with the fluid, which exchanges substances with neurons.

cerebrum (suh-REE-bruhm) Part of the vertebrate forebrain governing responses to olfactory input and, in mammals, the most complex information-encoding and information-processing center. Divided into two cerebral hemispheres; in humans, the left hemisphere deals generally with spoken language skills and the right, with abstract, nonverbal skills.

channel protein Type of transport protein that serves as a pore through which ions or other water-soluble substances move across the plasma membrane. Some channels remain open; others are gated, and open and close in controlled ways.

chemical bond A union between the electron structures of two or more atoms or ions.

chemical synapse (SIN-aps) [Gk. *synapsis*, union] A junction where a small gap, the synaptic cleft, separates two neurons (or a neuron and a muscle cell or gland cell). The presynaptic neuron releases a transmitter substance into the cleft, and this may have an excitatory or inhibitory effect on the postsynaptic cell.

chemiosmotic theory (kim-ee-OZ-MOT-ik) Theory that an electrochemical gradient across a cell membrane drives ATP synthesis. Metabolic reactions cause hydrogen ions (H^+) to accumulate in some type of membrane-bound compartment. The combined force of the resulting concentration and electric gradients propels hydrogen ions down the gradient, through channel proteins. Enzyme action at those proteins links ADP with inorganic phosphate to form ATP.

chemoreceptor (KEE-moe-ree-sep-tur) Sensory receptor that detects chemical energy (ions or molecules) dissolved in body fluids next to the cell.

chemosynthetic autotroph (KEE-moe-sin-THET-ik) One of a few kinds of bacteria able to synthesize all the organic molecules it requires using carbon dioxide as the carbon source and certain inorganic substances (such as sulfur) as the energy source.

chlorophyll (KLOR-uh-fill) [Gk. *chloros*, green, + *phyllon*, leaf] Photosynthetic pigment molecule that absorbs light of blue and red wavelengths and transmits green. Special chlorophyll pigments of photosystems give up electrons used in photosynthesis.

chloroplast (KLOR-uh-plast) Of plants and certain protistans, a membrane-bound organelle that specializes in photosynthesis.

chordate An animal having a notochord, a dorsal hollow nerve cord, a pharynx, and gill slits in the pharynx wall for at least part of its life cycle.

chorion (CORE-ee-on) Of land vertebrates, a protective membrane surrounding an embryo and other extraembryonic membranes; in placental mammals it becomes part of the placenta.

chromatid The name applied to each of the two parts of a duplicated eukaryotic chromosome for as long as the two parts remain attached at the centromere. Each chromatid consists of a DNA double helix and associated proteins, and it has the same gene sequence as its "sister" chromatid.

chromosome (CROW-moe-some) [Gk. *chroma*, color, + *soma*, body] In eukaryotes, a DNA molecule and many associated proteins. A chromosome that has undergone duplication prior to nuclear division consists of two DNA molecules and associated proteins; the two are called *sister chromatids*. A bacterial chromosome does not have a comparable profusion of proteins associated with the DNA.

cilium (SILL-ee-um), plural **cilia** [L. *cilium*, eyelid] Short, hairlike process extending from the plasma membrane and containing a regular array of microtubules. Cilia are typically more profuse than flagella. Cilia serve as motile structures, help create currents of fluids, or are part of sensory structures.

circadian rhythm (ser-KAYD-ee-un) [L. *circa*, about, + *dies*, day] Of many organisms, a cycle of physiological events that is completed every 24 hours or so, even when environmental conditions remain constant.

circulatory system Of multicelled animals, an organ system consisting of a muscular pump (heart, most often), blood vessels, and blood; the system transports materials to and from cells and often helps stabilize body temperature and pH.

cladistics An approach to biological systematics in which organisms are grouped according to similarities that are derived from a common ancestor.

cladogram Branching diagram that represents patterns of relative relationships between organisms based on discrete morphological, physiological, and behavioral traits that vary among taxa being studied.

cleavage Stage of animal development when mitotic cell divisions convert a zygote to a ball of cells, the *blastula*. Different cytoplasmic components end up in different daughter cells, and this *cytoplasmic localization* helps seal the developmental fate of their descendants.

cleavage furrow Of animal cells undergoing cytokinesis, a depression that forms at the cell surface as contractile microfilaments pull the plasma membrane inward; defines where the cell will be cut in two.

climate Prevailing weather conditions for an ecosystem, including temperature, humidity, wind speed, cloud cover, and rainfall.

climax community Of an ecosystem, a more or less stable array of species that results from the process of succession.

clonal selection theory Theory of immune system function stating that lymphocytes activated by a specific antigen rapidly multiply, giving rise to descendants (clones) that all retain the parent cell's specificity against that antigen.

cloned DNA Multiple, identical copies of DNA fragments contained within plasmids or some other cloning vector.

codominance Condition in which two alleles of a pair are not identical yet the expression of both can be discerned in heterozygotes. Each gives rise to a different phenotype.

codon One of a series of base triplets in an mRNA molecule that code for a series of amino acids that will be strung together during protein synthesis. Different codons specify different amino acids; a few serve as a stop signal and one type as a start signal.

coelom (SEE-lum) [Gk. *koilos*, hollow] Of many animals, a type of body cavity that occurs between the gut and body wall and that has a lining, the *peritoneum*.

coenzyme An organic molecule that serves as a carrier of electrons or atoms in metabolic reactions and that is necessary for proper functioning of many enzymes. NAD^+ is an example.

coevolution The joint evolution of two or more closely interacting species; when one evolves, the change affects selection pressures operating between the two species so the other also evolves.

cofactor A metal ion or coenzyme that either helps catalyze a reaction or serves briefly as an agent that transfers electrons, atoms, or functional groups from one substrate to another.

cohesion Condition in which molecular bonds resist rupturing when under tension.

cohesion theory of water transport Theory that water moves up through vascular plants due to hydrogen bonding among water molecules confined inside the xylem pipelines. The collective cohesive strength of those bonds allows water to be pulled up as columns in response to transpiration (evaporation from leaves).

collenchyma Of vascular plants, a ground tissue that helps strengthen the plant body.

colon (CO-lun) The large intestine.

commensalism [L. *com*, together, + *mensa*, table] Two-species interaction in which one species benefits significantly while the other is neither helped nor harmed to any notable extent.

communication signal Of social animals, an evolved action or cue that transfers information, to the benefit of both the member of the species sending the signal *and* the member receiving it.

community The populations of all species occupying a habitat; also applies to groups of organisms with similar life-styles in a habitat (such as the bird community).

comparative morphology [Gk. *morph*, form] Detailed study of differences and similarities in body form and structural patterns among major taxa.

competition, exploitation Interaction in which both species have equal access to a required resource, but differ in how fast or how efficiently they exploit it.

competition, interference Interaction in which one species may limit another species' access to some resource regardless of whether the resource is abundant or scarce.

competition, interspecific Two-species interaction in which both species can be harmed due to overlapping niches.

competition, intraspecific Interaction among individuals of the same species that are competing for the same resources.

competitive exclusion The theory that populations of two species competing for a limited resource cannot coexist indefinitely in the same habitat; the population better adapted to exploit the resource will enjoy a competitive (hence reproductive) edge and will eventually exclude the other species from the habitat.

complement system A group of about twenty proteins circulating in blood plasma that are activated during both general responses and immune responses to a foreign agent in the body; part of the *inflammatory response*.

concentration gradient A difference in the number of molecules or ions of a substance between one region and another, as in a given volume of fluid. In the absence of other forces, the molecules tend to move down their gradient.

condensation reaction Enzyme-mediated reaction leading to the covalent linkage of small molecules and, often, the formation of water.

cone cell In the vertebrate eye, a type of photoreceptor that responds to intense light and contributes to sharp daytime vision and color perception.

conjugation [L. *conjugatio*, a joining] Of some bacterial species, the transfer of DNA between two different mating strains that have made cell-to-cell contact.

consumer [L. *consumere*, to take completely] A heterotrophic organism that obtains energy and raw materials by feeding on the tissues of other organisms. Herbivores, carnivores, omnivores, and parasites are examples.

continuous variation For many traits, small degrees of phenotypic variation that occur over a more or less continuous range.

contractile vacuole (kun-TRAK-till VAK-you-ohl) [L. *contractus*, to draw together] In some protistans, a membranous chamber that takes up excess water in the cell body, then contracts, expelling the water through a pore to the outside.

control group In a scientific experiment, a group used to evaluate possible side effects of the manipulation of the experimental group. Ideally, the experimental group differs from the control group only with respect to the key factor, or variable, being studied.

convergence, morphological Resemblance of body parts between dissimilar and only distantly related species, an evolutionary outcome of their ancestors having adopted a similar way of life and having used those body parts for similar functions.

cork cambium Of woody plants, a type of lateral meristem that produces a tough, corky replacement for the epidermis on older plant parts.

corpus callosum (CORE-pus ka-LOW-sum) In the human brain, a band of 200 million axons that functionally links the two cerebral hemispheres.

corpus luteum (CORE-pus LOO-tee-um) A glandular structure; it develops from cells of a ruptured ovarian follicle and secretes progesterone and some estrogen, both of which maintain the lining of the uterus (endometrium).

cortex [L. *cortex*, bark] In general, a rindlike layer; the kidney cortex is an example. In vascular plants, ground tissue that makes up most of the primary plant body, supports plant parts, and stores food.

cotyledon A so-called seed leaf that develops as part of a plant embryo; cotyledons provide nourishment for the germinating seedling.

covalent bond (koe-VAY-lunt) [L. *con*, together, + *valere*, to be strong] A sharing of one or more electrons between atoms or groups of atoms. When electrons are shared equally, the bond is *nonpolar*. When electrons are shared unequally, the bond is *polar*—slightly positive at one end and slightly negative at the other.

crossing over During prophase I of meiosis, an event in which nonsister chromatids of a pair of homologous chromosomes break at one or more sites along their length and exchange corresponding segments at the breakage points. As a result, new combinations of alleles replace old ones in a chromosome.

culture The sum total of behavior patterns of a social group, passed between generations by learning and by symbolic behavior, especially language.

cuticle (KEW-tih-kull) A body covering. In plants, a cuticle consisting of waxes and lipid-rich cutin is deposited on the outer surface of epidermal cell walls. Annelids have a thin, flexible cuticle. Arthropods have a thick, protein- and chitin-containing cuticle that is flexible, lightweight, and protective.

cyclic AMP, cyclic adenosine monophosphate (SIK-lik ah-DEN-uh-seen mon-oh-FOSS-fate) A nucleotide present in cytoplasm that serves as a mediator of the cell's response to hormonal signals; a type of second messenger.

cyclic photophosphorylation (SIK-lik foe-toe-FOSS-for-ih-LAY-shun) Photosynthetic pathway in which electrons excited by sunlight energy move from a photosystem to a transport chain, then back to the photosystem. Operation of the transport chain helps produce electrochemical gradients that lead to ATP formation. (Compare *chemiosmotic theory*.)

cytochrome (SIGH-toe-krome) [Gk. *kytos*, hollow vessel, + *chrōma*, color] Iron-containing protein molecule present in the electron transport systems used in photosynthesis and aerobic respiration.

cytokinesis (SIGH-toe-kih-NEE-sis) [Gk. *kinesis*, motion] The actual splitting of a parental cell into two daughter cells; also called cytoplasmic division.

cytokinin (SIGH-tow-KY-nun) Any of the class of plant hormones that stimulate cell division, promote leaf expansion, and retard leaf aging.

cytomembrane system [Gk. *kytos*, hollow vessel] The membranous system in the cytoplasm in which proteins and lipids take on their final form and are distributed. Components of the system include the endoplasmic reticulum, Golgi bodies, lysosomes, and a variety of vesicles.

cytoplasm (SIGH-toe-plaz-um) [Gk. *plassein*, to mold] All cellular parts, particles, and semifluid substances enclosed by the plasma membrane *except* the nucleus (in eukaryotes) or the nucleoid (in prokaryotes).

cytosine (SIGH-toe-seen) A pyrimidine; one of the nitrogen-containing bases in nucleotides.

cytoskeleton Of eukaryotic cells, an internal "skeleton" that structurally supports the cell, organizes its components, and often moves components about. Microtubules, microfilaments, and intermediate filaments are the most common cytoskeletal elements.

decomposer [L. *de-*, down, away, + *companere*, to put together] Generally, any of the heterotrophic bacteria or fungi that obtain energy by chemically breaking down the remains, products, or wastes of other organisms. Their activities help cycle nutrients back to producers.

degradative pathway A metabolic pathway by which molecules are broken down in stepwise reactions that lead to products of lower energy.

deletion Loss of a chromosome segment, nearly always resulting in a genetic disorder.

demographic transition model Model of human population growth in which changes in the growth pattern correspond to different stages of economic development. These are a preindustrial stage, when birth and death rates are both high, a transitional stage, an industrial stage, and a postindustrial stage, when the death rate exceeds the birth rate.

denaturation (deh-NAY-chur-AY-shun) Disruption of bonds holding a protein in its three-dimensional form, such that its polypeptide chain(s) unfolds partially or completely.

dendrite (DEN-drite) [Gk. *dendron*, tree] A short, slender extension from the cell body of a neuron.

denitrification (DEE-nite-rih-fih-KAY-shun) The conversion of nitrate or nitrite, by certain bacteria, to gaseous nitrogen (N_2) and a small amount of nitrous oxide (N_2O).

density-dependent controls Factors such as predation, parasitism, disease, and competition for resources, which limit population growth by reducing the birth rate, increasing the rates of death and dispersal, or all of these.

density-independent controls Factors such as storms or floods that increase a population's death rate more or less independently of its density.

dentition (den-TIH-shun) The type, size, and number of an animal's teeth.

dermis The layer of skin underlying the epidermis, consisting mostly of dense connective tissue.

desertification (dez-urt-ih-fih-KAY-shun) Conversion of grasslands, rain-fed cropland, or irrigated cropland to desertlike conditions, with a drop of agricultural productivity of 10 percent or more.

detrital food web A network of interlinked food chains in which energy flows from plants through decomposers and detritivores.

detritivore (dih-TRY-tih-vore) [L. *detritus*; after *deterere*, to wear down] An earthworm, crab, nematode, or other heterotroph that obtains energy by feeding on partly decomposed particles of organic matter.

deuterostome (DUE-ter-oh-stome) [Gk. *deuteros*, second, + *stoma*, mouth] Any of the bilateral animals, including echinoderms and chordates, in which the first indentation in the early embryo develops into the anus.

diaphragm (DIE-uh-fram) [Gk. *diaphragma*, to partition] Muscular partition between the thoracic and abdominal cavities, the contraction and relaxation of which contribute to breathing. Also, a contraceptive device used temporarily to prevent sperm from entering the uterus during sexual intercourse.

dicot (DIE-kot) [Gk. *di*, two, + *kotylēdōn*, cup-shaped vessel] Short for dicotyledon; class of flowering plants characterized generally by seeds having embryos with two cotyledons (seed leaves); net-veined leaves; and floral parts arranged in fours, fives, or multiples of these.

differentiation Of the cells of multicelled organisms, differences in composition, structure, and function that arise through selective gene expression. All the cells inherit the same genes but become specialized by activating or suppressing some fraction of those genes in different ways.

diffusion Tendency of molecules or ions of the same substance to move from a region of greater concentration to a region where they are less concentrated. See *concentration gradient*.

digestive system Of most animals, an internal tube or cavity in which ingested food is reduced to molecules small enough to be absorbed into the internal environment; often divided into regions specialized for food transport, processing, and storage.

dihybrid cross An experimental cross between two organisms, each of which breeds true (is homozygous) for forms of *two* traits that are distinctly different from those displayed by the other organism. For each trait, the first-generation offspring inherit a pair of nonidentical alleles.

diploid (DIP-loyd) Of sexually reproducing species, having two chromosomes of each type (that is, homologous chromosomes) in somatic cells. Except for sex chromosomes, the two homologues of a pair resemble each other in length, shape, and which genes they carry. Compare *haploid*.

directional selection A shift in allele frequencies in a population in a steady, consistent direction in response to a new environment (or a directional change in the old one), so that forms of traits at one end of a range of phenotypic variation become more common than the intermediate forms.

disaccharide (die-SAK-uh-ride) [Gk. *di*, two, + *sakcharon*, sugar] A type of simple carbohydrate, of the class called oligosaccharides; two monosaccharides covalently bonded.

disruptive selection Selection that favors forms of traits at both ends of a range of phenotypic variation in a population, and operates against intermediate forms.

distal tubule The tubular section of a nephron most distant from the glomerulus; a major site of water and sodium reabsorption.

divergence Accumulation of differences in allele frequencies between reproductively isolated populations of a species.

divergence, morphological Among similar and evolutionarily related species, decreased resemblance in one or more aspects of body patterning or function, usually corresponding to divergences in life-styles.

diversity, organismic Sum total of variations in form, function, and behavior that have accumulated in different lineages. Those variations generally are adaptive to prevailing conditions or were adaptive to conditions that existed in the past.

DNA Deoxyribonucleic acid (dee-OX-ee-RYE-bow-new-CLAY-ik) Usually, two strands of nucleotides twisted together in the shape of a double helix. The nucleotides differ only in their nitrogen-containing bases (adenine, thymine, guanine, cytosine), but *which* ones follow others in a DNA strand represents instructions for assembling proteins, and, ultimately, new organisms.

DNA library A collection of DNA fragments produced by restriction enzymes and incorporated into plasmids.

DNA ligase (LYE-gaze) Enzyme that links together short stretches of nucleotides on a parent DNA strand during replication; also used by recombinant DNA technologists to join base-paired DNA fragments to cut plasmid DNA.

DNA polymerase (poe-LIM-uh-raze) Enzyme that assembles a new strand on a parent DNA strand during replication; also "proofreads" for mismatched base pairs, which are replaced with correct bases.

dominance hierarchy Form of social organization in which some members of the group are subordinate to other members, which in turn are dominated by others.

dominant allele In a diploid cell, an allele whose expression masks the expression of its partner on the homologous chromosome.

dormancy [L. *dormire*, to sleep] Of plants, the temporary, hormone-mediated cessation of growth under conditions that might appear to be quite suitable for growth.

double fertilization Of flowering plants only, the fusion of one sperm nucleus with the egg nucleus (to produce a zygote), *and* fusion of a second sperm nucleus with the nuclei of the endosperm mother cell, which gives rise to nutritive tissue.

duplication Type of chromosome rearrangement in which a gene sequence occurs in excess of its normal amount in a chromosome.

ecology [Gk. *oikos*, home, + *logos*, reason] Study of the interactions of organisms with one another and with their physical and chemical environment.

ecosystem [Gk. *oikos*, home] A whole complex of organisms and their environment, all of which interact through a one-way flow of energy and a cycling of materials.

ecosystem modeling Method of identifying pieces of information about different components of an ecosystem, then combining that information with computer programs and models in order to predict the outcome of a disturbance to the system.

ectoderm [Gk. *ecto*, outside, + *derma*, skin] Of animal embryos, the outermost primary tissue layer, or *germ layer*, that gives rise to the outer portion of the integument and to tissues of the nervous system.

effector A muscle (or gland) that responds to nerve signals by producing movement (or chemical change) that helps adjust the body to changing conditions.

egg A female gamete; of complex animals, a mature ovum.

electron Negatively charged particle occupying one of the orbitals around the nucleus of an atom.

electron transport phosphorylation (FOSS-for-ih-LAY-shun) Final stage of aerobic respiration, in which ATP forms after hydrogen ions and electrons (from the Krebs cycle) are sent through a transport system that gives up the electrons to oxygen. (Compare *chemiosmotic theory* and *electron transport system*).

electron transport system An organized array of membrane-bound enzymes and cofactors that accept and donate electrons in sequence. Operation of such systems leads to the flow of hydrogen ions (H^+) across a cell membrane, and this flow results in ATP formation and other reactions.

element Any substance that cannot be decomposed into substances with different properties.

embryo (EM-bree-oh) [Gk. *en*, in, + probably *bryein*, to swell] Of animals generally, the early stages of development (cleavage, gastrulation, organogenesis, and morphogenesis). In most plants, a young sporophyte, from the first cell divisions after fertilization until germination.

embryo sac In flowering plants, the female gametophyte.

endergonic reaction (en-dur-GONE-ik) Chemical reaction showing a net gain in energy.

endocrine gland Ductless gland that secretes hormones into interstitial fluid, after which they are distributed by way of the bloodstream.

endocrine system System of cells, tissues, and organs that is functionally linked to the nervous system and that helps control body functions with its hormones and other chemical secretions.

endocytosis (EN-doe-sigh-TOE-sis) A process by which part of the plasma membrane encloses substances (or cells, in the case of phagocytes) at or near the cell surface, then pinches off to form a vesicle that transports the substance into the cytoplasm.

endoderm [Gk. *endon*, within, + *derma*, skin] Of animal embryos, the inner primary tissue layer, or *germ layer*, that gives rise to the inner lining of the gut and organs derived from it.

endodermis In roots, a sheetlike wrapping of single cells around the vascular cylinder; functions in controlling the uptake of water and dissolved nutrients. An impermeable barrier (*Casparian strip*) prevents water from passing between the walls of abutting endodermal cells.

endometrium (EN-doh-MEET-ree-um) [Gk. *metrios*, of the womb] Inner lining of the uterus, consisting of connective tissues, glands, and blood vessels.

endoplasmic reticulum or **ER** (EN-doe-PLAZ-mik reh-TIK-yoo-lum) System of membranous channels, tubes, and sacs in the cytoplasm in which many newly formed proteins become modified and the protein and lipid components of most organelles are manufactured. Rough ER has ribosomes on the surface facing the cytoplasm; smooth ER does not.

endoskeleton [Gk. *endon*, within, + *skléros*, hard, stiff] In chordates, the internal framework of bone, cartilage, or both. Together with skeletal muscle, supports and protects other body parts, helps maintain posture, and moves the body.

endosperm (EN-doe-sperm) Nutritive tissue that surrounds and serves as food for a flowering plant embryo and, later, for the germinating seedling.

endospore Of certain bacteria, a resistant body that forms around the DNA and some cytoplasm under unfavorable conditions; it germinates and gives rise to new bacterial cells when conditions become favorable.

energy The capacity to make things happen, to do work.

energy pyramid A pyramid-shaped representation of the trophic structure of an ecosystem, based on the decreasing energy flow at each upward transfer to a different trophic level.

entropy (EN-trow-pee) A measure of the degree of disorder in a system—that is, how much energy in the system has become so dispersed (usually as low-quality heat) that it is no longer available to do work.

enzyme (EN-zime) One of a special class of proteins that greatly speed up (catalyze) reactions involving specific substrates.

epidermis The outermost tissue layer of a multicelled plant or animal.

epistasis (eh-PISS-tih-sis) An absence of an expected phenotype owing to a masking of the expression of one gene pair by another gene pair.

epithelium (EP-ih-THEE-lee-um) Of multicelled animals, one or more layers of adhering cells having one free surface; the opposite surface rests on a basement membrane that intervenes between it and an underlying connective tissue. Epithelia cover external body surfaces and line internal cavities and tubes.

equilibrium, dynamic [Gk. *aequus*, equal, + *libra*, balance] The point at which a chemical reaction runs forward as fast as it runs in reverse, so that there is no net change in the concentrations of products or reactants.

erythrocyte (eh-RITH-row-site) [Gk. *erythros*, red, + *kytos*, vessel] Red blood cell.

esophagus (ee-SOF-uh-gus) Tubular portion of a digestive system that receives swallowed food and leads to the stomach.

essential amino acid Any of eight amino acids that human cells cannot synthesize and must obtain from food.

essential fatty acid Any of the fatty acids that the human body cannot synthesize and must obtain from food.

estrogen (ESS-trow-jun) A sex hormone required in egg formation, preparing the uterine lining for pregnancy, and maintaining secondary sexual traits; also influences growth and development.

estrus (ESS-truss) [Gk. *oistrus*, frenzy] For mammals generally, the cyclic period of a female's sexual receptivity to the male.

estuary (EST-you-ary) A partly enclosed coastal region where seawater mixes with freshwater from rivers or streams and runoff from the land.

ethylene (ETH-il-een) Plant hormone that stimulates fruit ripening and triggers abscission.

eukaryotic cell (yoo-CARRY-oht-ik) [Gk. *eu*, good, + *karyon*, kernel] A cell that has a "true nucleus" and many other membrane-bound organelles; any cell except bacteria.

evaporation [L. *e-*, out, + *vapor*, steam] Changes by which a substance is converted from a liquid state into vapor.

evolution [L. *evolutio*, act of unrolling] Change within a line of descent over time; entails successive changes in allele frequencies in a population as brought about by mutation, natural selection, genetic drift, and gene flow.

excitatory postsynaptic potential or **EPSP** One of two competing signals at an input zone of a neuron; a graded potential that brings the neuron's plasma membrane closer to threshold.

excretion Any of several processes by which excess water, excess or harmful solutes, or waste materials are passed out of the body. Compare *secretion*.

exergonic reaction (EX-ur-GONE-ik) A chemical reaction that shows a net loss in energy.

exocrine gland (EK-suh-krin) [Gk. *es*, out of, + *krinein*, to separate] Glandular structure that secretes products, usually through ducts or tubes, to a free epithelial surface.

exocytosis (EK-so-sigh-TOE-sis) A process by which substances are moved out of cells. A vesicle forms inside the cytoplasm, moves to the plasma membrane, and fuses with it, so that the vesicle's contents are released outside.

exodermis Layer of cells just inside the root epidermis in most flowering plants; functions in controlling the uptake of water and dissolved nutrients.

exon Any of the portions of a newly formed mRNA molecule that are spliced together to form the mature mRNA transcript and that are ultimately translated into protein.

exoskeleton [Gk. *exo*, out, + *skléros*, hard, stiff] An external skeleton, as in arthropods.

exponential growth (EX-po-NEN-shul) Pattern of population growth that occurs when *r* (the net reproduction per individual) holds constant; then, the number of individuals increases in doubling increments (from 2 to 4, then 8, 16, 32, 64, 128, and so on).

extinction, background A steady rate of species turnover that characterizes lineages through most of their histories.

extinction, mass An abrupt increase in the rate at which major taxa disappear, with several taxa being wiped out simultaneously.

extracellular fluid In animals generally, all the fluid not inside cells; includes blood plasma and interstitial fluid, which occupies the spaces between cells and tissues.

extracellular matrix Of animals, a meshwork of fibrous proteins and other components in a ground substance that helps hold many tissues together in certain shapes and that influences cell metabolism by virtue of its composition.

facilitated diffusion The passive transport of specific solutes through the inside of a channel protein or carrier protein that spans the lipid bilayer of a cell membrane; the solutes simply move in the direction that diffusion would take them.

family pedigree A chart of the genetic relationships of the individuals in a family through a number of generations.

fat A lipid with one, two, or three fatty acid tails attached to a glycerol backbone.

fatty acid A compound having a long, unbranched carbon backbone (a hydrocarbon) with a —COOH group at the end.

feedback inhibition A mechanism of enzyme control in which the output of the reaction (such as a particular molecule) works in a way that inhibits further output.

fermentation [L. *fermentum*, yeast] Type of anaerobic pathway of ATP formation that begins with glycolysis and ends with electrons being transferred back to one of the breakdown products or intermediates. Glycolysis yields two ATP; the rest of the pathway serves to regenerate NAD^+.

fertilization [L. *fertilis*, to carry, to bear] Fusion of sperm nucleus with egg nucleus. See also *double fertilization*.

fibrous root system Adventitious roots and their branchings.

filtration In urine formation, the process by which blood pressure forces water and solutes out of glomerular capillaries and into the cupped portion of a nephron wall (Bowman's capsule).

first law of thermodynamics [Gk. *therme*, heat, + *dynamikos*, powerful] Law stating that the total amount of energy in the universe remains constant. Energy cannot be created or destroyed, but can only be converted from one form to another.

flagellum (fluh-jell-um) plural **flagella**, [L. whip] Motile structure of many free-living eukaryotic cells; has a 9 + 2 microtubule array.

flower The often showy reproductive structure that distinguishes angiosperms from other seed plants.

fluid mosaic model Model of membrane structure in which proteins are embedded in a lipid bilayer or attached to one of its surfaces. Lipids impart structure to the membrane as well as impermeability to water-soluble molecules. Packing variations and movements of lipids impart fluidity to the membrane. Proteins carry out most membrane functions, such as transport, enzyme action, and reception of chemical signals.

follicle (FOLL-ih-kul) In a mammalian ovary, a primary oocyte (immature egg) together with the surrounding layer of cells.

food chain A linear sequence of who eats whom in an ecosystem.

food web A network of crossing, interlinked food chains, encompassing primary producers and an array of consumers, detritivores, and decomposers.

forebrain Brain region that includes the cerebrum and cerebral cortex, the olfactory lobes, and the hypothalamus.

fossil Recognizable evidence of an organism that lived in the distant past. Most fossils are skeletons, shells, leaves, seeds, and tracks that were buried in rock layers before they could be decomposed.

fossil fuel Coal, petroleum, or natural gas; formed in sediments by the compression of carbon-containing plant remains over hundreds of millions of years.

founder effect An extreme case of genetic drift in which a few individuals leave a population and establish a new one. Simply by chance, allele frequencies for many traits may differ from those in the original population.

fruit [L. after *frui*, to enjoy] In flowering plants, the ripened ovary of one or more carpels, sometimes with accessory structures incorporated.

functional group An atom or group of atoms covalently bonded to the carbon backbone of an organic compound, contributing to its structure and properties.

Fungi The kingdom of fungi.

fungus A heterotroph that secretes enzymes able to break down an external food source into molecules small enough to be absorbed by cells (extracellular digestion and absorption). Saprobic types feed on nonliving organic matter; parasitic types feed on living organisms. Fungi as a group are major decomposers.

gall bladder Organ of the digestive system that stores bile secreted from the liver.

gamete (GAM-eet) [Gk. *gametēs*, husband, and *gametē*, wife] Haploid cell (sperm or egg) that functions in sexual reproduction.

gametophyte (gam-EET-oh-fite) [Gk. *phyton*, plant] The haploid, multicelled, gamete-producing phase in the life cycle of most plants.

ganglion (GANG-lee-un), plural **ganglia** [Gk. *ganglion*, a swelling] A clustering of cell bodies of neurons into a distinct structure in regions other than the brain or spinal cord.

gastrulation (gas-tru-LAY-shun) Stage of embryonic development in which cells become arranged into two or three primary tissue layers (germ layers); in humans, the layers are an inner endoderm, an intermediate mesoderm, and a surface ectoderm.

gene [short for German *pangan*, after Gk. *pan*, all + *genes*, to be born] Any of the units of instruction for heritable traits. Each gene is a linear sequence of nucleotides that calls for the assembly of a sequence of specific amino acids into a polypeptide chain.

gene flow Microevolutionary process whereby allele frequencies in a population change due to immigration, emigration, or both.

gene frequency More precisely, allele frequency: the relative abundances of different alleles carried by the individuals of a population.

gene locus Particular location on a chromosome for a given gene.

gene mutation [L. *mutatus*, a change] Change in DNA due to the deletion, addition, or substitution of one to several bases in the nucleotide sequence.

gene pair In diploid cells, the two alleles at a given gene locus on a pair of homologous chromosomes.

gene pool Sum total of all genotypes in a population. More accurately, allele pool.

gene therapy Inserting one or more normal genes into existing cells of an organism as a way to correct some genetic defect.

genetic code [After L. *genesis*, to be born] The correspondence between nucleotide triplets in DNA (then in mRNA) and a specific sequence of amino acids in the resulting polypeptide chains; the basic language of protein synthesis.

genetic drift Microevolutionary process whereby allele frequencies in a population change randomly over time, as a result of chance events.

genetic engineering Altering the information content of DNA through use of recombinant DNA technology.

genetic equilibrium Hypothetical state in a population in which allele frequencies for a trait remain stable through the generations; a reference point for measuring rates of evolutionary change.

genetic recombination Presence of a new combination of alleles in a DNA molecule compared to the parental genotype; the result of processes such as crossing over at meiosis, chromosome rearrangements, gene mutation, and recombinant DNA technology.

genome All the DNA in a haploid number of chromosomes of a species.

genotype (JEEN-oh-type) Genetic constitution of an individual. Can mean a single gene pair or the sum total of the individual's genes. Compare *phenotype*.

genus, plural **genera** (JEEN-us, JEN-er-ah) [L. *genus*, race, origin] A taxon into which all

species exhibiting certain phenotypic similarities and evolutionary relationship are grouped.

germ cell Animal cell that may give rise to gametes. Compare *somatic cell*.

germ layer Of animal embryos, one of two or three primary tissue layers that form during gastrulation and that gives rise to certain tissues of the adult body. Compare *ectoderm*, *mesoderm*, and *endoderm*.

germination (jur-min-AY-shun) Of plants, the time at which an embryo sporophyte breaks through its seed coat and resumes growth.

gibberellin (JIB-er-ELL-un) Any of a class of plant hormones that promote stem elongation.

gill A respiratory organ, typically with a moist, thin, vascularized layer of epidermis that functions in gas exchange.

glomerulus (glow-MARE-you-luss) [L. *glomus*, ball] Region where a nephron wall balloons around a cluster of capillaries and where water and solutes are filtered from blood.

glucagon (GLUE-kuh-gone) Type of hormone, secreted by alpha cells of the pancreas, that stimulates conversion of glycogen and amino acids to glucose.

glyceride (GLISS-er-eyed) A molecule having one, two, or three fatty acid tails attached to a backbone of glycerol. Glycerides—fats and oils—are the body's most abundant lipids and its richest source of energy.

glycerol (GLISS-er-ol) [Gk. *glykys*, sweet, + L. *oleum*, oil] A three-carbon molecule with three hydroxyl groups attached; combines with fatty acids to form fat or oil.

glycogen (GLY-kuh-jen) In animals, a storage polysaccharide that is a main food reserve; can be readily broken down into glucose subunits.

glycolysis (gly-CALL-ih-sis) [Gk. *glykys*, sweet, + *lysis*, loosening or breaking apart] Initial stage of both aerobic and anaerobic pathways by which glucose (or some other organic compound) is partially broken down to pyruvate, with a net yield of two ATP.

Golgi body (GOHL-gee) Organelle in which many newly forming proteins and lipids undergo final processing, then are sorted and packaged in vesicles.

gonad (GO-nad) Primary reproductive organ in which gametes are produced.

graded potential Of neurons, a local signal that slightly changes the voltage difference across a small patch of the plasma membrane. Such signals vary in magnitude, depending on the stimulus. With prolonged or intense stimulation, graded potentials may spread to a trigger zone of the membrane and initiate an *action potential*.

granum, plural **grana** Within many chloroplasts, any of the stacks of flattened membranous compartments that incorporate chlorophyll and other light-trapping pigments and reaction sites for ATP formation.

gravitropism (GRAV-i-TROPE-izm) [L. *gravis*, heavy, + Gk. *trepein*, to turn] The tendency of a plant to grow directionally in response to the earth's gravitational force.

gray matter Of vertebrates, the dendrites, neuron cell bodies, and neuroglial cells of the spinal cord and cerebral cortex.

grazing food web A network of interlinked food chains in which energy flows from plants to herbivores, then through some array of carnivores.

greenhouse effect Warming of the lower atmosphere due to the buildup of so-called greenhouse gases—carbon dioxide, methane, nitrous oxide, ozone, water vapor, and chlorofluorocarbons.

green revolution In developing countries, the use of improved crop varieties, modern agricultural practices (including massive inputs of fertilizers and pesticides), and equipment to increase crop yields.

ground meristem (MARE-ih-stem) [Gk. *meristos*, divisible] Of vascular plants, a primary meristem that produces ground tissue, hence the bulk of the plant body.

guard cell Either of two adjacent cells having roles in the movement of gases and water vapor across leaf or stem epidermis. An opening (stoma) forms when both cells swell with water and move apart; it closes when they lose water and collapse against each other.

gut A body region where food is digested and absorbed; of complete digestive systems, the portions from the stomach onward.

gymnosperm (JIM-noe-sperm) [Gk. *gymnos*, naked, + *sperma*, seed] A plant that bears seeds at exposed surfaces of reproductive structures, such as cone scales. Pine trees are examples.

habitat [L. *habitare*, to live in] The type of place where an organism normally lives, characterized by physical features, chemical features, and the presence of certain other species.

hair cell Type of mechanoreceptor that may give rise to action potentials when bent or tilted.

haploid (HAP-loyd) Of sexually reproducing species, having only one of each pair of homologous chromosomes that were present in the nucleus of a parent cell; an outcome of meiosis. Compare *diploid*.

heart Muscular pump that keeps blood circulating through the animal body.

hemoglobin (HEEM-oh-glow-bin) [Gk. *haima*, blood, + L. *globus*, ball] Iron-containing, oxygen-transporting protein that gives red blood cells their color.

hemostasis (HEE-mow-STAY-iss) [Gk. *haima*, blood, + *stasis*, standing] Stopping of blood loss from a damaged blood vessel through coagulation, blood vessel spasm, platelet plug formation, and other mechanisms.

herbivore [L. *herba*, grass, + *vovare*, to devour] Plant-eating animal.

heterocyst (HET-er-oh-sist) Of some filamentous cyanobacteria, a type of thick-walled, nitrogen-fixing cell that forms when nitrogen is scarce.

heterotroph (HET-er-oh-trofe) [Gk. *heteros*, other, + *trophos*, feeder] Organism that cannot synthesize its own organic compounds and must obtain them by feeding on plants or other autotrophs. Animals, fungi, many

protistans, and most bacteria are heterotrophs.

heterozygous condition (HET-er-oh-ZYE-guss) [Gk. *zygoun*, join together] For a given trait, having nonidentical alleles at a particular locus on homologous chromosomes.

hindbrain Of the vertebrate brain, the medulla oblongata, cerebellum, and pons; includes reflex centers for respiration, blood circulation, and other basic functions; also coordinates motor responses and many complex reflexes.

histone Of eukaryotic chromosomes, any of a class of structural proteins intimately associated with the DNA.

homeostasis (HOE-me-oh-STAY-sis) [Gk. *homo*, same, + *stasis*, standing] Of multicelled organisms, a physiological state in which the physical and chemical conditions of the internal environment are stabilized within tolerable ranges.

hominid [L. *homo*, man] All species on the evolutionary branch leading to modern humans. *Homo sapiens* is the only living representative.

hominoid Apes, humans, and their recent ancestors.

homologous chromosome (huh-MOLL-uh-gus) [Gk. *homologia*, correspondence] One of a pair of chromosomes that resemble each other in length, shape, and the genes they carry, and that pair with each other at meiosis. X and Y chromosomes differ in these respects but still function as homologues.

homologous structures Similarity in some aspect of body form or patterning between different species; evolutionary outcome of descent from a common ancestor.

homozygous condition (HOE-moe-ZYE-guss) For a given trait, having two identical alleles at a particular locus on homologous chromosomes.

homozygous dominant An individual having two dominant alleles at a given gene locus (on a pair of homologous chromosomes).

homozygous recessive An individual having two recessive alleles at a given gene locus (on a pair of homologous chromosomes).

hormone [Gk. *hormon*, to stir up, set in motion] Any of the signaling molecules secreted from endocrine glands, endocrine cells, and some neurons and that travel the bloodstream to nonadjacent target cells.

hydrogen bond Type of chemical bond in which an atom of a molecule interacts weakly with a hydrogen atom already taking part in a polar covalent bond.

hydrogen ion A hydrogen atom that has lost its electron and so bears a positive charge (H^+); a "naked" proton.

hydrologic cycle A biogeochemical cycle in which hydrogen and oxygen move, in the form of water molecules, through the atmosphere, on or through the uppermost layers of land masses, to the oceans, and back again; driven by solar energy.

hydrolysis (high-DRAWL-ih-sis) [L. *hydro*, water, + Gk. *lysis*, loosening or breaking apart] Enzyme-mediated reaction that breaks covalent bonds in a molecule, which splits

into two or more parts; at the same time, H^+ and OH^- (derived from a water molecule) become attached to the exposed bonding sites.

hydrophilic substance [Gk. *philos*, loving] A polar substance that is attracted to water molecules and so dissolves easily in water.

hydrophobic substance [Gk. *phobos*, dreading] A nonpolar substance that is repelled by water molecules and so does not readily dissolve in water. Oil is an example.

hydrosphere All liquid or frozen water on or near the earth's surface.

hypha, plural **hyphae** (HIGH-fuh) [Gk. *hyphe*, web] Of fungi, a generally tube-shaped filament with chitin-reinforced walls and, often, reinforcing cross-walls; component of the mycelium.

hypodermis A subcutaneous layer having stored fat that helps insulate the body; although not part of skin, it anchors skin while allowing it some freedom of movement.

hypothalamus [Gk. *hypo*, under, + *thalamos*, inner chamber or possibly *tholos*, rotunda] Of vertebrate forebrains, a brain center that monitors visceral activities (such as salt-water balance, temperature control, and reproduction) and that influences related forms of behavior (as in hunger, thirst, and sex).

hypothesis A plausible answer, or "educated guess," concerning a question or problem. In science, predictions drawn from hypotheses are tested by making observations, developing models, and performing repeatable experiments.

immune system White blood cells (macrophages, T lymphocytes, and B lymphocytes) and their interactions and products; the system shows specificity in response to a particular foreign agent, and memory—the ability to mount a more rapid attack if that specific agent returns.

immunization Deliberate introduction into the body of an antigen that can provoke an immune response and the production of memory lymphocytes.

imprinting Category of learning in which an animal that has been exposed to specific key stimuli early in its behavioral development forms an association with the object.

incomplete dominance Of heterozygotes, a condition in which one allele of a pair only partially dominates expression of its partner.

independent assortment Mendelian principle that each gene pair tends to assort into gametes independently of other gene pairs located on nonhomologous chromosomes.

indirect selection A theory in evolutionary biology that self-sacrificing individuals can indirectly pass on their genes by helping relatives survive and reproduce.

induced-fit model Model of enzyme action whereby a bound substrate induces changes in the shape of the enzyme's active site, resulting in a more precise molecular fit between the enzyme and its substrate.

inflammatory response A series of events involving many cells, complement proteins, and other substances that destroy foreign agents in the body and that restore tissues and internal operating conditions to normal. Occurs during both nonspecific defense responses and immune responses.

inheritance The transmission, from parents to offspring, of structural and functional patterns that have a genetic basis and are characteristic of each species.

inhibiting hormone A signaling molecule produced and secreted by the hypothalamus that controls secretions by the anterior lobe of the pituitary gland.

inhibitor A substance that can bind with an enzyme and interfere with its functioning.

inhibitory postsynaptic potential, or **IPSP** Of neurons, one of two competing types of graded potentials at an input zone; tends to drive the resting membrane potential away from threshold.

instinctive behavior The capacity of an animal to complete fairly complex, stereotyped responses to particular environmental cues without having had prior experience with those cues.

insulin Hormone that lowers the glucose level in blood; it is secreted from beta cells of the pancreas and stimulates cells to take up glucose; also promotes protein and fat synthesis and inhibits protein conversion to glucose.

integration, neural [L. *integrare*, to coordinate] Moment-by-moment summation of all excitatory and inhibitory synapses acting on a neuron; occurs at each level of synapsing in a nervous system.

integument Of animals, a protective body covering such as skin. Of flowering plants, a protective layer around the developing ovule; when the ovule becomes a seed, its integument(s) harden and thicken into a seed coat.

integumentary exchange (in-teg-you-MEN-tuh-ree) Of some animals, a mode of respiration in which oxygen and carbon dioxide diffuse across a thin, vascularized layer of moist epidermis at the body surface.

interleukin A type of communication signal that sensitizes specific cells of the immune system to the presence of a foreign agent and that stimulates them into action.

interneuron Any of the neurons in the vertebrate brain and spinal cord that integrate information arriving from sensory neurons and that influence other neurons in turn.

internode In vascular plants, the stem region between two successive nodes.

interphase Of cell cycles, the time interval between nuclear divisions in which a cell increases its mass, roughly doubles the number of its structures and organelles, and replicates its DNA. The interval is different for different species.

interstitial fluid (IN-ter-STISH-ul) [L. *interstitus*, to stand in the middle of something] In multicelled animals, that portion of the extracellular fluid occupying spaces between cells and tissues.

intertidal zone Generally, the area on a rocky or sandy shoreline that is above the low water mark and below the high water mark; organisms inhabiting it are alternately submerged, then exposed, by tides.

intervertebral disk One of a number of disk-shaped structures containing cartilage that serve as shock absorbers and flex points between bony segments of the vertebral column.

intron A noncoding portion of a newly formed mRNA molecule.

inversion Type of chromosome rearrangement in which a segment that has become separated from the chromosome is reinserted at the same place but in reverse, so the position and sequence of genes are altered.

invertebrate Animal without a backbone.

ion, negatively charged (EYE-on) An atom or a compound that has gained one or more electrons, hence has acquired an overall negative charge.

ion, positively charged An atom or a compound that has lost one or more electrons, hence has acquired an overall positive charge.

ionic bond An association between ions of opposite charge.

isotonic condition Equality in the relative concentrations of solutes in two fluids; for two fluids separated by a cell membrane, there is no net osmotic (water) movement across the membrane.

isotope (EYE-so-tope) An atom that contains the same number of protons as other atoms of the same element, but that has a different number of neutrons.

karyotype (CARRY-oh-type) Cut-and-paste micrograph of a cell's metaphase chromosomes, arranged according to length, shape, banding patterns, and other features.

keratin A tough, water-insoluble protein manufactured by most epidermal cells.

kidney In vertebrates, one of a pair of organs that filter mineral ions, organic wastes, and other substances from the blood, and help regulate the volume and solute concentrations of extracellular fluid.

kinetochore At the centromere of a chromosome, a specialized group of proteins and DNA that serves as an attachment point for several spindle microtubules during mitosis or meiosis. Each chromatid of a duplicated chromosome has its own kinetochore.

Krebs cycle Stage of aerobic respiration in which pyruvate is completely broken down to carbon dioxide and water. Resulting hydrogen ions and electrons are shunted to the next stage, which yields most of the ATP produced in aerobic respiration.

lactate fermentation Anaerobic pathway of ATP formation in which pyruvate from glycolysis is converted to the three-carbon compound lactate, with a net yield of two ATP.

large intestine The colon; a region of the gut that receives unabsorbed food residues from the small intestine and concentrates and stores feces until they are expelled from the body.

larva, plural **larvae** Of animals, a sexually immature, free-living stage between the embryo and the adult.

larynx (LARE-inks) A tubular airway that leads to the lungs. In humans, contains vocal cords, where sound waves used in speech are produced.

lateral meristem Of vascular plants, a type of meristem responsible for secondary growth; either vascular cambium or cork cambium.

leaf For most vascular plants, a structure having chlorophyll-containing tissue that is the major region of photosynthesis.

learning The adaptive modification of behavior in response to neural processing of information that has been gained from specific experiences.

lichen (LY-kun) A symbiotic association between a fungus and a captive photosynthetic partner such as a green alga.

life cycle A recurring, genetically programmed frame of events in which individuals grow, develop, maintain themselves, and reproduce.

light-dependent reactions First stage of photosynthesis, in which sunlight energy is absorbed and converted to the chemical energy of ATP alone (by the cyclic pathway) or to ATP and NADPH (by the noncyclic pathway).

light-independent reactions Second stage of photosynthesis, in which sugar phosphates are assembled with the help of the ATP and NADPH produced during the first stage.

lignification Of mature land plants, a process by which lignin is deposited in secondary cell walls. The deposits impart strength and rigidity, stabilize and protect other wall components, and form a waterproof barrier around the cellulose. Probably a key factor in the evolution of vascular plants.

lignin An inert substance containing different sugar alcohols in amounts that vary among plant species.

limbic system Brain regions that, along with the cerebral cortex, collectively govern emotions.

lineage (LIN-ee-age) A line of descent.

linkage Tendency of genes located on the same chromosome to stay together during meiosis and to end up together in the same gamete.

lipid A compound of mostly carbon and hydrogen that generally does not dissolve in water, but that does dissolve in nonpolar substances. Some lipids serve as energy reserves; others are components of membranes and other cell structures.

lipid bilayer Of cell membranes, two layers of mostly phospholipid molecules, with all the fatty acid tails sandwiched between the hydrophilic heads as a result of hydrophobic interactions.

liver Glandular organ with roles in storing and interconverting carbohydrates, lipids, and proteins absorbed from the gut; maintaining blood; disposing of nitrogen-containing wastes; and other tasks.

local signaling molecules Secretions from cells in many different tissues that alter chemical conditions in the immediate vicinity where they are secreted, then are swiftly degraded.

locus (LOW-cuss) The specific location of a particular gene on a chromosome.

logistic growth (low-JIS-tik) Pattern of population growth in which the growth rate of a low-density population goes through a rapid growth phase and then levels off.

loop of Henle The hairpin-shaped, tubular region of a nephron that functions in reabsorption of water and solutes.

lung An internal respiratory surface in the shape of a cavity or sac.

lymph (LIMF) [L. *lympha*, water] Tissue fluid that has moved into the vessels of the lymphatic system.

lymphatic system System of lymphoid organs (which function in defense responses) and lymph vessels (which return excess tissue fluid to the bloodstream and also transport absorbed fats to it).

lymphocyte Any of various white blood cells that take part in vertebrate immune responses.

lymphoid organs The lymph nodes, spleen, thymus, tonsils, adenoids, and patches of tissue in the small intestine and appendix.

lysis [Gk. *lysis*, a loosening] Gross induced leakage across a plasma membrane that leads to cell death. Examples are lysis of a virus-infected cell or of a pathogen under chemical attack by complement proteins.

lysosome (LYE-so-sohm) In eukaryotic cells, an organelle containing digestive enzymes that can break down polysaccharides, proteins, nucleic acids, and some lipids.

lytic pathway A mode of viral replication; the virus quickly takes over a host cell's metabolic machinery, the viral genetic material is replicated, and new virus particles are produced and then released as the cell undergoes lysis.

macroevolution The large-scale patterns, trends, and rates of change among major taxa.

macrophage A phagocytic white blood cell that develops from circulating monocytes and defends tissues; takes part in inflammatory responses and in both cell-mediated and antibody-mediated immune responses.

mass extinction An abrupt rise in extinction rates above the background level; a catastrophic, global event in which major taxa are wiped out simultaneously.

mass number The total number of protons and neutrons in an atom's nucleus. The relative masses of atoms are also called atomic weights.

mechanoreceptor Sensory cell or cell part that detects mechanical energy associated with changes in pressure, position, or acceleration.

medulla oblongata Part of the vertebrate brainstem with reflex centers for respiration, blood circulation, and other vital functions.

medusa (meh-DOO-sah) [Gk. *Medusa*, one of three sisters in Greek mythology having snake-entwined hair; this image probably evoked by the tentacles and oral arms extending from the medusa] Free-swimming, bell-shaped stage in cnidarian life cycles.

megaspore Of seed-bearing plants, a type of spore that develops into a female gametophyte.

meiosis (my-OH-sis) [Gk. *meioun*, to diminish] Two-stage nuclear division process in which the parental number of chromosomes in each daughter nucleus becomes haploid (with one of each type of chromosome that was present in the parent nucleus). Basis of gamete formation, also of spore formation in plants. Compare *mitosis*.

membrane excitability A membrane property of any cell that can produce action potentials in response to appropriate stimulation.

memory The storage and retrieval of information about previous experiences; underlies the capacity for learning.

memory lymphocyte Any of the various B or T lymphocytes of the immune system that are formed in response to invasion by a foreign agent and that circulate for some period, available to mount a rapid attack if the same type of invader reappears.

menopause (MEN-uh-pozz) [L. *mensis*, month, + *pausa*, stop] Physiological changes that mark the end of a human female's potential to bear children.

menstrual cycle The cyclic release of oocytes and priming of the endometrium (lining of the uterus) to receive a fertilized egg; the complete cycle averages about 28 days in female humans.

menstruation Periodic sloughing of the blood-enriched lining of the uterus when pregnancy does not occur.

mesoderm (MEH-so-derm) [Gk. *mesos*, middle, + *derm*, skin] In most animal embryos, a primary tissue layer (germ layer) between ectoderm and endoderm; gives rise to muscle, organs of circulation, reproduction, and excretion, most of the internal skeleton (when present), and connective tissue layers of the gut and body covering.

messenger RNA A linear sequence of ribonucleotides transcribed from DNA and translated into a polypeptide chain; the only type of RNA that carries protein-building instructions.

metabolic pathway A linear or cyclic series of breakdown or synthesis reactions in cells, the steps of which are catalyzed by the action of specific enzymes.

metabolism (meh-TAB-oh-lizm) [Gk. *meta*, change] All those chemical reactions by which cells acquire and use energy as they synthesize, accumulate, break apart, and eliminate substances in ways that contribute to growth, maintenance, and reproduction.

metamorphosis (met-uh-MOR-foe-sis) [Gk. *meta*, change, + *morphe*, form] Transformation of a larva into an adult form by way of major tissue reorganization.

metaphase Stage of mitosis when spindle has fully formed, sister chromatids of each chromosome become attached to opposite spindle poles, and all chromosomes lie at the spindle equator.

metaphase I Stage of meiosis when all pairs of homologous chromosomes are aligned with their partners at the spindle equator.

metaphase II Stage of meiosis when the chromosomes, already separated from their homologous partner but still in the duplicated state, are aligned at the spindle equator.

metazoan Any multicelled animal.

MHC marker Any of the surface receptors that mark an individual's cells as self; except for identical twins, the markers are unique to each individual.

microevolution Changes in allele frequencies brought about by mutation, genetic drift, gene flow, and natural selection.

microfilament [Gk. *mikros*, small, + L. *filum*, thread] Component of the cytoskeleton; involved in cell shape, motion, and growth.

microspore Of seed-bearing plants, a type of spore that develops into an immature male gametophyte called a pollen grain.

microtubular spindle Of eukaryotic cells, a bipolar structure composed of organized arrays of microtubules; it forms during mitosis or meiosis and moves the chromosomes.

microtubule Hollow cylinder of mainly tubulin subunits; a cytoskeletal element with roles in cell shape, motion, and growth and in the structure of cilia and flagella.

microtubule organizing center, or **MTOC** Small mass of proteins and other substances in the cytoplasm; the number, type, and location of MTOCs determine the organization and orientation of microtubules.

microvillus (MY-crow-VILL-us) [L. *villus*, shaggy hair] A slender, cylindrical extension of the animal cell surface that functions in absorption or secretion.

midbrain Of vertebrates, a brain region that evolved as a coordinating center for reflex responses to visual and auditory input; together with the pons and medulla oblongata, part of the brainstem, which includes the reticular formation.

migration Of certain animals, a cyclic movement between two distant regions at times of year corresponding to seasonal change.

mimicry (MIM-ik-ree) Situation in which one species (the mimic) bears deceptive resemblance in color, form, and/or behavior to another species (the model) that enjoys some survival advantage.

mineral Any of a number of small inorganic substances required for the normal functioning of body cells.

mitochondrion, plural **mitochondria** (MY-toe-KON-dree-on) Organelle in which the second and third stages of aerobic respiration occur; those stages are the Krebs cycle and preparatory conversions in it, as well as electron transport phosphorylation.

mitosis (my-TOE-sis) [Gk. *mitos*, thread] Type of nuclear division that maintains the parental number of chromosomes for daughter cells. It is the basis of bodily growth and, in some cases, of asexual reproduction of eukaryotes.

molecule A unit of two or more atoms of the same or different elements, bonded together.

molting The shedding of hair, feathers, horns, epidermis, or exoskeleton in a process of growth or periodic renewal.

Monera The kingdom of single-celled prokaryotes; bacteria.

monocot (MON-oh-kot) Short for monocotyledon; a flowering plant in which seeds have only one cotyledon, whose floral parts generally occur in threes (or multiples of threes), and whose leaves typically are parallel-veined. Compare *dicot*.

monohybrid cross [Gk. *monos*, alone] An experimental cross between two parent organisms that breed true for distinctly different forms of a single trait; heterozygous offspring result.

monomer A simple sugar or some other small organic compound that can serve as one of the individual units of *polymers*.

monophyletic group A set of independently evolving lineages that share a common evolutionary heritage.

monosaccharide (MON-oh-SAK-ah-ride) [Gk. *monos*, along, single, + *sakharon*, sugar] A sugar monomer; the simplest carbohydrate. Glucose is an example.

monosomy Abnormal condition in which one chromosome of diploid cells has no homologue.

morphogenesis (MORE-foe-JEN-ih-sis) [Gk. *morphe*, form, + *genesis*, origin] Processes by which differentiated cells in an embryo become organized into tissues and organs, under genetic controls and environmental influences.

motor neuron Nerve cell that relays information away from the brain and spinal cord to the body's effectors (muscles, glands, or both), which carry out responses.

mouth An oral cavity; in human digestion, the site where polysaccharide breakdown begins.

multicelled organism An organism that has differentiated cells arranged into tissues, and often into organs and organ systems.

multiple allele system Three or more alternative molecular forms of a gene (alleles), any of which may occur at the gene's locus on a chromosome.

muscle tissue Tissue having cells able to contract in response to stimulation, then passively lengthen and so return to their resting state.

mutagen (MEW-tuh-jen) An environmental agent that can permanently modify the structure of a DNA molecule. Certain viruses and ultraviolet radiation are examples.

mutation [L. *mutatus*, a change, + *-ion*, result of a process or an act] A heritable change in the molecular structure of DNA.

mutualism [L. *mutuus*, reciprocal] A type of community interaction in which members of two species each receive benefits from the association. When the interaction is intimate and involves a permanent dependency, it is called *symbiosis*.

mycelium (my-SEE-lee-um), plural **mycelia** [Gk. *mykes*, fungus, mushroom, + *helos*, callus] A mesh of tiny, branching filaments (hyphae) that is the food-absorbing part of a multicelled fungus.

mycorrhiza (MY-coe-RISE-uh) "Fungus-root," a symbiotic arrangement between fungal hyphae and young roots of many vascular plants, in which the fungus obtains carbohydrates from the plant and in turn releases dissolved mineral ions to the plant roots.

myelin sheath Of many sensory and motor neurons, an axonal sheath that affects how fast action potentials travel; formed from the plasma membranes of Schwann cells that are wrapped repeatedly around the axon and are separated from each other by a small node.

myofibril (MY-oh-FY-brill) One of many threadlike structures inside a muscle cell; composed of actin and myosin molecules arranged as sarcomeres, the fundamental units of contraction.

myosin (MY-uh-sin) One of two types of protein filaments that make up sarcomeres, the contractile units of a muscle cell; the other is actin.

NAD$^+$ Nicotinamide adenine dinucleotide, a large organic molecule that serves as a cofactor in enzyme reactions. When carrying electrons and protons (H$^+$) from one reaction site to another, it is abbreviated NADH.

NADP$^+$ Nicotinamide adenine dinucleotide phosphate. When carrying electrons and protons (H$^+$) from one reaction site to another, it is abbreviated NADPH.

natural selection A microevolutionary process; a measure of the differences in survival and reproduction that have occurred among individuals of a population that differ from one another in one or more traits.

negative feedback mechanism A homeostatic feedback mechanism in which an activity changes some condition in the internal environment and so triggers a response that reverses the changed condition.

nematocyst (NEM-ad-uh-sist) [Gk. *nema*, thread, + *kystis*, pouch] Of cnidarians only, a stinging capsule that assists in prey capture and possibly protection.

nephridium, plural, **nephridia** (neh-FRID-ee-um) Of earthworms and some other invertebrates, a system of regulating water and solute levels.

nephron (NEFF-ron) [Gk. *nephros*, kidney] Of the vertebrate kidney, a slender tubule in which water and solutes filtered from blood are selectively reabsorbed and in which urine forms.

nerve Cordlike communication line of the peripheral nervous system, composed of axons of sensory neurons, motor neurons, or both packed within connective tissue. In the brain and spinal cord, similar cordlike bundles are called nerve pathways or tracts.

nerve cord Of many animals, a cordlike communication line consisting of axons of neurons.

nerve impulse See *action potential*.

nerve net Cnidarian nervous system.

nervous system System of neurons oriented relative to one another in precise message-conducting and information-processing pathways.

neuroendocrine control center Those portions of the hypothalamus and pituitary gland that interact to control many body functions.

neuroglial cell (NUR-oh-GLEE-uhl) Of vertebrates, one of the cells that provide structural and metabolic support for neurons and that collectively represent about half the volume of the nervous system.

neuromodulator Type of signaling molecule that influences the effects of transmitter substances by enhancing or reducing membrane responses in target neurons.

neuromuscular junction Chemical synapses between the axon terminals of a motor neuron and a muscle cell.

neuron A nerve cell; the basic unit of communication in nervous systems. Neurons collectively sense environmental change, integrate sensory inputs, then activate muscles or glands that initiate or carry out responses.

neutral mutation Mutation in which the altered allele has no more measurable effect on survival and reproduction than do other alleles for the trait.

neutron Subatomic particle of about the same size and mass as a proton but having no electric charge.

niche (NITCH) [L. *nidas*, nest] Of a species, the full range of physical and biological conditions under which its members can live and reproduce.

nitrification (nye-trih-fih-KAY-shun) Process by which certain soil bacteria strip electrons from ammonia or ammonium, releasing nitrite (NO$_2$) that other soil bacteria break down, releasing nitrate (NO$_3$).

nitrogen cycle Cycling of nitrogen atoms between living organisms and the environment, through nitrogen fixation, assimilation and biosynthesis of nitrogen-containing compounds, decomposition, ammonification, nitrification, and denitrification.

nitrogen fixation Process by which a few kinds of bacteria convert gaseous nitrogen (N$_2$) to ammonia, which dissolves rapidly in water to produce ammonium. Other organisms as well as the bacteria use the fixed nitrogen.

nociceptor A sensory receptor, such as a free nerve ending, that detects any stimulus causing tissue damage.

node In vascular plants, a point on a stem where one or more leaves are attached.

noncyclic photophosphorylation (non-SIK-lik foe-toe-FOSS-for-ih-LAY-shun) [L. *non*, not, + Gk. *kylos*, circle] Photosynthetic pathway in which new electrons derived from water molecules flow through two photosystems and two transport chains, and ATP and NADPH form.

nondisjunction Failure of one or more chromosomes to separate during meiosis.

notochord (KNOW-toe-kord) Of chordates, a rod of stiffened tissue (not cartilage or bone) that serves as a supporting structure for the body.

nuclear envelope A double membrane (two lipid bilayers and associated proteins) that is the outermost portion of a cell nucleus.

nucleic acid (new-CLAY-ik) A single- or double-stranded chain of nucleotide units; DNA and RNA are examples.

nucleoid Of bacteria, a region in which DNA is physically organized apart from other cytoplasmic components.

nucleolus (new-KLEE-oh-lus) [L. *nucleolus*, a little kernel] Within the nucleus of a nondividing cell, a mass of proteins, RNA, and other material used in ribosome synthesis.

nucleosome (NEW-klee-oh-sohm) Of eukaryotic chromosomes, an organizational unit consisting of a segment of DNA looped twice around a core of histone molecules.

nucleotide (NEW-klee-oh-tide) A small organic compound having a five-carbon sugar (deoxyribose), nitrogen-containing base, and phosphate group. Nucleotides are the structural units of adenosine phosphates, nucleotide coenzymes, and nucleic acids.

nucleotide coenzyme A protein that transports hydrogen atoms (free protons) and electrons from one reaction site to another in cells.

nucleus (NEW-klee-us) [L. *nucleus*, a kernel] In atoms, the central core of one or more positively charged protons and (in all but hydrogen) electrically neutral neutrons. In eukaryotic cells, a membranous organelle containing the DNA.

obesity An excess of fat in the body's adipose tissues, caused by imbalances between caloric intake and energy output.

oligosaccharide A carbohydrate consisting of a small number of covalently linked sugar monomers. One subclass, disaccharides, consists of two sugar monomers. Compare *monosaccharide* and *polysaccharide*.

omnivore [L. *omnis*, all, + *vovare*, to devour] An organism able to obtain energy from more than one source rather than being limited to one trophic level.

oncogene (ON-coe-jeen) Any gene having the potential to induce cancerous transformations in a cell.

oogenesis (oo-oh-JEN-uh-sis) Formation of a female gamete, from a germ cell to a mature haploid ovum (egg).

operator A short base sequence between a promoter and the start of a gene; interacts with regulatory proteins to control transcription.

operon Of transcription, a promoter-operator sequence that services more than a single gene. The lactose operon of *E. coli* is an example.

orbitals Volumes of space around the nucleus of an atom in which electrons are likely to be at any instant.

organ A structure of definite form and function that is composed of more than one tissue.

organ formation Stage of development in which primary tissue layers (germ layers) split into subpopulations of cells, and different lines of cells become unique in structure and function; foundation for growth and tissue specialization, when organs acquire specialized chemical and physical properties.

organ system Two or more organs that interact chemically, physically, or both in performing a common task.

organelle Any of various membranous sacs, envelopes, and other compartmented portions of cytoplasm. Organelles separate different, often incompatible metabolic reactions in the space of the cytoplasm and in time (by allowing certain reactions to proceed only in controlled sequences).

organic compound In biology, a compound assembled in cells and having a carbon backbone, often with carbon atoms arranged as a chain or ring structure.

osmosis (oss-MOE-sis) [Gk. *osmos*, act of pushing] Of cell membranes, the passive movement of water through the interior of membrane-spanning proteins in response to solute concentration gradients, a pressure gradient, or both.

ovary (OH-vuh-ree) In female animals, the primary reproductive organ in which eggs form. In seed-bearing plants, the portion of the carpel where eggs develop, fertilization takes place, and seeds mature. A mature ovary and sometimes other plant parts is a fruit.

oviduct (OH-vih-dukt) Duct through which eggs travel from the ovary to the uterus. Formerly called Fallopian tube.

ovulation (AHV-you-LAY-shun) During each turn of the menstrual cycle, the release of a secondary oocyte (immature egg) from an ovary.

ovule (OHV-youl) [L. *ovum*, egg] Any of one or more structures that form on the inner wall of the ovary of seed-bearing plants and that, at maturity, are seeds. An ovule contains the female gametophyte with its egg, surrounded by nutritive and protective tissues.

ovum (OH-vum) A mature female gamete (egg).

oxidation-reduction reaction An electron transfer from one atom or molecule to another. Often hydrogen is transferred along with the electron or electrons.

pancreas (PAN-cree-us) Glandular organ that secretes enzymes and bicarbonate into the small intestine during digestion, and that also secretes the hormones insulin and glucagon.

pancreatic islets Any of the 2 million clusters of endocrine cells in the pancreas, including alpha cells, beta cells, and delta cells.

parasite [Gk. *para*, alongside, + *sitos*, food] An organism that obtains nutrients directly from the tissues of a living host, which it lives on or in and may or may not kill.

parasitoid An insect larva that grows and develops inside a host organism (usually another insect), eventually consuming the soft tissues and killing it.

parasympathetic nerve Of the autonomic nervous system, any of the nerves carrying signals that tend to slow the body down overall and divert energy to basic tasks; such nerves also work continually in opposition with *sympathetic nerves* to bring about minor adjustments in internal organs.

parathyroid glands (PARE-uh-THY-royd) In vertebrates, endocrine glands embedded in the thyroid gland that secrete parathyroid hormone, which helps restore blood calcium levels.

parenchyma Most abundant ground tissue of root and shoot systems. Its cells function in photosynthesis, storage, secretion, and other tasks.

parthenogenesis Development of an embryo from an unfertilized egg.

passive immunity Temporary immunity conferred by deliberately introducing antibodies into the body.

passive transport Movement of a solute across a cell membrane in response to its concentration gradient, through the interior of

proteins that span the membrane. No energy expenditure is required.

pathogen (PATH-oh-jen) [Gk. *pathos*, suffering, + *-genēs*, origin] Disease-causing organism.

pattern formation Of animals, mechanisms responsible for specialization and positioning of tissues during embryonic development.

PCR *See* polymerase chain reaction.

pelagic province The entire volume of ocean water; subdivided into *neritic zone* (relatively shallow waters overlying continental shelves) and *oceanic zone* (water over ocean basins).

penis A male organ that deposits sperm into a female reproductive tract.

perennial [L. *per-*, throughout, + *annus*, year] A plant that lives for three or more growing seasons.

pericycle (PARE-ih-sigh-kul) [Gk. *peri-*, around, + *kyklos*, circle] Of a root vascular cylinder, one or more layers just inside the endodermis that give rise to lateral roots and contribute to secondary growth.

periderm Of vascular plants showing secondary growth, a protective covering that replaces epidermis.

peripheral nervous system (per-IF-ur-uhl) [Gk. *peripherein*, to carry around] Of vertebrates, the nerves leading into and out from the spinal cord and brain, and the ganglia along those communication lines.

peristalsis (pare-ih-STAL-sis) A rhythmic contraction of muscles that moves food forward through the animal gut.

peritoneum A lining of the coelom that also covers and helps maintain the position of internal organs.

permafrost A permanently frozen, water-impenetrable layer beneath the soil surface in arctic tundra.

PGA Phosphoglycerate (FOSS-foe-GLISS-er-ate); a key intermediate in glycolysis as well as the Calvin-Benson cycle.

PGAL Phosphoglyceraldehyde; a key intermediate in glycolysis as well as the Calvin-Benson cycle.

pH scale A scale used in measuring the concentration of free (unbound) hydrogen ions in solutions; pH 0 is the most acidic, 14 the most basic, and 7, neutral.

phagocyte (FAG-uh-sight) [Gk. *phagein*, to eat, + *kytos*, hollow vessel] A macrophage or certain other white blood cells that engulf and destroy foreign agents in body tissues.

phagocytosis (FAG-uh-sigh-TOE-sis) [Gk. *phagein*, to eat, + *kytos*, hollow vessel] Engulfment of foreign cells or substances by amoebas and some white blood cells, by means of endocytosis.

pharynx (FARE-inks) A muscular tube by which food enters the gut and, in land vertebrates, the windpipe (trachea).

phenotype (FEE-no-type) [Gk. *phainein*, to show, + *typos*, image] Observable trait or traits of an individual; arises from interactions between genes, and between genes and the environment.

pheromone (FARE-oh-moan) [Gk. *phero*, to carry, + *-mone*, as in hormone] A type of signaling molecule secreted by exocrine glands that serves as a communication signal between individuals of the same species.

phloem (FLOW-um) Of vascular plants, a tissue with living cells that interconnect and form the tubes through which sugars and other solutes are conducted.

phospholipid A type of lipid with a glycerol backbone, two fatty acid tails, and a phosphate group to which an alcohol is attached; the main lipid of plant and animal cell membranes.

phosphorylation (FOSS-for-ih-LAY-shun) The attachment of inorganic phosphate to a molecule; also the transfer of a phosphate group from one molecule to another, as when ATP phosphorylates glucose.

photolysis (foe-TALL-ih-sis) [Gk. *photos*, light, + *-lysis*, breaking apart] First step in noncyclic photophosphorylation, when water is split into oxygen, hydrogen, and associated electrons; photon energy indirectly drives the reaction.

photoreceptor Light-sensitive sensory cell.

photosynthesis The trapping and conversion of sunlight energy to chemical energy (ATP, NADPH, or both), followed by synthesis of sugar phosphates that become converted to sucrose, cellulose, starch, and other end products.

photosynthetic autotroph An organism able to synthesize all organic molecules it requires using carbon dioxide as the carbon source and sunlight as the energy source. All plants, some protistans, and a few bacteria are photosynthetic autotrophs.

photosystem Of photosynthetic membranes, a light-trapping unit having organized arrays of pigment molecules and enzymes.

photosystem I A type of photosystem that operates during the cyclic pathway of photosynthesis.

photosystem II A type of photosystem that operates during both the cyclic and noncyclic pathways of photosynthesis.

phototropism [Gk. *photos*, light, + *trope*, turning, direction] Adjustment in the direction and rate of plant growth in response to light.

phylogeny Evolutionary relationships among species, starting with the most ancestral forms and including all the branches leading to all their descendants.

phytochrome Light-sensitive pigment molecule, the activation and inactivation of which trigger plant hormone activities governing leaf expansion, stem branching, stem lengthening, and often seed germination and flowering.

phytoplankton (FIE-toe-PLANK-tun) [Gk. *phyton*, plant, + *planktos*, wandering] A freshwater or marine community of floating or weakly swimming photosynthetic autotrophs, such as diatoms, green algae, and cyanobacteria.

pineal gland (py-NEEL) Of vertebrates, a light-sensitive endocrine gland that secretes melatonin, a hormone that influences the development of reproductive organs and reproductive cycles.

pioneer species Typically small plants with short life cycles that are adapted to growing in exposed, often windy areas with intense sunlight, wide swings in air temperature, and soils deficient in nitrogen and other nutrients. By improving conditions in areas they colonize, pioneers invite their replacement by other species.

pituitary gland Of vertebrate endocrine systems, a gland that interacts with the hypothalamus to coordinate and control many physiological functions, including the activity of many other endocrine glands. Its *posterior lobe* stores and secretes hypothalamic hormones; the *anterior lobe* produces and secretes its own hormones.

placenta (play-SEN-tuh) Of a uterus, an organ composed of maternal tissues and extraembryonic membranes (chorion especially); delivers nutrients to and carries away wastes from the embryo.

plankton [Gk. *planktos*, wandering] Any community of floating or weakly swimming organisms, mostly microscopic, in freshwater and saltwater environments. See *phytoplankton* and *zooplankton*.

plant Most often, multicelled autotroph able to build its own food molecules through photosynthesis.

Plantae The kingdom of plants.

plasma (PLAZ-muh) Liquid component of blood; consists of water, various proteins, ions, sugars, dissolved gases, and other substances.

plasma cell Of immune systems, any of the antibody-secreting daughter cells of a rapidly dividing population of B cells.

plasma membrane Of cells, the outermost membrane that separates internal metabolic events from the environment but selectively permits passage of various substances. Composed of a lipid bilayer and proteins that carry out most functions, including transport across the membrane and reception of outside signals.

plasmid Of many bacteria, a small, circular DNA molecule that carries some genes and replicates independently of the bacterial chromosome.

plasmodesma (PLAZ-moe-DEZ-muh) Of multicelled plants, a junction between linked walls of adjacent cells through which nutrients and other substances flow.

plasticity Of the human species, the ability to remain flexible and adapt to a wide range of environments, rather than becoming narrowly adapted to one specific environment.

plate tectonics Arrangement of the earth's outer layer (lithosphere) in slablike plates, all in motion and floating on a hot, plastic layer of the underlying mantle.

platelet (PLAYT-let) Any of the cell fragments in blood that release substances necesary for clot formation.

pleiotropy (PLEE-oh-troh-pee) [Gk. *pleon*, more, + *trope*, direction] Form of gene expression in which a single gene exerts multiple effects on seemingly unrelated aspects of an individual's phenotype.

pollen grain [L. *pollen*, fine dust] Of gymnosperms and flowering plants, an immature male gametophyte (gamete-producing body).

pollen sac In anthers of flowers, any of the chambers in which pollen grains develop.

pollen tube A tube formed after a pollen grain germinates; grows down through carpel tissues and carries sperm to the ovule.

pollination Of flowering plants, the arrival of a pollen grain on the landing platform (stigma) of a carpel.

pollutant Any substance with which an ecosystem has had no prior evolutionary experience, in terms of kinds or amounts, and that can accumulate to disruptive or harmful levels. Can be naturally occurring or synthetic.

polymer (POH-lih-mur) [Gk. *polus*, many, + *meris*, part] A molecule composed of three to millions of small subunits that may or may not be identical.

polymerase chain reaction DNA amplification method; DNA having a gene of interest is split into single strands, which enzymes (polymerases) copy; the enzymes also act on the accumulating copies, multiplying the gene sequence by the millions.

polymorphism (poly-MORE-fizz-um) [Gk. *polus*, many, + *morphe*, form] Of a population, the persistence through the generations of two or more forms of a trait, at a frequency greater than can be maintained by new mutations alone.

polyp (POH-lip) Vase-shaped, sedentary stage of cnidarian life cycles.

polypeptide chain Three or more amino acids joined by peptide bonds.

polyploidy (POL-ee-PLOYD-ee) A condition in which offspring end up with three or more of each type of chromosome characteristic of the parental stock.

polysaccharide [Gk. *polus*, many, + *sakcharon*, sugar] A straight or branched chain of hundreds of thousands of covalently linked sugar monomers, of the same or different kinds. The most common polysaccharides are starch, cellulose, and glycogen.

polysome Of protein synthesis, several ribosomes all translating the same messenger RNA molecule, one after the other.

population A group of individuals of the same species occupying a given area.

positive feedback mechanism Homeostatic mechanism by which a chain of events is set in motion and intensifies an original condition.

post-translational controls Of eukaryotes, controls that govern modification of newly formed polypeptide chains into functional enzymes and other proteins.

predator [L. *prehendere*, to grasp, seize] An organism that feeds on and may or may not kill other living organisms (its *prey*); unlike parasites, predators do not live on or in their prey.

pressure flow theory Of vascular plants, a theory that organic compounds move through phloem because of gradients in solute concentrations and pressure between source regions (such as photosynthetically active leaves) and sink regions (such as growing plant parts).

primary growth Plant growth originating at root tips and shoot tips.

primary immune response Actions by white blood cells and their products, elicited by a first-time encounter with an antigen; includes both antibody-mediated and cell-mediated responses.

primary productivity, gross Of an ecosystem, the total rate at which the producers capture and store a given amount of energy, as by photosynthesis, during a specified interval.

primary productivity, net Of an ecosystem, the rate of energy storage in the tissues of producers in excess of their rate of aerobic respiration.

procambium (pro-KAM-bee-um) Of vascular plants, a primary meristem that gives rise to the primary vascular tissues.

producer, primary An organism such as a plant that directly or indirectly nourishes consumers, decomposers, and detritivores.

progesterone (pro-JESS-tuh-rown) Female sex hormone secreted by the ovaries.

prokaryote (pro-CARRY-oht) [L. *pro*, before, + Gk. *karyon*, kernel] Single-celled organism that has no nucleus or other membrane-bound organelles; only bacteria are prokaryotes.

promoter Of transcription, a base sequence that signals the start of a gene; the site where RNA polymerase initially binds.

prophase First stage of mitosis, when each duplicated chromosome becomes condensed into a thicker, rodlike form.

prophase I Stage of meiosis when each duplicated chromosome condenses and pairs with its homologous partner, followed by crossing over and genetic recombination among nonsister chromatids.

prophase II Brief stage of meiosis after interkinesis during which each chromosome still consists of two chromatids.

protein Organic compound composed of one or more chains of amino acids (polypeptide chains).

Protista The kingdom of protistans.

protistan (pro-TISS-tun) [Gk. *prōtistos*, primal, very first] Single-celled eukaryote.

proton Positively charged unit of energy in the atomic nucleus.

proto-oncogene A gene sequence similar to an oncogene but that codes for a protein required in normal cell function; may trigger cancer, generally when specific mutations alter its structure or function.

protostome (PRO-toe-stome) [Gk. *proto*, first, + *stoma*, mouth] A bilateral animal in which the first indentation in the early embryo develops into the mouth. Includes mollusks, annelids, and arthropods.

proximal tubule Of a nephron, the tubular region that receives water and solutes filtered from the blood.

pulmonary circuit Blood circulation route leading to and from the lungs.

Punnett-square method A diagramming technique for predicting the possible outcome of a mating or an experimental cross.

purine Nucleotide base having a double ring structure. Adenine and guanine are two examples.

pyrimidine (pih-RIM-ih-deen) Nucleotide base having a single ring structure. Cytosine and thymine are examples.

pyruvate (PIE-roo-vate) Three-carbon compound produced by the initial breakdown of a glucose molecule during glycolysis.

radial symmetry Body plan having four or more roughly equivalent parts arranged around a central axis.

rain shadow A reduction in rainfall on the leeward side of high mountains, resulting in arid or semiarid conditions.

reabsorption Of urine formation, the diffusion or active transport of water and usable solutes out of a nephron and into capillaries leading back to the general circulation; regulated by ADH and aldosterone.

receptor Of cells, a molecule at the cell surface or within the cytoplasm that may be activated by a specific hormone, virus, or some other outside agent. Of nervous systems, a sensory cell or cell part that may be activated by a specific stimulus.

receptor protein Protein that binds a signaling molecule such as a hormone, then triggers alterations in cell behavior or metabolism.

recessive allele [L. *recedere*, to recede] In heterozygotes, an allele whose expression is fully or partially masked by expression of its partner; fully expressed only in the homozygous recessive condition.

recognition protein Protein at cell surface recognized by cells of like type; helps guide the ordering of cells into tissues during development and functions in cell-to-cell interactions.

recombinant technology Procedures by which DNA (genes) from different species may be isolated, cut, spliced together, and the new recombinant molecules multiplied in quantity.

red blood cell Erythrocyte; an oxygen-transporting cell in blood.

red marrow A substance in the spongy tissue of many bones that serves as a major site of blood cell formation.

reflex [L. *reflectere*, to bend back] A simple, stereotyped, and repeatable movement elicited by a sensory stimulus.

reflex arc [L. *reflectere*, to bend back] Type of neural pathway in which signals from sensory neurons can be sent directly to motor neurons, without intervention by an interneuron.

refractory period Of neurons, the period following an action potential at a given patch of membrane when sodium gates are shut and potassium gates are open, so that the patch is insensitive to stimulation.

releasing hormone A hypothalamic signaling molecule that stimulates or slows down secretion by target cells in the anterior lobe of the pituitary gland.

repressor protein Regulatory protein that provides negative control of gene activity by preventing RNA polymerase from binding to DNA.

reproduction, asexual Production of new individuals by any mode that does not involve gametes.

reproduction, sexual Mode of reproduction that begins with meiosis, proceeds through gamete formation, and ends at fertilization.

reproductive isolating mechanism Any aspect of structure, functioning, or behavior that prevents successful interbreeding (hence gene flow) between populations or between local breeding units within a population.

resource partitioning A community pattern in which similar species generally share the same kind of resource in different ways, in different areas, or at different times.

respiration [L. *respirare*, to breathe] In most animals, the overall exchange of oxygen from the environment and carbon dioxide wastes from cells by way of circulating blood. Compare *aerobic respiration*.

resting membrane potential Of neurons and other excitable cells that are not being stimulated, the steady voltage difference across the plasma membrane.

restriction enzymes Class of enzymes that cut apart foreign DNA that enters a cell, as by viral infection; also used in recombinant DNA technology.

reticular formation Of the vertebrate brainstem, a major network of interneurons that helps govern activity of the whole nervous system.

reverse transcriptase Viral enzyme required for reverse transcription of mRNA into DNA; also used in recombinant DNA technology.

reverse transcription Assembly of DNA on a single-stranded mRNA molecule by viral enzymes.

RFLPs (restriction fragment length polymorphisms) Slight but unique differences in the banding pattern of DNA fragments from different individuals of a species; result from individual differences in the number and location of DNA sites that restriction enzymes can recognize and cut.

ribosomal RNA (rRNA) Type of RNA molecule that combines with proteins to form ribosomes, on which polypeptide chains are assembled.

ribosome Of cells, a structure having two subunits, each composed of RNA and protein molecules; the site of protein synthesis.

RNA Ribonucleic acid; a category of nucleotides used in translating the genetic message of DNA into protein.

rod cell A vertebrate photoreceptor sensitive to very dim light and that contributes to coarse perception of movement.

root hair Of vascular plants, an extension of a specialized root epidermal cell; root hairs collectively enhance the surface area available for absorbing water and solutes.

RuBP Ribulose bisphosphate, a five-carbon compound required for carbon fixation in the Calvin-Benson cycle of photosynthesis.

salivary gland Any of the glands that secrete saliva, a fluid that initially mixes with food in the mouth and starts the breakdown of starch.

salt An ionic compound formed when an acid reacts with a base.

saltatory conduction In myelinated neurons, rapid, node-to-node hopping of action potentials.

saprobe Heterotroph that obtains its nutrients from nonliving organic matter. Most fungi are saprobes.

sarcomere (SAR-koe-meer) Of skeletal and cardiac muscles, the basic unit of contraction; a region of organized myosin and actin filaments between two Z lines of a myofibril inside a muscle cell.

sarcoplasmic reticulum (sar-koe-PLAZ-mik reh-TIK-you-lum) A calcium-storing membrane system surrounding myofibrils of a muscle cell.

Schwann cells Specialized neuroglial cells that grow around neuron axons, forming a *myelin sheath.*

sclerenchyma Of vascular plants, a ground tissue that provides mechanical support and protection in mature plant parts.

second law of thermodynamics The spontaneous direction of energy flow is from high-quality to low-quality forms. With each conversion, some energy is randomly dispersed in a form that is not as readily available to do work.

second messenger A molecule inside a cell that mediates and generally triggers amplified response to a hormone.

secondary immune response Rapid, prolonged immune response by white blood cells, memory cells especially, to a previously encountered antigen.

secretion Generally, the release of a substance for use by the organism producing it. (Not the same as *excretion,* the expulsion of excess or waste material.) Of kidneys, a regulated stage in urine formation, in which ions and other substances move from capillaries into nephrons.

sedimentary cycle A biogeochemical cycle without a gaseous phase; the element moves from land to the seafloor, then returns only through long-term geological uplifting.

seed Of gymnosperms and flowering plants, a fully mature ovule (contains the plant embryo), with its integuments forming the seed coat.

segmentation Of earthworms and many other animals, a series of body units that may be externally similar to or quite different from one another.

segregation (Mendelian principle of) [L. *se-,* apart, + *grex,* herd] The principle that diploid organisms inherit a pair of genes for each trait (on a pair of homologous chromosomes) and that the two genes segregate during meiosis and end up in separate gametes.

selective gene expression Of multicelled organisms, activation or suppression of a fraction of the genes in unique ways in different cells, leading to pronounced differences in structure and function among different cell lineages.

selfish behavior Form of behavior by which an individual protects or increases its own chance of producing offspring, regardless of the consequences for the group to which it belongs.

semen (SEE-mun) [L. *serere,* to sow] Sperm-bearing fluid expelled from a penis during male orgasm.

semiconservative replication [Gk. *hēmi,* half, + L. *conservare,* to keep] Reproduction of a DNA molecule when a complementary strand forms on each of the unzipping strands of an existing DNA double helix, the outcome being two "half-old, half-new" molecules.

senescence (sen-ESS-cents) [L. *senescere,* to grow old] Sum total of processes leading to the natural death of an organism or some of its parts.

sensory neuron Any of the nerve cells that act as sensory receptors, detecting specific stimuli (such as light energy) and relaying signals to the brain and spinal cord.

sessile animal Animal that remains attached to a substrate during some stage (often the adult) of its life cycle.

sex chromosomes Of most animals and some plants, chromosomes that differ in number or kind between males and females but that still function as homologues during meiosis. Compare *autosomes.*

sexual dimorphism Phenotypic differences between males and females of a species.

sexual reproduction Production of offspring by way of meiosis, gamete formation, and fertilization.

sexual selection Natural selection based on a trait that provides a competitive edge in mating and producing offspring.

shoot system Stems and leaves of vascular plants.

sieve tube member Of flowering plants, a cellular component of the interconnecting conducting tubes in phloem.

sink region In a vascular plant, any region using or stockpiling organic compounds for growth and development.

sliding filament model Model of muscle contraction, in which actin filaments physically slide over myosin filaments toward the center of the sarcomere. The sliding requires ATP energy and the formation of cross-bridges between actin and myosin filaments.

small intestine Of vertebrates, the portion of the digestive system where digestion is completed and most nutrients absorbed.

smog, industrial Gray-colored air pollution that predominates in industrialized cities with cold, wet winters.

smog, photochemical Form of brown, smelly air pollution occurring in large cities in warm climates.

social behavior Tendency of individual animals to enter into cooperative, interdependent relationships with others of their kind; based on the ability to use communication signals.

social parasite Animal that depends on the social behavior of another species to gain food, care for young, or some other factor to complete its life cycle.

sodium-potassium pump A transport protein spanning the lipid bilayer of the plasma membrane. When activated by ATP, its shape changes and it selectively transports sodium ions out of the cell and potassium ions in.

solute (SOL-yoot) [L. *solvere,* to loosen] Any substance dissolved in a solution. In water, this means its individual molecules are surrounded by spheres of hydration that keep their charged parts from interacting, so the molecules remain dispersed.

solvent Fluid in which one or more substances are dissolved.

somatic cell (SO-MAT-ik) [Gk. *somā,* body] Of animals, any cell that is not a germ cell (which develops by meiosis into sperm or eggs).

somatic nervous system Those nerves leading from the central nervous system to skeletal muscles.

source region Of vascular plants, any of the sites of photosynthesis.

speciation (spee-cee-AY-shun) The time at which a new species emerges, as by divergence or polyploidy.

species (SPEE-sheez) [L. *species,* a kind] Of sexually reproducing species in nature, one or more populations composed of individuals that are interbreeding and producing fertile offspring, and that are reproductively isolated from other such groups.

sperm [Gk. *sperma,* seed] Mature male gamete.

spermatogenesis (sperm-AT-oh-JEN-ih-sis) Formation of mature sperm following meiosis in a germ cell.

sphere of hydration Through positive or negative interactions, a clustering of water molecules around the individual molecules of a substance placed in water. Compare *solute.*

sphincter (SFINK-tur) Ring of muscle between regions of a tubelike system (as between the stomach and small intestine).

spinal cord Of central nervous systems, the portion threading through a canal inside the vertebral column and providing direct reflex connections between sensory and motor neurons as well as communication lines to and from the brain.

spindle apparatus A bipolar structure composed of microtubules that forms during mitosis or meiosis and that moves the chromosomes.

sporangium, plural sporangia (spore-AN-gee-um) [Gk. *spora,* seed] The protective tissue layer that surrounds haploid spores in a sporophyte.

spore Of fungi, a walled, resistant cell or multicelled structure, produced by mitosis or meiosis, that can germinate and give rise to a new mycelium. Of land plants, a reproductive cell formed by meiosis that can develop into a gametophyte (gamete-producing body).

sporophyte [Gk. *phyton,* plant] Diploid, spore-producing stage of plant life cycles.

stabilizing selection Mode of natural selection in which the most common phenotypes in a population are favored, and the underlying allele frequencies persist over time.

stamen (STAY-mun) Of flowering plants, a male reproductive structure; commonly consists of pollen-bearing structures (anthers) on single stalks (filaments).

start codon Of protein synthesis, a base triplet in a strand of mRNA that serves as the start signal for mRNA translation.

steroid (STAIR-oid) A lipid with a backbone of four carbon rings. Steroids differ in the number and location of double bonds in the backbone and in the number, position, and type of functional groups.

stimulus [L. *stimulus*, goad] A specific form of energy, such as light, heat, and mechanical pressure, that the body can detect through sensory receptors.

stoma (STOW-muh), plural **stomata** [Gk. *stoma*, mouth] A controllable gap between two guard cells in stems and leaves; any of the small passageways across the epidermis through which carbon dioxide moves into the plant and water vapor moves out.

stomach A muscular, stretchable sac that receives ingested food; of vertebrates, an organ between the esophagus and intestine in which considerable protein digestion occurs.

stop codon Of protein synthesis, a base triplet in a strand of mRNA that serves as the stop signal for translation, so that no more amino acids are added to the polypeptide chain.

stroma [Gk. *strōma*, bed] Of chloroplasts, the semifluid matrix surrounding the thylakoid membrane system; the zone where sucrose, starch, and other end products of photosynthesis are assembled.

substrate Specific molecule or molecules that an enzyme can chemically recognize, bind briefly to itself, and modify in a specific way.

substrate-level phosphorylation Enzyme-mediated reaction in which a substrate gives up a phosphate group to another molecule, as when an intermediate of glycolysis donates phosphate to ADP, producing ATP.

succession, primary (suk-SESH-un) [L. *succedere*, to follow] Orderly changes from the time pioneer species colonize a barren habitat through replacements after replacements by various species; the changes lead to a *climax community*, when the composition of species remains steady under prevailing conditions.

succession, secondary Orderly changes in a community or patch of habitat toward the climax state after having been disturbed, as by fire.

surface-to-volume ratio A mathematical relationship in which volume increases with the cube of the diameter, but surface area increases only with the square. Of growing cells, the volume of cytoplasm increases more rapidly than the surface area of the plasma membrane that must service the cytoplasm. Because of this constraint, cells generally remain small or elongated, or have elaborate membrane foldings.

symbiosis (sim-by-OH-sis) [Gk. *sym*, together, + *bios*, life, mode of life] A form of mutualism in which interacting species have become intimately and permanently dependent on each other for survival and reproduction.

sympathetic nerve Of the autonomic nervous system, any of the nerves generally concerned with increasing overall body activities during times of heightened awareness, excitement, or danger; such nerves also work continually in opposition with *parasympathetic nerves* to bring about minor adjustments in internal organs.

sympatric speciation [Gk. *syn*, together, + *patria*, native land] Speciation that follows after ecological, behavioral, or genetic barriers arise within the geographical boundaries of a single population.

synaptic integration (sin-AP-tik) Moment-by-moment combining of excitatory and inhibitory signals arriving at a trigger zone of a neuron.

systematics Branch of biology that deals with patterns of diversity among organisms in an evolutionary context; its three approaches include taxonomy, phylogenetic reconstruction, and classification.

systemic circuit (sis-TEM-ik) Circulation route in which oxygen-enriched blood flows from the lungs to the left half of the heart, through the rest of the body (where it gives up oxygen and picks up carbon dioxide), then back to the right side of the heart.

taproot system A primary root and its lateral branchings.

taxonomy (tax-ON-uh-mee) Approach in biological systematics that involves identifying organisms and assigning names to them.

telophase (TEE-low-faze) Of mitosis, final stage when chromosomes decondense into threadlike structures and two daughter nuclei form.

telophase I Of meiosis, stage when one of each type of duplicated chromosome has arrived at one or the other end of the spindle pole.

telophase II Of meiosis, final stage when four daughter nuclei form.

temperate pathway Mode of viral replication in which the virus enters latency instead of killing the host cell outright; viral genes remain inactive and, if integrated into the bacterial chromosome, may be passed on to any of the cell's descendants—which will be destroyed when the viral genes do become activated.

testcross Experimental cross in which hybrids of the first generation of offspring (F_1) are crossed with an individual known to be true-breeding for the same recessive trait as the recessive parent.

testis, plural **testes** Male gonad; primary reproductive organ in which male gametes and sex hormones are produced.

testosterone (tess-TOSS-tuh-rown) In male mammals, a major sex hormone that helps control male reproductive functions.

theory A related set of hypotheses that, taken together, form a broad-ranging explanation about some aspect of the natural world; differs from a scientific hypothesis in its breadth of application. In modern science, only explanations that have been extensively tested and can be relied upon with a very high degree of confidence are accorded the status of theory.

thermal inversion Situation in which a layer of dense, cool air becomes trapped beneath a layer of warm air; can cause air pollutants to accumulate to dangerous levels close to the ground.

thermoreceptor Sensory cell that can detect radiant energy associated with temperature.

thigmotropism (thig-MOTE-ruh-pizm) [Gk. *thigm*, touch] Of vascular plants, growth oriented in response to physical contact with a solid object, as when a vine curls around a fencepost.

threshold Of neurons and other excitable cells, a certain minimum amount by which the steady voltage difference across the plasma membrane must change to produce an action potential.

thylakoid membrane Of chloroplasts, an internal membrane commonly folded into flattened channels and stacked disks (*grana*); contains light-absorbing pigments and enzymes used in the formation of ATP, NADPH, or both during photosynthesis.

thymine Nitrogen-containing base in some nucleotides.

thymus gland Of endocrine systems, a gland in which certain white blood cells multiply, differentiate, and mature, and which secretes hormones that affect their functions.

thyroid gland Of endocrine systems, a gland that produces hormones that affect overall metabolic rates, growth, and development.

tissue Of multicelled organisms, a group of cells and intercellular substances that function together in one or more specialized tasks.

tonicity The relative concentrations of solutes in two fluids, such as inside and outside a cell. When solute concentrations are *isotonic* (equal in both fluids), water shows no net osmotic movement in either direction. When one fluid is *hypotonic* (has less solutes than the other), the other is *hypertonic* (has more solutes) and is the direction in which water tends to move.

trachea (TRAY-kee-uh), plural **tracheae** Of insects, spiders, and some other animals, a finely branching air-conducting tube that functions in respiration; of land vertebrates, the windpipe that carries air between the larynx and bronchi.

tracheid (TRAY-kid) Of flowering plants, one of two types of cells in xylem that conduct water and dissolved minerals.

transcript-processing controls Of eukaryotic cells, controls that govern modification of new mRNA molecules into mature transcripts before shipment from the nucleus.

transcription [L. *trans*, across, + *scribere*, to write] Of protein synthesis, the assembly of an RNA strand on one of the two strands of a DNA double helix; the base sequence of the resulting transcript is complementary to the DNA region on which it is assembled.

transcriptional controls Controls influencing when and to what degree a particular gene will be transcribed.

transfer RNA (tRNA) Of protein synthesis, any of the type of RNA molecules that bind and deliver specific amino acids to ribosomes *and* pair with mRNA code words for those amino acids.

translation Of protein synthesis, the conversion of the coded sequence of information in mRNA into a particular sequence of amino acids to form a polypeptide chain; depends on interactions of rRNA, tRNA, and mRNA.

translational controls Of eukaryotic cells, controls governing the rates at which mRNA transcripts that reach the cytoplasm will be translated into polypeptide chains at ribosomes.

translocation Of cells, the transfer of part of one chromosome to a nonhomologous chromosome. Of vascular plants, the conduction of organic compounds through the plant body by way of the phloem.

transmitter substance Any of the class of signaling molecules that are secreted from neurons, act on immediately adjacent cells, and are then rapidly degraded or recycled.

transpiration Evaporative water loss from stems and leaves.

transport control Of eukaryotic cells, controls governing when mature mRNA transcripts are shipped from the nucleus into the cytoplasm.

transposable element DNA element that can spontaneously "jump" to new locations in the same DNA molecule or a different one. Such elements often inactivate the genes into which they become inserted and give rise to observable changes in phenotype.

trisomy (TRY-so-mee) Of diploid cells, the abnormal presence of three of one type of chromosome.

trophic level (TROE-fik) [Gk. *trophos*, feeder] All organisms in an ecosystem that are the same number of transfer steps away from the energy input into the system.

tropism (TROE-pizm) Of vascular plants, a growth response to an environmental factor, such as growth toward light.

tumor A tissue mass composed of cells that are dividing at an abnormally high rate.

turgor pressure (TUR-gore) [L. *turgere*, to swell] Internal pressure applied to a cell wall when water moves by osmosis into the cell.

upwelling An upward movement of deep, nutrient-rich water along coasts to replace surface waters that winds move away from shore.

uracil (YUR-uh-sill) Nitrogen-containing base found in RNA molecules; can base-pair with adenine.

urinary system Of vertebrates, an organ system that regulates water and solute levels.

urine Fluid formed by filtration, reabsorption, and secretion in kidneys; consists of wastes and excess water and solutes.

uterus (YOU-tur-us) [L. *uterus*, womb] Chamber in which the developing embryo is contained and nurtured during pregnancy.

vagina Part of a female reproductive system that receives sperm, forms part of the birth canal, and channels menstrual flow to the exterior.

variable Of the factors characterizing or influencing an experimental group under study, the only one (ideally) that is *not* identical to those of a control group.

vascular bundle One of several to many strandlike arrangements of primary xylem and phloem embedded in the ground tissue of roots, stems, and leaves.

vascular cambium A lateral meristem that increases stem or root diameter of vascular plants showing secondary growth.

vascular cylinder Of plant roots, the arrangement of vascular tissues as a central cylinder.

vascular plant Plant having tissues that transport water and solutes through well-developed roots, stems, and leaves.

vein Of the circulatory system, any of the large-diameter vessels that lead back to the heart; of leaves, one of the vascular bundles that thread lacily through photosynthetic tissues.

vernalization Of flowering plants, stimulation of flowering by exposure to low temperatures.

vertebra, plural **vertebrae** Of vertebrate animals, one of a series of hard bones, arranged with intervertebral disks, into a backbone.

vertebrate Animal having a backbone of bony segments, the *vertebrae*.

vesicle (VESS-ih-kul) [L. *vesicula*, little bladder] Of cells, a small membranous sac that transports or stores substances in the cytoplasm.

villus (VIL-us), plural **villi** Any of several types of absorptive structures projecting from the free surface of an epithelium.

viroid An infectious nucleic acid that has no protein coat; a tiny rod or circle of single-stranded RNA.

virus A noncellular infectious agent, consisting of DNA or RNA and a protein coat; can replicate only after its genetic material enters a host cell and subverts that cell's metabolic machinery.

vision Precise light focusing onto a layer of photoreceptive cells that is dense enough to sample details concerning a given light stimulus, followed by image formation in the brain.

vitamin Any of more than a dozen organic substances that animals require in small amounts for normal cell metabolism but generally cannot synthesize for themselves.

water potential The sum of two opposing forces (osmosis and turgor pressure) that can cause the directional movement of water into or out of a walled cell.

watershed A region where all precipitation becomes funneled into a single stream or river.

wax A type of lipid with long-chain fatty acid tails; waxes help form protective, lubricating, or water-repellent coatings.

white blood cell Leukocyte; of vertebrates, any of the macrophages, eosinophils, neutrophils, and other cells which, together with their products, comprise the immune system.

white matter Of spinal cords, major nerve tracts so named because of the glistening myelin sheaths of their axons.

wild-type allele Of a population, the allele that occurs normally or with greatest frequency at a given gene locus.

wing Of birds, a forelimb of feathers, powerful muscles, and lightweight bones that functions in flight. Of insects, a structure that develops as a lateral fold of the exoskeleton and functions in flight.

X-linked gene Any gene on an X chromosome.

X-linked recessive inheritance Recessive condition in which the responsible, mutated gene occurs on the X chromosome.

Y-linked gene Any gene on a Y chromosome.

xylem (ZYE-lum) [Gk. *xylon*, wood] Of vascular plants, a tissue that transports water and solutes through the plant body.

yolk sac Of many vertebrates, an extraembryonic membrane that provides nourishment (from yolk) to the developing embryo; of humans, the sac does not include yolk but helps give rise to a digestive tube.

zooplankton A freshwater or marine community of floating or weakly swimming heterotrophs, mostly microscopic, such as rotifers and copepods.

zygote (ZYE-goat) The first diploid cell formed after fertilization (fusion of nuclei from a male and a female gamete).

CREDITS AND ACKNOWLEDGMENTS

Front Matter

Pages xiv–xv Stock Imagery / **Pages xvi–xvii** James M. Bell/Photo Researchers / **Pages xviii–xix** S. Stammers/SPL/Photo Researchers / **Pages xx–xxi** © 1990 Arthur M. Greene / **Pages xxii–xxiii** Thomas D. Mangelsen / **Pages xxiv–xxv** Lennart Nilsson from *A Child Is Born*, © 1966, 1977 Dell Publishing Company, Inc. / **Pages xxvi–xxvii** Jim Doran

Page 1 Tom Van Sant/The GeoSphere Project, Santa Monica, CA

Chapter 1

1.1 Frank Kaczmarek / **1.4** (left) Paul DeGreve/FPG; (right) Norman Meyers/Bruce Coleman, Inc. / **1.5** (a) Walt Anderson/Visuals Unlimited; (b) Edward S. Ross; (c) Gregory Dimijian/Photo Researchers; (d) Alan Weaving/Ardea, London / **1.6** Jack deConingh / **1.7** (a) Tony Brain/SPL/Photo Researchers; (b) M. Abbey/Visuals Unlimited; (c) Edward S. Ross; (d) Dennis Brokaw; (e) Edward S. Ross; (f) Pat & Tom Leeson/Photo Researchers / **1.8** Levi Publishing Company

Page 17 James M. Bell/Photo Researchers

Chapter 2

2.1 Martin Rogers/FPG / **2.2** Jack Carey / **Page 21** (a) Kingsley R. Stern; (b) Chip Clark/ **Page 22** (c) Stanford Medical Center; (d) (left) Hank Morgan/Rainbow; (right) Dr. Harry T. Chugani, M.D., UCLA School of Medicine / **2.8** Art by Palay/Beaubois / **2.9** (a) Richard Riley/FPG / **2.10** Colin Monteath, Hedgehog House New Zealand; art by Palay/Beaubois / **2.11** (a), (b) Paolo Fioratti; (c) H. Eisenbeiss/Frank Lane Picture Agency / **2.14** Michael Grecco/Picture Group

Chapter 3

3.1 (a) Field Museum of Natural History, Neg. #75400C, Chicago; (b) Brian Parker/Tom Stack & Associates / **3.2, 3.3, 3.5** Art by Palay/Beaubois / **3.8** Biophoto Associates/SPL/Photo Researchers / **3.9** Art by Jeanne Schreiber / **3.13** Clem Haagner/Ardea, London / **3.14** Larry Lefever/Grant Heilman / **Page 42** Lewis L. Lainey / **3.19** Art by Palay/Beaubois / **3.20** (a) CNRI/SPL/Photo Researchers; (b), (c) art by Robert Demarest / **3.22** (b) A. Lesk/SPL/Photo Researchers

Chapter 4

4.1 (a) Jan Hinsch/SPL/Photo Researchers; (b), (c) NASA / **4.2** (a) (left) National Library of Medicine; (right) Armed Forces Institute of Pathology; (b) The Bettmann Archive; (c) George Musil/Visuals Unlimited / **4.5** (a–d) Jeremy Pickett-Heaps, School of Botany, University of Melbourne / **4.7** (a) Gary Gaard and Arthur Kelman; (b) micrograph G. Cohen-Bazire; art by Palay/Beaubois / **4.8–4.9** Art by Leonard Morgan / **4.10** Micrograph M.C. Ledbetter, Brookhaven National Laboratory; art by D. & V. Hennings / **4.11** Micrograph G.L. Decker; art by D. & V. Hennings / **4.12** Micrograph D. Fawcett, *The Cell*, Philadelphia: W.B. Saunders Co.; art by D. & V. Hennings / **4.13** (a) Don W. Fawcett/Visuals Unlimited; (b) A.C. Faberge, *Cell and Tissue*

Research, 151:403–415, 1974 / **4.14** Art by A. Kasnot / **4.16** (a), (b) Micrographs Don W. Fawcett/Visuals Unlimited; (below, right) art by Robert Demarest / **4.17** (left) Art by Robert Demarest after a model by J. Kephart; micrograph Gary W. Grimes / **4.18** Gary W. Grimes / **4.19** Micrograph Keith R. Porter / **4.20** Micrograph L.K. Shumway; (below) art by Palay/Beaubois / **4.21** (a–c) Mark McNiven and Keith R. Porter; (d) Andrew S. Bajer; (e) J. Victor Small and Gottfried Rinnerthaler / **4.23** (a) Sidney L. Tamm; (b) art by D. & V. Hennings / **4.24** (b) U.W. Goodenough and J.W. Heuser / **4.25** (a) After Alberts et al., *Molecular Biology of the Cell*, Garland Publishing Co., 1983; (b) Dianne T. Woodrum and Richard W. Linck / **4.26** Sketch by D. & V. Hennings after P. Raven et al., *Biology of Plants*, third edition, Worth Publishers, 1981; micrograph P.A. Roelofsen

Chapter 5

5.1 (a) Runk/Schoenberger/Grant Heilman; (b) Inigo Everson/Bruce Coleman Ltd. / **5.3** Art by Leonard Morgan; micrograph H.C. Aldrich / **Pages 78–79** Art by Palay/Beaubois; micrograph P. Pinto da Silva and D. Branton, *Journal of Cell Biology*, 45:598, by copyright permission of The Rockefeller University Press / **5.5** Micrograph M. Sheetz, R. Painter, and S. Singer, *Journal of Cell Biology*, 70:193, by copyright permission of The Rockefeller University Press / **5.6** Frieder Sauer/Bruce Coleman Ltd. / **5.7** Photographs Frank B. Salisbury / **5.8** Art by Leonard Morgan after Alberts et al., *Molecular Biology of the Cell*, second edition, Garland Publishing Co., 1989 / **5.9–5.11** Art by Leonard Morgan / **5.12** From *Molecular Cell Biology* by James Darnell, Harvey Lodish, and David Baltimore. Copyright © 1986 by Scientific American Books. Reprinted with permission of W.H. Freeman and Company; art by Palay/Beaubois / **5.13** M.M. Perry and A.B. Gilbert / **5.14** Art by Palay/Beaubois

Chapter 6

6.1 Robert C. Simpson/Nature Stock / **6.3** (left) NASA; (right) Manfred Kage/Peter Arnold, Inc. / **6.4** Art by Palay/Beaubois / **6.8** Thomas A. Steitz / **6.9** Art by Palay/Beaubois / **6.12** Douglas Faulkner/Sally Faulkner Collection / **6.15** Art by L. Calver after B. Alberts et al., *Molecular Biology of the Cell*, Garland Publishing Co., 1983 / **Page 101** (a) Kathie Atkinson/Oxford Scientific Films; (b) Keith V. Wood

Chapter 7

7.1 Photograph Sam Zarember/Image Bank / **7.2** (a) Hans Reinhard/Bruce Coleman Ltd.; (b), (c) micrographs David Fisher; (b) art by Palay/Beaubois; (c), (d) art by K. Kasnot / **7.3** Photograph Barker-Blakenship/FPG; art by Victor Royer / **7.4** Larry West/FPG / **7.5** Photograph E.R. Degginger / **7.6** Art by Illustrious, Inc. / **7.10** Art by L. Calver

Chapter 8

8.1 Stephen Dalton/Photo Researchers / **8.2** (a), (b) NASA; (c) Janeart/Image Bank / **8.4** (right) Art by Palay/Beaubois / **8.5** (a) Keith R. Porter; (b), (c) art by L. Calver / **8.9** Photograph Adrian Warren/Ardea, London / **8.11** David M. Phillips/Visuals Unlimited / **Page 130** Ralph Pleasant/FPG / **Page 133** R. Llewellyn/Superstock, Inc.

Page 135 © Lennart Nilsson

Chapter 9

Page 136 (left) Chris Huss; (right) Tony Dawson / **Page 137** (above) Chris Huss; (below) Tony Dawson / **9.4** Andrew S. Bajer, University of Oregon / **9.5** Photographs Andrew S. Bajer, University of Oregon / **9.6** Micrographs Ed Reschke; art by K. Kasnot / **9.7** Art by Palay/Beaubois / **9.8** Micrographs H. Beams and R.G. Kessel, *American Scientist*, 64:279–290, 1976 / **9.9** B.A. Palevitz and E.H. Newcomb, University of Wisconsin/ BPS/Tom Stack & Associates / **9.10** (a–c), (e) Lennart Nilsson from *A Child Is Born*, © 1966, 1967 Dell Publishing Company, Inc.; (d) Lennart Nilsson from *Behold Man*, © 1974 by Albert Bonniers Forlag and Little, Brown and Company, Boston

Chapter 10

10.1 (a) Jane Burton/Bruce Coleman Ltd.; (b) Dan Kline/Visuals Unlimited / **10.2** Courtesy of Kirk Douglas/The Bryna Company / **10.3** Art by L. Calver / **10.4** Art by K. Kasnot / **10.5** Micrograph B. John / **10.6** CNRI/SPL/Photo Researchers / **10.7** Art by K. Kasnot / **10.12** (b) © David M. Phillips/Visuals Unlimited / **10.14** Art by K. Kasnot

Chapter 11

11.1 (a) David M. Phillips/Visuals Unlimited; (b) Bill Longcore/Photo Researchers; (c) Moravian Museum, Brno / **11.2** Photograph Jean M. Labat/Ardea, London; art by Jennifer Wardrip / **11.11** Photographs William E. Ferguson / **11.13** Tedd Somes / **11.14** (a), (b) Michael Stuckey/Comstock Inc.; (c) Russ Kinne/Comstock Inc. / **11.15** David Hosking / 11.16 Photograph Bill Longcore/Photo Researchers / **11.17** Photograph Jane Burton/Bruce Coleman Ltd.; art by D. & V. Hennings / **11.18** After John G. Torrey, *Development in Flowering Plants*, by permission of Macmillan Publishing Company, copyright © 1967 by John G. Torrey / **11.19** (top to bottom) Frank Cezus; Frank Cezus; Michael Keller; Ted Beaudin; Stan Sholik/all FPG / **11.20** (a) F. Blakeslee, *Journal of Heredity*, 1914 / **Page 181** Evan Cerasoli

Chapter 12

12.1 Eddie Adams/AP Photo / **12.3** Art by Palay/Beaubois; photograph Omikron/Photo Researchers / **Page 186** (a) Reprinted by permission from page 109 of *Human Heredity: Principles and Issues*, second edition. Copyright © 1991 by West Publishing Company. All rights reserved. Art by Robert Demarest / **Page 187** Art by Robert Demarest after Patten, Carlson, and others / **12.5** Photographs Carolina Biological Supply Company / **12.8** Photograph Dr. Victor A. McKusick / **12.12** After Victor A. McKusick, *Human Genetics*, second edition, copyright © 1969. Reprinted by permission of Prentice-Hall, Inc., Englewood Cliffs, NJ; photograph The Bettmann Archive / **12.13** (a) Courtesy of David D. Weaver, M.D. / **12.15** Photograph courtesy of B.R. Brinkley from D.E. Merry et al., *American Journal of Human Genetics*, 37:425–430, 1985. © 1985 by The American Society of Human Genetics. All rights reserved. Used by permission of The University of Chicago Press / **12.16** Art by K. Kasnot / **12.17** (a) Cytogenetics Laboratory, University of California, San Francisco; (b) after Collman and Stoller, *American Journal of Public Health*, 52, 1962 / **12.18** (top left) Used by permission of Carole Lafrate; (top right and below) courtesy of Peninsula Association for Retarded Children and Adults, San Mateo Special Olympics, Burlingame, CA / **Page 200** Art by Palay/Beaubois / **Page 203** (above) Bonnie Kamin/Stuart Kenter Associates; (below) Carolina Biological Supply Company

Chapter 13

13.1 Photograph A.C. Barrington Brown © 1968 J.D. Watson; model A. Lesk/SPL/Photo Researchers / **13.3** Micrograph Lee D. Simon, Waksman Institute of Microbiology / **13.5** Micrograph Biophoto Associates/SPL/Photo Researchers / **13.19** (a) Photograph W.C. Earnshaw. From the *Journal of Cell Biology*, 1985, 100:1716–1725 by copyright permission of The Rockefeller University Press; (b), (c) photographs U.K. Laemmli from *Cell*, 12:817–828, copyright 1977 by MIT/Cell Press / **13.10** (a) C.J. Harrison et al., *Cytogenetics and Cell Genetics* 35:21–27, copyright 1983 S. Karger A.G., Basel; (c) U.K. Laemmli; (d) B. Hamkalo; (e) O.L. Miller, Jr., and Steve L. McKnight; art by Palay/Beaubois

Chapter 14

14.1 (above) Kevin Magee/Tom Stack & Associates; (below) Dennis Hallinan/FPG / **14.4** Art by Palay/Beaubois / **14.7** Art by Palay/Beaubois / **14.8** Photograph courtesy of Thomas A. Steitz from *Science*, 246:1135–1142, December 1, 1989 / **14.10** Art by L. Calver / **14.11** Micrograph Dr. John E. Heuser, Washington University School of Medicine, St. Louis, MO; art by Palay/Beaubois / **14.12** Art by L. Calver / **14.14** Peter Starlinger

Chapter 15

15.1 (a) Lennart Nilsson © Boehringer Ingelheim International GmbH / **15.2** Art by Palay/Beaubois / **15.4** Art by Palay/Beaubois / **15.5** (a) M. Roth and J. Gall; (e) W. Beerman / **15.6** Stuart Kenter Associates / **15.7** Jack Carey

Chapter 16

16.1 Secchi-Lecague/Roussel–UCLAF/CNRI/SPL/Photo Researchers / **16.2** (a) Dr. Huntington Potter and Dr. David Dressler; (b) C.C. Brinton, Jr., and J. Carnahan / **16.6** Art by Jeanne Schreiber / **Page 252** Photograph Damon Biotech, Inc. / **Pages 252–253** Art by Palay/Beaubois / **16.9** Michael Maloney/San Francisco Chronicle / **16.10** Runk/Schoenberger/Grant Heilman / **16.11** W. Merrill / **16.12** Keith V. Wood / **16.13** (a), (b) Monsanto Company; (c), (d) Calgene, Inc. / **16.14** R. Brinster and R.E. Hammer, School of Veterinary Medicine, University of Pennsylvania

Page 259 S. Stammers/SPL/Photo Researchers

Chapter 17

17.1 (a) Art by Leonard Morgan after R.H. Dott, Jr., and R.L. Batten, *Evolution of the Earth*, third edition, McGraw-Hill, 1981. Reproduced by permission of McGraw-Hill, Inc.; (b) Werner Stoy/Bruce Coleman Ltd.; / **17.2** (a) Jen & Des Bartlett/Bruce Coleman Ltd.; (b) Kenneth W. Fink/Photo Researchers; (c) Dave Watts/A.N.T. Photo Library / **17.4** (a) Courtesy George P. Darwin, Darwin Museum, Down House; (b) Christopher Ralling / **17.5** (a) D. Barrett/Planet Earth Pictures; (b) (above) Field Museum of Natural History (Neg. No. CK21T), Chicago, and the artist Charles R. Knight; (below) Lee Kuhn/FPG / **17.6** (a) Photograph Dieter & Mary Plage/Survival Anglia; art by Leonard Morgan; (b) C.P. Hickman, Jr.; (c), (d) Heather Angel / **17.7** (a) George W. Cox; (b) David Cavagnaro; (c) Heather Angel; (d) Alan Root/Bruce Coleman Ltd; (e) David Steinberg / **17.8** Down House and The Royal College of Surgeons of England / **17.9** Photograph John H. Ostrom, Yale University

Chapter 18

18.1 Elliott Erwitt/Magnum Photos, Inc. / **18.2** Alan Solem / **18.3** (left) Eric Crichton/Bruce Coleman Ltd.; (right) William E. Ferguson / **18.5** After D. Futuyma, *Evolutionary Biology*, Sinauer, 1979 / **18.6** David Cavagnaro / **18.7** (above) David Neal Parks; (below) W. Carter Johnson / **18.9** (a) Thomas N. Taylor; (b) Edward S. Ross / **18.10** (c), (d) Alex Kerstitch / **18.11** After M. Karns and L. Penrose, *Annals of Eugenics*, 15:206–233, 1951 / **18.12** J.A. Bishop and L.M. Cook / **18.13** (a) Bruce Beehler; (b) Charles W. Fowler/National Marine Fisheries; (c) D. Avon/Ardea, London / **18.14** After F. Ayala and J. Valentine, *Evolving*, Benjamin-Cummings, 1979 / **18.15** After V. Grant, *Organismic Evolution*, W.H. Freeman and Co., 1977 / **18.16** Jen & Des Bartlett/Bruce Coleman Ltd. / **Page 289** (a) Nancy Sefton/Photo Researchers; (b) Daniel W. Gotshall / **18.17** After W. Jensen and F.B. Salisbury, *Botany: An Ecological Approach*, Wadsworth, 1972

Chapter 19

19.1 (left) Vatican Museums; (right) Martin Dohrn/SPL/Photo Researchers / **19.2** (a) Patricia G. Gensel; (b) A. Feduccia, *The Age of Birds*, Harvard University Press, 1980; (c) Jonathan Blair/Woodfin Camp & Associates; (d) Donald Baird, Princeton Museum of Natural History / **19.3** David Noble/FPG / **19.4** From T. Storer et al., *General Zoology*, sixth edition, McGraw-Hill, 1979. Reproduced by permission of McGraw-Hill, Inc. / **19.5** Art by Victor Royer / **19.6** Art by Joel Ito / **19.7** (top) Douglas P. Wilson/Eric & David Hosking; (center) Superstock, Inc.; (bottom) E.R. Degginger / **19.8** After "Reconstructing Bird Phylogeny by Comparing DNA's" by C.G. Sibley and J.E. Ahlquist, *Scientific American*, February 1986. Copyright © 1986 by Scientific American, Inc. All rights reserved / **19.9** Chesley Bonestell / **19.10** Gary Byerly, LSU / **19.12** (a) Sidney W. Fox; (b) W. Hargreaves and D. Deamer / **19.13** Art by Leonard Morgan / **19.14** (below) After S.M. Stanley, *Macroevolution: Pattern and Process*, W.H. Freeman and Co., 1979 / **19.15** After P. Dodson, *Evolution: Process and Product*, third edition, Prindle, Weber & Schmidt / **19.16** Art by Leonard Morgan / **19.17** Data from J.J. Sepkoski, Jr., *Paleobiology*, 7(1):36–53 and J.J. Sepkoski, Jr., and M.L. Hulver in Valentine, ed., *Phanerozoic Diversity Patterns: Profiles in Macroevolution*, Princeton University Press, 1985 / **19.18** (a) Stanley W. Awramik; (b) M.R. Walter / **19.19** (a), (b) Neville Pledge/South Australian Museum; (c), (d) Chip Clark / **19.21** (a) H.P. Banks; (b) Patricia G. Gensel / **19.22** From *Evolution of Life*, Linda Gamlin and Gail Vines (Eds.), Oxford University Press, 1987; art by D. & V. Hennings; (b) Rod Salm/Planet Earth Pictures / **Pages 312–313** (a) NASA; (b) © John Gurche 1989; (c) William K. Hartmann / **19.23** Jack Carey / **19.24** (a), (b) Field Museum of Natural History (Neg. Nos. CK46T & CK8T), Chicago, and the artist Charles R. Knight

Chapter 20

20.1 (top) Thomas D. Mangelsen/Images of Nature; (center) Jeffrey Sylvester/FPG; (bottom) Kjell Sandved/Visuals Unlimited / **20.2** Edward S. Ross / **20.3** Art by Raychel Ciemma after C.T. Regan and E. Trewavas, 1932 / **20.4** From F. Salisbury and C. Ross, *Plant Physiology*, fourth edition, Wadsworth, 1991 / **20.6** (a) Suzanne L. Collins and Joseph T. Collins; (b) from *The Amphibians and Reptiles of Missouri* by Tom R. Johnson. Copyright © 1987 by the Conservation Commission of the State of Missouri. Reprinted by permission / **20.7** Kevin Schafer/Tom Stack / **Page 324** Art by D. & V. Hennings / **Pages 325–327** Art by John & Judy Waller / **Page 326** (left to right) Larry Lefever/Grant Heilman; R.I.M. Campbell/Bruce Coleman Ltd.; Runk/Schoenberger/Grant Heilman; Bruce Coleman Ltd.

Chapter 21

21.1 (left) FPG; (right) Douglas Mazonowicz/Gallery of Prehistoric Art / **21.3** (a) Bruce Coleman Ltd.; (b)

Tom McHugh/Photo Researchers; (c) Larry Burrows/Aspect Picture Library / **21.4** Art by D. & V. Hennings / **21.6** © Time Inc. 1965/Larry Burrows Collection / **21.8** Art by D. & V. Hennings / **21.9** (a) Louise M. Robbins; (b) Dr. Donald Johanson, Institute of Human Origins / **21.10–21.11** Art by D. & V. Hennings / **21.12** Photographs by John Reader copyright 1981 / **Page 342** Art by Palay/Beaubois after "Emergence of Modern Humans" by Christopher B. Stringer, *Scientific American*, December 1990. Copyright © 1990 by Scientific American, Inc. All rights reserved

Page 345 © 1990 Arthur M. Greene

Chapter 22

22.1 Tony Brain and David Parker/SPL/Photo Researchers / **22.2** (b) Art by L. Calver / **22.3** Art by Palay/Beaubois / **22.4** (a) George Musil/Visuals Unlimited; (b) K.G. Murti/Visuals Unlimited / **22.5** Art by Palay/Beaubois / **22.6** Kenneth M. Corbett / **Page 352** Kent Wood/Photo Researchers / **22.7** Art by L. Calver / **22.8** L.J. LeBeau, University of Illinois Hospital/BPS / **22.9** (a) CNRI/SPL/Photo Researchers; (b) Stanley Flegler/Visuals Unlimited / **22.10** Micrograph J.J. Cardamone, Jr./BPS / **22.11** Paul A. Zahl, © 1967 National Geographic Society / **22.12** (a) John D. Cunningham/Visuals Unlimited; (b) Tony Brain/SPL/Photo Researchers; (c) P.W. Johnson and J. McN. Sieburth, University of Rhode Island/BPS / **22.13** Stanley W. Watson, *International Journal of Systematic Bacteriology*, 21:254–270, 1971 / **22.14** T.J. Beveridge, University of Guelph/BPS / **22.15** Richard Blakemore / **22.16** Hans Reichenbach, Gesellschaft für Biotechnologische Forschung, Braunschweig, Germany / **Page 361** (a) Art by Palay/Beaubois; (b) (above) H. Stolp; (below) L. Margulis / **22.17** (a) Edward S. Ross; (b) John Shaw/Bruce Coleman Ltd. / **22.18** (a) Art by Leonard Morgan; (b), (c) M. Claviez, G. Gerisch, and R. Guggenheim; (d) London Scientific Films; (e–g) Carolina Biological Supply Company; (h) photograph courtesy Robert R. Kay from R.R. Kay et al., *Development*, 1989 Supplement, pp. 81–90, © The Company of Biologists Ltd. 1989 / **22.19** (a) P.L. Walne and J.H. Arnott, *Planta*, 77:325–354, 1967; (b) T.E. Adams/Visuals Unlimited; art by Palay/Beaubois / **22.20** (a–c) Ronald W. Hoham, Dept. of Biology, Colgate University; (d) (above) G.A. Fryxell; (below) G. Shih and R.G. Kessel, *Living Images*, Jones and Bartlett Publishers, Inc., Boston © 1982 / **22.21** (left) C.C. Lockwood; (right) Florida Department of Natural Resources, Bureau of Marine Research / **22.22** (a) John D. Cunningham/Visuals Unlimited; (b) Jerome Paulin/Visuals Unlimited; (c) David M. Phillips/Visuals Unlimited / **22.23** (a) M. Abbey/Visuals Unlimited; (b) John Clegg/Ardea, London; (c) Manfred Kage/Bruce Coleman Ltd.; (d) T.E. Adams/Visuals Unlimited / **22.24** Micrograph Gary W. Grimes and Steven L'Hernault; art by Palay/Beaubois / **22.25** Gary W. Grimes and Steven L'Hernault / **22.26** Art by Leonard Morgan; micrograph Steven L'Hernault / **Page 369** (a) Richard W. Greene; (b) Laszlo Meszoly in L. Margulis, *Early Life*, Jones and Bartlett Publishers, Inc., Boston, © 1982

Chapter 23

23.1 Robert C. Simpson/Nature Stock / **23.2** Philip Springham from A.D.M. Rayner, *New Scientist*, November 19, 1988 / **23.3** (a) G.T. Cole, University of Texas, Austin/BPS / **23.5** M.S. Fuller, *Zoosporic Fungi in Teaching and Research*, M.S. Fuller and A. Jaworski (Eds.), 1987, Southeastern Publishing Company, Athens, GA / **Page 242** Heather Angel / **23.6** Heather Angel / **23.7** (a) John D. Cunningham/Visuals Unlimited; (b) David M. Phillips/Visuals Unlimited / **23.8** John Hodgin / **23.9** (a) After T. Rost et al., *Botany*, Wiley, 1979; (b), (c) Robert C. Simpson/Nature Stock; (d) Eric Crichton/Bruce Coleman Ltd. / **23.10** (a), (b) Robert C. Simpson/Nature Stock; (c) Victor Duran; (d) Jane Burton/Bruce Coleman Ltd.; (e), (f) Thomas J. Duffy / **23.11** (b) Martyn Ainsworth from A.D.M. Rayner, *New Scientist*, November 19, 1988; (c–e) G. Shih and R.G. Kessel, *Living Images*, Jones and Bartlett Publishers, Inc., Boston, © 1982 /

23.12 (a) G.T. Cole, University of Texas, Austin/BPS; (b) N. Allin and G.L. Barron; (c) G.L. Barron, University of Guelph / 23.13 (above) Mark Mattock/Planet Earth Pictures; (below) Ken Davis/Tom Stack & Associates / 23.14 After Raven, Evert, and Eichhorn, *Biology of Plants,* fourth edition, Worth Publishers, New York, 1986 / 23.15 © 1990 Gary Braasch / 23.16 F.B. Reeves

Chapter 24

24.1 (left) Raymond A. Mendez/Animals Animals; (above) Ken Lewis/Earth Scenes / 24.2 Art by Jeanne Schrieber / 24.3 (a) D.P. Wilson/Eric & David Hosking; (b) Douglas Faulkner/Sally Faulkner Collection / 24.4 (left) Steven C. Wilson/Entheos; (right) J.R. Waaland, University of Washington/BPS / 24.5 (a) Dennis Brokaw; (b) art by Jennifer Wardrip based on Gilbert M. Smith, *Marine Algae of the Monterey Peninsula,* Stanford University Press / 24.6 (a) Hervé Chaumeton/Agence Nature; (b) Alex Kerstitch/Tom Stack & Associates; (c), (d) Ronald W. Hoham, Dept. of Biology, Colgate University / 24.7 Carolina Biological Supply Company / 24.8 Photograph D.J. Patterson/Seaphot Limited: Planet Earth Pictures; art by D. & V. Hennings / 24.9 Hervé Chaumeton/Agence Nature / 24.10 Photograph Jane Burton/Bruce Coleman Ltd.; art by D. & V. Hennings / 24.11 (a), (b) Kingsley R. Stern; (c) John D. Cunningham/Visuals Unlimited / 24.12 Kingsley R. Stern / 24.13 Edward S. Ross / 24.14 (left) W.H. Hodge; (right) Kratz/ZEFA / 24.15 (a) Art by D. & V. Hennings; photograph A. & E. Bomford/Ardea, London; (b) Lee Casebere; (c) Jean Paul Ferrero/Ardea, London / 24.16 Art by Jennifer Wardrip / 24.17 Ed Reschke / 24.18 (a) John H. Gerard; (b) Kingsley R. Stern; (c) Edward S. Ross; (d) F.J. Odendaal, Duke University/BPS / 24.19 Photograph Edward S. Ross; art by D. & V. Hennings / 24.20 (a) Martin Grosnick/Ardea, London; (b) Hans Reinhard/Bruce Coleman Ltd.; (c) Edward S. Ross; (d) Heather Angel; (e) Dick Davis/Photo Researchers (f) Peter F. Zika/Visuals Unlimited; (g) L. Mellichamp/Visuals Unlimited / 24.21 Art by D. & V. Hennings

Chapter 25

25.1 (a) Chip Clark; (b) (above) Jim Stewart/Scripps Institution of Oceanography; (below) Chip Clark / 25.5 Art by D. & V. Hennings / 25.6 (a) (above) Douglas Faulkner/Sally Faulkner Collection; (below) David C. Haas/Tom Stack & Associates; (b) Marty Snyderman/Planet Earth Pictures / 25.7 Art by Palay/Beaubois / 25.8 (b), (c) Kim Taylor/Bruce Coleman Ltd. / 25.9 (a) Frieder Sauer/Bruce Coleman Ltd.; (b) Walter Deas/Seaphot Limited: Planet Earth Pictures; (c) Bill Wood/Seaphot Limited: Planet Earth Pictures; (d) Douglas Faulkner /Sally Faulkner Collection; (e) F. Stuart Westmorland /Tom Stack & Associates / 25.11 Photograph Andrew Mounter/Seaphot Limited: Planet Earth Pictures; art by Raychel Ciemma / 25.12 Photograph E.R. Degginger; art by Joan Carol after T. Storer et al., *General Zoology,* sixth edition, © 1979 McGraw-Hill / 25.13 Larry Madin/Planet Earth Pictures / 25.14 Photograph Kim Taylor/Bruce Coleman Ltd.; art by K. Kasnot / 25.15 (above) Cath Ellis, University of Hull/SPL/Photo Researchers; (below) Robert & Linda Mitchell / **Page 416** (a) Photograph Robert L. Calentine / **Page 417** (b) Photograph Carolina Biological Supply Company; art by K. Kasnot; (c) Lorus J. and Margery Milne; (d) Dianora Niccolini / 25.16 Kjell B. Sandved / 25.18 Photograph J. Solliday/BPS; art by Raychel Ciemma / 25.19 Art by Palay/Beaubois / 25.21 (c) Anthony & Elizabeth Bomford/Ardea, London; (d) Kjell B. Sandved / 25.22 Jeff Foott/Tom Stack & Associates / 25.24 (a) Rick M. Harbo; (b) Alex Kerstitch; (c) Hervé Chaumeton/Agence Nature / 25.25 Art by Laszlo Meszoly and D. & V. Hennings / 25.26 Hervé Chaumeton/Agence Nature / 25.27 J. Grossauer/ZEFA / 25.28 Photograph Douglas Faulkner/Sally Faulkner Collection; art by Laszlo Meszoly and D. & V. Hennings / 25.29 © Cabisco/Visuals Unlimited / 25.30 J.A.L. Cooke/Oxford Scientific Films / 25.31 (a) Hervé Chaumeton/Agence Nature; (b) Jon Kenfield/Bruce Coleman Ltd. / 25.32 Art by Raychel Ciemma / 25.33 C.B. and D.W. Frith/Bruce Coleman Ltd. / 25.34 (a) Peter Green/Ardea, London; (b) Angelo Giampiccolo/FPG; (c) Jane Burton/Bruce Coleman Ltd. / 25.35 (a) John H. Gerard; (b) Ken Lucas/Seaphot Limited: Planet Earth Pictures; (c) P.J. Bryant, University of California, Irvine/BPS / 25.37 (a) Frans Lanting/Bruce Coleman Ltd.; (b) photograph Hervé Chaumeton/Agence Nature; art by Laszlo Meszoly / 25.38 Fred Bavendam/Peter Arnold, Inc. / 25.39 Agence Nature / 25.40 (a) Z. Leszczynski/Animals Animals; (b) Steve Martin/Tom Stack & Associates / 25.41 Art by D. & V. Hennings / 25.42 (a) David Maitland/Seaphot Limited: Planet Earth Pictures; (b–e), (g–i), (k) Edward S. Ross; (f) Ralph A. Reinhold/FPG; (j) C.P. Hickman, Jr. / 25.43 (a) Ian Took/Biofotos; (b) Kjell B. Sandved; (c) John Mason/Ardea, London; (d) Chris Huss/The Wildlife Collection / 25.44 (a) Hervé Chaumeton/Agence Nature; (b), (c) art by L. Calver / 25.45 (a) Kjell B. Sandved; (b) Jane Burton/Bruce Coleman Ltd.

Chapter 26

26.1 (a) Tom McHugh/Photo Researchers; (b) Jean Phillipe Varin/Jacana/Photo Researchers / 26.2 Art by D. & V. Hennings / 26.3 Photographs (left) Rick M. Harbo; (above) Peter Parks/Oxford Scientific Films/Animals Animals; (a–d) from *Living Invertebrates,* V. & J. Pearse and M. & R. Buchsbaum, The Boxwood Press, 1987. Used by permission / 26.4 (a) C.R. Wyttenbach, University of Kansas/BPS; art by D. & V. Hennings / 26.5 Photograph Hervé Chaumeton/Agence Nature; art by Laszlo Meszoly and D. & V. Hennings / 26.6 Art by D. & V. Hennings / 26.7 After A.S. Romer and T.S. Parsons, *The Vertebrate Body,* sixth edition, Saunders College Publishing, © 1986 CBS College Publishing; art by Laszlo Meszoly and D. & V. Hennings / 26.8 After C.P. Hickman, Jr., and L.S. Roberts, *Integrated Principles of Zoology,* seventh edition, St. Louis: Times Mirror/Mosby College Publishing, 1984; art by Palay / Beaubois / 26.9 Heather Angel / 26.10 (a) Allan Power/Bruce Coleman Ltd.; (b) Erwin Christian/ZEFA; (c) Tom McHugh/Photo Researchers; (d) Patrice Ceisel/© 1986 John G. Shedd Aquarium; (e) Douglas Faulkner/Sally Faulkner Collection; (f) Robert & Linda Mitchell; (g) William H. Amos / 26.11 Photograph Bill Wood/Bruce Coleman Ltd.; art by Raychel Ciemma / 26.12 Peter Scoones/Seaphot Limited: Planet Earth Pictures / 26.13 Art by Laszlo Meszoly and D. & V. Hennings / 26.14 Art by D. & V. Hennings after Romer and others / 26.15 (a) From *The Vertebrate Body,* sixth edition, copyright © 1986 by Saunders College Publishing, reprinted by permission of the publisher; art by Leonard Morgan; (b) Jerry W. Nagel; (c) Stephen Dalton/Photo Researchers; (d) © John Serrano/Visuals Unlimited; (e) Juan M. Renjifo/Animals Animals / 26.17 (a) Zig Leszczynski/Animals Animals / 26.18 Art by D. & V. Hennings / 26.19 (a) D. Kaleth/Image Bank; (b) Peter Scoones/Seaphot Limited: Planet Earth Pictures / 26.20 (a) Andrew Dennis/A.N.T. Photo Library; (b) Kim Taylor/Bruce Coleman Ltd.; (c) Stephen Dalton/Photo Researchers; art by Raychel Ciemma; (d) Bob McKeever/Tom Stack & Associates; (e) C.B. & D.W. Frith/Bruce Coleman Ltd.; (f) W.J. Weber/Visuals Unlimited / 26.21 (a) Heather Angel; (b) Kevin Schafer/Tom Stack & Associates; (c) W.A. Banaszewski/Visuals Unlimited / 26.22 (a) Robert A. Tyrrell; (b) Rajesh Bedi; (c) J.L.G. Grande/Bruce Coleman Ltd.; (d) Thomas D. Mangelsen/Images of Nature / 26.23 (a) Gerard Lacz/A.N.T. Photo Library; (b) art by D. & V. Hennings / 26.24 D. & V. Blagden/A.N.T. Photo Library / 26.25 (a) Jack Dermid; (b) Douglas Faulkner/Photo Researchers; (c) Clem Haagner/Ardea, London; (d), (e) Leonard Lee Rue III/FPG; (f) Sandy Roessler/FPG

Chapter 27

27.1 (a) Roger Werth; (b) © 1980 Gary Braasch; (c) © 1989 Gary Braasch / 27.3 (left) Micrograph James D. Mauseth, *Plant Anatomy,* Benjamin-Cummings, 1988; (a–c) Biophoto Associates / 27.4 Thomas Eisner, Cornell University / 27.5 Art by Jennifer Wardrip / 27.6 Micrographs H.A. Core, W.A. Coté, and A.C. Day, *Wood Structure and Identification,* second edition, Syracuse University Press, 1979 / 27.7 (a) Chuck Brown; (b) G. Shih and R.G. Kessel, *Living Images,* Jones and Bartlett Publishers, Inc., Boston, © 1982 / 27.8 Art by D. & V. Hennings / 27.9 (a), (b) Edward S. Ross / 27.10 Art by D. & V. Hennings; (center) Carolina Biological Supply Company; (right) James W. Perry / 27.11 (left) Art by D. & V. Hennings: (center) Ray F. Evert; (right) James W. Perry / 27.12 (a) Robert & Linda Mitchell; (b), (c) Roland R. Dute / 27.13 (b–d) E.R. Degginger / 27.14 Art by D. & V. Hennings / 27.15 Heather Angel / 27.16 C.E. Jeffree et al., *Planta,* 172(1):20–37, 1987. Reprinted by permission of C.E. Jeffree and Springer-Verlag; (b) art by D. & V. Hennings / 27.17 John E. Hodgin / 27.18 Micrograph E.R. Degginger / 27.19 Sketch after T. Rost et al., *Botany: A Brief Introduction to Plant Biology,* second edition, © 1984, John Wiley & Sons; micrographs Chuck Brown / 27.20 Carolina Biological Supply Company / 27.21 Art by Palay/Beaubois / 27.22 Ripon Microslides, Inc. / 27.25 After Marian Reeve / 27.26 (b) Jerry D. Davis / 27.27 (a) Biophoto Associates; (b) H.A. Core, W.A. Coté, and A.C. Day, *Wood Structure and Identification,* second edition, Syracuse University Press, 1979

Chapter 28

28.1 (a) Robert & Linda Mitchell; micrograph John N.A. Lott, *Scanning Electron Microscope Study of Green Plants,* St. Louis: C.V. Mosby Company, 1976; (b) Robert C. Simpson/Nature Stock / 28.2 (a), (b) Art by Jennifer Wardrip; (c) Mark E. Dudley and Sharon R. Long; (d) Adrian P. Davies/Bruce Coleman Ltd.; (e) NifTAL Project, University of Hawaii, Maui / 28.3 Micrograph Jean Paul Revel / 28.4 Art by Leonard Morgan / 28.6 Art by Leonard Morgan / 28.7 (a) John Troughton and L.A. Donaldson; (b) W. Thomson, *American Journal of Botany,* 57(3):316, 1970 / 28.8 Micrograph Jeremy Burgess/SPL/Photo Researchers; art by Palay/Beaubois / 28.9 T.A. Mansfield / 28.11 Martin Zimmerman, *Science,* 133:73–79, © AAAS 1961 / 28.12 (b) David Fisher; (c–d) micrographs David Fisher; art by Palay/Beaubois

Chapter 29

29.1 (a) (above) Edward S. Ross; (below) Thomas D.W. Friedmann/Photo Researchers; (b) Jeffry Myers/FPG / 29.3 Photograph Bonnie Rauch/ Photo Researchers / 29.4 John Shaw/Bruce Coleman Ltd. / 29.5 (a), (b) David M. Phillips/Visuals Unlimited; (c) David Scharf/Peter Arnold, Inc. / 29.7 Art by Leonard Morgan / 29.8 Art by D. & V. Hennings / **Page 506** (a) Peter Steyn/Ardea, London/ (c) Thomas Eisner, Cornell University / **Page 507** (b) M.P.L. Fogden/Bruce Coleman Ltd.; (d) Edward S. Ross / **Page 508** (e) Ted Schwartz / 29.9 (a), (b) Patricia Schulz; (c), (d) Ray F. Evert; (e), (f) Ripon Microslides; (far right) Kingsley R. Stern / 29.10 F. Bracegirdle and P. Miles, *An Atlas of Plant Structure,* Heinemann Educational Books, 1977 / 29.11 Janet Jones / 29.12 (a) B.J. Miller, Fairfax, VA/BPS; (b) R. Carr/Bruce Coleman Ltd.; (c) Richard H. Gross, Motlow State Community College / 29.13 (a) Grant Heilman; (b) Kjell Sandved / **Pages 512–513** John Alcock

Chapter 30

30.1 (a) R. Lyons/Visuals Unlimited; (b) Michael A. Keller/FPG / 30.2 Photograph Carolina Biological Supply Company / 30.3 Photograph Hervé Chaumeton/Agence Nature / 30.4 Art by Palay/ Beaubois / 30.5 (a) Kingsley R. Stern / 30.6 Frank B. Salisbury /

30.7 John Digby and Richard Firn / 30.8 B.E. Juniper / 30.9 Cary Mitchell / 30.10 Frank B. Salisbury / 30.12 Frank B. Salisbury / 30.14 Jan Zeevart / 30.15 Photograph N.R. Lersten / 30.16 A.C. Leopold et al., *Plant Physiology*, 34:570, 1958 / 30.17 A.C. Leopold and M. Kawase, *American Journal of Botany*, 51:294–298, 1964 / 30.18 R.J. Downs in T.T. Kozlowski, ed., *Tree Growth*, The Ronald Press, 1962 / **Page 528** Edward S. Ross / **Page 529** Dennis Brokaw

Page 531 © Kevin Schafer

Chapter 31

31.1 David Macdonald / 31.2 Photographs (a) Lennart Nilsson from *Behold Man*, © 1974 Albert Bonniers Forlag and Little, Brown and Company, Boston; (b) Manfred Kage/Bruce Coleman Ltd.; (c) Ed Reschke/Peter Arnold Inc. / 31.3 (a) Art by Palay/Beaubois; (b) Focus on Sports; (inset) Manfred Kage/Bruce Coleman Ltd. / 31.4 Art by Palay/Beaubois / 31.5 Photographs Ed Reschke / 31.6 (left) Art by L. Calver / 31.7 Photographs Ed Reschke / 31.8 Lennart Nilsson from *Behold Man*, © 1974 Albert Bonniers Forlag and Little, Brown and Company, Boston / 31.9 Art by L. Calver / 31.10 Art by Palay/Beaubois / **Page 544** (a) Manfred Kage/Bruce Coleman Ltd.; (b–d) Ed Reschke

Chapter 32

32.1 Adrian Warren/Ardea, London / 32.2 Manfred Kage/Peter Arnold, Inc. / 32.3 (left) Art by Kevin Somerville; (right) art by L. Calver / 32.4, 32.6 Art by Leonard Morgan / 32.7 (b) A.L. Hodgkin, *Journal of Physiology*, vol. 131, 1956 / 32.8 (top) Art by Leonard Morgan; (bottom) art by Jeanne Schreiber / 32.9 (a) Carolina Biological Supply Company; art by Leonard Morgan / 32.10 Art by D. & V. Hennings; (c) J.E. Heuser and T.S. Reese / **Page 558** Painting by Sir Charles Bell, 1809, courtesy of Royal College of Surgeons, Edinburgh / 32.12 (a) Art by Robert Demarest; (b) from *Tissues and Organs: A Text-Atlas of Scanning Electron Microscopy*, by R.G. Kessel and R.H. Kardon. Copyright © 1979 by W.H. Freeman and Company. Reprinted with permission / 32.13 Art by K. Kasnot

Chapter 33

33.1 Comstock/Comstock Inc. / **Page 564** Art by D. & V. Hennings / 33.2 Photograph Francois Gohier/Photo Researchers; art by Raychel Ciemma / 33.3 Art by Palay/Beaubois / 33.4–33.5 Art by Kevin Somerville / 33.6 (b) Art by Kevin Somerville / 33.8 Art by Robert Demarest / 33.9 (a) Art by Kevin Somerville; (b) Manfred Kage/Peter Arnold, Inc. / 33.11 Art by Kevin Somerville / 33.12 C. Yokochi and J. Rohen, *Photographic Anatomy of the Human Body*, second edition, Igaku-Shoin Ltd., 1979 / **Page 573** Art by Palay/Beaubois / 33.13 Art by Joel Ito / 33.14 Art by Palay/Beaubois after Penfield and Rasmussen, *The Cerebral Cortex of Man*, copyright © 1950 Macmillan Publishing Company, Inc. Renewed 1978 by Theodore Rasmussen / 33.15 Art by Robert Demarest / 33.16 After H. Jasper, 1941

Chapter 34

34.1 Hugo van Lawick / 34.2–34.3 Art by Kevin Somerville / 34.4–34.7 Art by Robert Demarest / 34.8 (a) Mitchell Layton; (b) Syndication International (1986) Ltd. / 34.9 Photographs courtesy of Dr. William H. Daughaday, Washington University School of Medicine. From A.I. Mendelhoff and D.E. Smith, eds., *American Journal of Medicine*, 20:133 (1956) / 34.12 The Bettmann Archive / 34.13 Biophoto Associates /SPL/Photo Researchers / 34.14 Art by Leonard Morgan / **Page 592** Evan Cerasoli

Chapter 35

35.1 Merlin D. Tuttle, Bat Conservation International / 35.2 Eric A. Newman / 35.3 Art by Kevin Somerville / 35.4 Art by Palay/Beaubois after Penfield and Rasmussen, *The Cerebral Cortex of Man*, copyright © 1950 Macmillan Publishing Company, Inc. Renewed 1978 by Theodore Rasmussen / 35.5 From Hensel and Bowman, *Journal of Physiology*, 23:564–568, 1960 / 35.6 Art by Ron Ervin; photograph Ed Reschke / 35.7 Art by D. & V. Hennings / 35.8 Art by Robert Demarest; micrograph Omikron/SPL/Photo Researchers / 35.9 Art by Kevin Somerville / 35.10 Art by Robert Demarest / 35.11 (a), (b) Robert E. Preston, courtesy Joseph E. Hawkins, Kresge Hearing Research Institute, University of Michigan Medical School / 35.12 Photograph Edward W. Bower/ © 1991 TIB/West; art by Kevin Somerville / 35.13 (a) Keith Gillett/Tom Stack & Associates; (b) after M. Gardiner, *The Biology of Vertebrates*, McGraw-Hill, 1972 / 35.14 (a) E.R. Degginger / 35.15 G.A. Mazohkin-Porshnykov (1958). Reprinted with permission from *Insect Vision*, © 1969 Plenum Press / 35.16 Art by Robert Demarest / 35.17–35.18 Art by Kevin Somerville / 35.19 Micrograph Lennart Nilsson © Boehringer Ingelheim International GmbH / **Page 613** Photographs Gerry Ellis/The Wildlife Collection; art by Kevin Somerville / 35.20 Art by Robert Demarest / 35.21 Art by Palay/Beaubois after S. Kuffler and J. Nicholls, *From Neuron to Brain*, Sinauer, 1977 / **Page 616** Art by Robert Demarest

Chapter 36

36.1 Jeff Schultz/AlaskaStock Images / 36.2 Art by L. Calver / 36.3 (a) Robert & Linda Mitchell; (b) Jane Burton/Bruce Coleman Ltd. / 36.4 Chaumeton-Lanceau/Agence Nature / 36.5 Art by Robert Demarest / 36.6 Ed Reschke / 36.7 Michael Keller/FPG / 36.8 CNRI/SPL/Photo Researchers / 36.9 Linda Pitkin/Planet Earth Pictures / 36.10 Photograph Stephen Dalton/Photo Researchers; art by Raychel Ciemma / 36.11 D.A. Parry, *Journal of Experimental Biology*, 36:654, 1959 / 36.12 Art by D. & V. Hennings / 36.13 Art by Joel Ito; micrograph Ed Reschke / 36.14 Art by K. Kasnot / 36.15 National Osteoporosis Foundation / 36.16 (b) Art by Ron Ervin / **Page 629** Photograph C. Yokochi and J. Rohen, *Photographic Anatomy of the Human Body*, second edition, Igaku-Shoin Ltd., 1979 / 36.16 (b), (c) Art by L. Calver / 36.18 (a) Ed Reschke; (b) D. Fawcett, *The Cell*, Philadelphia: W.B. Saunders Co., 1966 / 36.21 Art by R.M. Jensen / 36.22 Adapted from R. Eckert and D. Randall, *Animal Physiology: Mechanisms and Adaptations*, second edition, W.H. Freeman and Co., 1983 / 36.23 Art by Kevin Somerville; (b) Ed Reschke / **Page 637** Photograph Michael Neveux / 36.25 N.H.P.A./A.N.T. Photo Library

Chapter 37

37.1 (a) D. Robert Franz/Planet Earth Pictures; (b) art by D. & V. Hennings adapted from *Mammalogy*, third edition, by Terry Vaughan, copyright © 1986 by Saunders College Publishing. Used by permission of the publisher / 37.2 (a) Kim Taylor/Bruce Coleman Ltd.; (b) Wardene Weisser/Ardea, London / 37.4 Art by Robert Demarest / 37.6 Art by Raychel Ciemma; (b) after A. Vander et al., *Human Physiology: Mechanisms of Body Function*, fifth edition, McGraw-Hill, 1990. Used by permission / 37.8 (a), (c) Lennart Nilsson © Boehringer Ingelheim International GmbH; (b) Biophoto Associates/SPL/Photo Researchers; art by Victor Royer / 37.9 Art by Robert Demarest / 37.10 Art by L. Calver / 37.11 (b) Steven Jones/FPG / **Page 650** Photograph CNRI/Phototake / 37.12 Photograph Ralph Pleasant/FPG / 37.13 Modified after A. Vander et al. *Human Physiology*, fourth edition, McGraw-Hill, 1985 / **Page 655** Photograph courtesy of David Steinberg

Chapter 38

38.1 (a) From A.D. Waller, *Physiology, The Servant of Medicine*, Hitchcock Lectures, University of London Press, 1910; (b) photograph courtesy of The New York Academy of Medicine Library / 38.3 (b) (below) After M. Labarbera and S. Vogel, *American Scientist*, 70:54–60, 1982 / **Page 662** Art by Palay/Beaubois / 38.4 (a) CNRI/SPL/Photo Researchers; (b) Lennart Nilsson from *Behold Man*, © 1974 by Albert Bonniers Forlag and Little, Brown and Company, Boston / 38.5 (left) Art by L. Calver and Victor Royer; (right) art by Victor Royer / 38.6 (a) Art by Leonard Morgan; (b) art by Kevin Somerville / 38.7 (a) Art by Joel Ito; (b) C. Yokochi and J. Rohen, *Photographic Anatomy of the Human Body*, second edition, Igaku-Shoin Ltd., 1979 / 38.11 Art by Robert Demarest based on A. Spence, *Basic Human Anatomy*, Benjamin-Cummings, 1982 / 38.14 Art by Raychel Ciemma / 38.15 After J. A. Gosling et al., *Atlas of Human Anatomy with Integrated Text*, copyright © 1985 by Gower Medical Publishing Ltd. / **Page 671** (a) (above) Ed Reschke; (below) F. Sloop and W. Ober/Visuals Unlimited / 38.16 Photograph Lennart Nilsson © Boehringer Ingelheim International GmbH / 38.17 (a) After F. Ayala and J. Kiger, *Modern Genetics*, © 1980 Benjamin-Cummings; (b) Lester V. Bergman & Associates, Inc. / 38.18 After Gerard J. Tortora and Nicholas P. Anagnostakos, *Principles of Anatomy and Physiology*, sixth edition, copyright © 1990 by Biological Sciences Textbooks, Inc., A & P Textbooks, Inc. and Elia-Sparta, Inc. Reprinted by permission of Harper Collins Publishers / 38.19 Art by Kevin Somerville

Chapter 39

39.1 (a) The Granger Collection, New York; (b) Lennart Nilsson © Boehringer Ingelheim International GmbH / 39.2 Lennart Nilsson © Boehringer Ingelheim International GmbH / 39.5 Art by Palay/Beaubois / 39.6 Art by L. Calver and Victor Royer / 39.7 Art by Palay/Beaubois after S. Tonegawa, *Scientific Ameican*, October 1965 / **Page 689** Photographs Dr. Gilla Kaplan / 39.8 Art by Palay/Beaubois / 39.9 Art by Palay/Beaubois after B. Alberts et al., *Molecualr Biology of the Cell*, Garland Publishing Company, 1983 / **Page 693** (a) Art by L. Calver / **Page 694** (b), (c) Micrographs Z. Salahuddin, National Institutes of Health

Chapter 40

40.1 Galen Rowell/Peter Arnold, Inc. / 40.2 (b) Steve Lissau/Rainbow; (c) Peter Parks/Oxford Scientific Films / 40.4 Ed Reschke / 40.5 Art by D. & V. Hennings after C. P. Hickman et al., *Integrated Principles of Zoology*, sixth edition, St. Louis: C. V. Mosby Co., 1979 / 40.6 After C. P. Hickman et al., *Integrated Principles of Zoology*, sixth edition, St. Louis: C. V. Mosby Co., 1979 / 40.7 Art by Palay/Beaubois adapted from H. Scharnke, *Z. vergl. Physiol.*, 25:548–583 (1938) in *Form and Function in Birds*, Vol. 4, A. King and J. McLelland, Eds., Academic Press, 1989; micrograph H.R. Duncker, Justus-Liebig University, Giessen, Germany / 40.8 Art by L. Calver / 40.9 Art by Kevin Somerville / 40.11 CNRI/SPL/Photo Researchers / 40.12 After A. Vander et al., *Human Physiology*, third edition, McGraw-Hill, 1980 / 40.13 Art by K. Kasnot / 40.14 From L.G. Mitchell, J.A. Mutchmor, and W.D. Dolphin, *Zoology*, © 1988 Benjamin-Cummings Publishing Company / 40.16 Art by Leonard Morgan / **Page 710** (a) Gerard D. McLane / **Page 711** (b) Lennart Nilsson from *Behold Man*, © 1974 by Albert Bonniers Forlag and Little, Brown and Company, Boston / **Page 713** Christian Zuber/Bruce Coleman Ltd. / 40.17 (b) Giorgio Gualco/Bruce Coleman Ltd.

Chapter 41

41.1 Claude Steelman/Tom Stack & Associates / 41.3 Art by Kevin Somerville / 41.4 Art by Robert Demarest / 41.8 Art by Joel Ito / 41.9 Thomas D. Mangelsen/Images of Nature / 41.10 (left) David Jennings/Image Works; (right) Evan Cerasoli / 41.11 Art by Kevin Somerville / 41.12 The Bettmann Archive / 41.13 Terry Vaughan / 41.14 Fred Bruemmer

Chapter 42

42.1 (a) Hans Pfletschinger; (b) Carolina Biological Supply Company; (c–e) John H. Gerard / **42.2** (a) Frieder Sauer/Bruce Coleman Ltd.; (b) Evan Cerasoli; (c) Fred McKinney/FPG; (d) Carolina Biological Supply Company; (e) Leonard Lee Rue III / **42.4** Art by Palay/Beaubois adapted from R.G. Ham and M.J. Veomett, *Mechanisms of Development*, St. Louis: C.V. Mosby Co., 1980 / **42.5** Photographs Carolina Biological Supply Company; sketch after M.B. Patten, *Early Embryology of the Chick*, fifth edition, McGraw-Hill, 1971 / **42.6** J.R. Whittaker / **42.8** Photographs Carolina Biological Supply Company / **42.9** (a), (b) Micrographs J.B. Morrill and N. Ruediger; (c), (d) Micrographs J.B. Morrill; art by Raychel Ciemma after V.E. Foe and B.M. Alberts, *Journal of Cell Science*, 61:32, © The Company of Biologists 1983 / **42.10** Micrographs F.R. Turner; art by Raychel Ciemma / **42.11** Sketches after B. Burnside, *Developmental Biology*, 26:416–441, 1971; micrograph K.W. Tosney / **42.12** (a–c) Adapted by permission of Macmillan Publishing Company from *Developmental Biology: Patterns, Problems, Principles* by John W. Saunders, Jr., Copyright © 1982 by John W. Saunders, Jr.; (d) S.R. Hilfer and J.W. Yang, *The Anatomical Record*, 197:423–433, 1980 / **42.13** (a) K.W. Tosney / **42.14** Art by Palay/Beaubois after Robert F. Weaver and Philip W. Hedrick, *Genetics*. Copyright © 1989 Wm. C. Brown Publishers, Dubuque, Iowa. All rights reserved. Reprinted by permission / **42.15** After J.W. Fristrom et al., in E.W. Hanly, ed., *Problems in Biology: RNA Development*, University of Utah Press / **42.16** Carolina Biological Supply Company / **42.17** Sketches after Willier, Weiss, and Hamburger, *Analysis of Development*, Philadelphia: W.B. Saunders Co., 1955; photograph Roger K. Burnard / **42.18** Art by Raychel Ciemma adapted from L.B. Arey, *Developmental Anatomy*, Philadelphia: W.B. Saunders Co., 1965 / **Page 755** Dennis Green/Bruce Coleman Ltd.

Chapter 43

43.1 Lennart Nilsson from *A Child Is Born*, © 1966, 1977 Dell Publishing Company, Inc. / **43.2** (left) Art by Ron Ervin; (right) art by L. Calver / **43.3** Art by L. Calver; (c) R.G. Kessel and R.H. Kardon, *Tissues and Organs: A Text-Atlas of Scanning Electron Microscopy*, W.H. Freeman and Co., copyright © 1979 / **43.4–43.5** Art by Ron Ervin / **43.7** Art by Ron Ervin; (right) art by L. Calver / **43.8** (top) Art by Robert Demarest; photograph Lennart Nilsson from *A Child Is Born*, © 1966, 1977 Dell Publishing Company, Inc. / **43.10** Art by Robert Demarest / **43.11** Art by Robert Demarest; (left) micrograph from Lennart Nilsson, *A Child Is Born*, © 1966, 1977 Dell Publishing Company, Inc.; (right) from Lennart Nilsson, *Behold Man*, © 1974 by Albert Bonniers Forlag and Little, Brown and Co., Boston / **43.12** Art by Robert Demarest / **43.13** Art by L. Calver; (c) after A.S. Romer and T.S. Parsons, *The Vertebrate Body*, sixth edition, Saunders College Publishing, © CBS College Publishing / **43.14** Art by L. Calver after Bruce Carlson, *Patten's Foundations of Embryology*, fourth edition, McGraw-Hill, 1981 / **43.15** From Lennart Nilsson, *A Child Is Born*, © 1966, 1977 Dell Publishing Company, Inc. / **Page 772** Modified from Keith L. Moore, *The Developing Human: Clinically Oriented Embryology*, fourth edition, Philadelphia: W.B. Saunders Co., 1988 / **Page 773** James W. Hanson, M.D. / **43.16–43.17** From Lennart Nilsson, *A Child Is Born*, © 1966, 1977 Dell Publishing Company, Inc. / **43.18** Art by Robert Demarest / **Page 776** Mills-Peninsula Hospitals / **43.19** Art by Ron Ervin / **Page 781** (a) Cheun-mo To and C.C. Brinton / **Page 782** (b) Joel B. Baseman

Page 785 Alan and Sandy Carey

Chapter 44

44.1 Antoinette Jongen/FPG / **44.2** (above) Fran Allan/Animals Animals; (below) E.R. Degginger / **44.3** (c) Stanley Flegler/Visuals Unlimited / **44.5** (b) E. Vetter/ZEFA / **Page 794** Photograph Eric Crichton /Bruce Coleman Ltd. / **44.6** (left) Jonathan Scott/ Planet Earth Pictures; (right) (above) Wisniewski/ ZEFA; (below) Fred Bavendam/Peter Arnold, Inc. / **44.7** (a), (b) John Endler; (c) art by Raychel Ciemma / **44.8** Photograph NASA / **44.11** After G.T. Miller, *Living in the Environment*, sixth edition, Wadsworth, 1990 / **44.12** Data from Population Reference Bureau

Chapter 45

45.1 (left) Edward S. Ross; (right) Dona Hutchins / **45.2** (a), (c) Harlo H. Hadow; (b) Bob and Miriam Francis/Tom Stack & Associates / **45.3** Robert A. Tyrrell / **45.4** After G. Gause, 1934 / **45.5** Stephen G. Tilley / **45.6** Photograph John Dominis, *Life Magazine*, © Time, Inc. / **45.7** (a) Ed Cesar/Photo Researchers / **45.8** Edward S. Ross / **45.9** (a) Roger T. Petersen/NAS/Photo Researchers; (b), (c) Thomas Eisner, Cornell University / **45.10** (a) Douglas Faulkner/Sally Faulkner Collection; (b) W.M. Laetsch; (c), (d) Edward S. Ross; (e) James H. Carmichael / **45.11** Data from P. Price and H. Tripp, *Canadian Entymology*, 104:1003–1016, 1972 / **45.12** After N. Weland and F. Bazazz, *Ecology*, 56:681–688, © 1975 Ecological Society of America / **45.13** (a), (b) Jane Burton/Bruce Coleman Ltd.; (c) Heather Angel; graph from Jane Lubchenco, *American Naturalist*, 112:23–29, © 1978 by The University of Chicago Press / **Page 818** R. Slavin/FPG / **45.14** (a–f), (i) Roger K. Bernard; (g), (h) E.R. Degginger / **45.15** Photograph Dr. Harold Simon/Tom Stack & Associates; (below) after S. Fridriksson, *Evolution of Life on a Volcanic Island*, Butterworth, London, 1975 / **45.16** After J.M. Diamond, *Proceedings of the National Academy of Sciences*, 69:3199–3201, 1972 / **45.17** After M.H. Williamson, *Island Population*, Oxford University Press, 1981 / **45.18** (a) After F.G. Stehli et al., *Geological Society of America Bulletin*, 78:455–466, 1967; (b) after M. Kusenov, *Evolution*, 11:298–299, 1957; (c) after T. Dobzhansky, *American Scientist*, 38:209–221, 1950

Chapter 46

46.1 Wolfgang Kaehler / **46.3** Photograph Sharon R. Chester / **46.7** (b) Photograph Steven D. Bach / **46.10** Photograph © 1991 Gary Braasch / **46.11** Art by Raychel Ciemma / **46.12** (a) Photograph by Gene E. Likens from G.E. Likens and F.H. Bormann, *Proceedings First International Congress of Ecology*, pp. 330–335, September 1974, Centre Agric. Publ. Doc. Wagenigen, The Hague, the Netherlands; (b), (c) photographs by Gene E. Likens from G.E. Likens et al., *Ecology Monograph*, 40(1):23–47, 1970 / **Pages 838–839** Art by Raychel Ciemma; photograph NASA / **46.14** Photograph William J. Weber/Visuals Unlimited

Chapter 47

47.1 (a) (left) Edward S. Ross; (right) David Noble/ FPG; (b) (left) Edward S. Ross; (right) Richard Coomber/Planet Earth Pictures / **47.3** (b) Art by L. Calver / **47.5** (b) Edward S. Ross / **47.7** Art by Raychel Ciemma / **47.9** Art by Joan Carol / **47.11** After Whittaker; Bland; and Tilman / **47.12** Harlo H. Hadow / **47.13** (a) Jack Wilburn/Earth Scenes; (b) John D. Cunningham / Visuals Unlimited / **47.14** Kenneth W. Fink/Ardea, London / **47.15** Ray Wagner/Save the Tall Grass Prairie, Inc. / **47.16** Jonathan Scott/ Planet Earth Pictures / **47.17** © 1991 Gary Braasch / **47.18** Thomas E. Hemmerly / **47.19** Dennis Brokaw / **47.20** Jack Carey / **47.21** Fred Bruemmer / **47.22** D.W. MacManiman / **47.24** Modified after Edward S. Deevy, Jr., *Scientific American*, October 1951 / **47.25** D.W. Schindler, *Science*, 184:897–899 / **47.27** (a) E.R. Degginger; (b) art by D. & V. Hennings / **47.28** Courtesy of J.L. Sumich, *Biology of Marine Life*, fourth edition, William C. Brown, 1988 / **47.29** (left and center) © 1991 Gary Braasch; (right) Phil Degginger / **47.31** (top right) Jim Doran; all other photographs Douglas Faulkner/Sally Faulkner Collection / **47.32** (a) McCutcheon / ZEFA; (b) Chuck Niklin; (c) Fred Grassle, Woods Hole Institution of Oceanography; (d) Robert Hessler / **Page 871** Photographs R. Legeckis/NOAA

Chapter 48

48.1 (a) Gerry Ellis/The Wildlife Collection; (b) Adolf Schmidecker/FPG; (c) Edward S. Ross / **48.2** Photograph John Lawlor/FPG / **48.3** After G.T. Miller, *Environmental Science: An Introduction*, Wadsworth, 1986, and the Environmental Protection Agency / **48.4** (a) USDA Forest Service; (b) Heather Angel / **48.5** (bottom left) National Science Foundation; (top left; right) NASA / **48.6** Dr. Charles Henneghien/Bruce Coleman Ltd. / **48.7** From Water Resources Council / **48.8** Data from G.T. Miller / **48.9** (above) R. Bieregaard/Photo Researchers; (below) after G.T. Miller, *Living in the Environment*, sixth edition, Wadsworth, 1990 / **48.10** NASA / **48.11** USDA Soil Conservation Service/ Thomas G. Meier / **48.12** Agency for International Development / **48.13** Data from G.T. Miller / **Page 889** J. McLoughlin/FPG / **Page 891** © 1983 Billy Grimes

Chapter 49

49.1 (left) Robert Maier/Animals Animals; (right) (above) John Bova/Photo Researchers; (below) photograph Jack Clark/Comstock Inc.; graph L. Clark, *Parasitology Today*, 6(11), 1990, Elsevier Trends Journals, Cambridge, U.K. / **49.2** (a) Eugene Kozloff; (b), (c) Stevan Arnold / **49.3** Photograph John S. Dunning/Ardea, London; sonogram J. Bruce Falls and Tom Dickson, University of Toronto / **49.4** Photograph Hans Reinhard/Bruce Coleman Ltd.; sonogram G. Pohl-Apel and R. Sussinka, *Journal for Ornithologie*, 123:211–214 / **49.5** (a) Eric Hosking; (b) Stephen Dalton/Photo Researchers / **49.6** (left) Evan Cerasoli; (right) from A.N. Meltzoff and M.K. Moore, "Imitation of Facial and Manual Gestures by Human Neonate," *Science*, 198:75–78. Copyright 1977 by the AAAS / **49.7** (a) Nina Leen in *Animal Behavior*, Life Nature Library; (b) F. Schultz / **49.8** Michael Francis/The Wildlife Collection / **49.9** (left) David C. Fritts/Animals Animals; (right) Ray Richardson/Animals Animals / **49.10** John Alcock / **Page 904** (left) Lincoln P. Brower; (right) Eric Hosking

Chapter 50

50.1 John Alcock / **50.2** Edward S. Ross / **50.3** (a) E. Mickleburgh/Ardea, London; (b–d) G. Ziesler/ ZEFA / **50.4** Art by D. & V. Hennings / **50.5** A.E. Zuckerman/Tom Stack & Associates / **50.6** Fred Bruemmer / **50.7** John Alcock / **50.8** Patricia Caulfield / **50.9** Timothy Ransom / **50.10** Frank Lane Agency/Bruce Coleman Inc. / **50.11** Kenneth Lorenzen / **Page 919** Gregory D. Dimijian/Photo Researchers

INDEX

Italic numerals refer to illustrations.

A

Abalone, 390
Abdomen, insect, 433
Abdominal cavity, 407, *541*
Abiotic, defined, 300
Abiotic component, of ecosystems, 788
Abiotic formation, of organic compounds, 300, *300*, 308
ABO blood group locus, 172
ABO blood typing, 172, 674–675
Abortion, 196, 779, 800
Abscisic acid, 494, 520–521, *520*, 527
Abscission, 521, 526
Abscission zone, *526*
Absorption
  in capillary bed, *669*
  by digestive system, 642
  fungal, 373
  by mycorrhizae, 490
  phosphorus, by plants, 382
  by photoreceptors, 614–615
  by roots, 489–490, *490*
  water–solute balance and, 720
Absorption spectrum, photosynthetic pigments, *108*
Abstention, sexual, 778
Abyssal zone, 870
Accutane, 772
Acer, 477, *526*
Acetabularia, 390
Acetaldehyde, 128, *129*
Acetylase, *235*
Acetylcholine, 556
Acetyl-CoA, *124–125*, 131
Achondroplasia, *192*, 193
Acid, defined, 30
Acid–base balance, 726–727
Acid deposition, 877–879, 887
Acidity
  of blood, 31
  hydrogen ion concentration and, 30, *30*
  soil, and flower color, *275*
Acidity, stomach, 645
Acid rain, 31, *31*, 382
Acid stomach, 30
Acne, 623
Acoelomate, 407, *407*
Acorn, 279, *279*, 528, 529
Acorn worm, *446*
Acoustical signaling, 909–910
Acquired immune deficiency syndrome (*See* AIDS)
Acrasiomycota, *360*
Acromegaly, 586, *586*
Acrosome, *761*, 763, 769
Actin, *67–68*, *631–632*, *634*, 635
Actinomycetes, 353, 356, 358
Action potential (nerve impulse) (*See also* Membrane potential)
  all-or-nothing nature of, 552
  defined, 549
  duration of, 554
  function, 549
  from hand pressure receptor, *602*
  mechanism of excitation, 552–553

and muscle contraction, 556–558, 630
propagation of, *552–554*, 554
recording, *552–553*
refractory period, 554
saltatory conduction of, 554
and sensory receptors, 602
threshold, 552, *552*
Activation energy, 97–98, *97*
Activator protein, 234
Active transport
  and ATP, 85, *85*, 490
  by calcium pump, 84–85
  defined, 84
  and intestinal absorption, 647, *647*
  mechanisms of, 85–86
  in nephron, *724*
  overview of, *84*
  in plant cells, 490
  by sodium-potassium pump, 85, 550–551, *551–552*, 554
Acuity, visual, 611
Acute pancreatitis, 99
Acyclovir, 353, 782
Adaptation, to stimuli, 603
Adaptive behavior, 899*ff*
Adaptive radiation
  and adaptive zones, 304–305, 310
  animals, 310–311, 404
  defined, 304
  dinosaurs, 312
  flowering plants, *307*
  hominid, 342
  hominoid, 336
  insects, *307*, 310–311
  mammals, 305, *305*, *307*, 463
  marsupials, 464
  occurrence of, 304
  plants, 310–311
  rabbits, *305*
  reptiles, 456–457
  whales, *305*
Adaptive trait, 9, 12
  and mass extinction, 304
Adaptive zone, 304–305, 310, 314
Adder's tongue fern, *139*
Addiction, 576–577
Adélie penguin, 827
Adenine, 101, 207, *207*, 209, *223*
Adenoids, 676
Adenosine
  diphosphate (*See* ADP)
  phosphate, 47, *48*
  triphosphate (*See* ATP)
Adenosine deaminase, 244, 251
Adenosine diphosphate (*See* ADP)
Adenosine monophosphate, 244
Adenosine triphosphate (*See* ATP)
Adenovirus, *348–349*
Adenylate cyclase, 594, *594*, 595
ADH (*See* Antidiuretic hormone)
Adhering junctions, 71
Adhesion
  cancer cell, 241
  cell, 80
Adhesive cue, 748
Adipose tissue, *537*, 538, 649, 654
Adolescence (human), defined, 777
Adoption, 920
ADP (adenosine diphosphate)
  formation by photophosphorylation, *111*
  ion flow and, *112–113*

Adrenal cortex, 588
Adrenal gland
  cortex, *452*, 588
  innervation, 568
  location, 582
  medulla, 587
  role in metabolism, 588, 655–656
Adrenalin (*See* Epinephrine)
Adrenal medulla, 588–589
Adrenocorticotropic hormone, *452*, 588, 656
Adult (developmental stage), 754, 777, *777*
Adventitious root, 478, *478*, 518
Aegyptopithecus, *336*
Aerobic respiration
  ATP yield from, *127*
  by bacteria, *109*
  and carbon cycle, 836–837
  defined, 5, 120
  and early atmosphere, 132
  equation for, 120
  and germination, 518
  links with photosynthesis, *104*, 105, 132
  and muscle contraction, 633, *633*
  net energy yield, 127–128, *127*
  origin of, 132
  overview of, 120, *121*
  and oxygen, *109*, *121*, 123–125, 308
  reactions of, 120–127
  stages of, 120–126, *120–127*
Aerosol spray, and ozone layer, 880
Aesculus, 477
Afferent nerve, *567*
African blood lily, *67*
Africanized bee, 118–119, *118*, 819
African sleeping sickness, 366
African violet, *513*
Afterbirth, 775
Agar, 388, *521*
Agaricus brunnescens, 379
Agave, 477
Agent Orange, 520
Age structure, population, *800–801*
Agglutination, blood, 675, *675*
Aggressive mimicry, 813
Aging
  and bone turnover, 627
  and cancer resistance, 688
  and cardiovascular disorders, 670
  and death, 754–755
  defined, 777, *777*
  and eye disorders, 612
  gene mutation effects, 182–183
  in Gilford progeria syndrome, 182–183, 192
  glandular secretions and, 623
  and human development, 777
  and posture, 628
  skin changes in, 623
  and sun tanning, 623
Agnatha, 448, *448*, 450
Agriculture
  animal-assisted, 883
  and deforestation, 883–884, *884*
  and desertification, 886, *886*
  and genetic engineering, 245, 254–255
  and green revolution, 882–883
  and hormones, 520–521
  intensive, 883

land available for, 882–883
and orchardists, 118, 520
and population growth, 798
and salination, 880
shifting cultivation (slash-and-burn), 883
and strip mining, 887
subsistence, 883
Agrobacterium, 356
Agrobacterium tumefaciens, 255
AIDS (*See also* Human immunodeficiency virus)
  carriers of, 695
  HIV virus, *348*
  and immune response, 693
  incidence of, 692, 694
  spread of, 353, 780–781
  testing for, 781
  transmission modes, 694, 780
  treatments for, 695, 780
AIDS-related complex (ARC), 780
Air circulation, 848, *848–849*
Air pollution, 283, 875, 876*ff*, 876–877, 885 (*See also* Industrial pollution)
Air sac, *704–705*
Alanine, 222
Albatross, courtship behavior, 909, *909*
Albinism, 174, *175*, *192*, 199
Albino, 174, *175*
Albumin, 661, *662*
  primary structure, 46
  protein denaturation, 46
Alcohol (ethyl alcohol)
  addiction, 577
  and blood-brain barrier, 571
  and brain function, 571
  degradation, 65
  effects on central nervous system, 577
  effects on heart, 672
  effects on liver, 576
  fetal alcohol syndrome, 773
  hydroxyl group, 37
  and low-density lipoprotein levels, 42, *42*
  peroxisome action on, 65
  and pregnancy, 772, 773
Alcoholic fermentation, 128–129, *129*
Aldehyde group, *36*, 48
Alder, 821
Aldosterone, 587, 725
Alfalfa, 489
  stem structure, *475*
Algae (*See also* specific type)
  blue-green, 358
  brown, 66, 396
  classification, 388
  evolution, *307*
  golden, *360*, 364
  green, *18*, *109*, 114, 308, 369, 381, 390–391, *390*, *391*, 396
  homosporous, 387–388
  of lichen, 381
  red, 308, 375, 396, *869*
  species of, *396*
  surface-to-volume ratio constraints, 56
  in tidepools, *817*
  uses of, 388–391
  yellow-green, *360*, 364
Algin, 388–390